Springer Series in Astrophysics and Cosmology

The series covers all areas of astrophysics and cosmology, including theory, observations, and instrumentation. It publishes monographs and edited volumes. All books are authored or edited by leading experts in the field and are primarily intended for researchers and graduate students.

Cosimo Bambi
Editor

Regular Black Holes

Towards a New Paradigm of Gravitational Collapse

Springer

Editor
Cosimo Bambi
Department of Physics
Fudan University
Shanghai, China

ISSN 2731-734X ISSN 2731-7358 (electronic)
Springer Series in Astrophysics and Cosmology
ISBN 978-981-99-1595-8 ISBN 978-981-99-1596-5 (eBook)
https://doi.org/10.1007/978-981-99-1596-5

This Springer imprint is published by the registered company Springer Nature Singapore Pte Ltd.
The registered company address is: 152 Beach Road, #21-01/04 Gateway East, Singapore 189721, Singapore

Preface

General Relativity is one of the pillars of modern physics. While all the available observational data are in good agreement with the predictions of Einstein's gravity, there are a few unresolved theoretical issues that strongly point out the existence of new physics. One of the most outstanding and longstanding problems in General Relativity is the inevitability of spacetime singularities in physically relevant solutions of the Einstein Equations. At a spacetime singularity, predictability is lost and standard physics breaks down. It is widely believed that the problem of spacetime singularities can be solved within a theory of quantum gravity. However, despite the significant efforts and progress in the past decades, we do not have yet a robust theory of quantum gravity.

In the absence of a robust theory of quantum gravity, we can investigate qualitatively different scenarios that can solve the problem of spacetime singularities and can be obtained by violating specific conditions or assumptions valid in General Relativity. In other words, instead of following the more traditional top-down approach, in which we start from a well-formulated theory and derive its predictions, we can try to follow a bottom-up strategy, in which we consider plausible predictions for a certain phenomenon and we try to derive the required key-ingredients of the fundamental theory.

Within this spirit, in the past years there has been an increasing interest in the study of regular black holes and singularity-free gravitational collapse models. Actually, the first attempts in this direction are not new, but initially they did not attract much interest from the community, and only in the past 10 years this line of research has grown significantly. The construction of the first regular black hole was presented by James Bardeen in 1968 at the *5th International Conference on Gravitation and the Theory of Relativity* in Tbilisi, and the first quantum gravity-inspired model of singularity-free gravitational collapse was discussed by Valeri Frolov and Grigorii Vilkovisky in 1979 at the *2nd Marcel Grossmann Meeting* in Trieste. In 1998, Eloy Ayon-Beato and Alberto Garcia showed that it is possible to obtain regular black hole solutions in General Relativity in the presence of a nonlinear electrodynamic field. It was only around 2012–2013 that regular black holes started becoming a hot topic, as we can clearly see from the rise in the number of citations per year of the first works as well as the rise in the number of new publications on the topic.

The idea to publish a book on regular black holes and singularity-free gravitational collapse models is motivated, in part, by the fact that this line of research has significantly grown in the past 10 years and there are a number of quite interesting recent results, and, at the same time, by the absence of a review covering all main models proposed so far. I am very grateful to all the contributed authors of this book, who accepted enthusiastically to be involved in this project and wrote high-quality contributions. I hope that this book can help young researchers to enter this exciting line of research and can contribute to progress further in our understanding of black holes and gravitational collapse models.

Shanghai, China
January 2023

Cosimo Bambi

Contents

Contributors

Abhay Ashtekar Physics Department, and, Institute for Gravitation & the Cosmos, Penn State, University Park, PA, USA

Alfio Bonanno INAF, Osservatorio Astrofisico di Catania, Catania, Italy;
INFN, Sezione di Catania, Catania, Italy

Kirill A. Bronnikov VNIIMS, Moscow, Russia;
Institute of Gravitation and Cosmology, Peoples' Friendship University of Russia (RUDN University), Moscow, Russia;
National Research Nuclear University "MEPhI", Moscow, Russia

Raúl Carballo-Rubio CP3-Origins, University of Southern Denmark, Odense, Denmark;
Florida Space Institute, University of Central Florida, Orlando, FL, USA

Tibério de Paula Netto Department of Physics, Southern University of Science and Technology, Shenzhen, China

Irina Dymnikova A.F. Ioffe Physico-Technical Institute of the Russian Academy of Sciences, St Petersburg, Russia

Astrid Eichhorn CP3-Origins, University of Southern Denmark, Odense M, Denmark

Francesco Di Filippo Center for Gravitational Physics, Yukawa Institute for Theoretical Physics, Kyoto University, Kyoto, Japan

Breno L. Giacchini Department of Physics, Southern University of Science and Technology, Shenzhen, China

Aaron Held Theoretisch-Physikalisches Institut, Friedrich-Schiller-Universität Jena, Jena, Germany;
The Princeton Gravity Initiative, Jadwin Hall, Princeton University, Princeton, NJ, United States

Stefano Liberati SISSA - International School for Advanced Studies, Trieste, Italy;
IFPU - Institute for Fundamental Physics of the Universe, Trieste, Italy;
INFN - Sez, Trieste, Italy

Daniele Malafarina Department of Physics, Nazarbayev University, Astana, Kazakhstan

Emil Mottola Department of Physics and Astronom, University of New Mexico, Albuquerque, NM, USA

Piero Nicolini Dipartimento di Fisica, Università degli Studi di Trieste, Trieste, Italy;
Institut für Theoretische Physik, Johann Wolfgang Goethe-Universität, Frankfurt am Main, Germany;
Frankfurt Institute for Advanced Studies (FIAS), Frankfurt am Main, Germany

Javier Olmedo Departamento de Física Teórica y del Cosmos, Universidad de Granada, Granada, Spain

Gonzalo J. Olmo Departamento de Física Teórica and IFIC, Centro Mixto Universidad de Valencia - CSIC. Universidad de Valencia, Valencia, Spain

Nikodem Popławski Department of Mathematics and Physics, University of New Haven, West Haven, CT, USA

Diego Rubiera-Garcia Departamento de Física Teórica and IPARCOS, Universidad Complutense de Madrid, Madrid, Spain

Frank Saueressig Institute for Mathematics, Astrophysics and Particle Physics (IMAPP), Radboud University, Nijmegen, The Netherlands

Parampreet Singh Department of Physics and Astronomy, Louisiana State University, Baton Rouge, LA, USA

Ramón Torres Department of Physics, Universitat Politècnica de Catalunya, Barcelona, Spain

Matt Visser School of Mathematics and Statistics, Victoria University of Wellington, Wellington, New Zealand

Chapter 1
Regular Rotating Black Holes and Solitons with the de Sitter/Phantom Interiors

Irina Dymnikova

1.1 Introduction

The basic features of regular rotating black holes (RRBHs) and spinning solitons G-lumps, which replace naked singularities, are determined by the Einstein equations which govern their geometry, and by the energy conditions which define the character of matter content in their interiors. G-lumps are the non-singular particle-like structures without the event horizons, defined as the physical solitons in the spirit of the Coleman lumps - nonsingular non-dissipative compact objects which keep themselves together by their own self-interaction [37]. In our case lumps are bound by their gravitational self-interaction balanced at the zero-gravity surface, where the Strong Energy Condition of the singularity theorems [87], the basic condition for the existence of a singularity, is violated, and gravitational attraction becomes gravitational repulsion. For this reason they have been qualified as G-lumps [51].

G-lumps differ from gravastars (for a review [124]) in that their internal structure is not constructed by imposed a priori certain models but is determined, as we shall see below, by geometry and by the energy conditions.

The Einstein equations admit the class of regular spherical solutions, generated by stress-energy tensors with the algebraic structure such that

$$T_t^t = T_r^r \ \ (p_r = -\rho) \tag{1.1}$$

and described by the metric

I. Dymnikova (✉)
A.F. Ioffe Physico-Technical Institute of the Russian Academy of Sciences,
Politekhnicheskaya 26, 194021 St Petersburg, Russia
e-mail: igd.ammp@mail.ioffe.ru

C. Bambi (ed.), *Regular Black Holes*, Springer Series in Astrophysics and Cosmology,
https://doi.org/10.1007/978-981-99-1596-5_1

$$ds^2 = g(r)dt^2 - \frac{dr^2}{g(r)} - r^2 d\Omega^2; \tag{1.2}$$

$$g(r) = 1 - \frac{2G\mathcal{M}(r)}{r}; \quad \mathcal{M}(r) = 4\pi \int_0^r \tilde{\rho}(x)x^2 dx; \quad M = 4\pi \int_0^\infty \tilde{\rho}(x)x^2 dx. \tag{1.3}$$

Here $\tilde{\rho}(r)$ is a spherical density profile. Spherically symmetric metrics (1.2) belong to the Kerr-Shild class of algebraically special metrics [99], which can be presented as $g_{\mu\nu} = \eta_{\mu\nu} + 2f(r)k_\mu k_\nu$, where $f(r) = r\mathcal{M}(r)$, $\eta_{\mu\nu}$ is the Minkowski metric and k_μ are principal null congruences (radial geodesics in the spherically symmetric case). For the Kerr-Schild metrics presented in the Lorentz covariant coordinate system, where the Einstein equations are linear, source terms satisfy $T^i_{k,i} = 0$ [83], and the eigenvalues of the stress-energy tensor, $T^t_t = \rho$, $T^r_r = -p_r$, $T^\theta_\theta = T^\phi_\phi = -p_\perp$ satisfy the $r-$dependent equation of state [47, 121]

$$p_r = -\rho; \quad p_\perp = -\rho - r\rho'/2. \tag{1.4}$$

We conclude that all the structures described by the metrics (1.2)–(1.3), are characterized by the anisotropy of pressures.

For regular structures the Strong Energy Condition, which requires $\rho + \sum p_k \geq 0$ in the comoving reference frame, is violated at the zero gravity surface [48] defined by $\rho + \sum p_k = 0$, where the gravitational acceleration, proportional to $\rho + \sum p_k$, changes its sign and becomes repulsive.

For RRBHs and G-lumps satisfying the Weak Energy Condition (WEC), which implies non-negativity of density as measured by a local observer on a time-like curve and requires $\rho \geq 0$ and $p_k + \rho \geq 0$ for any principal pressure [87], in our case $p_\perp + \rho \geq 0$, which leads to $\rho' \leq 0$ by virtue of (1.4), the density monotonically grows to the regular center, where the metric takes the de Sitter form [48, 51]

$$g(r) \to 1 - r^2/r_\Lambda^2; \quad r_\Lambda^2 = 3/\Lambda; \quad \Lambda = 8\pi G\rho_\Lambda; \quad T^i_k \to \rho_\Lambda \delta^i_k. \tag{1.5}$$

In the de Sitter center (1.5) density achieves its maximal value ρ_Λ, directly related to the cosmological constant Λ, in accordance with the underlying idea to associate cosmological constant with the energy density of self-interaction [138].

The gravitational mass M given in (1.3), is generically related to breaking of spacetime symmetry from the de Sitter group in the origin for an object specified by (1.1) [51, 54]. This provides for the intrinsic relation between gravity, spacetime symmetry and the the Higgs mechanism for a mass generation [67].

The de Sitter-Schwarzschild geometry presented by the metrics (1.2), asymptotically de Sitter as $r \to 0$ and asymptotically Schwarzschild as $r \to \infty$ [48], can match the Schwarzschild exterior with the de Sitter interior directly [121] or continuously [47, 50]. The energy scale of the interior de Sitter vacuum can be the Planck scale in the frame of the hypothesis of transition to the de Sitter interior due to self-regulation of geometry by vacuum polarization effects [121], or of existence of the limiting curvature of the Planck scale [75]. But it can be also the GUT scale, where the interior de Sitter vacuum appears due to restoration of the spacetime symmetry to the de Sitter group in the course of a gravitational collapse, corresponding to the process reversed with respect to the violation of spacetime symmetry in the process of spontaneous symmetry breaking in the early stage of the universe evolution [48, 49]. Appearance of the de Sitter core instead of the Schwarzschild singularity has been demonstrated in the frame of "renormalization group improving" [16], of the noncommutative geometry [113], of an ultraviolet quantum gravity [114] , and of the quadratic gravity [15].

The spherical metrics of the Kerr-Schild class can be transformed to the axially symmetric metrics by the complex Trautman-Newman translations of the general Gürses-Gürsey formalism [83], which include the Newman-Janis algorithm [117]

$$(r,t) \to (r,u); \quad u = t - \int \frac{dr}{g(r)}; \quad r \to r + ia\cos\theta; \quad u \to u = ia\cos\theta. \tag{1.6}$$

This procedure has been used for construction of the axially symmetric regular solutions [8–11, 28, 64, 71, 77, 93, 104, 106, 109, 116, 133, 135, 136]. More general approach based on the properties of the Kerr-Schild metrics, has been applied in the non-commutative geometry [5, 113, 118]. The extension and applications of the Kerr-Schild formalism have been presented in [23]).

The complex translation (1.6) transforms the de Sitter-Schwarzschild geomtery into the de Sitter-Kerr geometry, asymptotically de Sitter (1.5) as $r \to 0$, and asymptotically Kerr as $r \to \infty$, so that the regular rotating compact objects represent the de Sitter-Kerr black holes and spinning solitons replacing naked singularities [63].

The Kerr solution to the Einstein equations can be obtained from the Schwarzschild metric with using the Newman-Janis complex translation (1.6), although originally it was obtained as a particular case of the algebraically special metric [98].

In the Boyer-Lindquist coordinates the Kerr metric reads (in the units $c = G = 1$)

$$ds^2 = \frac{2Mr - \Sigma}{\Sigma}dt^2 + \frac{\Sigma}{\Delta}dr^2 + \Sigma d\theta^2 - \frac{4aMr\sin^2\theta}{\Sigma}dtd\phi \tag{1.7}$$

$$+\left(r^2 + a^2 + \frac{2Mra^2\sin^2\theta}{\Sigma}\right)\sin^2\theta d\phi^2; \ \Sigma = r^2 + a^2\cos^2\theta; \ \Delta = r^2 - 2Mr + a^2 \tag{1.8}$$

where M is the gravitational mass and a is the specific angular momentum.

The Boyer-Lindquist coordinates (r, θ, ϕ) are related to the Cartesian coordinates (x, y, z) by

$$x^2 + y^2 = (r^2 + a^2)\sin^2\theta; \quad z = r\cos\theta. \tag{1.9}$$

In the axially symmetric geometry the surfaces of constant r are the oblate confocal ellipsoids of revolution [34]

$$r^4 - (x^2 + y^2 + z^2 - a^2)r^2 - a^2z^2 = 0 \tag{1.10}$$

which degenerate, for $r = 0$, to the equatorial disk

$$x^2 + y^2 \leq a^2, \quad z = 0, \tag{1.11}$$

centered on the symmetry axis and bounded by the ring

$$x^2 + y^2 = a^2, \quad z = 0. \tag{1.12}$$

In the Kerr geometry the ring (1.12) comprises the Kerr ring singularity [34].

The Kerr metric represents the exterior gravitational field of a massive rotating body. Presented in the literature models of an internal material source for the Kerr exterior field, include disk-like [20, 84, 94], shell-like [36, 39], bag-like [21, 22], and string-like [24] models. The problem of matching the Kerr exterior to an internal source does not have a unique solution, because of a freedom in the choice of the boundary between them [94], as well as of the shape and matter content of a source.

Here we show that in regular geometry this freedom is strongly restricted by the dynamical equations and reduced to the choice of the density profile in (1.3). The shape of an interior of a regular compact object is strictly determined by geometry, while the type of its matter content is regulated by the energy conditions. The axial symmetry requires the de Sitter center $r = 0$ to become the de Sitter equatorial disk $r = 0$ which is the obligatory constituent of all compact objects described by regular solutions specified by (1.1). Algebraic structure of their stress-energy tensors (1.1) implies that the pressure is intrinsically anisotropic.

Regular axially symmetric geometry includes ergoregions where processes of energy extraction involve not only rotational energy as in the Kerr geometry [129], but also the de Sitter and phantom energy ([66] and references therein).

A regular rotating black hole has one ergoregion between the ergosphere and the event horizon. A spinning soliton can have two ergospheres and ergoregion between them, or one ergosphere and ergoregion behind it [63].

Energy conditions distinguish two kinds of interiors. The 1-st kind satisfies WEC and reduces to the de Sitter vacuum disk. The 2-nd kind consists of a two-dimensional closed S-surface of the de Sitter vacuum with the de Sitter disk as a bridge. In the cavities between the S-surface and the disk WEC is violated, they are filled with a phantom fluid, defined for the isotropic case by $p = w\rho, \ w < -1$ [26] (for a review [19]). For regular rotating black holes and G-lumps a phantom fluid is essentially anisotropic $p_r = w_r\rho, w_r = -1; \ p_\perp = w_\perp\rho, w_\perp < -1$ [63, 64].

For a regular rotating black hole information about nature and the energy scale of its interior can be extracted from its shadow ([65] and references therein). The shadow of a black hole presents its direct image over an image of a bright distant source of radiation. The black hole shadows can be observed, recently the Event Horizon Telescope Collaboration [44] reported the first observation of the black hole shadow in M87 [4]. The black hole shadow is formed by the photon gravitational capture cross-section confined by the innermost unstable photon orbits [34, 46]. For a regular rotating black hole an asymmetry of its shadow depends on its angular momentum, and essentially depends on the pace of decreasing of its density [65].

Primordial black holes and their remnants (end products of the Hawking evaporation) are considered as a source of dark matter (DM) [30, 108, 122] for more than thirty decades. Regular primordial black holes, their remnants and G-lumps can be qualified as heavy DM candidates generically related to a dark energy via their de Sitter/phantom interiors [55, 59, 62] (for a review [66]).

They are formed at the early inflationary stage(s) by the quantum collapse of density fluctuations related to primordial inhomogeneities [60]. Therefore their observational signatures as heavy DM candidates can be a source of information about inhomogeneity of the early universe [62]. In their GUT scale false vacuum interiors baryon and lepton numbers are not conserved, as a result they can induce proton decay in an underground detector like IceCUBE, which would present their observational signature in heavy dark matter searches [62].

For a G-lump with 2-nd kind interior a phantom energy is not screened by the event horizon. This makes possible extraction of a phantom energy from ergoregions in addition to rotational energy. This would provide an additional source of information about the scale and properties of the interior phantom fluid [63].

Primordial regular black hole remnants and G-lumps can form graviatoms—gravitationally bound quantum systems—by capturing charged particles ([60, 66] and references therein). Their specific observational signature is the electromagnetic radiation, whose frequency essentially depends on the energy density of the interior de Sitter vacuum and falls within the range available for observations [60].

In Sect. 1.2 we describe geometry including relation of the mass with spacetime symmetry. Section 1.3 is devoted to physics of interiors, and Sect. 1.4 to observational signatures of regular rotating black holes and G-lumps. Section 1.5 contains conclusions. In the estimates for pictures we apply the GUT scale od symmetry restoration, $E_{GUT} = 10^{15}$ GeV which gives the de Sitter radius $r_\Lambda = 2.4 \times 10^{-25}$ cm.

1.2 Geometry

In this Section we present geometry of regular rotating compact objects described by the axially symmetric solutions, obtained from spherical solutions of the Kerr-Schild class with using the Gürsey-Gürses approach which includes the Newman-Janis algorithm. We describe the asymptotic behavior of solutions to dynamical equations, their predictions concerning spacetime symmetry and its relation with the gravitational mass, and the number and location of the horizons and ergospheres.

1.2.1 Metric and Spacetime Symmetry

In the Boyer-Lindquist coordinates and in the units $c = G = 1$, the axially symmetric Gürses-Gürsey metric reads [83]

$$ds^2 = \frac{2f - \Sigma}{\Sigma}dt^2 + \frac{\Sigma}{\Delta}dr^2 + \Sigma d\theta^2 - \frac{4af\sin^2\theta}{\Sigma}dtd\phi$$

$$+\left(r^2 + a^2 + \frac{2fa^2\sin^2\theta}{\Sigma}\right)\sin^2\theta d\phi^2;\ \Delta = r^2 + a^2 - 2f(r);\ \Sigma = r^2 + a^2\cos^2\theta \tag{1.13}$$

and involves a master function

$$f(r) = r\mathscr{M}(r) = 4\pi r\int_0^r \tilde{\rho}(x)x^2dx \tag{1.14}$$

where $\tilde{\rho}(r)$ is the density profile of the original spherical solution related with the axial solution by (1.14). For spherical solutions satisfying WEC, the function $f(r)$ is everywhere positive. For $r \to 0$ it approaches the de Sitter asymptotic $2f(r) = r^4/r_\Lambda^2;\ r_\Lambda^2 = 3/\Lambda = 3/8\pi G\tilde{\rho}_\Lambda$, where $\tilde{\rho}_\Lambda = \tilde{\rho}(0)$, and monotonically increases as

$$f_{de\ S} = (4\pi\tilde{\rho}_\Lambda/3)r^4 \longleftarrow f(r) \longrightarrow f_{Kerr} = Mr.$$

For $r \to \infty$ the metric coincides with the Kerr metric (1.8). In the equatorial plane the function $2f(r)/\Sigma$ in (1.13) tends to $2f(r)/\Sigma = r^2/r_\Lambda^2$, and vanishes at $r = 0$. It follows that the equatorial disk $r = 0$ is totally intrinsically flat. But $\Lambda = 8\pi G\rho_\Lambda$ is non-zero throughout the disk where $\rho_\Lambda = (r^4/\Sigma^2)\tilde{\rho}_\Lambda$ [54]. The original spherical metrics are asymptotically de Sitter as $r \to 0$ which is asymptotically flat but with non-zero Λ. Rotation transforms the de Sitter center $r = 0$ into the de Sitter vacuum disk (1.11)). The metric (1.13) in the limit $r \to 0$ takes the form

$$ds^2 = \left(\frac{r^2}{r_\Lambda^2} - 1\right)dt^2 + \frac{r^2}{\Delta}dr^2 + r^2d\theta - \frac{r^2}{r_\Lambda^2}2adtd\phi + \left(r^2 + a^2 + \frac{r^2}{r_\Lambda^2}\right)d\phi^2. \tag{1.15}$$

and represents the rotating de Sitter vacuum with Λ spread over the disk (1.11).

Appearance of the de Sitter vacuum inside a regular rotating compact object implies that the mass of an object, M given in (1.3), is generically related to its interior de Sitter vacuum and breaking of spacetime symmetry from the de Sitter group in its origin to the Poincaré group at infinity in the asymptotically flat space-time [51, 54]. The direct consequence of the appearance of the de Sitter vacuum instead of a central singularity, is thus the generic relation of an object mass with gravity presented by the de Sitter geometry, and with breaking of spacetime symmetry.

This provides for the inherent relation between gravity, spacetime symmetry and the Higgs mechanism for a particle mass generation [67, 68]. The Higgs mechanism endows a particle with a mass via spontaneous symmetry breaking of intrinsically incorporated scalar fields [73, 82, 89] (for overview [123]). Invoking the spontaneous symmetry breaking for scalar fields, the Higgs mechanism incorporates the de Sitter vacuum as its basic ingredient, and, in consequence, incorporates the relation of a particle mass with gravity and spacetime symmetry, since spontaneous symmetry breaking of a scalar field from its false (de Sitter) vacuum state $p = -\rho$, to its true

(Minkowski) vacuum state generically involves breaking of spacetime symmetry from the de Sitter group to the Poincaré group [67].

The intrinsic relation of the Higgs mechanism with the spacetime symmetry can be verified by analysis of the experimental data on negative mass square differences for neutrino reported since 1991 [127]. In our analysis [2, 56] we take into account that the interaction vertex is gravito-electroweak due to intrinsic involvement of gravity of the de Sitter vacuum, and apply the Casimir operators in the de Sitter space for description of particle states in the vertex. The further evolution of particle states in the Minkowski background involves the change in spacetime symmetry. We find that the de Sitter symmetry in the gravito-electroweak vertex leads to the exact bi-maximal mixing which allows us to provide the explanation for the origin of the negative mass squares, and to evaluate the gravito-electroweak unification scale from the data on the solar and atmospheric neutrino as $M_{unif} \sim (6 - 15)$ TeV, in the same range as predicted by theories of gravito-electroweak unification [2, 3, 56].

The Higgs mechanism, as responsible for the first inflation, powers the universe dynamics via the spontaneous symmetry breaking from its false vacuum state with $p = -\rho_{GUT}$, corresponding to the Sitter vacuum with $\Lambda = 8\pi G\rho_\Lambda = 8\pi G\rho_{GUT}$, whose further decay provides the necessary energetic support.

The generic relation of the Higgs mechanism with spacetime symmetry breaking suggests the possibility of symmetry restoration of the Higgs field at the presently observed de Sitter vacuum scale, which would make it dynamically responsible for the universe evolution from the initial symmetry breaking at the GUT scale with $\rho_\Lambda = \rho_{GUT}$ to the final symmetry restoration at the presently observed de Sitter vacuum scale $\rho_\lambda = \rho_{today}$.

Analysis of this case in the frame of the minisuperspace model of quantum cosmology has shown that the cosmological constant today $\lambda = 8\pi G\rho_\lambda$ must be non-zero in principle, since it corresponds to an absolute lower limit in its quantum spectrum. The energy scale for the presently observed spacetime symmetry restoration, which generically involves the symmetry restoration of the Higgs field, is estimated as $E_\lambda \simeq 23 \times 10^{-4}$ eV [70].

1.2.2 Horizons and Ergoregions

The basic features of geometry of regular rotating objects are actually determined by properties of the related spherical metrics (1.2)–(1.3).

Horizons are defined by $\Delta(r) = 0$. The function $\Delta(r)$ is related to the spherical metric function $g(r)$ as

$$\Delta(r) = r^2 + a^2 - 2f(r) = a^2 + r^2 g(r). \tag{1.16}$$

It evolves from $\Delta = a^2$ as $r = 0$ to $\Delta \to \infty$ as $r \to \infty$, and takes the value $\Delta = a^2$ at zero points of the metric function $g(r)$.

A metric function $g(r)$ in (1.3) in the asymptotically flat case evolves between $g(r) = 1$ at $r = 0$ and at $r \to \infty$, and has one minimum in between. It can have two zero points $g(r) = 0$, or one zero point $g(r) = 0,\ g'(r) = 0$, either no zero points [51]. Typical behavior of a metric function $g(r)$ is shown in Fig. 1.1 (Left) [64].

Derivatives of Δ are $\Delta' = 2rg(r) + r^2 g'(r);\ \ \Delta'' = 2g(r) + 4rg'(r) + r^2 g''(r)$. At $r = 0,\ \ \Delta' = 0;\ \Delta'' > 0$ and the function Δ has the minimum, $\Delta = a^2$. Next it grows and can have maximum at a certain value $r_m < r_1$ where r_1 is the 1-st zero of the function $g(r)$. At the maximum $\Delta'(r_m) = 0$ and hence $g'(r_m) = -2g(r_m)/r_m < 0$, and $\Delta''(r_m) = -6g(r_m) + r_m^2 g''(r_m) < 0$. After passing the maximum, decreasing Δ achieves $\Delta = a^2$ at the 1-st zero of $g(r)$, while $g(r)$ decreases to its minimum. In this region $g'(r)$ is first negative, then passes zero and becomes positive, hence $g'' \geq 0$ everywhere between zeros of $g(r)$; as a result $\Delta'' \geq 0$ everywhere in this region, so that the function $\Delta(r)$ can have only minimum and only one. At $r > r_2$ we have $g(r) > 0$ and $g' > 0$ so that Δ cannot vanish. Since the function $g(r)$ has at most two zero points, axially symmetric space-time can have at most two horizons, the event horizon r_+ and the internal Cauchy horizon $r_- < r_+$ [61, 63].

Horizons are defined by equation $r^2 + a^2 - 2r\mathscr{M}(r) = 0$ which gives

$$r_{+,-} = \mathscr{M}(r) \pm \sqrt{\mathscr{M}^2 - a^2}. \tag{1.17}$$

The extreme black hole is defined by the double horizon $r_\pm = \mathscr{M}(r_\pm)$, and by the angular momentum

$$a = a_{dh} = \mathscr{M}(r_\pm) = r_\pm. \tag{1.18}$$

Ergosphere is a surface of a static limit $g_{tt} = 0$ given by

$$g_{tt}(r, \theta) = r^2 + a^2 \cos^2\theta - 2f(r) = 0. \tag{1.19}$$

Ergoregions are defined by $g_{tt} < 0$ which makes possible extraction of rotational energy (see, e.g. [136] and references therein).

Existence of the ergospheres depends on the mass function $\mathscr{M}(r)$. At the z-axis ergosphere coincides with the event horizon presented by the ellipsoid $r_+ = const$. In the equatorial plane the equation of ergosphere is $r^2 - 2r\mathscr{M}(r) = 0$, for monotonically growing mass function $\mathscr{M}(r)$ there can exist at most two zero points.

The basic condition for the existence of ergospheres is $W_e > W_h$ where $W_e = \sqrt{2f(r_e)}$ is the maximal width of ergosphere and $W_h = \sqrt{2f(r_+)}$ is the width of the event horizon in the equatorial plane. Each point of the ergosphere belongs to some of confocal ellipsoids (1.10). According to (1.19), it satisfies $z^2 = (2r^2 f(r) - r^4)/a^2$. The cases for the existence of ergospheres are shown in Fig. 1.1 (Right) [64].

The function $f(r)$ (curves 3,4 in Fig. 1.1 (Right)) is everywhere non-negative and growing from $f(r) = 4\pi\rho_\Lambda r^4/3$ as $r \to 0$ to $f(r) = rM$ as $r \to \infty$ [51].

The width of the ergoregion at a certain z is $x^2 + y^2 = (a^2 + r^2)(1 - z^2/r^2)$. In the equatorial plane $x^2 + y^2 = a^2 + r^2$ and $z^2 = 0 \to 2f(r) - r^2 = 0$. Ergosphere exists when the curve $u = 2f(r)$ intersects or touches the parabola $u = r^2$ (curves 2

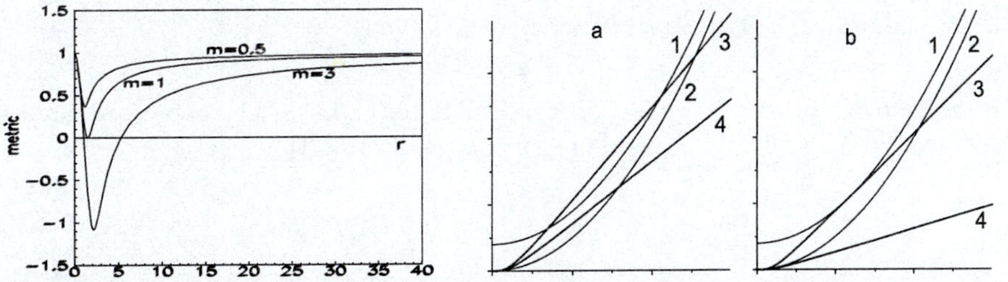

Fig. 1.1 Left: Typical behavior of spherical metric function $g(r)$. The mass parameter m is the mass M normalized to its critical value corresponding to the double-horizon state. Right: Four cases for the existence of ergospheres and ergoregions

in Fig. 1.1 (Right)). It is evident that in this case the curve $u = 2f(r)$ intersects also the parabola $u = r^2 + a^2 \cos^2\theta$ for a given θ (curves 1 in Fig. 1.1 (Right)).

In the black hole case, ergosphere and ergoregion exist for any density profile (the curve 3a in Fig. 1.1 (Right) for the case of the black hole with two horizons, the curve 3b for the extreme black hole with the double horizon).

In the case of G-lump the existence of ergospheres depends on the density profile. G-lumps can have two ergospheres and ergoregion between them (the curve 4a in Fig. 1.1 (Right)), one ergosphere and ergoregion involving the whole interior(the curve 4b in Fig. 1.1 (Right)), or no ergospheres [61, 63].

The difference is that in the case of a black hole the existence of ergosphere is restricted by the condition for coincidence of its upper and lower point with the event horizon at the z-axis, additional to the equation (1.19).

Geometry of all regular rotating structures described by solutions of the Kerr-Schild class with $T_t^t = T_r^r$ ($p_r = -\rho$) and satisfying WEC, contains the de Sitter vacuum disk $r = 0$. Regular rotating black hole can have at most two horizons, and ergospheres around the event horizon, which confines ergoregions and exist for any density profile. For spinning G-lumps the existence of ergospheres depends on the density profile. They can have two ergosperes and ergoregion between them, one ergosphere which confines the whole interior as the ergoregion, or no ergospheres.

1.3 Interiors of Regular Rotating Black Holes and Spinning Solitons

In this Section we present the matter content responsible for two kinds of interiors of the regular rotating black holes and spinning solitons, dependently on the form of the density profile, and the light rings surrounding these objects.

1.3.1 Stress-Energy Tensor

The anisotropic stress-energy tensor responsible for (1.13) can be expressed via the density $\rho(r,\theta)$ and the transversal pressure $p_\perp(r,\theta)$ as [83]

$$T_{\mu\nu} = (\rho + p_\perp)(u_\mu u_\nu - l_\mu l_\nu) + p_\perp g_{\mu\nu} \tag{1.20}$$

in the orthonormal tetrad

$$u^\mu = \frac{1}{\sqrt{\pm\Delta\Sigma}}[(r^2+a^2)\delta_0^\mu + a\delta_3^\mu], \quad l^\mu = \sqrt{\frac{\pm\Delta}{\Sigma}}\delta_1^\mu,$$
$$n^\mu = \frac{1}{\sqrt{\Sigma}}\delta_2^\mu, \quad m^\mu = \frac{-1}{\sqrt{\Sigma}\sin\theta}[a\sin^2\theta\,\delta_0^\mu + \delta_3^\mu]. \tag{1.21}$$

The sign plus refers to the regions outside the event horizon and inside the internal horizon, in which the vector u^μ is time-like. The sign minus refers to the regions between the horizons where the vector l^μ is time-like. The vectors m^μ and n^μ are space-like in all regions. The eigenvalues of the stress-energy tensor (1.20) in the co-rotating frame, where each of ellipsoidal layers rotates with the angular velocity $\omega(r) = u^\varphi/u^t = a/(r^2+a^2)$ [22], are given by

$$\rho(r,\theta) = T_{\mu\nu}u^\mu u^\nu; \quad p_r(r,\theta) = T_{\mu\nu}l^\mu l^\nu = -\rho(r,\theta); \quad p_\perp(r,\theta) = T_{\mu\nu}n^\mu n^\nu = T_{\mu\nu}m^\mu m^\nu \tag{1.22}$$

outside the event horizon and inside the internal horizon where the density is defined as the eigenvalue associated with the time-like eigenvector u^μ.

The density and pressures depend on a related spherical function $f(r)$ as $8\pi\Sigma^2\rho(r,\theta) = 2(f'r - f)$; $8\pi\Sigma^2 p_\perp(r,\theta) = 2(f'r - f) - f''\Sigma$ [83]. Applying the definition of the function $f(r)$ (1.14), we obtain

$$\rho(r,\theta) = \frac{r^4}{\Sigma^2}\tilde{\rho}(r); \quad p_r = -\rho: \quad p_\perp(r,\theta) = \left(\frac{r^4}{\Sigma^2} - \frac{2r^2}{\Sigma}\right)\tilde{\rho}(r) - \frac{r^3}{2\Sigma}\tilde{\rho}'(r) \tag{1.23}$$

where $\tilde{\rho}(r)$ and $\tilde{p}_\perp(r)$ are density and pressure for an original spherical solution, and the prime denotes the differentiation with respect to r.

The equation of state in the co-rotating frame can be presented in the form [54]

$$p_r(r,\theta) = -\rho(r,\theta); \quad p_\perp(r,\theta) = -\rho - \frac{\Sigma}{2r}\frac{\partial\rho(r,\theta)}{\partial r}. \tag{1.24}$$

A medium with anisotropic pressures (1.23) can be identified as an anisotropic quintessence [66]. Originally quintessence was defined by the equation of state $p = w\rho$ with $-1 < w < 0$ [25, 120]. For anisotropic quintessence the definition extends to $p_\alpha = w_\alpha\rho$, $w_\alpha = -1$ and $p_\beta = w_\beta\rho$, $\beta \neq \alpha$, where $w_\beta > -1$. In our case $p_r = w_r\rho,\ w_r = -1;\ p_\perp = w_\perp\rho,\ \ w_\perp = -1 - (\Sigma/2r\rho)(\partial\rho(r,\theta)/\partial r)$, where $w_\perp$ is evidently coordinate-dependent. It follows that in the considered case we have anisotropic and $r-$dependent quintessence.

At the ellipsoidal layers

$$\frac{r^2}{\Sigma} = \frac{(x^2+y^2+z^2-a^2)r^2+a^2z^2}{(x^2+y^2+z^2-a^2)r^2+2a^2z^2}$$

In the limit $z \to 0, r^2/\Sigma \to 1$ in the whole equatorial plane. Taking the limit $z \to 0$ we find from (1.23)

$$\rho(r,\theta) = \tilde{\rho}(r);\ \ p_\perp + \rho = -\frac{r}{2}\tilde{\rho}'(r) \tag{1.25}$$

in the whole equatorial plane including the disk, the ring and the origin. For spherical solutions regularity requires $r\tilde{\rho}'(r) \to 0$ as $r \to 0$ [51]. As a result we find that on the disk, $r \to 0,\ \ z \to 0$, the equation of state becomes

$$p_r = p_\perp = -\rho \tag{1.26}$$

and represents rotating de Sitter vacuum in the co-rotating frame [54, 61, 63].

1.3.2 Two Types of Regular Interiors

In the paper [22] it was suggested that rotation should inevitably lead to violation of the weak energy condition in interior regions, except the equatorial plane. Violation of WEC was found for regular rotating solutions, obtained with the Newman-Janis algorithm from particular spherical metrics [10, 136], and for a solution found by postulating a metric $g_{\mu\nu}$ and calculating $T_{\mu\nu}$ from the Einstein equations [116].

General analysis of dynamical equations has shown that the condition for WEC violation essentially depends on the density profile, and the regular rotating objects can have two kinds of interiors, one preserving and the other violating WEC [61, 63]. WEC is satisfied if and only if $\rho \geq 0$ and $p_k + \rho \geq 0$ [87]. In our case the first condition is satisfied by virtue of the basic property (1.1). Below we consider the conditions for the transversal pressure $p_\perp$, which is responsible for the WEC satisfaction/violation [61, 63].

As follows from (1.23), in the co-rotating frame

$$p_\perp + \rho = 2\left(\frac{r^4}{\Sigma^2} - \frac{r^2}{\Sigma}\right)\tilde{\rho}(r) - \frac{r^3}{2\Sigma}\tilde{\rho}'(r) \tag{1.27}$$

This equation can be written in the form [61, 63]

$$p_\perp + \rho = \frac{2r^2}{\Sigma^2}\left(\frac{\Sigma r}{4}|\tilde{\rho}'| - \tilde{\rho}a^2\cos^2\theta\right) \qquad (1.28)$$

which evidently implies a possibility for generic violation of the weak energy condition, dependently on the form of the spherical density profile, beyond a surface with $p_\perp + \rho = 0$, where the right-hand side in (1.28) would change its sign.

The simple condition can be obtained by writing equation (1.27) dependently on both density and tangential pressure $\tilde{p_\perp}$ for original spherical solution, which gives

$$p_\perp + \rho = \frac{r^2}{\Sigma}\left(\frac{2r^2}{\Sigma}\tilde{\rho} - (\tilde{\rho} - \tilde{p_\perp})\right) \qquad (1.29)$$

and immediately leads to the following conclusions:

(**1**) WEC is satisfied in the whole equatorial plane (in agreement with [22]) where $r^2/\Sigma = 1$ and, as a result, $p_\perp + \rho = \tilde{p_\perp} + \tilde{\rho}$. On the equatorial disk (1.11) the equation of state takes the form $p_\perp = p_r = -\rho$ and represents the rotating de Sitter vacuum as we have seen above. The existence of the interior de Sitter vacuum disk is the generic property of all regular rotating objects specified by (1.1) and described by axial solutions obtained from spherical solutions satisfying WEC.

(**2**) WEC is satisfied for rotating objects if the dominant energy condition (DEC) is violated for original spherical solutions. DEC requires ($\tilde{\rho} \geq \tilde{p}_k$ for principal pressures [87]. When it is violated, $(\tilde{\rho} - \tilde{p_\perp}) \leq 0$, and $p_\perp + \rho \geq 0$.

(**3**) When DEC is satisfied for a spherical solution, there are two cases:

(**3A**) WEC is still satisfied for a rotating object as long as

$$(\tilde{\rho} - \tilde{p_\perp}) \leq \frac{2r^2}{\Sigma}\tilde{\rho}. \qquad (1.30)$$

In all these cases the interior reduces to the de Sitter equatorial disk only. This is the 1-st type interior, shown in Fig. 1.2 (Left) [64], where we plotted also the black hole horizons and ergosphere.

(**3B**) WEC is violated for a rotating object if

$$(\tilde{\rho} - \tilde{p_\perp}) > \frac{2r^2}{\Sigma}\tilde{\rho} \qquad (1.31)$$

and there exists an additional surface where $p_\perp + \rho = 0$, which incorporates the de Sitter disk as a bridge, and beyond which WEC is violated [61, 63]

For detailed analysis of the the 2-nd type interior (**3B**) we present Eq. (1.28) in the form

$$p_\perp + \rho = \frac{r|\tilde{\rho}'|}{2\Sigma^2}S(r,z);\quad S(r,z) = r^4 - z^2P(r);\quad P(r) = \frac{2a^2}{r|\tilde{\rho}'|}(\tilde{\rho} - \tilde{p_\perp}). \quad (1.32)$$

This is the master equation, basic for investigation of energy conditions for the class of regular rotating objects, presented by stress-energy tensors satisfying (1.1), and, as a result, described by the metrics (1.13), obtained from the metrics (1.2) and (1.3) with using the Newman-Janis algorithm [61, 63, 64].

According to (1.32), WEC can be violated beyond the S-surface, on which $S(r,z)=0$. We see that when DEC is violated for a spherical solution, $\tilde{\rho}<\tilde{p}_\perp$, then $P(r)\le 0$, and the function $S(r,z)$ in (1.32) vanishes only at approaching the disk $r=0$, and we have the 1-st type interior satisfying WEC. When DEC is satisfied for a related spherical solution, then $P(r)\ge 0$, and there can exist an S-surface, at which $p_\perp+\rho=0$, beyond which WEC is violated, and we have the 2-nd type interior.

Below we present the detailed analysis of the 2-nd type interior for the spinning soliton replacing a naked singularity, when analysis can be carried out within one causal domain, for the whole region from $r=0$ to $r\to\infty$. The case of the black hole requires further research of energy conditions in different causal domains.

As follows from (1.32), the S-surface exists, if $P(r)>r^4/z^2$, which results in the condition on z-axis $\tilde{\rho}(z)-\tilde{p}_\perp\ge(r^5|\tilde{\rho}'(z)|)/(2a^2z^2)$. With taking into account that $\tilde{p_\perp}=-\tilde{\rho}-r\tilde{\rho}'/2$ and $\tilde{\rho}'\le 0$ [51] for spherical solutions satisfying WEC, we obtain the condition for the existence of S-surface

$$\left(\frac{z^2}{a^2}+1\right)|\tilde{\rho}'|\le\tilde{\rho}<\frac{16}{3x_g^2z^3};\quad x_g=\frac{r_g}{r_\Lambda}. \tag{1.33}$$

On the S-surface $r^2=|z|\sqrt{P(r)}$. Then the squared width of the S-surface, $W_S^2=(x^2+y^2)_S=(a^2+|z|\sqrt{P(r)})(1-|z|/\sqrt{P(r)})$. The squared width for the ellipsoid specified by $r=|z|_{max}=r_e$ is $W_e^2=(a^2+r_e^2)(1-z^2/r_e^2)$, and the difference $W_e^2-W_S^2=(a^2|z|/\sqrt{P(r)}+r_e^2)(1-|z|\sqrt{P(r)}/r_e^2)$, is characteristic for location of S-surface with respect to the ellipsoid with $r=|z|_{max}=r_e$. Assume that $W_S>W_e$ for some value of z, it means that there exists a ring specified by $|z|<|z|_{max}$ which belongs to the ellipsoid $r=r_e$, where these two surfaces intersect and where thus $|z|\sqrt{P(r_e)}/r_e^2=1$. As a result $|z|=r_e^2/\sqrt{P(r_e)}=|z|_{max}$, and hence these two surfaces can only touch each other on the poles and thus cannot cross. It follows that the S-surface is always entirely confined within the ellipsoid, defined by $|z|_{max}=r_e=r_e^2/\sqrt{P(r_e)}$, and contains the disk $r=0$ as a bridge.

For regular spherical solutions $r\tilde{\rho}'\to 0,\ \tilde{p}_\perp\to-\tilde{\rho}$ as $r\to 0$, and in the most general case $P(r)\to A^2r^{-(n+1)}$ with the integer $n\ge 0$ as $r\to 0$. The derivative of $W_S(z)$ near $z\to 0$ behaves as $z^{-(n+3)/(n+5)}$, goes to $\pm\infty$ as $z\to 0$, and the function $W_S(z)$ has the cusp at approaching the disk, and two symmetric maxima between $z=\pm|z_{max}|$ and $z=0$.

Typical form of the S-surface for the 2-nd type interior is shown in Fig. 1.2 (Right) for the case of a spinning soliton [64].

Relation between the width of the S-surface in the equatorial plane $W_S=a$ and its height $H_S=|z|_{max}=\sqrt{ar_v}$ defines the form of the S-surface by the oblateness parameter $\eta=H_S/W_S$. For a spinning G-lump $\eta<1$ and the S-surface is oblate (as shown in 1.2 (Right)) [63].

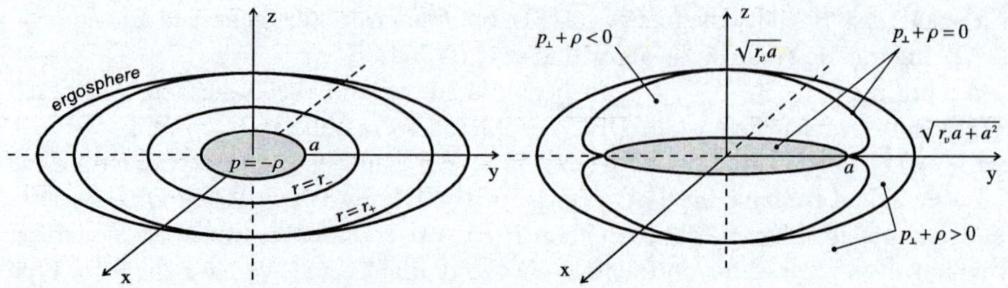

Fig. 1.2 Left: Horizons, ergosphere and de Sitter disk for the 1-st type of interior. Right: Vacuum S-surface incorporating the de Sitter disk in the 2-nd type interior

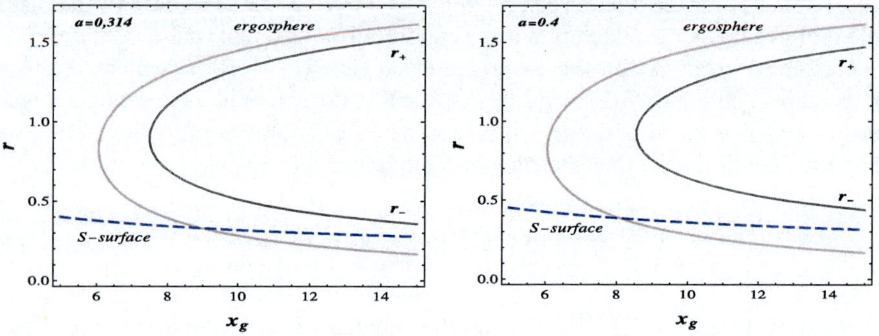

Fig. 1.3 $S-$surface, horizons and ergospheres dependently on the basic parameter $x_g = r_g/r_\Lambda$ for $a = 0.314$ (Left) and $a = 0.4$ (Right)

In the cavities between the upper and down boundaries of S-surface and the disk, WEC is violated, $p_\perp + \rho < 0$, and the cavities are filled with an anisotropic phantom fluid defined by $p_r = w_r\rho$, $w_r = -1$; $p_\perp = w_\perp\rho$, where $w_\perp < -1$ and is coordinate-dependent in accordance with (1.24).

In Fig. 1.3 [69] we show the location of the S-surface with respect to horizons and ergosphere, dependently on the basic characteristic parameter $x_g = r_g/r_\Lambda$, for two values of the specific angular momentum a.

The S-surface is located within the ergoregion, and the processes of energy extraction can involve the phantom energy. For a spinning soliton absence of the horizons makes the phantom region open to the outside region.

As we have seen above, the form of the de Sitter S-surfaces is generic. The pictures, illustrating the basic properties of regular rotating objects, are plotted for the exact axially symmetric solution, originated from the spherical solution with the phenomenologically regularized Newtonian profile [61, 63]

$$\tilde{\rho} = \frac{B^2}{(r^2+r_v^2)^2};\ r_v = \frac{\pi^2 B^2}{M};\ B^2 = \tilde{\rho}_\Lambda r_v^4;\ r_v = \left(\frac{4r_\Lambda^2 r_g}{3\pi}\right)^{1/3} = \left(\frac{4x_g}{3\pi}\right)^{1/3} r_\Lambda \quad (1.34)$$

where $\tilde{\rho}_\Lambda$ is the density of the de Sitter vacuum on the disk $r = 0$ and $x_g = r_g/r_\Lambda$.

The mass function $\mathcal{M}(r)$, and the metric function $g(r)$ are given by

$$\mathcal{M}(r) = \frac{r_g}{\pi}\left(\arctan\frac{r}{r_v} - \frac{rr_v}{r^2+r_v^2}\right);\ \ g(r) = 1 - \frac{2r_g}{\pi r}\left(\arctan\frac{r}{r_v} - \frac{rr_v}{r^2+r_v^2}\right). \tag{1.35}$$

The cut-off length scale r_v in (1.34) is proportional to the de Sitter radius r_Λ. The parameter B is directly related to ρ_Λ as $B^2/r_v^4 = \tilde{\rho}_\Lambda$, where $\tilde{\rho}_\Lambda$ is the maximal density at $r = 0$, which represents the density of self-interaction [63], and clearly verifies the underlying idea [138] to associate the cosmological constant with self-interaction. The cut-off scale r_v corresponds to the zero gravity surface, at which the strong energy condition is violated and the gravitational acceleration changes its sign, $\tilde{\rho} + \sum_k \tilde{p}_k = 0 \rightarrow \tilde{\rho}_\Lambda r_v^4(r^2 - r_v^2)/(r^2 + r_v^2)^3 = 0$ [63].

The mass M is related with the density $\tilde{\rho}_\Lambda$ by the formula $M = (4\pi/3)\tilde{\rho}_\Lambda r_\Lambda^2 r_g = (4\pi/3)\tilde{\rho}_\Lambda r_*^3$. The characteristic length scale $r_* = (r_\Lambda^2 r_g)^{1/3}$, appeared here in the natural way, is typical for geometry with the de Sitter interior and the Schwarzschild exterior [47, 50, 94]. The mass M can be also related to the cut-off parameter r_v by the relation $M = \pi^2\tilde{\rho}_\Lambda r_v^3$, where $\pi^2 r_v^3$ corresponds to the volume in the closed de Sitter world, which confirms the relation of the mass with the interior de Sitter vacuum [51].

For the density profile (1.34) $P(r) = r_*^4/r^2$ $(r_*^2 = r_v a)$, and the condition $S = 0$ in (1.32) reads $r^6 - r_*^4 z^2 = 0$. The width of the S-surface $W_S^2 = (x^2 + y^2)_S$ as the function of z has two maxima at $z_m = \pm r_*(1 - a^2/r_*^2)/2$ [61, 63]. The relation between the width of the S-surface in the equatorial plane $W_S = a$ and its height $H_S = |z|_{max} = \sqrt{ar_v}$ determines the form of the S-surface by the oblateness parameter $\eta = H_S/W_S$. For the density profile (1.34), the parameter $B^2 = Mr_v/\pi^2$ can be presented as $B^2 = \beta^2 GM^2$, where the dimensionless parameter $\beta^2 = (2/\pi^2)(4/3\pi)^{1/3}(x_g)^{-2/3}$. Detailed form of the S-surface depends on the parameters β and on the angular momentum a. The quantities H_S, W_S and η are given by

$$H_S = \sqrt{a/r_v};\ \ W_S = a;\ \ \eta = \sqrt{r_v/a} = \pi\beta/\sqrt{a}. \tag{1.36}$$

The case $a > \pi^2\beta^2$ defines the oblate S-surface shown in Fig. 1.2 (Right).

1.3.3 Light Rings

Light rings, formed by the innermost closed photon orbits, encompass the rotating compact objects and comprise their structure.

Photon orbits for the Kerr and Kerr-like black holes were examined in [95], and for the extreme asymptotically Kerr-Newman-AdS black holes in [134]. Stable photon orbits in the stationary axisymmetric electrovacuum spacetimes are considered in [45]. For the Hayward black hole where $g(r) = 1 - (2mr^2)/(r^3 + 2l^2m)$ and (l, m) are positive constants [88], the null geodesics are studied in [35]. The question of stability of light rings around spherical ultracompact objects has been addressed in [91, 92], and for axisymmetric case in [40, 42, 78, 81].

General analysis and classification of the closed equatorial photon orbits for RRBHs and spinning solitons with the de Sitter/phantom interiors are presented in [69], with the special attention to the innermost orbits and their location with respect to ergospheres.

Investigation of orbits is typically carried out in terms of the variable $u = 1/r$. The dynamical equations for timelike geodesics, representing the circular equatorial orbits, read [34]

$$u^{-4}\dot{u}^2 = F(u) = 0; \quad F' = 0; \tag{1.37}$$

$$F(u) = 2\mathcal{M}x^2u^3 - (x^2 + 2aEx)u^2 - (a^2u^2 - 2\mathcal{M}u + 1) + E^2 = 0; \quad x = L - aE \tag{1.38}$$

where E is the energy of a particle on the orbit, and L is the projection of its angular momentum on the $z-$axis, related to the Killing vector $k_\varphi = \delta^\alpha_\varphi$ in the axially symmetric geometry (1.13). The prime denotes differention with respect to u. The general equation, governing the photon orbits, can be obtained by taking the limit $E \to \infty$ in the general expressions for E and L, derived from Eqs. (1.37) and (1.38) [34].

In the regular geometry [69]

$$F'(u) = [3\mathcal{M}(u)u^2 + \mathcal{M}'(u)u^3 - u]x^2 - 2aEux - a^2u + (\mathcal{M}(u) + \mathcal{M}'(u)u) = 0. \tag{1.39}$$

From Eqs. (1.37)–(1.39) we obtain

$$E^2 = 1 - (\mathcal{M}(u)u - \mathcal{M}'(u)u^2) + (\mathcal{M}(u)u^3 + \mathcal{M}'(u)u^4)x^2; \tag{1.40}$$

$$2aExu = (3\mathcal{M}(u)u^2 + \mathcal{M}'(u)u^3)x^2 - x^2u - a^2u + (\mathcal{M}(u) + \mathcal{M}'(u)u). \tag{1.41}$$

Excluding E from (1.40) and (1.41), we obtain the equation for x^2

$$Ax^4 + Bx^2 + C = 0; \tag{1.42}$$

where $A = u^2[(3\mathcal{M}(u)u + \mathcal{M}'(u)u^2 - 1)^2 - 4a^2(\mathcal{M}(u)u^3 + \mathcal{M}'(u)u^4)]$;
$B = 2u[(\mathcal{M}(u) + \mathcal{M}'(u)u - a^2u)(3\mathcal{M}(u)u + \mathcal{M}'(u)u^2 - 1) - 2a^2u(1 - \mathcal{M}(u)u/ + \mathcal{M}'(u)u^2)]$; $C = [\mathcal{M}(u) + \mathcal{M}'(u)u - a^2u]^2$.

Determinant of the Eq. (1.42) is given by

$$B^2 - 4AC = 16u^2a^2(\mathcal{M}(u)u + \mathcal{M}'(u)u^2)\Delta_u^2; \quad \Delta_u = a^2u^2 - 2\mathcal{M}(u) + 1, \tag{1.43}$$

and the Eq. (1.42) yields

$$x^2u^2 = \frac{\Delta_u}{Q_{\mp}} - 1; \quad Q_{\pm} = 1 - 3\mathcal{M}(u)u - \mathcal{M}'(u)u^2 \pm 2a\sqrt{\mathcal{M}(u)u^3 + \mathcal{M}'(u)u^4}, \tag{1.44}$$

and $x = -(a\sqrt{u} \pm \sqrt{\mathcal{M}(u) + \mathcal{M}'(u)u})(\sqrt{uQ_{\mp}})^{-1}$.

The integrals of motion E and L on the orbits are

$$E = \frac{1}{\sqrt{Q_{\mp}}}\left[1 - 2\mathcal{M}(u)u \mp au\sqrt{\mathcal{M}(u)u + \mathcal{M}'(u)u^2}\right]; \tag{1.45}$$

$$L = \mp\frac{\sqrt{\mathcal{M}(u) + \mathcal{M}'(u)u}}{\sqrt{uQ_{\mp}}}\left[1 + a^2u^2 \pm \frac{2au\mathcal{M}(u)\sqrt{u}}{\sqrt{\mathcal{M}(u) + \mathcal{M}'(u)u}}\right]. \tag{1.46}$$

An upper sign corresponds to the retrograde orbits, and a lower sign to the direct orbits. For $\mathcal{M} = const$ the integrals L and E coincide with those in the Kerr geometry [34]. For analysis of behavior of photon orbits we apply the Eq. (1.45) in the limit $E \to \infty$ ($Q_{\mp} = 0$), which gives the general equation for the circular equatorial photon orbits [69]

$$1 - 3\mathcal{M}(u)u - \mathcal{M}'(u)u^2 \mp 2a\sqrt{\mathcal{M}(u)u^3 + \mathcal{M}'(u)u^4} = 0. \tag{1.47}$$

The detailed characteristics of particular orbits are obtained by analysis of the null geodesics which describe the orbits of photons with the wavelength less than the characteristic scale of geometry, and are described by ([69] and references therein)

$$\dot{r}^2 = E^2 - V_{\gamma}(r) = 0; \quad V_{\gamma}(r) = -\frac{2\mathcal{M}(r)}{r^3}(L - aE)^2 + \frac{(L^2 - a^2E^2)}{r^2}. \tag{1.48}$$

The key point here is that potentials $V_{\gamma}(r)$ in the regular geometry decrease from $V(r) \to \infty$ at $r = 0$, and the first extremum is the minimum, which provides the existence of the innermost stable orbits, unlike the Kerr geometry, where the potentials increase from $V(r) \to -\infty$ at $r = 0$, so that the first extremum is the maximum and the innermost orbits are unstable.

Introducing the parameter $\xi = L/E$ we obtain the equation for the potential, normalized on E^2, on the orbits (1.48)

$$V_{\gamma}(r) = \left[\frac{(\xi^2 - a^2)}{r^2} - \frac{2\mathcal{M}(r)}{r^3}(\xi - a)^2\right] = 1 \tag{1.49}$$

which immediately give the basic constraint on the orbital moment ξ

$$\xi^2 > a^2. \tag{1.50}$$

Orbits are determined by zeros of the potential derivative

$$V'_\gamma(r) = \left[-\frac{2}{r^3}(\xi^2 - a^2) + \frac{6\mathscr{M}(r)}{r^4}(\xi - a)^2 - \frac{2\mathscr{M}'(r)}{r^3}(\xi - a)^2 \right] = 0. \quad (1.51)$$

Here the prime denotes differentiation with respect to r. Equation (1.51) gives the relation

$$r_\gamma = [3\mathscr{M} - r\mathscr{M}'] \frac{(\xi - a)^2}{(\xi^2 - a^2)}. \quad (1.52)$$

which coincides with the relevant relation in the Kerr geometry for $\mathscr{M} = const = M$ [34].

As follows from the Eqs. (1.49) and (1.51), the integrals of motion on the orbits satisfy

$$(\xi - a)^2 = \frac{r^3}{(\mathscr{M}(r) - r\mathscr{M}'(r))}; \quad (\xi^2 - a^2) = \frac{r^2(3\mathscr{M} - r\mathscr{M}'(r))}{(\mathscr{M}(r) - r\mathscr{M}'(r))}. \quad (1.53)$$

These relations impose the evident (taking into account (1.50)) constraints on the mass function

$$(\mathscr{M}(r) - r\mathscr{M}'(r)) > 0; \quad (3\mathscr{M}(r) - r\mathscr{M}'(r)) \geq 0. \quad (1.54)$$

Introducing the variable $y = \xi + a$ in Eq. (1.52), we obtain for the orbit radius

$$r_\gamma = \left(3\mathscr{M} - \mathscr{M}'r\right)(1 - 2a/y), \quad (1.55)$$

and putting it into (1.49) we derive the equation for the integral of motion y

$$y^3 + 3py + 2q = 0; \quad p = -\frac{1}{3}\frac{(3\mathscr{M} - \mathscr{M}'r)^3}{(\mathscr{M} - \mathscr{M}'r)}; \quad q = a\frac{(3\mathscr{M} - \mathscr{M}'r)^3}{(\mathscr{M} - \mathscr{M}'r)} \quad (1.56)$$

which for $\mathscr{M} = const$ coincides with the relevant equation in the Kerr geometry [34].

For an extreme black hole its double horizon, $r = r_\pm$ presents the lower boundary of a manifold for an outside observer. For $r = r_\pm$ the involved functions are fixed, and we can apply the standard procedure relating the roots of Eq. (1.56) with the function $D = p^3 + q^2$ which gives

$$D(r_\pm) = \frac{(3\mathscr{M} - \mathscr{M}'r_\pm)^6}{(\mathscr{M} - \mathscr{M}'r_\pm)^2}\left[a^2 - \frac{(3\mathscr{M} - \mathscr{M}'r_\pm)^3}{27(\mathscr{M} - \mathscr{M}'r_\pm)}\right] \propto \left(a^2 - \frac{1}{27}\frac{(r_\pm^2 + 2a^2)^3}{r_\pm^2 a^2}\right). \quad (1.57)$$

On the double horizon, $r_\pm = \mathscr{M}(r_\pm) = a_{dh}$ we obtain $D(r_\pm) = 0$, and the appropriate value of y given by the double root of Eq. (1.56)

$$y = 2\sqrt{|p|}\cos(\pi/3) = \sqrt{\frac{(3\mathscr{M} - \mathscr{M}'r_\pm)^3}{3(\mathscr{M} - \mathscr{M}'r_\pm)}}. \quad (1.58)$$

Expressing the mass function and its derivative in terms of the function $\Delta(r)$ and its derivatives

$$3\mathcal{M}(r) - \mathcal{M}'(r)r = r + 2a^2/r - 2\Delta/r + \Delta'/2; \quad \mathcal{M}(r) - \mathcal{M}'r = a^2/r - \Delta/r + \Delta'/2, \tag{1.59}$$

we get on the double horizon

$$\sqrt{3}ar_{\pm}y = (r_{\pm}^2 + 2a^2)^{3/2}. \tag{1.60}$$

At the same time the equation for r_γ (1.55) takes the form $3\sqrt{3}a^2 r_\gamma = (r_\gamma^2 + 2a^2)^{3/2}$ and is satisfied for $r_\gamma = r_\pm = \mathcal{M}(r_\pm)$ on the double horizon, where $a_{dh} = \mathcal{M}(r_\pm)$. The limiting (innermost) direct orbit

$$r_\gamma = r_\pm = \mathcal{M}(r_\pm) = a_{dh}; \; y = 3\mathcal{M}(r_\pm); \quad \xi = 2\mathcal{M}(r_\pm) = 2r_\pm = 2a_{dh} \tag{1.61}$$

coincides for $\mathcal{M} = const$ with the unstable orbit $\xi = 2M, \;\; r_\gamma = M$ in the Kerr geometry [34].

The retrograde limiting orbit is given by

$$\xi = -7a; \;\; r_\gamma = \frac{4}{3}((3\mathcal{M} - \mathcal{M}'r)) \tag{1.62}$$

and coincides, for $\mathcal{M} = const = M$, with the unstable retrograde orbit $r = 4M$ in the Kerr geometry [34].

For analysis of the sequence of photon orbits we apply the general Eq. (1.47), written in the r-variable

$$r^{3/2} + \mathcal{M}'r^{3/2} - 3\mathcal{M}r^{1/2} \mp 2a\sqrt{\mathcal{M} - \mathcal{M}'r} = 0 \tag{1.63}$$

where an upper sign applies to a retrograde orbit and a lower sign to a direct orbit.

To study the character of orbits, dependently on the angular momentum a, their location with respect to a rotating object and its nature, we consider the curve $f(r, a) = 0$ defined by the orbit Eq. (1.63). For the direct orbits

$$f(r, a) = r^{3/2} + \mathcal{M}'r^{3/2} - 3\mathcal{M}r^{1/2} + 2a\sqrt{\mathcal{M} - \mathcal{M}'r} = 0. \tag{1.64}$$

Differentiating $f(r, a)$ we obtain

$$\frac{dr}{da} = -\frac{2\sqrt{\mathcal{M} - \mathcal{M}'r}}{3r - 3\mathcal{M} - 3\mathcal{M}'r + 2\mathcal{M}''r^2 - (2a\mathcal{M}''r\sqrt{r})/(\sqrt{\mathcal{M} - \mathcal{M}'r})} \tag{1.65}$$

which can be reduced to the form

$$\frac{dr}{da} = -\frac{4\sqrt{\mathcal{M}r - \mathcal{M}'r^2}}{3\Delta' + 4\mathcal{M}''r^2 - (4a\mathcal{M}''r^2)/(\sqrt{\mathcal{M}r - \mathcal{M}'r^2})}. \tag{1.66}$$

The innermost stable direct orbits in the minimum of the potential V_γ represent the stable light rings $r < r_\pm$ around a spinning soliton with $a > a_{dh}$. In this region the curve $r(a)$ is directed towards decreasing r and increasing a, and $dr/da < 0$. At approaching the double horizon, corresponding to the orbit (1.61), the derivative $dr/da \to -4\mathcal{M}(r_\pm)/3\Delta' \to -\infty$. Since then two cases are possible, dependently on the density profile and hence on the mass function.

Case A—The curve $r(a)$ arrives at the asymptotically Kerr branch of unstable orbits without changing its direction, $dr/da < 0$ on this branch as on the preceding branch, the curve $r(a)$ is still directed to increasing r at decreasing a. For $a < a_{dh}$ photon orbits form the light rings for each value of a.

Case B—The second possibility is that the derivative dr/da changes its sign at the double horizon, the curve $r(a)$ changes its direction, and an additional branch appears with $dr/da > 0$, starting from the branching point at $r = r_\pm$ and $a = a_{dh}$, and directed to increasing $r > r_\pm$ with increasing $a > a_{dh}$. To arrive at the asymptotically Kerr branch with $dr/da < 0$, the curve $r(a)$ must meet the second branching point $r_{br} > r_\pm$ at $a = a_{br}$, defined by the equation

$$3\Delta' + 4\mathcal{M}''r_{br}^2 - \frac{4a\mathcal{M}''r_{br}^2}{\sqrt{\mathcal{M}r_{br} - \mathcal{M}'r_{br}^2}} = 0. \tag{1.67}$$

In Fig. 1.4 [69] we show typical sequence of photon orbits in the **Case B** dependently on a (Left), and the enlarged image of their behavior near the double horizon (Right). Direct orbits are shown in the blue color, and retrograde orbits in the orange color.

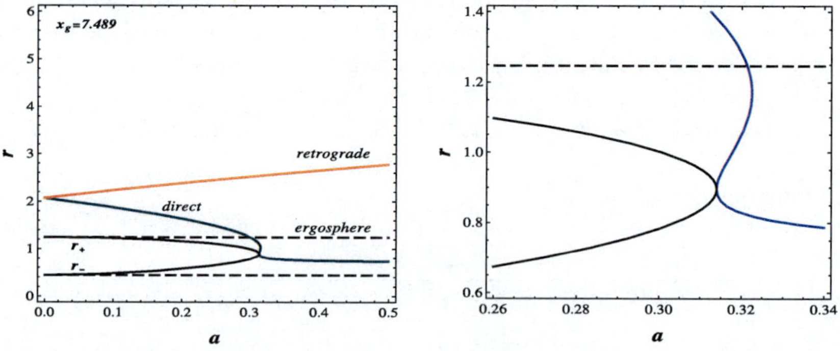

Fig. 1.4 Left: Photon orbits, horizons and ergospheres dependently on a. Right: The enlarged image near the double horizon

We see that the appearance of the additional branching point in the soliton region at $a = a_{br}$, leads to the existence of three branches around a spinning G-lump in the range $a_{dh} < a < a_{br}$, and two branches for $a = a_{br}$.

The type of an orbit is determined by the second derivative of the potential V_γ on the orbit satisfying (1.49) and (1.51), given by

$$r^3 V''_\gamma(\mathcal{M} - \mathcal{M}'r) = -6\mathcal{M}r + 6\mathcal{M}'r^2 - 2\mathcal{M}''r^3. \tag{1.68}$$

Presenting V''_γ as a polynomial with the positively defined coefficients, we get, taking into account (1.14) and (1.25)

$$4r^3 V''_\gamma(\mathcal{M} - \mathcal{M}'r) = [3\tilde{\rho} + 6(\tilde{\rho} + \tilde{p_\perp})]r^4 - 24\mathcal{M}r \tag{1.69}$$

which suggests that V''_γ can change its sign only once for the considered class of axially symmetric solutions originated from spherical solutions satisfying WEC. According to the Cartesius rule, which states that the number of real positive roots equals to the number of sign changes, V''_γ has one zero point, and behavior of the curve $r(a)$ depends now on where this point is located. On the double horizon

$$V''_\gamma r_\pm^2 = \Delta'' - 8; \quad \Delta''(r_\pm) = 2 - 2\mathcal{M}''r_\pm = 2 - (3/2)x_g^2(\tilde{\rho} - \tilde{p_\perp})r_\pm^2. \tag{1.70}$$

It follows

$$2r_\pm^2 V''_\gamma = -12 - 3x_g^2(\tilde{\rho} - \tilde{p_\perp})r_\pm^2. \tag{1.71}$$

For the case when a related spherical solution satisfies DEC, ($\tilde{\rho} > \tilde{p_\perp}$), it gives $V''_\gamma < 0$, and the orbit on the double horizon (1.61) is unstable. The option when a spherical solution satisfies DEC and an axial solution violates WEC, corresponds to a rotating objects with the 2-nd type interior. In this option the curve $r(a)$ starts in the soliton region from stable orbits in the minimum of V_γ, corresponding to the innermost stable light rings, unless $V''_\gamma = 0$ on the marginally (un)stable photon orbit r_{ms} for $a_{ms} > a_{dh}$, corresponding to the marginally (un)stable light ring, and then arrives at the unstable orbit $r_\gamma = r_\pm$ for $a = a_{dh}$. This branch includes thus the unstable innermost orbits around a soliton for $a_{dh} < a < a_{ms}$. Then in the **Case A** it follows asymptotically Kerr branch of unstable orbits after crossing $r_\pm$ corresponding to the unstable light ring. In the **Case B** for $a_{dh} < a < a_{ms}$ there appears an additional branch with $dr/da > 0$ of unstable orbits, meets the branching point $r_{br} > r_\pm$ and arrives at the asymptotically Kerr branch of unstable light rings.

In the case when a related spherical solution violates DEC ($\tilde{\rho} < \tilde{p_\perp}$), an axial solution satisfies WEC and describes an object with the 1-st type interior. In accordance with (1.71), the sign of V''_γ at the double horizon, in principle, can be any. When $V''_\gamma(r_\pm) < 0$, the behavior of the curve $r(a)$ is the same as described above. When $V''_\gamma(r_\pm) = 0$, the orbit on the double horizon is marginally (un)stable. In the **Case A** it is followed by the asymptotically Kerr branch of unstable orbits for $a < a_{dh}$, which correspond to the unstable light rings. In the **Case B** it is followed by an additional branch of unstable orbits with $dr/da > 0$, meets a branching point $r_{br} > r_\pm$ at

$a = a_{br}$ and go to the asymptotically Kerr branch with $dr/da < 0$. It is also possible that $V''_\gamma(r_\pm) > 0$, and the orbit (1.61) is stable. In the **Case A** a certain part of the next branch contains stable orbits and presents a set of the stable light rings around a regular rotating black hole. Next $V''_\gamma = 0$ at the marginally (un)stable orbit r_{ms} for $a = a_{ms}$, followed by the asymptotically Kerr branch of unstable orbits. In the **Case B** an additional branch continues as the stable orbits around a soliton and transforms to unstable asymptotically Kerr orbits at the marginally (un)stable orbit, eventually at the second branching point r_{br}.

For retrograde orbits the curve $r(a)$ has the form

$$f(r,a) = r^{3/2} + \mathcal{M}' r^{3/2} - 3\mathcal{M} r^{1/2} - 2a\sqrt{\mathcal{M} - \mathcal{M}' r} = 0 \tag{1.72}$$

which differs from (1.66) by the sign of dr/da, and by the sign in the denominator, denoted below as Den_r, which can be written as [69]

$$Den_r = 3\Delta' + 4\mathcal{M}'' r^2 + \frac{4a\mathcal{M}'' r^2}{\sqrt{\mathcal{M} r - \mathcal{M}' r^2}} = 6\mathcal{M} - 6\mathcal{M}' r + 4\mathcal{M}'' r^2 + \frac{4a\mathcal{M}'' r^2}{\sqrt{\mathcal{M} r - \mathcal{M}' r^2}}. \tag{1.73}$$

This function is positive on the unstable orbits in the asymptotically Kerr region, in accordance with the behavior of V''_γ given by (1.68). As a result $dr/da > 0$, the curve $r(a)$ continues monotonically from the asymptotically Kerr branch $dr/da = (4\sqrt{Mr})/(3\Delta')$ for $r \gg a$ towards the soliton region [69].

As we have seen, around a spinning soliton there can exist three branches of direct orbits, which correspond to the co-rotating light rings. The innermost co-rotating light rings are stable, and two additional can be stable or unstable dependently on the density profile. Around a black hole there exists one co-rotating branch of direct orbits, the innermost part of which can be stable for a certain class of density profiles. Around both black holes and G-lumps there exists one branch of unstable retrograde orbits, corresponding to the unstable counter-rotating light rings.

The innermost orbits in the field of a soliton have radii $r < r_\pm$, and hence exist inside the ergoregion, since its upper boundary goes on the level $r = r_+ > r_\pm$.

Existence of the light rings (closed photon orbits) around RRBHs and spinning G-lumps allows to identify them as the ultracompact objects in accordance with their definition as the self-gravitating systems with the light rings [29].

Stress-energy tensors responsible for regular rotating objects specified by $T^t_t = T^r_r$, *represent an intrinsically anisotropic medium with* $p_r = w_r\rho$; $w_r = -1$, $p_\perp = w_\perp \rho$, *where the parameter* $w_\perp$ *is coordinate-dependent. All regular rotating objects of this class have the equatorial de Sitter vacuum disk* $r = 0$. *Energy conditions imply the existence of two kinds of regular interiors. The 1-st kind interiors satisfy WEC and are filled with the anisotropic quintessence with* $w_\perp > -1$. *Interiors of the 2-nd type contain an additional closed de Sitter S-surface, with the de Sitter disk as the bridge. WEC is violated in the cavities between the S-surface and the disk, which are filled with an anisotropic phantom fluid with* $w_\perp < -1$. *Around a RRBH there exist one set of co-rotating light rings, which can include the branch of the innermost stable orbits for a certain class of density profiles, followed by a marginally (un)stable orbit and then by unstable light rings at the asymptotically Kerr branch. Around a soliton there exists the set of the innermost stable co-rotating light rings, and there can exist one or two additional sets of co-rotating light rings, stable of unstable dependently on the density profile. Counter-rotating unstable light rings exist around both regular rotating black holes and spinning solitons.*

1.4 Observational Signatures

In this Section we consider the identification of RRBHs by their shadows, and the observational signatures of the primordial RRBHs, their remnants, and spinning G-lumps as heavy dark matter candidates with the dark (de Sitter/phantom) interiors.

1.4.1 Identification of RRBH by its Shadow

The shadow of a black hole looks as a dark spot over an image of a remote source of radiation ([34, 46] and references therein). At present the black hole shadows can be directly observed [74, 132], giving information on the position and motion of an observer with respect to the black hole [110], and on the black hole parameters [90, 103]. The observational possibilities provided by the Event Horizon Telescope Collaboration [44] have been demonstrated by the first spectacular M87 result - observation of the black hole shadow in M87 [4].

The sensitivity of the shape for the Kerr black hole shadow to its angular momentum and orientation with respect to an observer, has been considered in [79, 103, 125]. Theoretical analysis of the black hole shadows was carried out in [103] for several black holes including the regular rotating black hole, described by the axial solutions, obtained from the Bardeen [7] and Hayward [88] spherical solutions. It was shown that the detection of an object by its shadow can give certain essential constraints on its nature [103]. Similar results have been obtained in the frame of different gravity theories [112], of general relativistic transfer [137] and hydrodynamical simulations of black hole accretion [33, 126], and for a dilaton black hole [76] (for a review [41]).

The black hole shadow presents its direct image, whose boundary is determined by the photons gravitational capture cross-section, which for a regular rotating black hole is less than that for the Kerr black hole, and the difference depends essentially on the form of the density profile of the RRBH [65]. This allows for identification of a regular black hole by the difference of its shadow from that for the Kerr black hole. Comparison of a predicted shadow for a regular rotating black hole shadow with the observed shadow and with the shadow of the Kerr black hole of the same mass would provide information about the interior content of a regular black hole (for a detailed analysis [65]). Below we present the results of the comparative numerical analysis

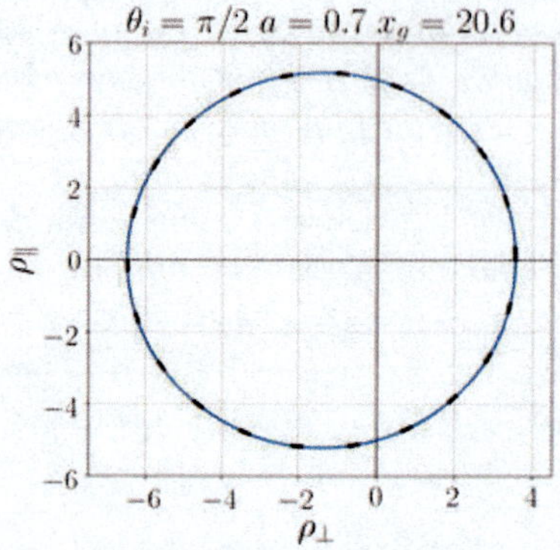

Fig. 1.5 Comparison of shadows for the case of the density profile (1.74) (Left) and for the density profile (1.34) for $a = 0.7$. An observer angular coordinate $\theta_i = \pi/2$

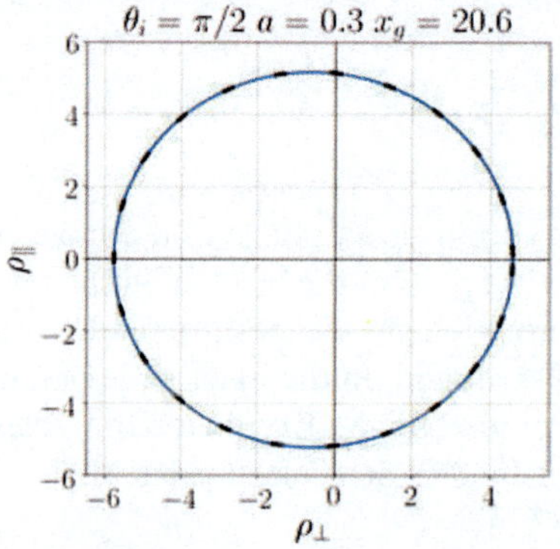

Fig. 1.6 Comparison of shadows for the case of the density profile (1.74) (Left) and for the density profile (1.34) for $a = 0.3$. An observer angular coordinate $\theta_i = \pi/2$

for two RRBHs, one with the phenomenologically regularized Newtonian density profile (1.74), the other with the density profile from the semiclassical model of the vacuum polarization in the spherical gravitational field [47]

$$\rho(r) = \rho_\Lambda e^{-r^3/(r_\Lambda^2 r_g)}; \quad \mathscr{M}(r) = M(1 - e^{-r^3/(r_\Lambda^2 r_g)}) \tag{1.74}$$

based on the hypothesis of symmetry restoration in the course of a gravitational collapse when all fields contribute to vacuum polarization and thus to gravity [47, 48, 51].

In Figs. 1.5 and 1.6 [65] we compare the RRBH shadows with that for the Kerr black hole (plotted with the dashed line), for the density profiles (1.74) and (1.34). The angle θ_i is the angular coordinate of an observer.

We see that the difference of the RRBH shadow from the Kerr shadow is much more substantial for the slowly decreasing density (1.34) than for a quickly decreasing density (1.74), which testifies for the essential dependence of RRBH shadows on the pace of the density decreasing. Difference related to spin is not as essential as the difference due to the pace of the density decreasing [65].

1.4.2 Primordial RRBHs and G-Lumps as Dark Matter Candidates

Primordial black holes are considered as reliable dark matter candidates for more than three decades [14, 31, 38, 100, 108, 122].

Primordial black holes with the masses exceeding 10^{15} g can survive to the present time and contribute to heavy dark matter. Primordial black holes with the masses less than 10^{15} g, had time to evaporate and can contribute to dark matter as remnants provided that they are stable. Characteristic scales for astrophysical structures originated from primordial quantum black holes have been considered in [27].

The primordial black holes arise from the primordial density inhomogeneities, and their population (including remnants) can serve as the pronounced signature for the inhomogeneity of the early Universe, responsible for the formation of these objects, whose contribution is constrained by the observed dark matter density [62].

The existence of black hole remnants and their role in the universe dynamics have been widely discussed since they were predicted in [86, 139]. In particular, it was shown, that the hybrid inflation can produce their abundance sufficient to be the primary source of dark matter [38, 107]. The production of black hole remnants in the very early universe can induce a matter-dominated stage before the onset of inflation. In [128] it has been shown that they can be responsible for the quadrupole anomaly of the CMB power spectrum. The ways of detection of remnants and the relevant cosmological constraints have been discussed in [119].

In particle physics, dark matter production from black hole remnants has been studied at CERN LHC [115]. Possible signatures of black hole events in *pp* collisions were investigated in the framework of the low scale gravity for the hypothesis that black holes do not decay completely into SM particles [43] but leave behind metastable remnants [101]. The production of black hole remnants has been predicted to occur with the rate of 10^8 per year [130]. For a reference stable remnant scenario with the Monte Carlo generator CHARYBDIS2, in which the remnant behavior is determined by the kinematic constraints and the conservation of quantum numbers, such as the baryon charge, it was shown that electrically neutral remnants are highly favored and that a significantly larger amount of the missing transverse momentum is to be expected, than in the case of a complete decay [13].

At the same time, for a singular black hole the existence itself of a remnant as the end-point of its evaporation, remains open question [32, 85, 102]. In this case the Generalized Uncertainty Principle requires existence of a remnant as a Planck size black hole [1, 12, 111]. Arguments for remnants based on the analysis of the Hawking evaporation and relevant causal domains have been presented in [72]. In the

Palatini framework a stable remnant with the Planckian mass can arise as a geon-like solitonic object, supported by the gravitational and electromagnetic fields [105].

Analysis of the process of evaporation in the course of formation of a black hole has shown that the end-point depends essentially on details of a collapse [97].

On the other hand, no evident symmetry or quantum number was found which would prevent complete evaporation [131]. Character and scale of uncertainty concerning an endpoint of the Hawking evaporation of a singular black hole, are clearly evident in the case of a multi-horizon space-time ([57] and references therein).

Complete evaporation of a black hole in the de Sitter space (in a sense actual for our universe with domination of the cosmological constant at the level of the order of 75% contribution to the total density), results in the problem, clearly formulated by Aros [6]: In the Schwarzschild-de Sitter space-time the cosmological horizon is not observer-dependent as in the de Sitter space, but the real horizon due to presence of the black hole, which breaks the global symmetries involving the radial direction. A serious doubt concerns therefore a causal structure of space-time: the fate of energy radiated once a black hole disappears, leaving behind the de Sitter space with nothing beyond the cosmological horizon but the de Sitter space itself, so energy can not be hidden there [6]. Complete evaporation would create also an additional and quite serious problem – how to evaporate a singularity? [57].

This problem, as well as the existential problem itself, does not arise for regular black holes, since their evaporation leaves behind thermodynamically stable double-horizon remnants with zero temperature and positive specific heat [48, 55, 58].

For the density profile (1.74) [47], which describes vacuum polarization effects [48], the mass of the double horizon remnant is given by [48, 58]

$$M_r = 0.3 M_{pl} \sqrt{\rho_{pl}/\rho_{int}} \tag{1.75}$$

where ρ_{int} is the density of interior de Sitter vacuum.

In the frame of the hypothesis of arising the de Sitter interior due to symmetry restoration in a collapse at the GUT scale [49], $\rho_{int} = \rho_{GUT}$ and $E_{int} = E_{GUT} \simeq 10^{15}$ GeV, the mass of the regular double-horizon remnant and its gravitational radius are [48, 58]

$$M_r \simeq 0.6 \times 10^3 \ g, \ \ r_g \simeq 10^{-25} \ cm. \tag{1.76}$$

In the frame of the hypothesis of self-regulation of geometry due to vacuum polarization effects near the Planck scale [121] or of the existence of the limiting curvature [75], $E_{int} = E_{Pl}$, the mass of the regular remnant is of the order of M_{Pl}.

The most general mechanism of formation of primordial black holes in the early Universe involves primordial density inhomogeneities leading to appearance of the overdense regions, which can stop expansion and collapse [86, 139]. Primordial black holes can form provided that the mass contained under its gravitational radius is sufficient for a collapse into a black hole [100, 122].

Regular primordial black holes with the de Sitter interior appear when a quantum collapse of a primordial fluctuation does not lead to formation of a central singularity but stops at achieving the de Sitter state $p = -\rho$, at which the acceleration proportional to $(\rho + 3p)$ changes its sign, and the gravitational attraction transforms by the de Sitter geometry to the gravitational repulsion [60].

Probability of formation of an object with mass M and the de Sitter interior in a quantum collapse of a quantum fluctuation at a phase transition involving an inflationary vacuum of the scale E_{infl}, is estimated as [60]

$$D > \exp\left[-4\left(\frac{M}{M_{Pl}}\right)^{3/4}\left(\frac{E_{Pl}}{E_{infl}}\right)\right]. \tag{1.77}$$

The basic constraint on the mass of M reads [60]

$$\frac{M}{M_{Pl}} > \left(\frac{E_{infl}}{E_{Pl}}\right)^4\left(\frac{E_{Pl}}{E_{int}}\right)^8. \tag{1.78}$$

In the case of the GUT scale for both inflationary and interior de Sitter vacuum, $E_{int} = E_{infl} = E_{GUT} \simeq 10^{15} GeV$, it gives the constraint $M > 10^{11}$ g. Primordial regular black holes have thus enough time to evaporate to stable remnants.

One more possibility for production of regular black holes and G-lumps from primordial quantum fluctuations appears at the second inflationary stage, predicted by the standard model of particle physics for the phase transition at the QCD scale $E_{infl} = E_{QCD} \simeq (100 - 200)$ MeV ([17] and references therein). At this stage supermassive remnants and G-lumps with M_r given by (1.75) with $\rho_{int} = \rho_{QCD}$, can be produced, although with the smaller probability, according to (1.77).

Regular primordial black holes, their remnants and G-lumps can be considered as heavy dark matter candidates, generically related to dark energy through their obligatory de Sitter interiors, representing dark ingredients in one drop [55, 59].

In the course of the universe evolution regular primordial black holes, their remnants and G-lumps can capture charged particles and form gravitationally bound ($\alpha_G = GMm/(\hbar c)$ where m is the mass of a captured particle) quantum systems called graviatoms ([60] and references therein). In particular, graviatoms formed to the end of the first inflationary stage, can, in principle, capture the GUT particles [60], whose binding energy in graviatoms is comparable with their mass, which makes graviatoms stable, and could survive to the present epoch as the constituents of graviatoms [62]. They can also capture leptoquarks, which arise at the GUT epoch and further decay into quarks and leptons, but a certain part of them can survive, and this part is accumulated in the galactic halos [80]. These leptoquarks can be captured by any of considered here objects with de Sitter interior: by near extreme primordial black holes, by remnants and by G-lumps, and become the constituents of graviatoms [60].

Graviatoms contribute, as heavy dark matter candidates, to atomic dark matter generically related to a dark energy through its de Sitter interiors. Their observational signatures as dark matter candidates include the electromagnetic radiation from graviatoms with the electrically charged captured particles.

The specific observational signatures of primordial black holes, their remnants, G-lumps and graviatoms are not influenced by the values of their angular momenta, because they are determined by the physical properties of their deep interiors.

All regular objects with the GUT scale interiors [47, 49], where baryon and lepton numbers are not conserved, may induce a proton decay in the matter of an underground detector. If the relevant cross section is determined by the geometrical size of a nucleon $\sigma_i \sim 10^{-26}$ cm^2, one can expect up to 300 events per year in the 1 km^3 detector, like IceCUBE, which would be their observational signature in heavy DM searches at the IceCUBE experiment [62]. In the most promising case, which would need a special study in the framework of e.g. bag models, the very penetration of a remnant or G-lump inside the nucleon may induce its decay.

In any case the key problem is how to distinguish such events, in which only 1 GeV energy is released. Search for this effect needs special study in the analysis of the IceCUBE data.

For the graviatom with the captured electrically charged particle, its nucleus component with the false vacuum interior component would induce the nucleon decay, while the charged component provides the enhancement of the relevant cross section. It makes the searches for graviatoms challenging for the IceCUBE experiment.

The specific observational signature of the graviatom with the electrically charged particle is its electromagnetic radiation of the oscillatory type, which bears information about the fundamental symmetry scale of its interior de Sitter vacuum. For the density profile (1.74) its frequency is given by [60]

$$\hbar\omega = 0.678\ \hbar c/r_{int} = 0.678 \times 10^{11} GeV (E_{int}/E_{GUT})^2 \tag{1.79}$$

where r_{int} is the de Sitter radius $r_{int}^2 = 3c^2/8\pi G\rho_{int}$.

Current experiments allow detection of photons up to $10^{11.5}$ GeV [96], so that the electromagnetic radiation falls within the range of observational possibilities for graviatoms with the GUT scale de Sitter interior, $E_{int} = E_{GUT}$.

Present observational possibilities prefer thus graviatoms with the GUT scale interiors, for a graviatom with the Planck scale interior the characteristic frequency, $\hbar\omega \simeq 0.7 \times 10^{19}$ GeV, is far from the today observational range.

Regular rotating black holes can be identified by their shadows. Comparison of an RRBH shadow with the shadow of the Kerr black hole testifies for the essential dependence of the RRBH shadow on the pace of its density decreasing. Restoring the density profile of an RRBH from its observed shadow would give a certain information about its interior. Primordial RRBHs, their remnants and G-lumps as heavy dark matter candidates provide a signature for inhomogeneity of the early Universe. Their predicted nontrivial observational signature is the induced proton decay in an underground detector due to non-conservation of baryon and lepton numbers in their GUT scale false vacuum interiors. These objects can form graviatoms by capturing charged particles. In graviatoms with the electrically charged particles the cross section of the induced proton decay is strongly enhanced. Their additional observational signature is the electromagnetic radiation with the frequencies within the range available for observations.

1.5 Conclusions

In this Chapter we overviewed the basic generic properties and observational signatures of regular rotating black holes and spinning solitons G-lumps replacing naked singularities, described by the axially symmetric asymptotically flat solutions, obtained in the frame of the Gürses-Gürsey approach, which includes the Newman-Janis algorithm, from spherical solutions of the Kerr-Schild class specified by $T_t^t = T_r^r$ $(p_r = -\rho)$, and satisfying the weak energy condition.

Regular rotating black holes can have two horizons or one double horizon, and one ergosphere and hence one ergoregion for any density profile. G-lumps can have two ergospheres and ergoregion between them, or one ergosphere and ergoregion involving the whole interior, or no ergosphere, dependently on the density profile [63].

The algebraic structure of stress-energy tensors, $T_t^t = T_r^r$ $(p_r = -\rho)$, determines the interior matter content as the intrinsically anisotropic fluid, $p_r = w_r \rho$, $w_r = -1$; $p_\perp = w_\perp \rho$, where the equation of state parameter $w_\perp$ is coordinate-dependent. For $w_\perp > -1$ it represents an anisotropic $r-$dependent quintessence [66].

The main generic constituent of any regular rotating object is the rotating de Sitter vacuum disk $r = 0$ in its deep interior. For a certain class of objects, distinguished by the dominant energy condition for original spherical solutions, there can exist an additional closed surface of the de Sitter vacuum, S-surface with the de Sitter disk as a bridge. The cavities between the S-surface and the disk are filled with the anisotropic $r-$dependent phantom fluid with $w_\perp < -1$ [63], which violates the weak energy condition. It follows that violation of WEC, which was reported for certain regular axially symmetric solutions [10, 116, 136] and predicted as inevitable for all rotating objects [22], takes place for those of them, which are described by the axial solutions, originated from a certain type of spherical solutions, distinguished by the dominant energy condition. Location of the S=surface with respect to ergospheres testifies for the important fact, that the internal energy, available in the ergoregions for processes of energy extraction, includes the vacuum and phantom energy [63].

Information on the interior content of a regular rotating black hole can be obtained from the direct observation of its shadow [74, 132], which provides information on its parameters [90, 103], and on its position and motion with respect to of an observer [110]. Current observational possibilities are provided by the Event Horizon Telescope Collaboration [4, 44]. Theoretical investigation of shadows for regular black holes by comparing them with each other and with the shadow of the Kerr black hole, reveals the essential dependence of their shadows on the pace of density decreasing and rather low dependence on differences, related to their spins. Recovery of the RRBH density profile from its observed shadow, would thus give a certain information on the nature of the material content of its interior [65].

Primordial RRBHs, their remnants and G-lumps are confined by the light rings formed by the closed photon orbits. They include the asymptotically Kerr unstable co-rotating light rings (direct photon orbits) and counter-rotating light rings (retrograde photon orbits) [69], in agreement with the results presented in [81], that co-rotating and counter-rotating light rings "always appear in pairs". Existence of light rings as the basic constituents of regular rotating black holes, their remnants and spinning G-lumps, qualify them as the ultracompact objects [69], in accordance with their definition as the self-gravitating systems with the light rings [29].

Existence and location of the light rings are uniquely defined by the sequence and behavior of the closed photon orbits [69]. The generic behavior of the potentials $V_\gamma(r)$ in the regular geometry, which, unlike those in the Kerr geometry, decrease from $V_\gamma(r) \to \infty$ at $r \to 0$, ensures the existence of the innermost stable $(V_\gamma(r)'' > 0)$ orbits in their first minima.

As a result, around the spinning G-lumps there exist the innermost stable co-rotating light rings. There can also exist the additional co-rotating light rings on the way to the asymptotically Kerr branch of unstable orbits, stable or unstable dependently on the density profile, and the marginally (un)stable $(V_\gamma(r)'' = 0)$ orbits corresponding to the degenerate co-rotating light rings, necessary as transitional from the stable to unstable orbits. Around regular rotating black holes there exist the unstable asymptotically Kerr direct orbits forming the unstable co-rotating light rings. Dependently on density profile, there can also exist the set of the stable co-rotating light rings, starting from the stable orbit on the double horizon, $r_\gamma = r_\pm$, $a = a_{dh}$, and followed by the degenerate light ring for $a < a_{dh}$ on the way to the asymptotically Kerr unstable co-rotating light rings [69]. This confirms the prediction of the existence of degenerate light rings for a certain class of spherical ultracompact objects [91], and extends it to the case of the regular rotating objects.

Counter-rotating light rings are presented by the continuous branch of the retrograde unstable photon orbits $r_\gamma(a)$, started in the black hole region with $a < a_{dh}$ and continued in the soliton region with $a > a_{dh}$.

Primordial RRBHs, their remnants and G-lumps can be considered as heavy dark matter candidates, generically related to dark energy via their deSitter/phantom interiors. They originate from primordial density inhomogeneities in the early universe and provide a signature for its inhomogeneity [62]. These ultracompact objects can form graviatoms capturing charged particles, which represent the atomic heavy dark matter candidates with the dark energy interiors [60, 62].

The fundamental observational signature, predicted for all dark matter candidates with the dark energy interiors of the GUT scale, is the induced by them proton decay in the matter of an underground detector, due to non-conservation of the baryon and lepton numbers in their GUT scale false vacuum interiors, which, in principle, can be detected in the IceCUBE experiment. For graviatoms the induced proton decay is strongly enhanced due to participation of charged particles, which makes promising their detection in heavy dark matter searches at the IceCUBE experiment [62].

The additional observational signature of graviatoms is their electromagnetic radiation whose frequencies depend on the scale of the interior de Sitter vacuum and fall within the range accessible to observations [60].

References

1. R.J. Adler et al., Gen. Rel. Grav. **33**, 2101–2108 (2001). 10.1023/A
2. D.V. Ahluwalia, I. Dymnikova, v2 (2003), arXiv:hep-ph/0305158
3. D.V. Ahluwalia, I. Dymnikova, Int. J. Mod. Phys. **D 12**, 1787–1794 (2003). https://doi.org/10.1142/S0218271803004328
4. K. Akiyama et al., The Event Horizon Telescope Collaboration, Astrophys. J. Lett. **875**, L1 (2019). https://doi.org/10.3847/2041-8213/ab0ec7
5. S. Ansoldi et al., Phys. Lett. **B 645**, 261–266 (2007). https://doi.org/10.1016/j.physletb.2006.12.020
6. R. Aros, Phys. Rev. **D 77**, 104013 (2008). https://doi.org/10.1103/PhysRevD.77.104013
7. A. Ayon-Beato, A. Garcia, Phys. Lett. B **493**, 5056–5059 (2000). https://doi.org/10.1016/S0370-2693
8. M. Azreg-Aïnou, Phys. Rev. **D 90**, 064041 (2014). https://doi.org/10.1103/PhysRevD.90.064041
9. L. Balart, E.C. Vagenas, Phys. Lett. **B 730**, 14–17 (2014). https://doi.org/10.1016/j.physletb.2014.01.024
10. C. Bambi, L. Modesto, Phys. Lett. **B 721**, 329–334 (2013). https://doi.org/10.1016/j.physletb.2013.03.025
11. C. Bambi et al., JCAP **05**, 003 (2017). https://doi.org/10.1088/1475-7516/2017/05.003
12. K. Banerjee, S. Ghosh, Phys. Lett. **B 688**, 224–229 (2010). https://doi.org/10.1016/j.physletb.2010.04.008
13. L. Bellagamba et al., Europ. Phys. J. C. **72**, 1957 (2012)
14. L. Bergstrom, AIP Conference Proceedings **1241**, 49 (2010). https://doi.org/10.1063/1.3462677
15. W. Berej et al., Gen. Rel. Grav. **38**, 885–906 (2006). https://doi.org/10.1007/s10714-006-0270-9
16. A. Bonanno, M. Reuter, Phys. Rev. **D 73** 083005 (2006). https://doi.org/10.1103/PhysRevD.73.083005
17. D. Boyanovski et al., Annu. Rev. Nucl. Part. Sci. **56**, 441 (2006). https://doi.org/10.1146/annurev.nucl.56.080805.140539
18. K.A. Bronnikov et al., Class. Quantum Gravity. **29**, 095025-095047 (2012). https://doi.org/10.1088/0264-9381/29/9/095025
19. K.A. Bronnikov, S.G. Rubin, *Black Holes* (Singapore, Cosmology and Extra Dimensions; World Scientific, 2013)
20. A. Ya. Burinskii, Sov. Phys. JETP. **39**, 193–217 (1974)
21. A. Burinskii, Gravit. Cosmol. **8**, 261–271 (2002)
22. A. Burinskii et al., Phys. Rev. **D 65**, 064039 (2002). https://doi.org/10.1103/PhysRevD.65.064039
23. A. Burinskii, Gravit. Cosmol. **12**, 119–125 (2006). https://doi.org/10.48550/arXiv.gr-qc/0610007
24. A. Burinskii, J. Phys. Conf. Ser. **532**, 012004 (2014). https://doi.org/10.1088/1742-6596/532/1/012004
25. R.R.Caldwell et al., Phys. Rev. Lett. **80**, 1582–1585 (1998). https://doi.org/10.1103/PhysRevLett.80.1582
26. R.R. Caldwell, Phys. Lett. **B 545**, 23–29 (2002). https://doi.org/10.1016/S0370-2693(02)02589-3
27. S. Capozziello et al., Eur. Phys.J. **C 69**, 293–303 (2010). https://doi.org/10.1140/epjc/s10052-010-1387-2
28. F. Caravelli, L. Modesto, Class. Quantum Gravity. **27**, 245022 (2010). https://doi.org/10.1088/0264-9381/27/24/245022
29. V. Cardoso et al., Phys. Rev. **D 90**, 044069 (2014). https://doi.org/10.1103/PhysRevD.90.044069

30. B.J. Carr et al., Phys. Rev. **D 50**, 4853–4867 (1994). https://doi.org/10.1103/PhysRevD.50.4853
31. B.J. Carr, Proc. of 22nd Texas Symp., ECONFCO **41213** v1 (2004) 0204. v1 (2004), arXiv.org/abs/astro-ph/0504034. https://doi.org/10.48550/arXiv.astro-ph/0504034
32. A. Casanova, E. Spallucci, Class. Quantum Gravity. **23**, R45 (2006). https://doi.org/10.1088/0264-9381/23/3/R01
33. C.-K. Chan et al., Astrophys. J. **799**, 1 (2015). https://doi.org/10.1088/0004-637X/799/1/1
34. S. Chandrasekhar, *The Mathematical Theory of Black Holes* (Clarendon Press, Oxford, UK, 1983)
35. T. Chiba, M. Kimura, Progres Theor. Exp. Phys. **2017**, 043E01 (2017). https://doi.org/10.1093/ptep/ptx037
36. J.M. Cohen, J. Math. Phys. **8**, 1477–1478 (1967). https://doi.org/10.1063/1.1705382
37. S. Coleman, *New Phenomena in Subnuclear Physics*, ed. by A. Zichichi, vol. Part A (New York, Plenum, 1977), pp. 297–421
38. E.J. Copeland et al., Phys. Rev. **D 49**, 6410–6433 (1994). https://doi.org/10.1103/PhysRevD.49.6410
39. V. de la Cruz et al., Phys. Rev. Lett. **24**, 423–426 (1970). https://doi.org/10.1103/PhysRevLett.24.423
40. P.V.P. Cunha et al., Phys. Rev. Lett. **119**, 251102 (2017). https://doi.org/10.1103/PhysRevLett.119.251102
41. P.V.P. Cunha, C.A.R. Herdeiro, Gen. Relativ. Gravit. **50**, 42 (2018). https://doi.org/10.1007/s10714-018-2361-9
42. P.V.P. Cunha, C.A.R. Herdeiro, Phys. Rev. Lett. **124**, 181101 (2020). https://doi.org/10.1103/PhysRevLett.124.181101
43. S. Dimopoulos, G.Landsberg, Phys. Rev. Lett. **87**, 161602 (2001). https://doi.org/10.1103/PhysRevLett.87.161602
44. S.S. Doeleman et al., Nat. **455**, 78–80 (2008). https://doi.org/10.1038/nature07245
45. S.R. Dolan, J.O. Shipley, Phys. Rev. **D 94**, 044038 (2016). https://doi.org/10.1103/PhysRevD.94.044038
46. I. Dymnikova, Sov. Phys. Uspekhi. **29**, 215–237 (1988). https://doi.org/10.1070/PU1986v029n03ABEH003178
47. I. Dymnikova, Gen. Relativ. Gravit. **24**, 235–242 (1992). doi:0001-7701/9/0300-0235506.50/0
48. I. Dymnikova, Int. J. Mod. Phys. D. **5**(5), 529–540 (1996)
49. I. Dymnikova, Internal Structure of Black Holes and Spacetime Singularities. ed. by M. Burko, A. Ori, Bristol In-t of Physics Publishing, Bristol, TN, USA; Philadelphia, PA, USA; Annals of the Israel Physical Society **13**, 422–440 (1997)
50. I. Dymnikova, Phys. Lett. **B 472**, 33–38 (2000). https://doi.org/10.1016/S0370-2693(99)01374-X
51. I. Dymnikova, Class. Quantum Gravity. **19**, 725–739 (2002). https://stacks.iop.org/CQG/19/725
52. I. Dymnikova, Int. J. Mod. Phys. **D 12**, 1015–1034 (2003). https://doi.org/10.1142/S021827180300358X
53. I. Dymnikova, Class. Quantum Gravity. **21**, 4417–4428 (2004). https://doi.org/10.1088/1361-6382/21/18/009
54. I. Dymnikova, Phys. Lett. **B 639**, 368–372 (2006). https://doi.org/10.1016/j.physletb.2006.06.035
55. I. Dymnikova, E. Galaktionov, Phys. Lett. **B 645**, 358–364 (2007). https://doi.org/10.1016/j.physletb.2006.12.047
56. I. Dymnikova, J. Phys. Math. & Theor. **41**, 304033 (2008). https://doi.org/10.1088/1751-8113/41/30/304033
57. I. Dymnikova, AIP Conference Proceedings **1241**, 361 (2010). https://doi.org/10.1063/1.3462656

58. I. Dymnikova, M. Korpusik, Phys. Lett. **B 685**, 12 (2010). https://doi.org/10.1016/j.physletb.2010.01.044
59. I. Dymnikova, E. Galaktionov, Central Europ. J. Phys. **9**, 644–653 (2011). https://doi.org/10.2478/s11534-010-0071-3
60. I. Dymnikova, M. Fil'chenkov, AHEP **2013**, 746894 (2013). https://doi.org/10.1155/2013/746894
61. I. Dymnikova, E. Galaktionov, Class. Quantum Gravity **32**, 165015 (2015). https://doi.org/10.1088/0264-9381/32/16/165015
62. I. Dymnikova, M. Khlopov, Int. J. Mod. Phys. **D 24**, 1545002 (2015). https://doi.org/10.1142/S0218271815450029
63. I. Dymnikova, E. Galaktionov, Class. Quantum Gravity. **33**, 145010 (2016). https://doi.org/10.1088/0264-9381/33/14/145010
64. I. Dymnikova, E. Galaktionov, Adv. Math. Phys. **2017**, 1035381 (2017). https://doi.org/10.1155/2017/1035381
65. I. Dymnikova, K. Kraav, Universe. **5**, 163 (2019). https://doi.org/10.3390/universe5070163
66. I. Dymnikova, Symmetry. **12**, 12 (2020). https://doi.org/10.3390/sym12040662
67. I. Dymnikova, Symmetry. **12**, 634 (2020). https://doi.org/10.3390/sym12040634
68. I. Dymnikova, Universe. **6**, 179 (2020). https://doi.org/10.3390/universe6100179
69. I. Dymnikova et al., Universe. **8**, 65 (2022). https://doi.org/10.3390/universe8020065
70. I. Dymnikova, Universe. **8**, 305 (2022). https://doi.org/10.3390/universe8060305
71. A. Eichhorn, A. Held, Eur. Phys. J. **C 81**, 933 (2021). https://doi.org/10.1140/epjc/s10052-021-09716-2
72. G.F.R. Ellis, (2013), arXiv:1310.4771
73. F. Englert, R. Brout, Phys. Rev. Lett. **13**, 321 (1964). https://doi.org/10.1103/PhysRevLett.13.321
74. H. Falcke et al., Astrophys. J. Lett. **528**, L13 (2000). https://doi.org/10.1086/312423
75. V.P. Frolov et al., Phys. Rev. **D 41**, 383–394 (1990). https://doi.org/10.1103/PhysRevD.41.383
76. A. Garcia et al., Phys. Rev. Lett. **74**, 1276 (1995). https://doi.org/10.1103/PhysRevLett.74.1276
77. S.G. Ghosh, Eur. Phys. J. **C 75**, 532 (2015). https://doi.org/10.1140/epjc/s10052-015-3740-y
78. R. Ghosh, S. Sarkar, Phys. Rev. **D 104**, 044019 (2021). https://doi.org/10.1103/PhysRevD.104.044019
79. A. Grenzebach et al., Phys. Rev. **D 89**, 124004 (2014). https://doi.org/10.1103/PhysRevD.89.124004
80. A.A. Grib, Yu.V. Pavlov, Mod. Phys. Lett. **A 23**, 1151–1159 (2008). https://doi.org/10.1142/S0217732308027072
81. M. Guo, S. Gao, Phys. Rev. **D 103**, 104031 (2021). https://doi.org/10.1103/PhysRevD.103.104031
82. G.Ś. Guralnik et al., Phys. Rev. Lett. **13**, 585–587 (1964). https://doi.org/10.1103/PhysRevLett.13.585
83. M. Gürses, F. Gürsey, J. Math. Phys. **16**, 2385 (1975). https://doi.org/10.1063/1.522480
84. V.H. Hamity, Phys. Lett. A. **56**, 77–78 (1976)
85. C.M. Harris et al., JHEP **0505**, 053 (2005). https://doi.org/10.1088/1126-6708/2005/05/053
86. S.W. Hawking, Mon. Not. R. Astron. Soc. **152**, 75 (1971). https://doi.org/10.1093/mnras/152.1.75
87. S.W. Hawking, G.F.R. Ellis, *The Large Scale Structure of Space-Time*, (Cambridge University Press, 1973)
88. S.A. Hayward, Phys. Rev. Lett. **96**, 031103 (2006). https://doi.org/10.1103/PhysRevLett.96.031103
89. P.W. Higgs, Phys. Rev. Lett. **13**, 508–509 (1964). https://doi.org/10.1103/PhysRevLett.13.508

90. K. Hioki, K. Maeda, Phys. Rev. D **80**, 024042 (2009). https://doi.org/10.1103/PhysRevD.80.024042
91. S. Hod, Phys. Lett. **B 776**, 1–4 (2018). https://doi.org/10.1016/j.physletb.2017.11.021
92. S. Hod, Europ. Phys. J. **C 78**, 417 (2018). https://doi.org/10.1140/epjc/s10052-018-5905-y
93. S. Iso et al., Phys. Rev. **D 7**, 044017 (2006). https://doi.org/10.1103/PhysRevD.74.044017
94. W. Israel, Phys. Rev. **D 2**, 641–646 (1970). https://doi.org/10.1103/PhysRevD.2.641
95. T. Johannsen, Astrophys. J. **777**, 170 (2013). https://doi.org/10.1088/0004-637X/777/2/170
96. O.E. Kalashev et al., Phys. Rev. **D 80**, 103006 (2009). https://doi.org/10.1103/PhysRevD.80.103006
97. H. Kawai et al., Int. J. Mod. Phys. **A 28**, 1350050 (2013). https://doi.org/10.1142/S0217751X13500504
98. R.P. Kerr, Phys. Rev. Lett. **11**, 237–238 (1963). https://doi.org/10.1103/PhysRevLett.11.237
99. R.P. Kerr, A. Schild, Proceedings of Symposia in Applied Mathematics (Amer. Math. Soc., Providence, R. I.) **17**, 199 (1965)
100. M.Y. Khlopov, Res. Astron. Astrophys. **10**, 495 (2010). https://doi.org/10.1088/1674-4527/10/6/001
101. B. Koch et al., JHEP **0510**, 053 (2005). https://doi.org/10.1088/1126-6708/2005/10/053
102. G. Landsberg, J. Phys. **G 32**, R337 (2006). https://doi.org/10.1088/0954-3899/32/9/R02
103. Z. Li, C. Bambi, J. Cosmol. & Astropart. Phys. **2014**, 041 (2014). https://doi.org/10.1088/1475-7516/2014/01/041
104. Y. Ling, M.-H. Wu, (2022), arXiv:2205.08919 [gr-qc] . https://doi.org/10.48550/arXiv.2205.08919
105. F.S.N. Lobo et al., JCAP **07**, 011 (2013). https://doi.org/10.1088/1475-7516/2013/07/011
106. T. De Lorenzo et al., Gen. Rel. Grav. **48**, 31–36 (2016). https://doi.org/10.1007/s10714-016-2026-5
107. D.H. Lyth, A. Riotto, Phys. Rep. **314**, 1 (1999). https://doi.org/10.1016/S0370-1573(98)00128-8
108. J.H. MacGibbon, Nat. **329**, 308–309 (1987). https://doi.org/10.1038/329308a0
109. A.D. Masa, V.T. Zanchin, (2022), arXiv:2204.08113 [gr-qc]. https://doi.org/10.48550/arXiv.2204.08113
110. K. Mars et al., Class. Quantum Gravity. **35**, 025005 (2018). https://doi.org/10.1088/1361-6382/aa97ff
111. M. Maziashvili, Phys. Lett. **B 635**, 232 (2006). https://doi.org/10.1016/j.physletb.2006.03.009
112. Y. Mizuno et al., Nat. Astron. Lett. **2**, 585 (2018). https://doi.org/10.1038/s41550-018-0449-5
113. L. Modesto, P. Nicolini, Phys. Rev. **D 82**, 104035 (2010). https://doi.org/10.1103/PhysRevD.82.104035
114. L. Modesto et al., Phys. Lett. **B 695**, 397–400 (2011). https://doi.org/10.1016/j.physletb.2010.11.046
115. G.C. Nayak, Phys. Part. Nucl. Lett. **8**, 337 (2011). https://doi.org/10.1134/S1547477111040121
116. J.C.S. Neves, A. Saa, Phys. Lett. **B 734**, 44–48 (2014). https://doi.org/10.1016/j.physletb.2014.05.026
117. E.T. Newman, A.J. Janis, J. Math. Phys. **6**, 915–917 (1965). https://doi.org/10.1063/1.1704350
118. P. Nicolini et al., Phys. Lett. **B 632**, 547–551 (2006). https://doi.org/10.1016/j.physletb.2005.11.004
119. K. Nozari, S.H. Mehdipour, Mod. Phys. Lett. **A 20**, 2937 (2005). https://doi.org/10.1142/S0217732305018050
120. S. Pameli, D. Ujjal, Europ. Phys. J. **C 79**, 919 (2019). https://doi.org/10.1140/epjc/s10052-019-7427-7
121. E. Poisson, W. Israel, Class. Quantum Gravity. **5**, L201 (1988). https://doi.org/10.1088/0264-9381/5/12/002

122. A.G. Polnarev, M.Yu. Khlopov, Sov. Phys. Uspekhi. **28**, 213–232 (1985). https://doi.org/10.3367/UFNr.0145.198503a.0369
123. C. Quigg, *Gauge Theories of the Strong, Weak and Electromagnetic Interactions*, (Addison-Wesley Publishing Company, Redwood City, CA, US, 1983)
124. S. Ray et al., Intern. J. Mod. Phys. **D 29**, 2030004 (2020). https://doi.org/10.1142/S0218271820300049
125. S.V. Repin et al., (2018), arXiv:1802.04667. https://doi.org/10.48550/arXiv.1802.04667
126. L. Rezzolla, A. Zhidenko, Phys. Rev. **D 90**, 084009 (2014). https://doi.org/10.1103/PhysRevD.90.084009
127. R.G.H. Robertson et al., Phys. Rev. Lett. **67**, 957 (1991). https://doi.org/10.1103/PhysRevLett.67.957
128. F. Scardigli et al., Phys. Rev. **D 83**(2011), 063507 (2011). https://doi.org/10.1103/PhysRevD.83.063507
129. J.D. Schnittman, Gen. Rel. Grav. **50**, 77 (2018). https://doi.org/10.1007/s10714-018-2373-5
130. H. Stoecker, Int. J. Mod. Phys. **D 16**, 185 (2007). https://doi.org/10.1142/S0218271807009930
131. L. Susskind, J. Math. Phys. **36**, 6377 (1995). https://doi.org/10.1063/1.531249
132. R. Takahashi, Astrophys. J. **611**, 996 (2004). https://doi.org/10.1086/422403
133. S. Takeuchi, Class. Quantum Gravity. **33**, 225016 (2016). https://doi.org/10.1088/0264-9381/33/22/225016
134. Zi.-Yu. Tang, Class. Quantum Gravity. **34** 245006 (2017). https://doi.org/10.1088/1361-6382/aa95ff
135. R. Torres, F. Fayos, (2016), arXiv:1611.03654 [gr-qc]. https://doi.org/10.48550/arXiv.1611.03654
136. B. Toshmatov et al., Phys. Rev. **D 89**, 104017 (2014). https://doi.org/10.1103/PhysRevD.89.104017
137. Z. Younsi et al., Astron. Astrophys. **545**, A13 (2012). https://doi.org/10.1051/0004-6361/201219599
138. Ya. B. Zel'dovich, JETP Lett. **6**, 316–317 (1967)
139. Ya.B. Zeldovich, I.D. Novikov, Sov. Astron. **10**, 602 (1967)

Chapter 2
Regular Black Holes Sourced by Nonlinear Electrodynamics

Kirill A. Bronnikov

2.1 Introduction

Nonlinear electrodynamics (NED) as a generalization of Maxwell's theory was proposed in the 1930s: M. Born and L. Infeld formulated a theory able to remove the central singularity of the electromagnetic field of a point charge as well as its energy divergence [12]. Another version of NED was put forward by W. Heisenberg and H. Euler while taking into consideration high-energy quantum processes with photons, such as pair creation [36]. Much later, J. Plebanski [61] developed a more general formulation of NED in special relativity, admitting an arbitrary function of the electromagnetic invariants.

More recently, the interest in NED received a new support when it was discovered that a Born-Infeld-like theory appears in the weak-field limit of some models of string theory [31, 53, 65]. It has also turned out that NED can be a material source of gravity able to lead to nonsingular geometries of interest, such as regular black holes (BHs) and solitonlike configurations without horizons in the framework of general relativity (GR) and various alternative theories. Let us also mention one more recent application of NED, namely, using it as one of the sources of gravity in Simpson-Visser-like (black-bounce) space-times [32, 50, 67] that are regular models simulating some expected effects of quantum gravity on the classical level [20, 25, 27].

K. A. Bronnikov (✉)
VNIIMS, Ozyornaya ul. 46, Moscow 119361, Russia
e-mail: kb20@yandex.ru

Institute of Gravitation and Cosmology, Peoples' Friendship University of Russia (RUDN University), ul. Miklukho-Maklaya 6, Moscow 117198, Russia

National Research Nuclear University "MEPhI", Kashirskoe sh. 31, Moscow 115409, Russia

C. Bambi (ed.), *Regular Black Holes*, Springer Series in Astrophysics and Cosmology,
https://doi.org/10.1007/978-981-99-1596-5_2

This chapter is devoted to NED application for obtaining regular BHs and solitons (monopoles) in GR. We will reproduce a number of well-known results in a somewhat pedagogical manner and also present some new observations. We will restrict ourselves to the simplest models assuming spherical symmetry, and also mostly focus on NED theories with Lagrangians of the form $L = L(f)$, where $f = F_{\mu\nu}F^{\mu\nu}$, and $F_{\mu\nu}$ is the electromagnetic field tensor. Then we will briefly discuss similar problems in some extensions of $L(f)$ theories: those with $L(f, h)$, where $h = {}^*F_{\mu\nu}F^{\mu\nu}$, where ${}^*F_{\mu\nu} = \frac{1}{2}\sqrt{-g}\varepsilon_{\mu\nu\rho\sigma}F^{\rho\sigma}$ is the Hodge dual of $F_{\mu\nu}$, and those with $L(f, J)$, where J is quartic with respect to $F_{\mu\nu}$ [28, 68]: $J = F_{\mu\nu}F^{\nu\rho}F_{\rho\sigma}F^{\sigma\mu}$.

When considering the NED-GR system in spherical symmetry, there are only two possible kinds of electromagnetic fields: radial electric fields and radial (monopole) magnetic ones. Two important circumstances should be taken into account. The first one is (in general) the absence of duality between electric and magnetic fields, so that solutions to the field equations containing these fields in the framework of the same NED theory will be quite different. Instead, there emerges the so-called FP duality that connects electric and magnetic solutions for *different* NED theories but involving the same space-time metric. The second circumstance is that it is insufficient to require finite values of the electric field itself and the electric field energy of a point charge in order to obtain a regular space-time: its regularity imposes more stringent requirements on NED, which cannot be satisfied, for example, by the Born-Infeld theory.

NED-GR solutions with electric or magnetic fields are currently widely discussed, probably beginning with finding a general form of an electric solution by Pellicer and Torrence [60]. Later on, a no-go theorem was proved [21, 23], showing that if NED is specified by a Lagrangian function $L(f)$ having a Maxwell weak-field limit ($L \sim f$ as $f \to 0$), a static, spherically symmetric solution of GR with an electric field cannot have a regular center. This theorem was extended to include static dyonic configurations, involving both electric and magnetic fields [17], and it was also proved [16, 17] that in any electric solutions describing systems with or without horizons (i.e., BH or solitonic ones), containing a regular center and a flat infinity with a Reissner-Nordström (RN) asymptotic behavior, different NED theories are valid at large and small r. The present paper describes this issue in detail.

It was also shown [17] that purely magnetic regular configurations, both BH and solitonic ones, can exist and are easily obtained if $L(f)$ tends to a finite limit as $f \to \infty$. Electric solutions with the same metric can also be found, but they suffer multivaluedness of $L(f)$ and inevitably exhibit infinite blueshifts of traveling photons on some surfaces [17].

Many further results of interest are known. In particular, the properties and examples of static, spherically symmetric dyonic NED-GR space-times were studied [18, 43–45, 54, 73]; a kind of phase transition was discussed, allowing one to circumvent the above no-go theorem on electric solutions [26]; the static, spherically symmetric solutions were extended to include a nonzero cosmological constant Λ [52]; the thermodynamic properties of regular NED BHs were investigated (see [4, 13, 30, 41, 46] and references therein); cylindrically [24] and axially [6, 29, 34, 47, 70] symmetric (rotating) NED-GR configurations were found and studied, as well as evolving wormhole models [1, 2, 9, 19]. (Note that static wormhole models with NED as a source are impossible because this kind of matter respects the weak energy condition.) Furthermore, the stability properties of NED BHs were investigated in [14, 49, 55], and quantum effects in their fields in [8, 51]. One should also mention a number of studies of special cases of both electric and magnetic solutions, their potential observational properties like gravitational lensing, particle motion and matter accretion in the fields of NED BHs as compared to their counterparts in scalar-tensor, $f(R)$ and multidimensional theories of gravity, consideration of NED with dilaton-like interactions, non-Abelian fields, different constructions with thin shells, etc., but the corresponding list of references would be too long. For a recent brief review on NED with and without relation to gravitational theories see [69].

The most relevant to the present subject, regular BHs, are the recent results obtained by Bokulić, Smolić and Jurić [10, 11] who have proved a number of no-go theorems in NED-GR solutions with NED Lagrangians of the form $L(f, h)$. Let us mention that this wide class of theories contains, among others, the Born-Infeld and Heisenberg-Euler theories. With all these no-go theorems, it seems that regular magnetic BHs with $L = L(f)$ are the only kind of regular BHs that can be found among NED-GR solutions with an asymptotically Maxwell NED, although some opportunities are still remaining unexplored.

In this chapter, we will discuss in detail the existence and main properties of static, spherically symmetric regular black holes and solitons with $L(f)$ NED theory, and more briefly consider the same with Lagrangians depending on two invariants, either $L(f, h)$ or $L(f, J)$. We begin with discussing the particular form of regularity and asymptotic conditions to be used (Sect. 2.2). Then, in Sect. 2.3, we discuss the $L(f)$ NED-GR field equations in static, spherically symmetric space-times. Section 2.4 is devoted to the regularity properties of black hole and soliton solutions to these equations, their compatibility with the known NED unitarity and causality [66] and stability [55] conditions as well as photon propagation in these space-times. Some particular examples known in the literature are also discussed. Section 2.5 presents some no-go theorems with $L(f, h)$ due to [10, 11] and with $L(f, J)$, and the latter results seem to be new. Section 2.6 is a brief conclusion.

We use the following conventions: the units with $c = 8\pi G = 1$; the metric signature $(+\,-\,-\,-)$; the curvature tensor $R^{\sigma}{}_{\mu\rho\nu} = \partial_\nu \Gamma^{\sigma}_{\mu\nu} + \ldots$; the Ricci tensor $R_{\mu\nu} = R^{\sigma}{}_{\mu\sigma\nu}$, so that the Ricci scalar $R = g^{\mu\nu} R_{\mu\nu} > 0$ for de Sitter space-time. The Einstein equations are written in the form

$$G^{\nu}_{\mu} \equiv R^{\nu}_{\mu} - \tfrac{1}{2}\delta^{\nu}_{\mu} R = -T^{\nu}_{\mu}, \tag{2.1}$$

where T^{ν}_{μ} is the stress-energy tensor (SET) of matter, such that T^{t}_{t} is the energy density.

2.2 Static Spherically Symmetric Space-Times. Regularity and Asymptotic Conditions

Before dealing with NED-Einstein equations, it makes sense to recall the conditions to be fulfilled by the desirable solutions to these equations.

Spherical symmetry is the simplest and natural assumption for descriptions of isolated bodies when their precise shape and possible rotation are regarded insignificant. The physical fields of any island-like objects are approximately spherically symmetric far from these objects.

In the general case, one can write a spherically symmetric metric in the form (see, e.g., [48])

$$ds^2 = e^{2\gamma} dt^2 - e^{2\alpha} dx^2 - r^2 d\Omega^2, \qquad d\Omega^2 = d\theta^2 + \sin^2\theta d\phi^2. \tag{2.2}$$

In general, α, γ, r are functions of the radial coordinate x and the time coordinate t. The quantity r has the geometric meaning of the radius of a coordinate sphere $x = \text{const}$, $t = \text{const}$, the so-called spherical radius, or it is sometimes called the areal radius since the area of a coordinate sphere is equal to $4\pi r^2$. Let us note that in curved space-time this radius r has nothing to do with a distance from the center (as happens in flat space-time), and there are many spherically symmetric space-times that contain no center at all, for example, wormholes.

In what follows we restrict ourselves to static space-times, such that α, γ, r depend on x only. There still remains the freedom of choosing the radial coordinate x and the possibility of its reparametrizations by replacing $x = x(x_{\rm new})$. The choice of the radial coordinate can be fixed by postulating a relation between the functions α, γ, r or by choosing some of them (or a function of some of them) as the coordinate. For example, very often the radius r is used as a coordinate, it is then called the Schwarzschild (or curvature) radial coordinate.

The convenient "exponential" notations in the metric (2.2), which simplify the appearance of many relations without fixing the radial coordinate, assume positive values of the corresponding quantities. However, the coefficients g_{tt} and g_{xx} can change their sign, in particular, this happens at black hole horizons. In such cases, it is helpful to use the so-called quasiglobal coordinate condition $\alpha + \gamma = 0$, and with the notation $e^{2\gamma} = e^{-2\alpha} = A(x)$, the metric is written as

$$ds^2 = A(x)dt^2 - \frac{dx^2}{A(x)} - r^2(x)d\Omega^2. \tag{2.3}$$

Regularity A Riemannian space-time is generally called regular at a particular point X if the Riemann tensor is well defined at X (hence the metric functions must be at least twice differentiable at X), and all algebraic curvature invariants are finite. (Other definitions of regularity, involving differential invariants of the Riemann tensor, are sometimes used, but the above definition is sufficient for our purposes.) Hence, the metric (2.2) or (2.3) is manifestly regular at any point where $r \neq 0$ as long as the functions $\alpha(x)$, $\gamma(x)$ and $r(x)$ (or $A(x)$ and $r(x)$) are sufficiently smooth. A point where $r = 0$ requires special attention because the metric becomes degenerate there, hence it is a singular point of the spherical coordinate system used in (2.2) or (2.3). Furthermore, a space-time as a whole (and in particular, a black hole space-time) is called regular if all its points are regular.

Very often, to verify regularity of a particular metric of the form (2.2) or (2.3), one directly calculates its basic invariants: the scalar curvature R, the Ricci tensor squared $R_{\mu\nu}R^{\mu\nu}$, and the Kretschmann scalar (the Riemann tensor squared) $\mathcal{K} = R_{\alpha\beta\gamma\delta}R^{\alpha\beta\gamma\delta}$. However, for a static metric (2.2), it is quite sufficient and much easier to verify finiteness of the four independent components $R_{\alpha\beta}{}^{\gamma\delta}$ of the Riemann tensor with two upper and two lower indices:

$$
\begin{aligned}
K_1 &= -R_{01}{}^{01} = \mathrm{e}^{-\alpha-\gamma}(\gamma' \mathrm{e}^{\gamma-\alpha})' = \frac{1}{2}A'', \\
K_2 &= -R_{02}{}^{02} = -R_{03}{}^{03} = \mathrm{e}^{-2\alpha}\frac{\gamma' r'}{r} = \frac{A'r'}{2r}, \\
K_3 &= -R_{12}{}^{12} = -R_{13}{}^{13} = \frac{\mathrm{e}^{-\alpha}}{r}(\mathrm{e}^{-\alpha}r')' = \frac{1}{2r}(2\,Ar'' - A'r'), \\
K_4 &= -R_{23}{}^{23} = \frac{1}{r^2}(1 - \mathrm{e}^{-2\alpha}r'^2) = \frac{1}{r^2}(1 - Ar'^2).
\end{aligned}
\tag{2.4}
$$

where the prime stands for d/dx (K_i in terms of the metric (2.3) are given in each line after the last equality sign). The point is that for static, spherically symmetric metrics, as well as and in many other important cases, the tensor $R_{\alpha\beta}{}^{\gamma\delta}$ is pairwise diagonal. Therefore, all algebraic curvature invariants are linear, quadratic, cubic, etc., combinations of K_i from (2.4) and are manifestly finite if K_i are finite. Moreover, the Kretschmann scalar is a sum of squares:

$$
\mathcal{K} = 4K_1^2 + 8K_2^2 + 8K_3^2 + 4K_4^2, \tag{2.5}
$$

hence it is finite if and only if *each* K_i is finite. Thus finiteness of all K_i is both necessary and sufficient condition of space-time regularity [22].

It is important to note that all K_i in (2.4) are invariant (behave as scalars) under reparametrizations of the x coordinate, and the same is true for mixed components of second-rank tensors, including the Ricci tensor R_μ^ν and the Einstein tensor $G_\mu^\nu = R_\mu^\nu - \frac{1}{2}\delta_\mu^\nu R$. Thus the space-time regularity can be verified using K_i in terms of any radial coordinate x.

As follows from (2.4), regularity at $r = 0$ requires not only finite values and smoothness of α and γ, but also, due to the expression for K_4,

$$
\mathrm{e}^{-2\alpha}r'^2 - 1 = \mathcal{O}(r^2) \quad \text{as } r \to 0. \tag{2.6}
$$

It is actually the local flatness condition, requiring a circumference to radius ratio of 2π for small circles around the center.

One more important observation follows from Eq. (2.6) for black hole space-times, in which $A(x)$ can become negative. The condition (2.6), rewritten as $Ar'^2 - 1 = \mathcal{O}(r^2)$, cannot be satisfied if $A(x) < 0$, which happens in nonstatic regions of spherically symmetric black holes (also called T-regions) beyond their horizons. We see that *the metric cannot be regular in the limit $r \to 0$ in T-regions of spherically symmetric black holes.* A regular center can only occur in a static region where $A > 0$.

It also follows from (2.4) that the metric (2.3) is regular at apparent horizons that correspond to regular zeros of the function $A(x)$ under the condition $r(x) > 0$.

Asymptotics For an island-like system, it is natural to assume that the space-time is asymptotically flat, and far from the source of gravity there is an approximately Schwarzschild gravitational field characterized by a certain mass m. In terms of an arbitrary radial coordinate x it means that in the metric (2.2), under the appropriate choice of the time scale,

$$e^{2\gamma(x)} = 1 - \frac{2m}{r(x)} + o(1/r) \qquad \text{as } r \to \infty \tag{2.7}$$

In addition, one should require a correct circumference to radius ratio for large circles around the source of gravity, which leads to a condition similar to (2.6),

$$e^{-2\alpha} r'^2 \to 1 \qquad \text{as } r \to \infty. \tag{2.8}$$

A limit other than unity in (2.8) leads to a deficit or excess of the solid angle at infinity, characterizing a global monopole space-time [72].

In the presence of a nonzero cosmological constant Λ, the gravitational field far from its island-like source as asymptotically de Sitter (if $\Lambda > 0$) or anti-de Sitter (if $\Lambda < 0$), well described by the metric (2.3) with $r = x$ and $A = 1 - \Lambda r^2/3$.

2.3 $L(f)$ NED Coupled to General Relativity. FP Duality

2.3.1 Field Equations

Let us now consider self-gravitating electromagnetic fields with the Lagrangian $L(f)$ in the framework of GR, so that the total action has the form

$$S = \frac{1}{2} \int \sqrt{-g} d^4x [R - L(f)], \tag{2.9}$$

where R is the Ricci scalar, the invariant f has the standard form $f = F_{\mu\nu}F^{\mu\nu} = 2(\mathbf{B}^2 - \mathbf{E}^2)$, where the 3-vectors $\mathbf{E}$ and $\mathbf{B}$ are the electric field strength and magnetic induction, and $L(f)$ is an arbitrary function. The electromagnetic tensor $F_{\mu\nu}$ obeys the Maxwell-like equations, obtained from (2.9) by variation with respect to the 4-vector potential A_μ, and the Bianchi identities for the dual field ${}^*F^{\mu\nu}$, following from the definition $F_{\mu\nu} = \partial_\mu A_\nu - \partial_\nu A_\mu$:

$$\nabla_\mu (L_f F^{\mu\nu}) = 0, \qquad \nabla_\mu {}^*F^{\mu\nu} = 0. \tag{2.10}$$

The corresponding SET is given by $(L_f \equiv dL/df)$

$$T^\nu_\mu = -2L_f F_{\mu\alpha} F^{\nu\alpha} + \tfrac{1}{2}\delta^\nu_\mu L(f). \tag{2.11}$$

Let us assume spherical symmetry, with a metric of the general form (2.2). The only nonzero components of $F_{\mu\nu}$ compatible with this symmetry are $F_{tr} = -F_{rt}$, representing a radial electric field, and $F_{\theta\phi} = -F_{\phi\theta}$, corresponding to a radial magnetic field. From (2.10) it follows

$$r^2 \,\mathrm{e}^{\alpha+\gamma} L_f F^{tr} = q_e, \qquad F_{\theta\phi} = q_m \sin\theta, \tag{2.12}$$

where $q_e =$ const has the meaning of an electric charge, and $q_m =$ const is a magnetic charge. Accordingly, the only nonzero SET components have the form

$$T^t_t = T^r_r = \tfrac{1}{2}L + f_e L_f, \qquad T^\theta_\theta = T^\phi_\phi = \tfrac{1}{2}L - f_m L_f, \tag{2.13}$$

where

$$f_e = 2E^2 = 2F_{tr}F^{rt} = \frac{2q_e^2}{L_f^2 r^4}, \qquad f_m = 2B^2 = 2F_{\theta\phi}F^{\theta\phi} = \frac{2q_m^2}{r^4}, \tag{2.14}$$

so that $f = f_m - f_e$. Here, $E = |\mathbf{E}|$ and $B = |\mathbf{B}|$ are the absolute values of the electric field strength and magnetic induction, measured by an observer at rest in our static space-time.

The SET (2.13) has two important properties $T^x_t = 0$ and $T^t_t = T^x_x$. The first one means the absence of radial energy flows, related to the absence of monopole electromagnetic radiation. The second one, due to the Einstein Eqs. (2.1), leads to $G^t_t = G^x_x$, and this equation is easily integrated if we use the Schwarzschild radial coordinate, $x \equiv r$ (see, e.g., [48]), leading to the relation $\alpha(r) + \gamma(r) =$ const. With a proper choice of the time scale, we have $\alpha + \gamma = 0$, and the metric can be rewritten as

$$ds^2 = A(r)dt^2 - \frac{dr^2}{A(r)} - r^2 d\Omega^2. \tag{2.15}$$

The other Einstein equation, $G^t_t = -T^t_t$, then reads

$$A + A'r = 1 - \rho r^2 \tag{2.16}$$

and can be rewritten in the integral form as

$$A(r) = 1 - \frac{2M(r)}{r}, \qquad M(r) = \frac{1}{2}\int \rho(r) r^2 dr, \tag{2.17}$$

where $\rho(r) \equiv T^t_t$ is the energy density, and $M(r)$ is called the mass function, such that $M(\infty)$ is the Schwarzschild mass in an asymptotically flat space-time. It is a solution for $A(r)$ if $\rho(r)$ is known. Note, however, that a complete solution for the system under consideration requires a knowledge of $L(f)$ and both electric and magnetic fields as functions of r.

2.3.2 *FP Duality*

NED with a Lagrangian function $L(f)$ is known to admit an alternative representation obtained from the original one by a Legendre transformation [5, 60, 64]: to this end, the new tensor $P_{\mu\nu} = L_f F_{\mu\nu}$ is defined, with its invariant $p = P_{\mu\nu}P^{\mu\nu}$. Then one considers the Hamiltonian-like quantity

$$H(p) = 2fL_f - L = -2T^t_t \tag{2.18}$$

as a function of p. It is possible to use the function $H(p)$ to specify the whole theory. The following relations are valid:

$$L = 2pH_p - H, \qquad L_f H_p = 1, \qquad f = pH_p^2, \qquad p = fL_f^2, \tag{2.19}$$

where $H_p \equiv dH/dp$. In terms of H and $P_{\mu\nu}$, the SET reads

$$T^\nu_\mu = -2H_p P_{\mu\alpha}P^{\nu\alpha} + \delta^\nu_\mu (pH_p - \tfrac{1}{2}H). \tag{2.20}$$

In a spherically symmetric space-time with the metric (2.15), Eqs. (2.12) are rewritten in the P framework as

$$r^2 P^{tr} = q_e, \qquad H_p P_{\theta\phi} = q_m \sin\theta. \tag{2.21}$$

Let us also introduce the quantities p_e and p_m quite similar to f_e and f_m:

$$p_e = 2P_{tr}P^{rt} = \frac{2q_e^2}{r^4} \geq 0, \qquad p_m = 2P_{\theta\phi}P^{\theta\phi} = \frac{2q_m^2}{H_p^2 r^4} \geq 0, \tag{2.22}$$

so that $p = p_m - p_e$, and then the SET (2.20) is transformed to

$$T^t_t = T^r_r = -\tfrac{1}{2}H + p_m H_p, \qquad T^\theta_\theta = T^\phi_\phi = -\tfrac{1}{2}H - p_e H_p. \tag{2.23}$$

One can notice that the F and P formulations of the same theory are not always equivalent [16, 17]. More precisely, a theory initially specified by $L(f)$ is equivalently reformulated in the P framework only in a range of f where $f(p)$ is a monotonic function. In any case, the main and physically preferred formulation is the Lagrangian one since it directly follows from the least action principle. Later on we will confirm this statement in the discussion of photon motion in magnetic and electric solutions with the same metric.

Now, comparing (2.13) and (2.23), one can see that they coincide up to the substitution

$$\{F_{\mu\nu},\ f,\ L(f)\} \quad \longleftrightarrow \quad \{{}^*P_{\mu\nu},\ -p,\ -H(p)\}, \tag{2.24}$$

where ${}^*P_{\mu\nu}$ is the Hodge dual of $P_{\mu\nu}$, such that ${}^*P_{\theta\phi} = P_{tx}$. As long as the SETs coincide, all possible metrics satisfying the Einstein equations (2.1) should also coincide. This coincidence was described in [17] for static systems and was named FP duality. In [55] this kind of duality was extended to general space-times and used for studying the stability of static solutions, and in [19] it was used while obtaining nonstatic spherically symmetric solutions to the Einstein-NED equations.

It should be stressed that the FP duality connects solutions with the same metric but belonging to *different* NED theories. Only in the Maxwell theory, in which $L = f = H = p$, the FP duality is the same as the conventional electric-magnetic duality.

2.4 Regular Black Holes with $L = L(f)$

2.4.1 Magnetic, Electric and Dyonic Solutions

Magnetic solutions ($q_e = 0, q_m \neq 0$) can be found most easily. If the Lagrangian $L(f)$ is specified, then, since now $f = 2q_m^2/r^4$, the density $\rho(r) = L/2$ is known according to (2.13), and the metric function $A(r)$ is found by integration in (2.17).

If, on the contrary, we know $A(r)$ (or choose it by hand), then $\rho = L(f)/2$ is found from (2.17), leading to

$$L(f(r)) = \frac{2}{r^2}[1 - (rA)'], \tag{2.25}$$

and $L(f)$ is restored since $f = 2q_m^2/r^4$.

Electric solutions ($q_e \neq 0, q_m = 0$) can be obtained in quite a similar manner if we use the Hamiltonian-like form of NED, see Eqs. (2.19)–(2.23). In this case, $p = -2q_e^2/r^4$, and if we specify $H(p) = -2\rho$, the mass function $M(r)$ is directly found, while $A(r)$ is obtained by integration in (2.17). If $A(r)$ is specified, then Eq. (2.17) allows for finding $\rho(r) = -H(p)/2$.

However, if one starts with the Lagrangian $L(f)$ and seeks electric solutions, a separate problem is the transition to the P framework, which is equivalent to the F framework only if $f(p)$ is a monotonic function, or only in such ranges of f and p in which $f(p)$ is monotonic. There is also a technical problem of expressing H as a function of p after its obtaining as a function of f according to (2.18).

For example, consider the simple rational function [45]

$$L(f) = \frac{f}{1+2\beta f}, \qquad \beta = \text{const} > 0. \tag{2.26}$$

The quantity (2.18) is easily found,

$$H = 2fL_f - L(f) = \frac{f(1-2\beta f)}{(1+2\beta f)^2}, \tag{2.27}$$

but finding the dependence $f(p)$ to be substituted to (2.27) requires solving a fourth-order algebraic equation:

$$f = p(1+2\beta f)^4. \tag{2.28}$$

It is therefore not surprising that the numerous existing electric solutions either start from a specific function $H(p)$ or postulate the metric function $A(r)$, as is actually done in [3] and a few other papers by the same authors.

Dyonic solutions with both nonzero charges q_e and q_m can be obtained with more effort. Neither $f(r)$ nor $p(r)$ is known explicitly now. Thus, in particular,

$$f(r) = \frac{2}{r^4}\left(q_m^2 - \frac{q_e^2}{L_f^2}\right). \tag{2.29}$$

Comparing the expressions for $\rho(r)$ from (2.13) and from (2.17), we can write

$$\frac{1}{2}L(f) + \frac{2q_e^2}{L_f r^4} = \frac{2M'(r)}{r^2} = \rho(r). \tag{2.30}$$

If $L(f)$ is known, Eq. (2.29) can be treated either (A) as an (in general, transcendental) equation for the function $f(r)$ or (B) as an expression of r as a function of f.

In case (A), if we can find explicitly $f(r)$, integration of Eq. (2.30) gives the metric function $A(r)$.

The scheme (B) gives a solution in quadratures expressed in terms of f that can be now chosen as a new radial coordinate. Indeed, if $L(f)$ and $r(f)$ are known and monotonic, so that $L_f \neq 0$ and $r_f \neq 0$, we can rearrange Eq. (2.30) as

$$M_f = \frac{r^2 r_f}{2}\left[\frac{L}{2} + \frac{q_e^2}{L_f r^4}\right] \tag{2.31}$$

(as before, the subscript f denotes d/df). Since the r.h.s. of (2.31) is known, we can calculate $M(f)$ and $A(r)$ and also rewrite the metric in terms of the coordinate f. Thus we obtain *a general scheme of finding dyonic solutions* under the above conditions [18]. '

As a trivial example of using the scheme (A), we can consider the Maxwell theory, $L = f$. Substituting $L = f$ and $L_f = 1$ to Eq. (2.30), we obtain $2M' = (q_e^2 + q_m^2)/r^2$, whence $2M(r) = 2m - (q_e^2 + q_m^2)/r$ and

$$A(r) = 1 - \frac{2m}{r} + \frac{q_e^2 + q_m^2}{r^2}, \qquad m = \text{const}, \tag{2.32}$$

that is, the dyonic Reissner-Nordström solution, as should be the case.

Another example is obtained [18] if we assume that Eq. (2.29) is linear in f. Then we have to put $L_f^{-2} = c_1 f + c_2$ with $c_{1,2} = \text{const}$, which yields after integration $L = L_0 + (2/c_1)\sqrt{c_1 f + c_2}$. Assuming a Maxwell behavior, $L \approx f$, at small f, we find $c_2 = 1$, $L_0 = -2/c_1$, and denoting $2/c_1 = b^2$, we arrive at the truncated Born-Infeld Lagrangian,

$$L(f) = b^2\Big(-1 + \sqrt{1 + 2f/b^2}\Big), \quad b = \text{const} \tag{2.33}$$

(the full Born-Infeld Lagrangian also involves the other electromagnetic invariant $h^2 = ({}^*F_{\mu\nu}F^{\mu\nu})^2$). With (2.33), we obtain

$$\begin{aligned} f(r) &= \frac{2b^2(q_m^2 - q_e^2)}{4q_e^2 + b^2 r^4}, \\ \rho(r) &= -\frac{b^2}{2} + \left(\frac{b^2}{2} + \frac{2q_e^2}{r^4}\right)\sqrt{\frac{4q_m^2 + b^2 r^4}{4q_e^2 + b^2 r^4}}. \end{aligned} \tag{2.34}$$

In the special case of a self-dual electromagnetic field, $q_e^2 = q_m^2$, we find simply $f = 0$ and $\rho(r) = 2q^2/r^4$, as in the Maxwell theory, and the dyonic solution for $A(r)$ coincides with (2.32). For arbitrary charges, Eq. (2.17) leads to a long expression with the Appel hypergeometric function F_1, not to be presented here.

Other examples of dyonic NED-GR solutions are found and discussed in [43–45, 54, 73].

2.4.2 *Regularity and no-go Theorems*

Magnetic solutions. According to (2.6), a regular center requires $A(r) = 1 + \mathcal{O}(r^2)$ at small r. In magnetic solutions with $f = 2q_m^2/r^4 \to \infty$ the metric regularity then requires $L \to L_0 < \infty$ as $f \to \infty$ [17] because the density that should be finite is now $T_t^t = \rho = L/2$. Furthermore, asymptotic flatness requires $A(r) = 1 - 2m/r +$

$o(1/r)$, where m is the Schwarzschild mass. By (2.17), it is the case if $\rho \sim r^{-4}$ or smaller as $r \to \infty$, which happens if $L(f) \sim f$, i.e., it has a Maxwell asymptotic behavior at small f. The metric is then approximately Reissner-Nordström at large r.

The infinite magnetic induction $B \sim 1/r^2$ at the center might cause a problem, but as discussed in [17], a correct estimate of the force applied to a charged test particle moving in the nonlinear magnetic field under consideration, obtained along the lines of Refs. [62, 63], shows that such forces are finite for both electrically and magnetically charged test particles and even vanish at $r = 0$.

Thus invoking a smooth function $L(f)$ such that $L \sim f$ as $f \to 0$ and $L \to L_0 < \infty$ as $f \to \infty$ is an easy way to obtain globally regular configurations including magnetic black holes and solitons, used in many papers, probably beginning with Ref. [17].

In all such solutions, a general feature is that $A \to 1$ as both $r \to 0$ and $r \to \infty$. Moreover, the mass term $-2m/r$ contributes negatively to $A(r)$ as long as $m > 0$. Thus in regular solutions $A(r)$ should inevitably have a minimum, at which the value of A depends on the mass and charge values. Their relationship determines the existence of horizons located at regular zeros of $A(r)$. If the mass $m > 0$ is fixed, then at small charges (which contribute positively to $A(r)$ at least at large r) the minimum of A is negative because the solution is close to Schwarzschild's almost everywhere, and then any regular function $A(r)$ has two zeros, one of which should be close to $r = 2m$, while the other emerges since it is necessary to return to $A(r) > 0$ at small r to reach $A = 1$ at $r = 0$. At large charges, on the contrary, the mass term $-2m/r$ is only significant at large r, and a minimum of A should be positive, leading to a solitonic solution. Some value of q must be critical, leading to a double zero of $A(r)$, corresponding to a single extremal horizon.

This general picture is really observed in the known examples of regular static, spherically symmetric NED-GR solutions. Let us illustrate it with the behavior of $A(r)$ in the example from [17], where

$$L(f) = \frac{f}{\cosh^2\left(b|f/2|^{1/4}\right)}, \qquad b = \text{const} > 0. \tag{2.35}$$

In the magnetic solution, with $q = q_m > 0$ (for simplicity),

$$\rho = \frac{q^2/r^4}{\cosh^2(b\sqrt{q}/r)}, \qquad A(r) = 1 - \frac{2m}{r}\left(1 - \tanh\frac{q^2}{2mr}\right), \tag{2.36}$$

where the mass m is determined as $M(\infty)$. The behavior of $A(r)$ is shown in Fig. 2.1 for three values of q/m leading to qualitatively different geometries. The causal structures and Carter-Penrose diagrams of these space-times are the same as those for Reissner-Nordström ones, but the important difference is that now the lines $r = 0$ denote a regular center instead of a singularity.

Regular models with more than two horizons are also possible, see, e.g., [33, 56] for detailed studies of such solutions.

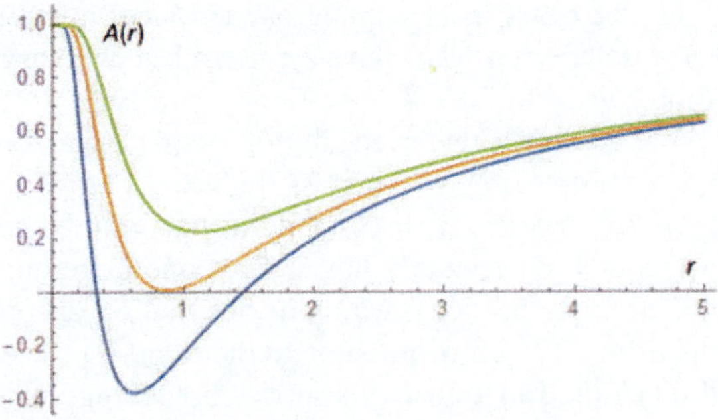

Fig. 2.1 The behavior of $A(r)$ according to Eq. (2.36) with $m = 1$ and $q = 0.9,\ 1.06,\ 1.2$ (bottom-up)

An important feature of regular solutions is that with given $L(f)$ the Schwarzschild mass m is uniquely fixed by the charge q. Indeed, to obtain a regular center, the integration in Eq. (2.17) must be carried out from $r = 0$ (where the density ρ is finite and determined by q) to arbitrary r, resulting in the Schwarzschild mass $m = M(\infty)$. It means that this mass is completely created by the electromagnetic field energy. Any additional mass m_1 that can appear in the solution as an integration constant in Eq. (2.17) would add the singular term $2m_1/r$ to $A(r)$.

Thus, in particular, returning to the solution (2.36) for the theory (2.35), it is easy to find that $m = q^{3/2}/(2b^{1/4})$, or on the contrary, the parameter b in $L(f)$ may be expressed in terms of m and q: $b = q^6/(16m^4)$.

Electric solutions with a regular center and a Reissner-Nordström asymptotic behavior can either be found in the same manner using the P formulation of NED (as is done in Refs. [3, 30] and many others), or obtained directly from the magnetic ones using the FP duality. However, as we saw above, solutions with the same metric correspond to quite different NED theories than those used in magnetic solutions, and this circumstance leads to their different physical properties. First of all, let us recall a theorem proved in [17, 21, 23]:

Theorem 2.1 *If a static, spherically symmetric electric solution ($q_e \neq 0, q_m = 0$) to the $L(f)$ NED-Einstein equation describes a space-time with a regular center, it cannot have a Maxwell behavior at small f ($L \approx f,\ L_f \to 1$).*

Proof To begin with, since the Ricci tensor for our metric is diagonal, the curvature invariant $R_{\mu\nu}R^{\mu\nu} = R^\nu_\mu R^\mu_\nu$ is a sum of squares of the components R^ν_μ, hence each of them taken separately must be finite at any regular point, including a center. It then follows that each of the components of T^ν_μ should be finite, as well as their any linear combination. In particular, by (2.13), we must have $|f_e L_f| < \infty$. But according to (2.14), $f_e L_f^2 = 2q_e^2/r^4 \to \infty$. These two conditions, taken together, lead to

$$f = -f_e \to 0, \qquad L_f \to \infty \qquad \text{as } r \to 0. \tag{2.37}$$

It means that we have a non-Maxwell function $L(f)$ at small f. □

On the other hand, regular asymptotically flat electric solutions obtained in the P formulation of NED have a correct Maxwell asymptotic behavior. How can it be combined with (2.37)?

The answer is that such solutions correspond to different Lagrangians $L(f)$ near $r = 0$ and at large r [16, 17]. Indeed, at a regular center $r = 0$ we have $-p = 2q_e^2/r^4 \to \infty$ and $f = 0$, while at flat infinity both $p \to 0$ and again $f \to 0$. It means that f inevitably has at least one extremum at some $p = p^*$, breaking the monotonicity of $f(p)$, which means that on different sides of p^*we have different functions $L(f)$ corresponding to the same $H(p)$. As shown in [17], at an extremum of $f(p)$ the function $L(f)$ suffers branching, at which the derivative L_f tends to the same finite limit as $p \to p^* + 0$ and $p \to p^* - 0$, while L_{ff} tends to infinities of opposite signs. This corresponds to a cusp in the plot of $L(f)$. Another form of branching of $L(f)$ takes place at extremum points of $H(p)$, if any, where the monotonicity of $f(p)$ also breaks down. The number of different Lagrangians $L(f)$ on the way from the center to infinity is equal to the number of monotonicity ranges of $f(p)$ [17].

To illustrate this unusual behavior of $L(f)$ let us use as an example the same metric function (2.36), where now $q = q_e$, as a solution corresponding to $H(p)$ dual to (2.35):

$$H(p) = -\frac{p}{\cosh^2\left(b|p/2|^{1/4}\right)}, \qquad b = \text{const} > 0. \tag{2.38}$$

Calculations reveal the behavior of the corresponding functions $f(p)$ and $L(f)$ shown in Figs. 2.2 and 2.3. It turns out that $L(f)$ has as many as four branches, in other words, there are four NED theories acting in different parts of space.

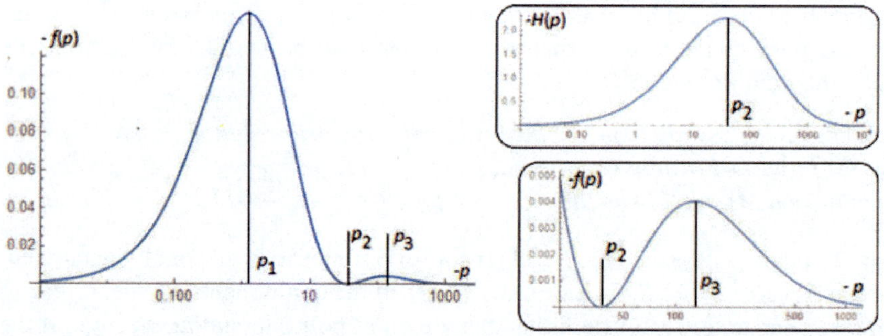

Fig. 2.2 The function $f(p)$ obtained from $H(p)$ given by Eq. (2.38) with $b = 1$. The points p_1 and p_3 show the maxima of $|f(p)|$ while p_2 shows its minimum corresponding to the maximum of $|H(p)|$. The upper inset shows the function $H(p)$, while the lower one is an enlarged view of the neighborhood of p_2 and p_3 in the plot of $f(p)$

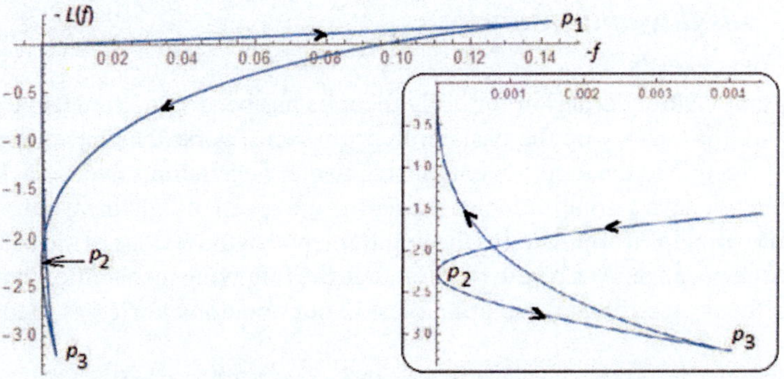

Fig. 2.3 The behavior of $L(f)$ in the electric solution for $H(p)$ from Eq. (2.38) with $b = 1$. The points p_1, p_2, p_3 correspond to the extrema of $f(p)$, at which the function $L(f)$ passes on from one branch to another. The inset shows more clearly the range close to p_2 and p_3. Arrows on the curves show the direction of growing $|p|$

Dyonic solutions If there are both nonzero q_e and q_m, then a combination of T_μ^ν components leads to the requirement

$$(f_e + f_m)|L_f| < \infty \tag{2.39}$$

that must hold at any regular point, including a regular center. Moreover, it must hold for each term separately because both f_e and f_m are positive. Applying it to f_e, we obtain, as before, that $L_f \to \infty$ at a regular center [17]. However, the inequality $f_m|L_f| < \infty$ leads to the requirement $L_f \to 0$, since $f_m = 2q_m^2/r^4 \to \infty$ as $r \to 0$. We arrive at a contradiction that leads to the general result:

Theorem 2.2 *Static spherically symmetric dyonic solutions to the NED-Einstein equations with arbitrary $L(f)$ cannot describe space-times with a regular center.*

Inclusion of a Cosmological Constant

If $\Lambda \neq 0$, asymptotically (A)dS solutions [52] are obtained by simply adding $-\Lambda r^2/3$ to $A(r)$ in (2.17). This new term does not affect the properties of the solutions near $r = 0$, therefore, all conclusions on the existence of a regular center and the necessary conditions for it, obtained with $\Lambda = 0$, remain valid with $\Lambda \neq 0$, although the latter drastically changes the global properties of space-time.

2.4.3 Causality and Unitarity

An important viability criterion for NED theories has been suggested by A. Shabad and V. Usov [66], partly on the basis of their previous work: they have used (i) the causality principle as the requirement that elementary excitations over a background field should not have a group velocity exceeding the speed of light in vacuum and (ii) the unitarity principle formulated as the requirement that the residue of the propagator should not be negative. As a result, there emerge the following inequalities that should hold for a theory satisfying these principles: in our notations, for $L(f)$ theories,

$$L_f > 0, \qquad L_{ff} \le 0, \qquad \Phi = L_f + 2fL_{ff} \ge 0. \tag{2.40}$$

One can notice that the third condition can be rewritten as $H_f \ge 0$, with the Hamiltonian-like quantity H given by (2.18), but does not directly concern the derivative H_p due to a possible complexity in the dependence $f(p)$. The quantity Φ also plays an important role in the effective metric for photon propagation and in the stability conditions, to be considered in the next subsections.

One immediate observation can be made about magnetic solutions with a regular center, both black hole and solitonic ones:

Theorem 2.3 *In static, spherically symmetric magnetic solutions to $L(f)$ NED-Einstein equations, the causality and unitarity conditions (2.40) are inevitably violated in a neighborhood of a regular center.*

Proof A regular center requires a finite limit of $L(f)$ as $f \to \infty$. If $L_f > 0$ (as required by the first inequality in (2.40)), then, to have a convergent integral $L = \int L_f df$, one has to require $L_f \ll 1/f$ at large f, hence the quantity $L_f\sqrt{f}$ is decreasing as $f \to \infty$, and its derivative in f is negative. On the other hand, we can write $\Phi = 2\sqrt{f}(L_f\sqrt{f})_f$, consequently, $\Phi < 0$ at large f, so that the first and third inequalities in (2.40) cannot hold simultaneously. □

2.4.4 Light Propagation and the Effective Metric

As we have seen, the same regular metric of the form (2.15) can be obtained with two kinds of sources, the electric and magnetic ones, described by different NED theories. It is thus natural to expect that the properties of electromagnetic fields will also be different in these two cases. Let us try to explore these differences using the effective metric formalism developed by M. Novello et al. [58] while studying the propagation of electromagnetic field discontinuities using Hadamard's approach [35]. According to [58, 59], photons governed by NED propagate along null geodesics of the effective metric

$$h^{\mu\nu} = g^{\mu\nu}L_f - 4L_{ff}F^{\mu}{}_{\alpha}F^{\alpha\nu}. \tag{2.41}$$

Electric Solutions

In the case of a purely electric field in the metric (2.15), $h^{\mu\nu}$ is diagonal, and we can write the effective metric as

$$ds^2_{\text{eff}} \equiv h_{\mu\nu}dx^\mu dx^\nu = \frac{1}{\Phi}\left[A(r)dt^2 - \frac{dr^2}{A(r)}\right] - \frac{r^2}{L_f}d\Omega^2, \qquad (2.42)$$

$$\Phi = L_f + 2fL_{ff} = \frac{H_p}{f_p}.$$

Consider the behavior of $h_{\mu\nu}$ at branching points of $L(f)$ that are inevitable in solutions with regular $g_{\mu\nu}$. At an extremum $p = p^*$ of $f(p)$ at which $f \neq 0$ (like points p_1 and p_3 in Fig. 2.2), we have $\Phi \to \infty$ since $f_p = 0$ while H_p is finite. This results in a curvature singularity of the effective metric due to blowing up of the quantity K_1 in (2.4).

Another kind of singularity of the metric (2.42) occurs at extrema of $H(P)$, those like point p_2 in Fig. 2.2: in this case, generically, Φ is finite but $L_f \to \infty$, which leads to a singular center in the auxiliary space-time with the metric $h_{\mu\nu}$ due to $h_{\theta\theta} \to 0$.

The changes in photon frequencies at their motion in space-time can be evaluated as outlined in [59]. Thus, if an emitter at rest at point X sends a photon with frequency ν_X, it comes to a receiver at rest at point Y with frequency ν_Y related to ν_X by

$$\frac{\nu_Y}{\nu_X} = \left[\frac{\sqrt{g_{tt}}}{h_{tt}}\right]_Y \left[\frac{\sqrt{g_{tt}}}{h_{tt}}\right]_X^{-1} = \left[\frac{\Phi}{\sqrt{A}}\right]_Y \left[\frac{\Phi}{\sqrt{A}}\right]_X^{-1}, \qquad (2.43)$$

where the second equality sign corresponds to the metric (2.42). If X is a regular point while Y is located at an inevitable branching point of $L(f)$ (like p_1 or p_3), then any photon arriving there is infinitely blueshifted, gaining an unlimited energy, which thus implies instability of the whole configuration.

The above reasoning used the assumption $A > 0$. In black hole solutions, the sphere where $L_f = 0$ may be located beyond the event horizon, where $A < 0$. In such a region, also called a T-region, r is a temporal coordinate, t is a spatial one, and in the redshift relation (2.43) we must replace g_{tt} with g_{rr}, or more specifically, $\sqrt{A}$ with $1/\sqrt{-A}$. However, as long as A is finite, this replacement does not affect the conclusion on an infinite blueshift on the sphere where $\Phi = \infty$.

Magnetic solutions For the same metric $g_{\mu\nu}$ with a magnetic source, we get, instead of (2.42),

$$ds^2_{\text{eff}} = \frac{1}{L_f}\left[A(r)dt^2 - \frac{dr^2}{A(r)}\right] - \frac{r^2}{\Phi}d\Omega^2, \qquad (2.44)$$

where, as before, $\Phi = L_f + 2fL_{ff}$. Then, for a photon traveling from point X to point Y, we find instead of (2.43):

$$\frac{\nu_Y}{\nu_X} = \left[\frac{L_f}{\sqrt{A}}\right]_Y \left[\frac{L_f}{\sqrt{A}}\right]_X^{-1}. \tag{2.45}$$

Now $L(f)$ has no branching points, while at a regular center ($r = 0,\ A = 1$) both L_F and Φ vanish, and the quantity $h_{22} \to \infty$, i.e., the spherical radius in the effective metric behaves as if in a wormhole, whereas $h_{tt} \to \infty$, which means an infinite redshift for photons. Also, all curvature invariants of the metric (2.44) vanish at $r = 0$. It is really a quiet place.

There still occurs something of interest between spatial infinity and the center of a regular magnetic model: there is necessarily a sphere $r = r^*$ on which $\Phi = 0$. Indeed, Φ can be presented as $\Phi = 2\sqrt{f}(\sqrt{f}L_f)_f$. The quantity $\sqrt{f}L_f$ tends to zero both at $r = 0$ (where $f \to \infty$ but $L \to$ const) and in the limit $r \to \infty$. Since $\sqrt{f}L_f$ is in general nonzero, it has at least one extremum at some $f \neq 0$, thus it is the value where $\Phi = 0$. The metric (2.44) is singular there due to $h_{22} \to \infty$, but this singularity seems to be unnoticed by the photons, as follows from an integral of their geodesic equation

$$L_f^{-2}\dot{r}^2 + [A(r)\Phi/r^2]\ell^2 = \varepsilon^2, \tag{2.46}$$

where the overdot denotes a derivative in an affine parameter, ε and ℓ are the photon's constants of motion characterizing its initial energy and angular momentum. Generically we have $L_f \neq 0$ at points where $\Phi = 0$, therefore the photon frequency remains finite. However, as we will see below, the photon velocities behave there in an unusual manner.

If $L_f = 0$ at some value of $f > 0$, it leads to another kind of singularity of the metric (2.44)), and this time it acts for NED photons as a potential wall, or a mirror, as is evident from (2.46) which then implies $\dot{r} = 0$. Also, from Eq. (2.45) it follows that the photons are infinitely redshifted there: ν_Y vanishes if $L_f(Y) = 0$. It means that in such a case no photon from outside can approach the center.

We thus observe a striking difference between the properties of photons moving in the same regular metric (2.15) in the cases where it is sourced by electric and magnetic fields. In the electric case, the photons inevitably "accelerate" to an infinite energy and destabilize the whole system, whereas in the magnetic case, even if they can approach the regular center (if $L_f \neq 0$, hence no mirror), they lose energy, being infinitely redshifted there.

The violent behavior of photons in electric regular black holes was discovered by Novello et al. [59] for a particular example of such a configuration. As shown in [17], it is quite a general property of NED-GR solutions.

Photon velocities A question of interest is the velocity of NED photons in regular or singular space-times. From (2.44) it follows that radially moving photons have the same velocity equal to c (=1) as the conventional Maxwell ones since the 2D metric of the (t, r) subspace in the effective metric (2.44) is conformal to that in the space-time metric (2.15), and their 1D light cones coincide. This is true for both electric and magnetic solutions. However, the situation is different for nonradial photon paths.

Consider a photon moving instantaneously in a tangential direction. Without loss of generality we can suppose that it moves along an equator of certain radius r in our coordinate system. For the corresponding null direction in terms of the effective metric we have $ds^2_{\text{eff}} = h_{tt}dt^2 - h_{\theta\theta}d\theta^2 = 0$, and for the photon's linear velocity $v_{\text{ph}} = r\, d\theta/dt$ we obtain:

$$\text{in an electric solution: } v^2_{\text{ph}} = AL_f/\Phi,$$
$$\text{in a magnetic solution: } v^2_{\text{ph}} = A\Phi/L_f, \tag{2.47}$$

For the Maxwell field, $L_f \equiv \Phi \equiv 1$, hence $v^2_{\text{ph}} = A$, and it would be equal to unity if we used the local time increment $dt_{\text{local}} = \sqrt{A}dt$ instead of the coordinate time increment dt, and the length element equal to $dr/\sqrt{A}$ instead of dr. Thus, as should be the case, Maxwell photons always travel in vacuum with the speed of light. The factor L_f/Φ or Φ/L_f changes the photons' velocity, working like a refractive index.

In particular, in electric solutions, at cusplike branching points where $\Phi \to \infty$ while L_f remains finite (like points p_1 and p_3 in Figs. 2.2 and 2.3), $v_{\text{ph}} \to 0$, in other words, tangentially moving photons have zero velocity at this value of r. On the contrary, at branching points like p_2, where $H_p = f_p = 0$ and Φ is finite but $L_f \to \infty$, we obtain $v_{\text{ph}} \to \infty$.

In magnetic solutions, at spheres where $\Phi = 0$ while L_f is finite, we have again $v_{\text{ph}} = 0$, a zero velocity of tangentially moving photons.

So far we were assuming $A(r) > 0$, while in black hole space-times there are T-regions where $A(r)$ is negative. However, the only change in Eq. (2.47) emerging in a T-region is the simple replacement $A \to 1/|A|$ because r is there a time coordinate instead of t, and in other respects our reasoning remains unaltered.

At intermediate directions between the radial and tangential ones, the NED photon velocities will obviously have intermediate values. We conclude altogether that these velocities can be both subluminal and superluminal, varying from zero to infinity.

We also observe that the conditions under which superluminal photon velocities are avoided ($L_f/\Phi \leq 1$ for electric solutions and $\Phi/L_f \leq 1$ for magnetic ones) do not coincide with the causality/unitarity conditions (2.40). Even more than that: any non-Maxwell NED, in which $\Phi/L_f \not\equiv 1$, predicts superluminal photon motion in either electric or magnetic space-times. Actually, this observation puts to doubt either any NED theory or the described straightforward interpretation of the effective metrics.

Also, the nonlinearity of NED is acting like a highly anisotropic medium, which, if one takes into account the wave properties of photons, naturally leads to such a phenomenon as birefringence, see the relevant recent studies in [37, 38] and references therein.

2.4.5 Dynamic Stability

Any static or stationary configuration may be regarded viable if it is stable under different kinds of perturbations, which always exist in nature, or at least if it decays slowly enough. Possible regular NED black holes do not make an exception, and their stability is discussed in a number of papers, e.g., [15, 55, 57, 71], see also references therein.

C. Moreno and O. Sarbach [55] have derived sufficient conditions for linear dynamic stability of the domain of outer communication of electric or magnetic black holes sourced by a general $L(f)$ NED. For magnetic black holes these conditions read (in the present notations)

$$L > 0, \qquad L_y > 0, \qquad L_{yy} > 0, \tag{2.48}$$

$$3L_y - A(r) y L_{yy} \geq 0, \tag{2.49}$$

where $y := \sqrt{q^2 f/2} = q^2/r^2$, and the index "$y$" stands for d/dy. In terms of f these conditions are rewritten as

$$L > 0, \qquad L_f > 0, \qquad \Phi \equiv L_f + 2fL_{ff} > 0, \tag{2.50}$$

$$[6 - A(r)]L_f - 2fL_{ff} \geq 0. \tag{2.51}$$

One can notice that the conditions (2.50) partly coincide with the causality and unitarity conditions (2.40). Moreover, if $L_f > 0$ and also the condition $L_{ff} \leq 0$ from (2.40) is valid, then the condition (2.51) holds automatically provided $f > 0$ (which is true for magnetic solutions) and $A(r) < 6$ (we can note that at least in regular black hole solutions, in general, $A(r) \leq 1$).

Thus Eq. (2.51) is not expected to make a problem, at least for regular magnetic solutions. Unlike that, by Theorem 2.3, the condition $\Phi > 0$ is always violated for such solutions near a regular center. This may be important for black hole solutions only if the range of r where $\Phi < 0$ extends to the domain of outer communication, which must be checked for each particular black hole solution.

The sufficient stability conditions for electric black holes have a form similar to (2.48), (2.49) in terms of the P-framework of the theory [55], which could be expected due to FP duality. Their reformulation to the F-framework is not possible in a general form due to problems with a relationship between f and p, see above.

More general stability conditions for NED-GR solutions involving both $F_{\mu\nu}$ and $^*F_{\mu\nu}$ have been recently obtained by K. Nomura, D. Yoshida and J. Soda in [57].

2.4.6 Examples

Let us enumerate some particular examples of the Lagrangians $L(f)$ discussed in the literature, along with their basic properties at $f > 0$ (that is, for their magnetic solutions): the existence of a correct Maxwell weak field (MWF) limit, a finite limit as $f \to \infty$, necessary for a regular center in magnetic solutions, and the validity of the causality, unitarity and stability conditions, (2.40) and (2.50).

It is convenient to do that in the form of a table, see Table 2.1. Among the conditions (2.40) and (2.50) we select there the inequalities $L_{ff} < 0$ and $\Phi > 0$ because the condition $L > 0$ holds in all examples, and $L_f > 0$ in all of them except the one with hyperbolic cosine.

The first line represents the truncated Born-Infeld Lagrangian which does not provide a regular center but satisfies the conditions (2.40) and (2.50). The next four lines correspond to different examples of NED considered by S. Kruglov, the first two of them provide regular magnetic black holes. The sixth line represents a special case from numerous examples considered by Fan and Wang in [30], selected there because it both has a correct MWF limit and provides a regular center. The last line is the special case of NED discussed above. The explicit form of the solutions can be found in the cited papers along with detailed discussions of their properties. This list certainly does not pretend to be complete, and many other solutions have been obtained and studied.

It can be observed from the table that in all NED theories that provide a regular center (those with "yes" in the column "finite as $f \to \infty$"), the inequality $\Phi > 0$ does not hold at sufficiently high values of f, in accordance with Theorem 2.3.

Table 2.1 Some examples of $L(f)$ NED theories: properties of magnetic solutions ($f \geq 0$)

References	Lagrangian[a]	Correct MWF limit	Finite as $f \to \infty$	Condition $L_{ff} < 0$	Condition $\Phi > 0$
[12], (2.33)	$\beta^2(-1+\sqrt{1+2f/\beta^2})$	yes	no	yes	yes
[41, 46], (2.26)	$\dfrac{f}{1+2\beta f}$	yes	yes	yes	partly[b]
[39]	$\beta^{-1}\arctan(\beta f)$	yes	yes	yes	partly
[40]	$\beta^{-1}\arcsin(\beta f)$	yes	no	no	yes
[42]	$\beta^2 \log\left(1+\dfrac{f}{\beta^2}\right)$	yes	no	yes	partly
[30]	$\dfrac{f}{(1+(\beta f)^{1/4})^4}$	yes	yes	yes	partly
[17], (2.35)	$\dfrac{f}{\cosh^2(\beta\|f/2\|^{1/4})}$	yes	yes	no	partly

[a] In all examples, $\beta = \text{const} > 0$

[b] Here and in other lines, "partly" means that $\Phi > 0$ at f smaller than some critical value

2.5 NED with More General Lagrangians

2.5.1 Systems with $L = L(f, h)$

Beginning with the paper by Born and Infeld [12], the researchers considered NED theories with Lagrangians more general than $L(f)$, depending on electromagnetic invariants other than f. The first and the most natural candidate is the pseudoscalar $h = {}^*F_{\mu\nu}F^{\mu\nu} = 2\mathbf{BE}$, where **E** and **B** are the the electric field strength and magnetic induction 3-vectors, respectively. Now the total action has the form

$$S = \frac{1}{2}\int \sqrt{-g}\, d^4x[R - L(f, h)]. \tag{2.52}$$

Special cases of $L(f, h)$ are the Born-Infeld Lagrangian

$$L^{\rm BI} = b^2\biggl(-1 + \sqrt{1 + \frac{f}{2b^2} - \frac{h^2}{16b^4}}\biggr), \qquad b > 0, \tag{2.53}$$

and the so-called modified Maxwell (ModMax) Lagrangian [7, 69]

$$L^{\rm MM} = \frac{1}{4}\Bigl(f\cosh\gamma - \sqrt{f^2 + h^2}\sinh\gamma\Bigr), \qquad \gamma \in \mathbb{R}. \tag{2.54}$$

Both these models are distinguished by their symmetry properties, in particular, the ModMax NED is conformally and duality invariant.[1]

The electromagnetic field equations due to (2.52) read

$$\nabla_\mu(L_f F^{\mu\nu} - L_h {}^*F^{\mu\nu}) = 0, \qquad \nabla_\mu {}^*F^{\mu\nu} = 0, \tag{2.55}$$

and the electromagnetic field SET has the form

$$T_\mu^\nu = -2L_f F_{\mu\alpha}F^{\nu\alpha} + \tfrac{1}{2}\delta_\mu^\nu(L - hL_h). \tag{2.56}$$

Assuming static spherical symmetry, hence having only radial electric and magnetic fields, we are again dealing with a SET with $T_t^r = 0$ and $T_t^t = T_r^r$, and the metric can be written in the form (2.15). We then have according to (2.55)

$$L_f F^{tr} - L_h {}^*F^{tr} = \frac{q_e}{r^2}, \qquad F_{\theta\phi} = q_m \sin\theta, \tag{2.57}$$

with the corresponding charges $q_e, q_m = \text{const}$.

---

[1] In our notations, see (2.52), some of the signs and factors are different from those in [11] and other papers. In particular, the Maxwell theory here corresponds to $L(f, h) = f$.

For static, spherically symmetric solutions to the NED-GR equations with $L = L(f, h)$, a number of no-go theorems have been proved in Ref. [11]. According to these theorems, such solutions with the metric (2.15) cannot describe a geometry with a regular center under the following assumptions on the electromagnetic field:

1. $q_e \neq 0$, $q_m = 0$ (electric), MWF limit.
2. $L(f, h) = L(f)$, $q_e \neq 0$, $q_m \neq 0$ (dyonic).
3. $L(f, h) = f + \eta(h)$, with an arbitrary function $\eta(h)$, $q_m \neq 0$ (magnetic or dyonic).
4. $L(f, h) = f + af^s h^u$, with real $a \neq 0$, positive integers $s > 1$, $u > 1$, and $q_e \neq 0$, $q_m \neq 0$ (dyonic).
5. $L(f, h) = f + af^2 + bfh + ch^2$, where $a, b, c \in \mathbb{R}$, $q_e \neq 0$, $q_m \neq 0$ (dyonic).
6. $L(f, h)$ given by (2.53) or (2.54), $q_m \neq 0$ (magnetic or dyonic).
7. $L(f, h) = f + af^2 + bfh + ch^2$, the pair $(b, c) \neq (0, 0)$, $q_e = 0$, $q_m \neq 0$ (magnetic).

The numbering here corresponds to the theorem numbers in Ref. [11]. Theorem 2 from this list coincides with our Theorem 2.2 presented in Sect. 2.4. We can notice that only two theorems, the first and the sixth ones, use the assumption of a correct MWF limit: in all other cases considered, a regular center is impossible irrespective of the weak field behavior of the theory.

On the other hand, there still remain some opportunities of obtaining regular BHs other than purely magnetic ones with $L = L(f)$. For example, both with $L(f)$ and $L(f, h)$, purely electric solutions with a regular center are possible with a theory having no MWF limit. Then, assuming a regular central region governed by such a theory, one can obtain an asymptotically flat electrically charged configuration by using a kind of phase transition, such that outside a certain sphere $r = r_{\text{crit}}$, another NED theory will be valid, having a correct MWF limit, as was suggested in [26].

2.5.2 Systems with $L = L(f, J)$

One more invariant, in addition to f,

$$J \equiv J_4 = F_{\mu\nu} F^{\nu\rho} F_{\rho\sigma} F^{\sigma\mu}, \tag{2.58}$$

has also been used for formulating an extended NED theory [28, 68]. With this invariant, the action reads

$$S = \frac{1}{2} \int \sqrt{-g} d^4x [R - L(f, J)], \tag{2.59}$$

the electromagnetic field equations are

$$\nabla_\mu Q^{\mu\nu} = 0, \qquad \nabla_\mu {}^*F^{\mu\nu} = 0, \tag{2.60}$$
$$Q^{\mu\nu} := 4L_f F^{\mu\nu} + 8L_J F^{\mu\rho} F_{\rho\sigma} F^{\sigma\nu}, \tag{2.61}$$

where $L_f = \partial L/\partial f$ and $L_J = \partial L/\partial J$. The SET has the form

$$T_\mu^\nu = -2(L_f + fL_J)F_{\mu\alpha}F^{\nu\alpha} + \frac{1}{2}\delta_\mu^\nu[(f^2 - 2J)L_J + L]. \tag{2.62}$$

An important subclass of the theories (2.59) is called conformal NED, or CNED, and is characterized by a zero trace of the SET (2.62) [28, 68], hence,

$$T = 2(L - fL_f - 2JL_J) = 0. \tag{2.63}$$

In this case, the SET as a whole is a multiple of the Maxwell field SET, and the field Eqs. (2.60) are invariant under general conformal mappings of the metric (conformally invariant) like the Maxwell equations.

The theory has a Maxwell asymptotic behavior at small fields if $L(f, J) \approx f$, so that $L_f \to 1$ and $|L_J < \infty|$ as $F_{\mu\nu} \to 0$ (the latter condition takes into account that $J \sim f^2$ at small $F_{\mu\nu}$).

Assuming static spherical symmetry, we have, as before, only radial electric and magnetic fields, and the SET has again the properties $T_t^r = 0$ and $T_t^t = T_r^r$, and the metric can be written in the form (2.3). Specifically,

$$T_t^t = T_r^r = 2(L_f + fL_J)E^2 + \tfrac{1}{2}L - 4B^2E^2L_J, \tag{2.64}$$
$$T_\theta^\theta = T_\phi^\phi = -2(L_f + fL_J)B^2 + \tfrac{1}{2}L - 4B^2E^2L_J,$$

with $E^2 = F^{tr}F_{rt}$ and $B^2 = F^{\theta\phi}F_{\theta\phi}$. The field Eqs. (2.60) lead to

$$Q^{tr} = 4(L_f + fL_J)F^{tr} = \frac{q_e}{r^2}, \qquad F_{\theta\phi} = q_m \sin\theta. \tag{2.65}$$

Let us prove that the same no-go theorems as in $L(f)$ theories coupled to GR, are valid in the theories (2.59).

Theorem 2.4 *The theories (2.59) do not admit static, spherically symmetric electric solutions ($q_e \neq 0,\ q_m = 0$) with a regular center and a correct MWF limit.*

Proof As with $L(f)$ theories in Sect. 2.4, assuming regularity at $r = 0$, we must require that all components of T_μ^ν should be finite, as well as their linear combinations. In particular, we require that

$$|T_t^t - T_\theta^\theta| = 2|L_f - 2E^2L_J|E^2 < \infty. \tag{2.66}$$

On the other hand, from (2.65) we obtain

$$Q^{tr}Q_{rt} = 16E^2(L_f - 2E^2L_J)^2 = q_e^2/r^4 \to \infty \quad \text{as } r \to 0. \tag{2.67}$$

The conditions (2.66) and (2.67) are only compatible if $E \to 0$ (that is, the field becomes weak) and $|L_f - 2E^2 L_J| \to \infty$, contrary to the desirable MWF limit. This completes the proof. □

Theorem 2.5 *The theories (2.59) do not admit static, spherically symmetric dyonic solutions ($q_e \neq 0,\ q_m \neq 0$) with a regular center.*

Proof The same requirement as in the previous theorem, $|T_t^t - T_\theta^\theta| < \infty$, necessary to be valid at a regular center, now reads

$$|T_t^t - T_\theta^\theta| = 2|L_f + fL_J|(E^2 + B^2) < \infty. \tag{2.68}$$

Moreover, since both $E^2 > 0$ and $B^2 > 0$, this inequality should hold with each of them taken separately. Applying it with E^2 together with the first equality (2.65) that now leads to

$$Q^{tr} Q_{rt} = 16(L_f + fL_J)^2 = q_e^2/r^4 \to \infty \quad \text{as } r \to 0, \tag{2.69}$$

we obtain, as before, $E \to 0$ and $L_f + fL_J \to \infty$. The same condition (2.68) with B^2 leads to $L_f + fL_J \to 0$ (since $B^2 = q_m^2/r^4 \to \infty$). The resulting contradiction proves the theorem. □

It is also evident than none of the CNED theories can produce a regular BH or, more generally, a solution with a regular center. Indeed, since the SET is proportional to that of Maxwell electrodynamics, it cannot lead to any static, spherically symmetric metric other than Reissner-Nordström, though certainly the interpretation of its constants m and q will be different.

As to purely magnetic solutions ($q_e \neq 0,\ q_m \neq 0$), in which $f = 2q_m^2/r^4$ and $J = 2q_m^4/r^8$, a regular center is possible under the condition that $L(f, J)$ tends to a finite constant as both f and J tend to infinity. The whole situation looks quite the same as with $L(f)$ theories.

Let us give a confirming example, taking as a basis Eq. (2.26) for $L(f)$:

$$L(f, J) = \frac{f}{1 + af/2} + \frac{bJ}{1 + cJ/2}, \qquad a, b, c = \text{const} > 0, \tag{2.70}$$

This Lagrangian has a correct MWF limit and tends to a finite limit at large f and J. Since with $q_e = 0$, according to (2.64), the density is simply $\rho = L/2$, the metric function $A(r)$ is found as

$$A(r) = 1 - \frac{2M(r)}{r}, \qquad M(r) = M_1(r) + M_2(r), \tag{2.71}$$

$$M_1(r) = \frac{q^2}{2}\int \frac{r^2 dr}{r^4 + aq^2}, \qquad M_2(r) = \frac{bq^4}{2}\int \frac{r^2 dr}{r^8 + cq^4},$$

where $q = q_m$. Integration gives

$$M_1(r) = \frac{q^2}{8h\sqrt{2}}\left[2\arctan\frac{h+\sqrt{2}r}{h} - 2\arctan\frac{h-\sqrt{2}r}{h} + \log\frac{h^2-\sqrt{2}hr+r^2}{h^2+\sqrt{2}hr+r^2}\right], \tag{2.72}$$

$$M_2(r) = \frac{bq^4}{16j^5}\left[2C\left(\arctan\frac{r+jC}{jS} + \arctan\frac{r-jC}{jS}\right) - 2S\left(\arctan\frac{r+jS}{jC} + \arctan\frac{r-jS}{jC}\right) + S\log\frac{j^2+2Cjr+r^2}{j^2-2Cjr+r^2} + C\log\frac{j^2-2Sjr+r^2}{j^2+2Sjr+r^2}\right],$$

where we have denoted $h = (aq^2)^{1/4}$, $j = (cq^4)^{1/8}$, $S = \sin(\pi/8)$, $C = \cos(\pi/8)$. At $r \to \infty$ we obtain

$$M_1(r) \to \frac{\pi q^{3/2}}{4\sqrt{2}a^{1/4}}, \qquad M_2(r) \to \frac{\pi b q^{3/2}(C-S)}{8c^{5/8}}, \qquad M = \lim_{r\to\infty}(M_1 + M_2), \tag{2.73}$$

where M is the Schwarzschild mass of completely electromagnetic origin. At the center, we have

$$M_1(r) \approx \frac{r^3}{6a}, \qquad M_2(r) \approx \frac{br^3}{6c} \quad \text{as } r \to 0, \tag{2.74}$$

which leads to $A(r) = 1 + \mathcal{O}(r^2)$, satisfying the regular center condition (2.6). We have obtained a regular asymptotically flat solution in $L(f, J)$ NED with a correct WMF limit.

Is it a black hole solution? To make it clear, let us fix the parameters $q = 1$, $a = 1$, $c = 1$, then the only remaining free parameter is b, and

$$M = \frac{\pi}{8}[\sqrt{2} + b(C - S)]. \tag{2.75}$$

The behavior of $A(r)$ at different values of b is shown in Fig. 2.4. An inspection shows that at small b the solution is of solitonic nature, at $b \approx 2.436$ (corresponding to $M \approx 1.073$) there emerges a single extremal horizon, and at larger b (large M) we obtain a regular black hole with two simple horizons.

2.6 Conclusion

We have discussed the opportunities of obtaining regular spherically symmetric black hole solutions in GR sourced by nonlinear electromagnetic fields governed by different NED theories. It happens that such NED black holes (as well as solitons)

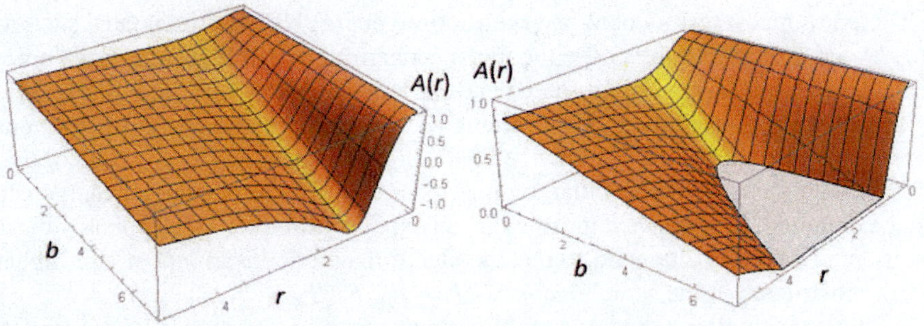

Fig. 2.4 The behavior of $A(r)$ in the magnetic solution for the theory (2.70) with $q = a = c = 1$ and different values of b. The left panel shows the full range of $A(r)$, while in the right one the same dependence is shown with a gray "floor" on the level $A = 0$, visualizing the places where A becomes zero or negative, and the solution then describes a black hole

with a regular center in $L(f)$ theories can exist with pure electric or pure magnetic charges, and only systems with a magnetic charge are compatible with Lagrangians having a correct Maxwell behavior at small f. Dyonic configurations, with $q_e \neq 0$ and $q_m \neq 0$, cannot contain a regular center, whatever be the function $L(f)$.

In Maxwell's electrodynamics there is the well-known symmetry (duality) between electric and magnetic fields, leading to the same symmetry between the corresponding solutions to the Einstein-Maxwell equations, at least in the absence of currents and charges. Unlike that, in NED, we only have FP duality that connects purely electric and purely magnetic configurations with the same metric but sourced by different NED theories. Accordingly, in a theory specified by a particular function $L(f)$, the properties of electric and magnetic solutions are quite different.

It turns out that magnetic solutions lead to completely regular configurations, while for their electric counterparts, obtained from them using FP duality and well-behaved in the framework of the "Hamiltonian" formulation of NED, the Lagrangian formulation is ill-defined, and the behavior of NED photons exhibits undesired features at some intermediate radii: they experience an infinite blueshift, indicating an instability of such a background configuration.

The dynamic stability of regular magnetic solutions is also questionable since one of the sufficient stability conditions ($\Phi > 0$) is inevitably violated near a regular center. Thus general stability results for regular black holes probably cannot be obtained, and stability studies of individual solutions seem to be necessary.

We here did not touch upon thermodynamic properties of NED black holes, this important issue is discussed in many papers, see, among others, [4, 13, 30, 41, 46] and references therein. Let us only remark here that a thermodynamic instability of black holes related to their negative heat capacity is implemented in the process of Hawking evaporation, which is very slow for sufficiently large black holes and can be practically ignored for black holes with stellar and larger masses, irrespective of their global regularity properties.

There are many results obtained with more general NED Lagrangians, such as $L(f, h)$ and $L(f, J)$. Many no-go theorems concerning possible regular black holes with $L(f, h)$ NED have been presented in Refs. [10, 11]. As to $L(f, J)$ NED, involving a fourth-order electromagnetic invariant, we have verified here that the restrictions on regular black hole existence obtained with $L(f)$ are extended to this class of theories without change. In particular, regular black holes with a magnetic charge can also be obtained, as we have confirmed by an explicit example. Very probably these results can be further extended to include electromagnetic invariants of still higher orders constructed in the same manner as $J = F_{\mu\nu}F^{\nu\rho}F_{\rho\sigma}F^{\sigma\mu}$.

Among the remaining theoretic problems deserving further studies let us mention the dynamic stability problem for regular magnetic black holes and the causality issue related to the predicted superluminal velocities of NED photons.

References

1. A.V.B. Arellano, F.S.N. Lobo, Evolving wormhole geometries within nonlinear electrodynamics. Class. Quantum Grav. **23**, 5811 (2006). (gr-qc/0608003)
2. A.V.B. Arellano, N. Bretón, R. Garcia-Salcedo, Some properties of evolving wormhole geometries within nonlinear electrodynamics. Gen. Rel. Grav. **41**, 2561 (2009). (arXiv: 0804.3944)
3. E. Ayon-Beato, A. Garcia Diaz, Regular black hole in general relativity coupled to nonlinear electrodynamics. Phys. Rev. Lett. **80**, 5056 (1998)
4. L. Balart, S. Fernando, Thermodynamics and heat engines of black holes with Born-Infeld-type electrodynamics. Mod. Phys. Lett. A **36**, 2150102 (2021). arXiv: 2103.15040. https://doi.org/10.1142/S0217732321501029
5. L. Balart, E.C. Vagenas, Regular black hole metrics and the weak energy condition. Phys. Lett. B. **730**, 14 (2014). (arXiv: 1401.2136)
6. C. Bambi, L. Modesto, Rotating regular black holes. Phys. Lett. B. **721**, 329 (2013). (arXiv: 1302.6075)
7. I. Bandos, K. Lechner, D. Sorokin, P. Townsend, (ModMax). Phys. Rev. D. **102**, 121703 (2020). arXiv: 2007.09092
8. W. Berej, J. Matyjasek, Vacuum polarization in the spacetime of charged nonlinear black hole. Phys. Rev. D. **66**, 024022 (2002). (gr-qc/0204031)
9. Ch.G. Boehmer, T. Harko, F.S.N. Lobo, Conformally symmetric traversable wormholes. Phys. Rev. D. **76**, 084014 (2007). (arXiv: 0708.1537)
10. A. Bokulić, I. Smolić, T. Jurić, Nonlinear electromagnetic fields in strictly stationary spacetimes. Phys. Rev. D. **105**, 024067 (2022). arXiv: 2111.10387
11. A. Bokulić, I. Smolić, T. Jurić, Constraints on singularity resolution by nonlinear electrodynamics. Phys. Rev. D. **106**, 064020 (2022). arXiv: 2206.07064
12. M. Born, L. Infeld, Foundations of the new field theory. Nat. **132**, 1004 (1933); Proc. R. Soc. Lond. **144**, 425 (1934)
13. N. Bretón, Smarr's formula for black holes with non-linear electrodynamics. Gen. Rel. Grav. **37**, 643 (2005). (gr-qc/0405116)
14. N. Bretón, Stability of nonlinear magnetic black holes. Phys. Rev. D. **72**, 044015 (2005). (hep-th/0502217)
15. N. Bretón, S.E.P. Bergliaffa, On the stability of black holes with nonlinear electromagnetic fields. arXiv: 1402.2922
16. K.A. Bronnikov, Comment on 'Regular black hole in general relativity coupled to nonlinear electrodynamics'. Phys. Rev. Lett. **85**, 4641 (2000)

17. K.A. Bronnikov, Regular magnetic black holes and monopoles from nonlinear electrodynamics. Phys. Rev. D. **63**, 044005 (2001). (gr-qc/0006014)
18. K.A. Bronnikov, Dyonic configurations in nonlinear electrodynamics coupled to general relativity. Grav. Cosmol. **23**, 343 (2017). arXiv: 1708.08125
19. K.A. Bronnikov, Nonlinear electrodynamics, regular black holes and wormholes. Int. J. Mod. Phys. D. **27**, 1841005 (2018). arXiv: 1711.00087
20. K.A. Bronnikov, Black bounces, wormholes, and partly phantom scalar fields. Phys. Rev. D. **106**, 064029 (2022). arXiv: 2206.09227
21. K.A. Bronnikov, V.N. Melnikov, G.N. Shikin, K.P. Staniukovich, Scalar, electromagnetic, and gravitational fields interaction: particlelike solutions. Ann. Phys. (N.Y.) **118**, 84 (1979)
22. K.A. Bronnikov, S.G. Rubin, *Black Holes, Cosmology, and Extra Dimensions* (World Scientific, 2012)
23. K.A. Bronnikov, G.N. Shikin, On the Reissner-Nordström problem with a nonlinear electromagnetic field, in *Class Quantum Gravity* (Trudy IF AN BSSR, Minsk, 1976), p. 88 (in Russian)
24. K.A. Bronnikov, G.N. Shikin, E.N. Sibileva, Self-gravitating stringlike configurations from nonlinear electodynamics. Grav. Cosmol. **9**, 169 (2003). (gr-qc/0308002)
25. K.A. Bronnikov, R.K. Walia, Field sources for Simpson-Visser space-times. Phys. Rev. D. **105**, 044039 (2022). arXiv: 2112.13198
26. A. Burinskii, S.R. Hildebrandt, New type of regular black holes and particlelike solutions from nonlinear electrodynamics. Phys. Rev. D. **65**, 104017 (2002). (hep-th/0202066)
27. P. Cañate, Black-bounces as magnetically charged phantom regular black holes in Einstein-nonlinear electrodynamics gravity coupled to a self-interacting scalar field. Phys. Rev. D. **106**, 024031 (2022). arXiv: 2202.02303
28. I.P. Denisova, B.D. Garmaev, V.A. Sokolov, Compact objects in conformal nonlinear electrodynamics. Eur. Phys. J. C. **79**, 531 (2019). arXiv: 1901.05318
29. I. Dymnikova, E. Galaktionov, Regular rotating electrically charged black holes and solitons in nonlinear electrodynamics minimally coupled to gravity. Class. Quantum Grav. **32**, 165015 (2015). arXiv: 1510.01353
30. Z-Y. Fan, X. Wang, Construction of regular black holes in general relativity. Phys. Rev. D. **94**, 124027 (2016). arXiv: 1610.02636
31. E.S. Fradkin, A.A. Tseytlin, Nonlinear electrodynamics from quantized strings. Phys. Lett. B. **163**, 123 (1985)
32. E. Franzin, S. Liberati, J. Mazza, A. Simpson, M. Visser, Charged black-bounce spacetimes. JCAP **07**, 036 (2021)
33. C. Gao, Black holes with many horizons in the theories of nonlinear electrodynamics. Phys. Rev. D. **104**, 064038 (2021). arXiv: 2106.13486. https://doi.org/10.1103/PhysRevD.104.064038
34. A.A. Garcia-Diaz, AdS-dS stationary rotating black hole exact solution within Einstein-nonlinear electrodynamics. Ann. Phys. **441**, 168880 (2022). arXiv: 2201.10682
35. J. Hadamard, in Leçons sur la propagation des ondes et les équations de l'hydrodynamique, ed. Hermann (Paris, 1903)
36. W. Heisenberg, H. Euler, Folgerungen aus der diracschen theorie des positrons. Z. Phys. **98**, 714 (1936)
37. C.M. Kim, S.P. Kim, Vacuum birefringence in a supercritical magnetic field and a subcritical electric field. arXiv: 2202.05477
38. S.I. Kruglov, Nonlinear electrodynamics with birefringence. Phys. Lett. A. **379**, 623 (2015). arXiv: 1504.03535. https://doi.org/10.1016/j.physleta.2014.12.026
39. S.I. Kruglov, Acceleration of universe by nonlinear electromagnetic fields. arXiv: 1603.07326. https://doi.org/10.1142/S0218271816400022
40. S.I. Kruglov, Nonlinear arcsin-electrodynamics and asymptotic Reissner-Nordström black holes. Ann. Physik. **528**, 588 (2016). arXiv: 1607.07726. https://doi.org/10.1002/andp.201600027
41. S.I. Kruglov, Asymptotic Reissner-Nordström solution within nonlinear electrodynamics. Phys. Rev. D. **94**, 044026 (2016). arXiv: 1608.04275

42. S.I. Kruglov, Dyonic black holes with nonlinear logarithmic electrodynamics. Grav. Cosmol. **25**, 190–195 (2019). arXiv: 1909.05674. https://doi.org/10.1134/S0202289319020105
43. S.I. Kruglov, Dyonic black holes in framework of Born–Infeld-type electrodynamics. Gen. Rel. Grav. **51**, 121 (2019). arXiv: 1909.11661
44. S.I. Kruglov, Dyonic and magnetic black holes with nonlinear arcsin-electrodynamics. Ann. Phys. **409** , 167937 (2019). arXiv: 1911.04253
45. S.I. Kruglov, Dyonic and magnetized black holes based on nonlinear electrodynamics. Eur. Phys. J. C. **80**, 250 (2020). arXiv: 2003.10845. https://doi.org/10.1140/epjc/s10052-020-7809-x
46. S.I. Kruglov, Remarks on nonsingular models of Hayward and magnetized black hole with rational nonlinear electrodynamics. Grav. Cosmol. **27**, 78–84 (2021). arXiv: 2103.14087. https://doi.org/10.1134/S0202289321010126
47. D. Kubiznak, T. Tahamtan, O. Svitek, Slowly rotating black holes in nonlinear electrodynamics. Phys. Rev. D. **105**, 104064 (2022). arXiv: 2203.01919
48. L. Landau, E. Lifshitz, *Classical Theory of Fields*, 3rd edn. (Pergamon, London, 1971)
49. Jin Li, Kai Lin, Nan Yang, Nonlinear electromagnetic quasinormal modes and Hawking radiation of a regular black hole with magnetic charge. Eur. Phys. J. C. **75**, 131 (2015). (arXiv: 1409.5988)
50. F.S.N. Lobo, M.E. Rodrigues, M.V.D.S. Silva, A. Simpson, M. Visser, Novel black-bounce spacetimes: wormholes, regularity, energy conditions, and causal structure. Phys. Rev. D. **103**, 084052 (2021)
51. J. Matyjasek, P. Sadurski, D. Tryniecki, Inside the degenerate horizons of regular black holes. Phys. Rev. D. **87**, 124025 (2013). (arXiv: 1304.6347)
52. J. Matyjasek, D. Tryniecki, M. Klimek, Regular black holes in an asymptotically de Sitter universe. Mod. Phys. Lett. A. **23**, 3377 (2009). (arXiv: 0809.2275)
53. R.R. Metsaev, M. Rakhmanov, A.A. Tseytlin, The Born-Infeld action as the effective action in the open superstring theory. Phys. Lett. B. **193**, 207 (1987)
54. K. Mkrtchyan, M. Svazas, Solutions in nonlinear electrodynamics and their double copy regular black holes. JHEP (9), 012 (2022). arXiv: 2205.14187. https://doi.org/10.1007/JHEP09(2022)012
55. C. Moreno, O. Sarbach, Stability properties of black holes in self-gravitating nonlinear electrodynamics. Phys. Rev. D. **67**, 024028 (2003). gr-qc/0208090. https://doi.org/10.1103/PhysRevD.67.024028
56. S. Nojiri, S.D. Odintsov, Regular multi-horizon black holes in modified gravity with nonlinear electrodynamics. Phys. Rev. D. **96**, 104008 (2017). arXiv: 1708.05226. https://doi.org/10.1103/PhysRevD.96.104008
57. K. Nomura, D. Yoshida, J. Soda, Stability of magnetic black holes in general nonlinear electrodynamics. Phys. Rev. D. **101**, 124026 (2020). arXiv: 2004.07560. https://doi.org/10.1103/PhysRevD.101.124026
58. M. Novello, V.A. de Lorenci, J.M. Salim, R. Klippert, Geometrical aspects of light propagation in nonlinear electrodynamics. Phys. Rev. D. **61**, 045001 (2000)
59. M. Novello, S.E. Perez Bergliaffa, J.M. Salim, Singularities in general relativity coupled to nonlinear electrodynamics. Class. Quantum Grav. **17**, 3821 (2000). gr-qc/0003052
60. R. Pellicer, R.J. Torrence, Nonlinear electrodynamics and general relativity. J. Math. Phys. **10**, 17+18 (1969)
61. J. Plebanski, *Non-Linear Electrodynamics–A Study* (C.I.E.A. del I.P.N, Mexico City, 1966)
62. N. Rosen, H.B. Rosenstock, The force between particles in a nonlinear field theory. Phys. Rev. **85**, 257 (1952). https://doi.org/10.1103/PhysRev.85.257
63. Yu.P. Rybakov, *Particle Structure in Nonlinear Field Theory* (Peoples' Friendship University Press, Moscow, 1985)
64. I.H. Salazar, A. Garcia, J. Plebanski, Duality rotations and type D solutions to Einstein equations with nonlinear electromagnetic sources. J. Math. Phys. **28**, 2171 (1987)
65. N. Seiberg, E. Witten, String theory and noncommutative geometry. J. High Energy Phys. **09**, 032 (1999). (hep-th/9908142)

66. Anatoly E. Shabad, Vladimir V. Usov, Effective Lagrangian in nonlinear electrodynamics and its properties of causality and unitarity. Phys. Rev. D. **83**, 105006 (2011). https://doi.org/10.1103/PhysRevD.83.105006
67. A. Simpson, M. Visser, Black bounce to traversable wormhole. JCAP **02**, 042 (2019)
68. V.A. Sokolov, Extended duality condition for conformal vacuum nonlinear electrodynamics. Phys. Rev. D. **104**, 124035 (2021). https://doi.org/10.1103/PhysRevD.104.124035
69. Dmitri P. Sorokin, Introductory notes on nonlinear electrodynamics and its applications. Fortschr. E Phys. **70**(7-8), 2200092 (2022). arXiv: 2112.12118
70. B. Toshmatov, Z. Stuchlik, B. Ahmedov, Generic rotating regular black holes in general relativity coupled to nonlinear electrodynamics. Phys. Rev. D. **95**, 084037 (2017). arXiv: 1704.07300
71. B. Toshmatov, Z. Stuchlík, B. Ahmedov, D. Malafarina, Relaxations of perturbations of spacetimes in general relativity coupled to nonlinear electrodynamics. Phys. Rev. D. **99**, 064043 (2019). arXiv: 1903.03778. https://doi.org/10.1103/PhysRevD.99.064043
72. A. Vilenkin, E.P.S. Shellard, *Cosmic Strings and Other Topological Defects* (Cambridge University Press, Cambridge, 1994)
73. Y. Yang, Dyonically charged black holes arising in generalized Born–Infeld theory of electromagnetism. Ann. Phys. **443**, 168996 (2022). arXiv: 2204.11313. https://doi.org/10.1016/j.aop.2022.168996

Chapter 3
How Strings Can Explain Regular Black Holes

Piero Nicolini

3.1 Introduction

Nowadays black holes are the focus of the attention of researchers working on a variety of topics in Physics and Mathematics. Astrophysicists have recently observed the shadow of black holes that presumably are harbored in the center of galaxies [1]. Gravitational waves due to black hole mergers have recently been detected at LIGO/Virgo facilities [2]. Mathematical physicists and mathematical relativists are interested in the properties of exact black hole solutions [3]. This activity intersects the work of those gravitational physicists that aim to circumvent the problem of dark sectors by means of theories alternative to general relativity [4–6]. The importance of black hole research, however, goes beyond the above research fields. It seems very likely that black holes are fated to be the cornerstone of our understanding of fundamental physics.

3.1.1 Three Facts About Evaporating Black Holes

To fully appreciate the significance of black holes, it is instructive to go back to the 1970's. At that time, theoretical physicists were interested in understanding nuclei and their phenomenology. Strings and dual models were formulated just few years earlier and they were already expected to die young due to the advent of QCD. Black holes and general relativity were topics of limited interest, because they were disconnected from the quantum realm. Astrophysicists, on the other hand, did not

P. Nicolini (✉)
Dipartimento di Fisica, Università degli Studi di Trieste, Strada Costiera 11, 34151 Trieste, Italy
e-mail: piero.nicolini@units.it

Institut für Theoretische Physik, Johann Wolfgang Goethe-Universität, Max-von-Laue-Str. 1, 60438 Frankfurt am Main, Germany

Frankfurt Institute for Advanced Studies (FIAS), Ruth-Moufang-Str. 1, 60438 Frankfurt am Main, Germany

C. Bambi (ed.), *Regular Black Holes*, Springer Series in Astrophysics and Cosmology,
https://doi.org/10.1007/978-981-99-1596-5_3

take seriously the existence of black holes, despite the growing evidence accumulated after the initial observation during the suborbital flight of the Aerobee rocket in 1964 [7]. Even the curvature singularity was considered just a mathematical problem, whose solution would never lead to physical consequences.

Hawking, however, radically changed this perspective. He actually set new goals for theoretical physics, by initiating the study of the Universe from a quantum mechanical view point. Along such a line of reasoning, Hawking showed that, in the vicinity of a black hole, quantum field theory is strongly disturbed by gravity. Particles become an ill-defined, coordinate dependent concept [8–10]. To an asymptotic observer black holes appear like black bodies emitting particles at a temperature $T \propto 1/M$, i.e. inversely proportional to their mass [11].

The existence of a thermal radiation offered the physical support for the thermodynamic interpretation of the laws governing black holes mechanics [12, 13]. It, however, left behind many open questions, such as the fate of an evaporating black hole[1] and the information loss paradox.[2] I list below some additional issues that are too often downplayed:

(i) If an horizon forms, Minkowski space cannot result from the Schwarzschild metric in the limit $M \to 0$, since it is forbidden by thermodynamics [14];
(ii) Quantum back reaction effects can tame a runaway temperature [15], but they can lead to mass inflation effects [16];
(iii) Quantum stress tensors imply violation of energy conditions [17].

In general, issues of this kind are mostly attributable to a breakdown of Hawking's semiclassical formalism. The last item of the above list is, however, intriguing. Without energy condition violation, standard matter would inevitably collapse into a curvature singularity [18]. As a result, already in the mid 1960's there were proposals, e.g. by Gliner [19] and Sakharov [20], to improve black hole spacetimes with energy violating source terms. Such proposals culminated with the work of Bardeen, who obtained the first regular black hole solution [21]. The related line elements reads:

$$ds^2 = -\left(1 - \frac{2Ml_P^2 r^2}{(r^2 + P^2)^{3/2}}\right) dt^2 + \left(1 - \frac{2Ml_P^2 r^2}{(r^2 + P^2)^{3/2}}\right)^{-1} dr^2 + r^2 d\Omega^2. \quad (3.1)$$

[1] By black hole evaporation one indicates the process of particle emission during the full life cycle. The $1/M$ dependence implies an increased emission rate as the hole loses mass. Such a nasty behavior is connected to the negative heat capacity of the black hole $C \equiv dM/dT < 0$.

[2] Microstates of a collapsing star are hidden behind an event horizon. The information is not lost but virtually not accessible. If the hole thermally radiates, it emits particles in a democratic way, de facto destroying the informational content of the initial star. This is the reason why the Hawking radiation is considered an effect that worsens the problem of the information in the presence of an event horizon.

Here the gravitational constant is written in terms of the Planck length $G = l_P^2$. At short scale, the singularity is replaced by a regular quantum vacuum region controlled by a magnetic monopole P [22].[3]

Against this background, the point (iii) in the above list represented a novelty. The violation was the direct consequence of a major principle, namely the combination of quantum and gravitational effects at short scale. Conversely, for the Bardeen metric, the energy condition violation is the result of an ad hoc choice e.g. the presence of a magnetic monopole. For this reason, already in the 1980's semiclassical gravity seemed to pave the way to a possible short scale completion of the spacetime [17].

3.2 Can One Probe Length Scales Smaller than $\sqrt{\alpha'}$?

As of today, Superstring Theory can be considered the major contender of the "quantum gravity war", namely the current debate about the formulation of a consistent quantum theory of gravity. The success of string theory is probably due to its wide spectrum, that covers a vast number of topics and paradigms, from particle physics to cosmology [31]. For what concerns black holes, string theory has been applied in a variety of situations, including thermodynamics [32] and derivation of new metrics [33]. The theory has also interesting spin-offs where black holes have a major role, e.g. large extra dimension paradigms [34–42] and the gauge/gravity duality [32]. There exist also proposals alternative to black holes like the fuzzball [43]. String theory is notoriously not free from problems. One of the major limitations is the identification of genuine effective theories, namely the string landscape [44].

For the present discussion, it is important to recall just one specific character of string theory: its intrinsic non-locality. Such a property should come as no surprise, because strings were introduced to replace quantum field theory and guarantee ultraviolet finiteness in calculations. To understand the nature of such a short scale convergence, string collisions at Planckian energies were extensively studied at the end of the 1980's [45–47]. The net result was simple and, at the same time, surprising. The particle Compton wavelength turned to be modified by an additional term, namely:

$$\Delta x \simeq \frac{1}{\Delta p} + \alpha' \Delta p. \tag{3.2}$$

Due the approximations for its derivation, the above uncertainty relation, known as generalized uncertainty principle (GUP), offers just the leading term of stringy corrections to quantum mechanics. Nevertheless, the GUP can capture several important new features. One can start by saying that the GUP depends on the combination of the conventional Compton wavelength λ and the gravitational radius r_g of the particle, being $\alpha' \sim G$. In practice, (3.2) is a genuine quantum gravity result. The GUP inher-

---

[3] Additional regular black hole metrics were proposed in the following years; see, e.g., [23–29] and Chaps. 1 and 2 in this book. For a review, see [30].

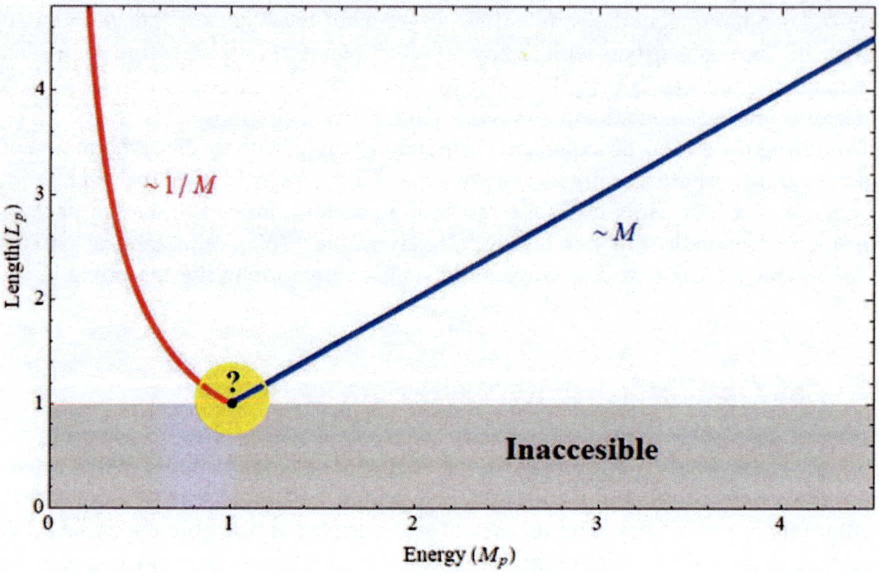

Fig. 3.1 The phase diagram of matter at the highest conceivable energy scale. The red curve represents the Compton wavelength, the blue curve the gravitational radius. The yellow spot represents the regime where string effects are expected to be dominant. The question mark indicates that the nature of the intersection is not known. The grey area is virtually inaccessible as a result of the ultraviolet self-complete nature of gravity

its the non-local character of string theory, being $\Delta x \geq \sqrt{\alpha'}$. For Planckian string tension $\sim 1/m_{\mathrm{P}}^2$, this is a equivalent to saying that the Planck length l_{P} is actually the smallest meaningful length scale in nature. The GUP also shows that quantum gravity has a peculiar characteristic: Quantum gravity effects show up *only* in the vicinity of the Planck scale. At energies lower than m_{P}, there is the conventional particle physics. At energies higher than m_{P}, there are conventional (classical) black holes with mass $M \sim \Delta p$. For this reason, one speaks of "classicalization" in the trans-Planckian regime [48, 49]. Particles (strings) and black holes are, therefore, two possible phases of matter. The relation between them is evident by the fact that black holes have a constant "tension", $M/r_{\mathrm{g}} \sim m_{\mathrm{P}}^2$, like a (Planckian) string [50]. In practice, the GUP suggests that matter compression has to halt due to the gravitational collapse into a Planckian black hole. Such a scenario is often termed gravity "ultraviolet self-completeness" and corresponds to the impossibility of probing length scale below l_{P} in any kind of experiment [51–56]. The diagram of self-completeness can be seen in Fig. 3.1.

Despite the great predictive power of the relation (3.2), many things remain unclear. For instance, the details of the collapse at the Planck scale is unknown. It is not clear whether the Lorentz symmetry is actually broken or deformed, prior, during and after the collapse [50]. Also the nature of the confluence of the two curves λ and r_{g} is debated. One could speculate that there exist a perfect symmetry between

λ and r_g and actually particles and black holes coincide [55]. Such a proposal, known as "Black Hole Uncertainty Principle Correspondence", is currently under investigation and requires additional ingredients for being consistent with observations [57]. Nevertheless, there could be some room for sub-Planckian black holes, as far as there exists a lower bound for the black hole mass [58, 59]. It is also possible to imagine that the confluence is non-analytic [59]. Finally, it has been shown that the number of the dimensions, charge and spin can drastically affect the self-completeness paradigm [59–62].

3.2.1 How to Derive a Consistent "Particle-Black Hole" Metric

There exists at least one thing one knows for sure about gravity self-completeness. The Schwarzschild metric simply does not fit in with the diagram in Fig. 3.1. The problem is connected to the possibility of having a black hole for any arbitrarily small mass, i.e., $M < m_P$. This implies a potential ambiguity since to a given mass, one could associate both a particle and a black hole. More importantly, Schwarzschild black holes for sub-Planckian masses have radii $\sim M/m_P^2$, smaller than l_P, a fact that is in contrast with the very essence of self-completeness. The formation of black holes in such a mass regime is the natural consequence of mass loss during the Hawking emission. Customarily, one circumvents the problem by saying that there is a breakdown of semiclassical gravity. Black holes would explode even before attaining sub-Planckian masses [63]. The problem, however, persists when one considers alternative formation mechanism, like early Universe fluctuations [64, 65] and quantum decay [66].

The most natural way to solve the puzzle is to postulate the existence of an extremal black hole at the confluence of λ and r_g. Degenerate horizons are zero temperature asymptotic states that can guarantee the switching off of black hole evaporation.[4] For instance, Denardo and Spallucci considered charged black holes and determined the parameters to obtain stable configurations [68]. Microscopic black holes can, however, share their charge and angular momentum very rapidly both via Hawking and Schwinger emissions [69, 70]. What one actually needs is a SCRAM phase following the Schwarzschild phase, similarly to what predicted by Balbinot and Barletta within the semiclassical approximation [15]. In conclusion, the issue can be solved only if one is able to derive a metric admitting a Planckian extremal horizon for $M = m_P$. Such a metric exists and it is known as holographic screen metric or simply holographic metric [71]. Its line element reads:

$$ds^2 = -\left(1 - \frac{2Ml_P^2 r}{r^2 + l_P^2}\right) dt^2 + \left(1 - \frac{2Ml_P^2 r}{r^2 + l_P^2}\right)^{-1} dr^2 + r^2 d\Omega^2. \tag{3.3}$$

[4] The switching off is also known as SCRAM phase, in analogy with the terminology in use for nuclear power plants [67].

Equation (3.3) is a prototype of a quantum gravity corrected black hole spacetime. Indeed, the holographic metric offers a sort of "preview" of the characteristics of string corrected black hole metrics, I will present in the next sections. In summary one notice that:

(a) For $M \gg m_P$, (3.3) becomes the Schwarzschild metric up to some corrections, that are consistent with the predictions of Dvali and Gomez's quantum N-portrait [72–74];
(b) For $M \simeq 2.06\, m_P$, the Hawking temperature reaches a maximum and the black hole undergoes a phase transition to a positive heat capacity cooling down (SCRAM phase);
(c) For $M = m_P$, one has $r_g = l_P$ and $T = 0$, namely the evaporation stops and leaves a *Planckian extremal black hole* as a remnant;
(d) For $M < m_P$, (3.3) describes a horizonless spacetime due to a particle sitting at the origin.

In practice (3.3) perfectly separates the two phases of matter, i.e. particles and black holes, and protects the region below l_P in Fig. 3.1 under any circumstances. In addition, (3.3) does not suffer from quantum back reaction, being $T/M \ll 1$ during the entire evaporation process. Also the issue of the mass inflation at point (ii) in Sect. 3.1.1 is circumvented. For $M > m_P$, there are actually an event horizon $r_g = r_+$ and a Cauchy horizon r_-,

$$r_\pm = l_P^2 \left(M \pm \sqrt{M^2 - m_P^2} \right), \tag{3.4}$$

but the latter falls behind the Planck length and it is actually not accessible. From (3.4), one notices that the horizon structure is the same as the Reissner-Nordström black hole, provided one substitutes the charge with the Planck mass

$$Q/G \longrightarrow m_P. \tag{3.5}$$

Equation (3.5) is another key aspect of self-completeness. Gravity does not need the introduction of a cut off. The completeness is achieved by exploiting the coupling constant G as a short scale regulator. At this point, there is, however, a caveat: The spacetime (3.3) does have a curvature singularity. The regularity was not the goal of the derivation of such a metric. The basic idea has been the introduction of fundamental surface elements (i.e. holographic screens), as building blocks of the spacetime. Each of such surface elements is a multiple of the extremal configuration, that becomes the basic information capacity or information bit. Indeed for the holographic metric the celebrated area law reads

$$S(\mathscr{A}_+) = \frac{\pi}{\mathscr{A}_0} (\mathscr{A}_+ - \mathscr{A}_0) + \pi \ln (\mathscr{A}_+/\mathscr{A}_0) \tag{3.6}$$

where $\mathscr{A}_0 = 4\pi l_P^2$ is the area of the extremal event horizon, and $\mathscr{A}_+ = n\mathscr{A}_0$.

If one accepts that surfaces (rather than volumes) are the fundamental objects, the question of the regularity of the gravitational field inside the minimal surface is no longer meaningful. The spacetime simply ceases to exist inside the minimal holographic screen. This interpretation is reminiscent of spacetime dissolution observed in quantum string condensates within Eguchi's areal quantization scheme [75].[5]

3.3 What T-duality can tell us About Black Holes

Suppose one has a physical system living on a compact space, whose radius is R. Suppose there exists another physical system defined on another compact space, whose radius is proportional to $1/R$. If the observables of the first system can be identified with that of the second system, one can say that such systems are equivalent or dual with respect to the transformation[6]

$$R \longrightarrow 1/R. \tag{3.7}$$

For example, by setting $R \sim 1/\Delta p$ in (3.2) one finds

$$\Delta x \simeq R + \frac{\alpha'}{R}. \tag{3.8}$$

The above relation actually maps length scales shorter than $\sqrt{\alpha'}$ to those larger that $\sqrt{\alpha'}$, being

$$\Delta x(R) = \Delta x(1/R), \tag{3.9}$$

for suitable values of α'. From this viewpoint, one can say that the GUP is a T-duality relation. This fact is per se intriguing because it offers an additional argument for a stringy interpretation of the holographic metric. The good part is that T-duality allows for an even more genuine contact between string theory and a short scale corrected metric. To do this, one needs to go back to a basic result due to Padmanabhan [76].

Standard path integrals can be thought as the sum of amplitudes over all possible particle trajectories. In the presence of gravity, the scenario is slightly modified. Indeed, there exist paths that cannot contribute to the path integral. If paths are shorter than the particle gravitational radius, they must be discarded in the computation of the amplitude. A simple way to achieve this is to introduce a damping term,

$$e^{-\sigma(x,y)/\lambda} \longrightarrow e^{-\sigma(x,y)/\lambda} e^{-r_g/\sigma(x,y)}, \tag{3.10}$$

[5] The classical spacetime is a condensate of quantum strings. At distances approaching $\sqrt{\alpha'}$, long range correlations of the condensate are progressively destroyed. For $\alpha' = G$, the whole spacetime boils over and no trace of the string/p-brane condensate is left over.

[6] The duality is termed T-duality, or target space duality.

for each path contribution in the sum over the paths.[7] The above relation implies that the path length $\sigma(x, y)$ admits a minimum. Interestingly, Padmanabhan performed the sum over the above path contributions and derived a modified propagator [77, 78]

$$G(x, y; m^2) = \int \frac{d^D p}{(2\pi)^D} e^{-ip\cdot(x-y)} G(p), \tag{3.11}$$

with

$$G(p) = -\frac{l_0}{\sqrt{p^2+m^2}} K_1\left(l_0\sqrt{p^2+m^2}\right), \tag{3.12}$$

where $K_1(x)$ is a modified Bessel function of the second kind, and l_0 is called "zero point length" [56]. Equation (3.12) is intrinsically non-local: The Bessel function has a damping term for momenta larger than $1/l_0$. Therefore l_0 is the minimal length that can be resolved over the manifold. Conversely, for small arguments one finds the conventional quantum field theory result.

The virtue of Padmanabhan's calculation (3.12) is two fold: The propagator is a robust result that descends from general considerations; The functional form in terms of the Bessel function exactly coincides with the correction of string theory to standard, "low energy" quantum field theory. To better understand such a crucial point we briefly sketch the line reasoning at the basis of series of papers authored by Spallucci and Padmanabhan in collaboration with Smailagic [79] and Fontanini [80, 81].

Let us start by considering a closed bosonic string in the presence of just one additional dimension, that is compactified on circle of length $l_0 = 2\pi R$. The string mass spectrum can be written as

$$M^2 = \frac{1}{2\alpha'}\left(n^2\frac{\alpha'}{R^2} + w^2\frac{R^2}{\alpha'}\right) + \text{harmonic excitations}, \tag{3.13}$$

where n labels the Kaluza-Klein excitations and w is the winding number of the string around the compact dimension.[8] As expected the above relation enjoys T-duality. It is invariant under simultaneous exchange $R \leftrightarrow \alpha'/R$ and $n \leftrightarrow w$ and leads to the identification of $\sqrt{\alpha'}$ as invariant length scale.

Strings are intrinsically non perturbative objects. As a result, any perturbative expansion destroys the very essence of the theory. The only way to extrapolate a nonpertubative character that can be "adapted" to the field theoretic concept of particle is the study of the string center of mass (SCM) dynamics. From the propagation kernel of the SCM in five dimensions, one can integrate out the fifth dimension to obtain an effective four dimensional propagator

---

[7] We temporarily assume Euclidean signature for the ease of presentation.

[8] In the process of path integral quantization, harmonic oscillators are irrelevant. Therefore we consider them frozen without unwanted consequences.

$$K(x-y, 0-nl_0; T) = \sum_n \int [\mathscr{D}z][\mathscr{D}p][\mathscr{D}x^5][\mathscr{D}p_5] \exp(\ldots) \rightarrow K_{\text{reg}}(x-y; T) \tag{3.14}$$

where $x-y$ and $0-nl_0$ are, respectively, the four dimensional interval and the separation along the fifth dimension. Already at this point, one can observe the regularity due to l_0, being

$$K_{\text{reg}}(x-y; T) \sim \sum_n e^{(i\mu_0/2T)[(x-y)^2+n^2l_0^2]} \tag{3.15}$$

where μ_0 is a parameter which will not appear in the final result. Additional integrations on T and w lead to Green's function

$$G_{\text{reg}}(x-y) \sim \sum_w \int dT\, e^{(iT/2\mu_0)m_0^2} e^{(\ldots w^2)} K_{\text{reg}}(x-y; T), \tag{3.16}$$

where m_0 is the mass of the particle in the limit $l_0 \rightarrow 0$. If one considers the leading term of the above expression, namely $n = w = 1$, one finds (3.11) upon the condition $l_0 = 2\pi\sqrt{\alpha'}$. In other words, the zero point length in four dimensions has a T-duality origin and coincides with the minimum length in string theory. The above result can be easily generalized to the case of more than one compact dimension. The conclusion is unaffected: (3.12) is both general and fundamental!

3.3.1 How to Implement T-duality Effects

Starting from (3.11), we expect important deviations from conventional Green's function equation

$$\{\text{Differential Operator}\}\ G(x, y) = \text{Dirac Delta}, \tag{3.17}$$

when $x \approx y$. For the specific case of black holes, we recall that, in the absence of spin, there are both spherical symmetry and static conditions. It is, therefore, instructive to consider the interaction potential between two static sources with mass m and M due to (3.12),

$$\begin{aligned} V(r) &= -\frac{1}{m}\frac{W[J]}{T} \\ &= -GM \int \frac{d^3k}{(2\pi)^3}\, G(k)|_{k^0=0} \exp(i\mathbf{k}\cdot\mathbf{r}) \\ &= -\frac{GM}{\sqrt{r^2+l_0^2}}. \end{aligned} \tag{3.18}$$

The fact that $V \approx -GM/l_0$ for $r \to 0$, is the first signal of a possible removal of the curvature singularity. To verify this is the case, one has to construct an effective energy momentum tensor for the r.h.s. of Einstein equations. The procedure is equivalent to the derivation of black hole solutions by means of non-local gravity actions

$$S = \frac{1}{2\kappa}\int \mathfrak{f}(R, \Box, \ldots)\sqrt{-g}\, d^4x + \int \mathfrak{L}\left(M, F^2, \Box, \ldots\right)\sqrt{-g}\, d^4x \tag{3.19}$$

with $\kappa = 8\pi G$, $\Box = \nabla_\mu \nabla^\mu$, F is the gauge field and $\ldots$ stand for higher derivative terms.

Equation (3.19) is a compact notation for a class of actions that have been studied to obtain ghost free, ultraviolet finite gravity field equations [82–85]. For the present discussion, the details of such an action are not relevant, since it is only an effective description of the full string dynamics. Accordingly, also the problem of the pathology of the action (e.g. ghosts, anomalies) is of secondary concerns, if one believes in the consistency of Superstring Theory. In conclusion, one can adopt a truncated version of the full non-local action [86–89] (see also Chap. 4 in this book) and derive the non-local Einstein equations. For $F^2 = 0$, they read

$$R_{\mu\nu} - \frac{1}{2} g_{\mu\nu} R = \kappa\, \mathfrak{T}_{\mu\nu} \tag{3.20}$$

where $\mathfrak{T}_{\mu\nu} = \mathcal{O}^{-1}(\Box) T_{\mu\nu}$, while the Einstein tensor and $T_{\mu\nu}$ are the conventional Einstein gravity tensors. The only thing that is important to know is the degree of ultraviolet convergence of the theory, encoded in the operator $\mathcal{O}(\Box)$. At this point, one can observe that (3.18) is consistent with Green's function equation for (3.11) [90], namely

$$\nabla^2 G\left(\mathbf{z}, \mathbf{z}'\right) = -l_0\sqrt{-\nabla^2}\, K_1\left(l_0\sqrt{-\nabla^2}\right)\delta^{(3)}\left(\mathbf{z} - \mathbf{z}'\right). \tag{3.21}$$

The operator can be simply read off from the above equation, taking into account that $\mathcal{O}(\Box) = \mathcal{O}(\nabla^2)$ if the source is static. In practice, the r.h.s. of (3.21) is equivalent to the ${}^t_t$ of the $\mathfrak{T}_{\mu\nu}$, namely

$$\mathfrak{T}^t_t \equiv -\rho(\mathbf{x}) = (4\pi)^{-1} M l_0 \sqrt{-\nabla^2}\, K_1\left(l_0\sqrt{-\nabla^2}\right)\delta^{(3)}(\mathbf{x}). \tag{3.22}$$

The effective energy density can be analytically derived and reads

$$\rho(\mathbf{x}) = \frac{3 l_0 M}{4\pi\left(|\mathbf{x}|^2 + l_0^2\right)^{5/2}}. \tag{3.23}$$

For large distances, the above density quickly dies off as $\sim 1/|\mathbf{x}|^5$. Conversely, at short scales $|\mathbf{x}| \lesssim l_0$, one finds the "Sea of Tranquility", i.e., a regular quantum region characterized by creation and annihilation of virtual particles at constant, finite energy. In such a sea, gravity becomes repulsive and prevents the full collapse of matter into a singularity. With a geometric description in terms of differential line elements, the quantum fluctuations of such a sea are not visible. One can only capture the average effect, namely a local de Sitter ball around the origin, whose cosmological constant is $\sim GM/l_0^3$. Local energy condition violations certify the correctness of such a scenario.

After the above prelude, one can analytically solve (3.20) and display the full metric [91]

$$ds^2 = -\left(1 - \frac{2Ml_P^2 r^2}{(r^2 + l_0^2)^{3/2}}\right) dt^2 + \left(1 - \frac{2Ml_P^2 r^2}{(r^2 + l_0^2)^{3/2}}\right)^{-1} dr^2 + r^2 d\Omega^2. \tag{3.24}$$

The magic of the above result is that it coincides with the Bardeen solution (3.1), provided

$$P \longrightarrow l_0. \tag{3.25}$$

This is reminiscent of the relation between the holographic metric and the Reissner-Nordström geometry (3.5): this time, however, one can say that the Dirac string has been traded with a closed bosonic string.

The general properties of the horizon structure and thermodynamics are similar to what seen in the context of the holographic metric–see (a)–(d) in Sect. 3.2.1. Horizon extremization allows for a SCRAM phase at the end of the evaporation, making the hole a stable system from a thermodynamic viewpoint. The Hawking temperature reads

$$T = \frac{1}{4\pi r_+}\left(1 - \frac{3l_0^2}{r_+^2 + l_0^2}\right), \tag{3.26}$$

while the entropy is

$$S = \frac{\mathscr{A}_+}{4}\left[\left(1 - \frac{8\pi l_0^2}{\mathscr{A}_+}\right)\sqrt{1 + \frac{4\pi l_0^2}{\mathscr{A}_+}} + \frac{12\pi l_0^2}{\mathscr{A}_+}\left(\operatorname{arsinh}\sqrt{\frac{\mathscr{A}_+}{4\pi l_0^2}} - \operatorname{arsinh}\sqrt{2}\right)\right], \tag{3.27}$$

with $\mathscr{A}_+ = 4\pi r_+^2$.

The great advantage of the metric (3.24) is the stability. This is a property in marked contrast to the case of the Bardeen metric, than can be, at the most a transient state. Even by postulating the existence of magnetic monopoles at some point of the history of the Universe [92], their coupling has to be much stronger than the QED coupling [93]

$$\alpha_m \gg \alpha_e \sim 137^{-1}. \tag{3.28}$$

This would imply for the Bardeen metric a sudden decay into the Schwarzschild black hole.

Charged and charged rotating regular T-duality black holes have recently been derived. The novelty is the replacement of the ring singularity with a finite tension rotating string–for further details see [94].

3.4 A Short Guide to Black Holes in Noncommutative Geometry

We are going to present a family of black hole solutions that represents a sort of coronation of the program of regular black holes in string theory. Indeed, after almost 20 years since their derivation, their good properties are still unmatched.

One should start by recalling that noncommutative geometry (NCG) is a field in Mathematics whose goal is the study of noncommutative algebras on certain topological spaces. In Physics, NCG has well known applications. For example, at the heart of quantum mechanics there is a noncommutative geometry, i.e., the algebra of quantum operators. The idea that further physically meaningful results can be obtained from NCG, however, remained dormant at least until the 1990's. At that time, Connes proposed the study of fundamental interactions from the spectral triple principle [95].[9] As a main goal, Connes aimed to construct a "quantum version" of the spacetime, by establishing a relation similar to that between quantum mechanics and classical phase space [96] (see also [97] for a pedagogical introduction).

The most simple way to construct a noncommutative geometry is based on the replacement of conventional coordinates with noncommutative operators

$$\left[x^i,\, x^j\right] = i\theta^{ij} \tag{3.29}$$

where θ^{ij} is a constant, real valued, antisymmetric $D \times D$ matrix. The above commutator implies a new kind of uncertainty

$$\Delta x^i \Delta x^j \geq \frac{1}{2}\left|\theta^{ij}\right|, \tag{3.30}$$

that can be used to improve the bad short distance behavior of fields propagating on the noncommutative geometry. To achieve this goal, one can deform field Lagrangians by introducing a suitable non-local product. For instance, a realization of noncommutive algebra of functions is based on the Moyal-produced (also known as star product or Weyl-Groenewold product)

$$f \star g \equiv e^{(i/2)\theta^{ij}\frac{\partial}{\partial \xi^i}\frac{\partial}{\partial \eta^j}} f(x+\xi)g(x+\eta)\Big|_{\xi=\eta=0}, \tag{3.31}$$

that can be used as a starting point to obtain a noncommutative field theory–for reviews see e.g. [98, 99]. Probably the biggest push to the popularity of noncommutative field theory was given by its connection to string theory. Open strings ending on D-branes display a noncommutative behavior in the presence of a non vanishing, (constant) Kalb-Ramond B-field [100]

[9] The spectral triple is made of three items, a real, associative, noncommutative algebra $\mathscr{A}$, a Hilbert space $\mathscr{H}$ and a self adjoint operator $\eth$ on it.

$$\theta^{ij} \sim (2\pi\alpha')^2 \left(\frac{1}{g + 2\pi\alpha' B} B \frac{1}{g - 2\pi\alpha' B} \right)^{ij} , \tag{3.32}$$

where g is the metric tensor.

Noncommutative gravity follows a similar procedure for the metric field, defined over the underlying noncommutative manifold. The program of noncommutative gravity is, however, still in progress. Apart from some specific examples, one still misses a consistent noncommutative version of general relativity. In addition, the existing attempts to derive noncommutative corrections to classical black hole solutions run into the general difficulty of improving curvature singularities–see [67].

In 2003, the possibility of obtaining from noncommutative geometry something meaningful for the physics of black hole physics was still perceived as quite remote. This was a time that followed the "explosive" predictions about the possibility of a plentiful production of mini black holes in particle detectors [101, 102]. Operations at the LHC, however, began only five year later. As a result, there was a huge pressure to predict the experimental signatures of such black holes. If the terascale quantum gravity paradigm was correct, it was expected to have repercussions on mini black hole cross section, evaporation and detection [103, 104].

Given this background, there was an unconventional attempt to study noncommutative geometry stripped of all elements, apart from its nonlocal character. From (3.31) one can guess that NCG introduces Gaussian damping terms. To prove this, Smailagic and Spallucci considered the average of noncommutative operators $\langle x^i \rangle$, on states of minimal uncertainty, namely coherent states similar to those introduced by Glauber in quantum optics [105]. Such averages were interpreted as the closest thing to the conventional concept of coordinate. Initial results for path integrals on the noncommutative plane led to the conclusion that Dirac delta distributions are smeared out and become Gaussian functions, whose width is controlled by the noncommutative parameter θ [106, 107].[10] The result was later formalized in terms of a nonlocal field theory formulation [108]. Green's function equation (3.17) was determined by applying a non-local operator to the source term namely[11]

$$\delta^{(D)}(x-y) \longrightarrow f_\theta(x,y) = e^{\theta\Box}\delta^{(D)}(x-y). \tag{3.33}$$

To derive a spacetime that account for noncommutative effects, one has to recall that the metric field can be seen as a "thermometer" that measures the average fluctuations of the manifold. From (3.33), one can derive the effective energy density,

$$\rho(\mathbf{x}) = \frac{M}{(4\pi\theta)^{3/2}}\, e^{-|\mathbf{x}|^2/4\theta} , \tag{3.34}$$

and follow the procedure presented in Sect. 3.3.1. There are, however, two caveats:

---

[10] The matrix θ^{ij} in (3.29) can be written as $\theta^{ij} = \theta\varepsilon^{ij}$. The parameter θ has the dimension of a length squared.

[11] Here the signature is Euclidean.

1. The resulting spacetime is an *effective description* that captures just one single character of NCG, i.e. non-locality.[12]
2. The matrix θ^{ij} is assumed to behave like a field to preserve Lorentz symmetry.[13]

At this point, one can display the central result [113]

$$ds^2 = -\left[1 - \frac{2Ml_P^2}{r}\frac{\gamma\left(\frac{3}{2}; \frac{r^2}{4\theta}\right)}{\Gamma\left(\frac{3}{2}\right)}\right] dt^2 + \left[1 - \frac{2Ml_P^2}{r}\frac{\gamma\left(\frac{3}{2}; \frac{r^2}{4\theta}\right)}{\Gamma\left(\frac{3}{2}\right)}\right]^{-1} dr^2 + r^2 d\Omega^2. \tag{3.35}$$

Here $\gamma(3/2, x) \equiv \int_0^x \frac{du}{u} u^{3/2} e^{-u}$ is the incomplete Gamma function. It guarantees regularity of the manifold and quick convergence to the Schwarzschild metric for $r \gg \sqrt{\theta}$. While the horizon structure and the thermodynamics are similar to those of the other quantum gravity improved metrics (3.3) and (3.24), the above result has some specific characters. The Gaussian function (3.33) is a non-polynomial smearing, in agreement to what found by Tseytlin in [114]. On the other hand, polynomial functions (like GUP, T-duality) can be seen as the result of a truncation of the expansion over the theta parameter [115]. The above metric has been obtained also in the context of non-local gravity actions [89] and has been extended to the case of additional spatial dimensions [116, 117], charged [118] and rotating [119, 120] solutions.

From the emission spectra of the higher dimensional extension of (3.35), one learns that mini black holes tend to radiate soft particles mainly on the brane. This is in marked contrast with results coming from the Schwarzschild-Tangherlini metric [121].

3.5 Conclusions

The very essence of the message I want to convey is the relation between particles and black holes in Fig. 3.1. It has already been noticed that strings and black holes share common properties [122]. In this work, however, the argument is reinforced and employed to improve classical black hole solutions. From this perspective the regularity of black hole metrics is the natural consequence of non-locality of particles, when described in terms of strings.

Another key point concerns the particle-black hole at the intersection of the curves for λ and r_g. The nature of this object is probably one of the most important topics in current research in quantum gravity. Indeed, the particle-black hole is essential to guarantee a self-complete character of gravity. Its mass and radius are related to the fundamental units of quantum gravity and string theory, along the common

[12] For this reasons, one speaks of "noncommutative geometry inspired" solution. Other authors have termed it as "minimalistic approach" [109].

[13] Lorentz violation associated to (3.29) is a debated issue in the literature, e.g., see [110–112].

denominator of non-locality. This is evident also from the correspondence between cut offs, $\sqrt{\alpha'}$ (string, GUP), l_0 (T-duality, GUP), l_P self completeness, $\sqrt{\theta}$ (NCG).

In this work, we have also mentioned some of the existing difficulties, e.g., the details of the collapse at the Planck scale, the absence of an actual "quantum manifold". This means that the program of quantum gravity is far from being complete. It is also not clear if the predictions emerging from string theory will have experimental corroboration in the future. The ideas here presented, however, tend to support a less pessimistic scenario. Black holes could offer a testbed for fundamental physics, that is alternative to conventional experiments in high energy particle physics.

Acknowledgements The work of P.N. has partially been supported by GNFM, Italy's National Group for Mathematical Physics. P.N. is grateful to Cosimo Bambi for the invitation to submit a contribution to the volume "Regular Black Holes: Towards a New Paradigm of Gravitational Collapse", Springer, Singapore. P.N. is grateful to Athanasios Tzikas for the support in drawing the picture.

References

1. *Astronomers Capture First Image of a Black Hole* (2019)
2. B.P. Abbott et al., Observation of gravitational waves from a binary black hole merger. Phys. Rev. Lett. **116**(6), 061102 (2016)
3. A.D. Rendall, Local and global existence theorems for the Einstein equations. Living Rev. Rel. **3**, 1 (2000)
4. T.P. Sotiriou, V. Faraoni, $f(r)$ theories of gravity. Rev. Mod. Phys. **82**, 451 (2010)
5. S. Capozziello, M. De Laurentis, Extended theories of gravity. Phys. Rept. **509**, 167 (2011)
6. T. Clifton, P.G. Ferreira, A. Padilla, C. Skordis, Modified gravity and cosmology. Phys. Rept. **513**, 1 (2012)
7. S. Bowyer, E.T. Byram, T.A. Chubb, H. Friedman, Cosmic X-ray sources. Sci. **147**(3656), 394 (1965)
8. S.A. Fulling, Nonuniqueness of canonical field quantization in Riemannian space-time. Phys. Rev. D. **7**(10), 2850 (1973)
9. P.C.W. Davies, Scalar production in Schwarzschild and Rindler metrics. J. Phys. Math. Gen. **8**(4), 609 (1975)
10. W.G. Unruh, Notes on black-hole evaporation. Phys. Rev. D. **14**(4), 870 (1976)
11. S.W. Hawking, Particle creation by black holes. Commun. Math. Phys. **43**, 199 (1975)
12. J.D. Bekenstein, Black holes and entropy. Phys. Rev. D. **7**, 2333 (1973)
13. J.M. Bardeen, B. Carter, S.W. Hawking, The four laws of black hole mechanics. Commun. Math. Phys. **31**, 161 (1973)
14. R.M. Wald, *General Relativity* (Chicago Univ. Pr, Chicago, USA, 1984)
15. R. Balbinot, A. Barletta, Switching off black hole evaporation? Class. Quant. Grav. **5**, L11 (1988)
16. R. Balbinot, E. Poisson, Mass inflation: the semiclassical regime. Phys. Rev. Lett. **70**, 13 (1993)
17. N.D. Birrell, P.C.W. Davies, *Quantum Fields in Curved Space, Cambridge Monographs on Mathematical Physics* (Cambridge Univ. Press, Cambridge, UK, 1984)
18. R. Penrose, Gravitational collapse and space-time singularities. Phys. Rev. Lett. **14**, 57 (1965)
19. E.B. Gliner, Algebraic Properties of the Energy-momentum Tensor and Vacuum-like States o^+ Matter. Sov. J. Exp. Theor. Phys. **22**, 378 (1966)

20. A.D. Sakharov, The initial stage of an expanding universe and the appearance of a nonuniform distribution of matter. Sov. Phys. JETP. **22**, 241 (1966)
21. J.M. Bardeen, Non-singular general-relativistic gravitational collapse. p.174 (1968)
22. E. Ayon-Beato, A. Garcia, The Bardeen model as a nonlinear magnetic monopole. Phys. Lett. B. **493**, 149 (2000)
23. I. Dymnikova, Vacuum nonsingular black hole. Gen. Rel. Grav. **24**, 235 (1992)
24. E. Ayon-Beato, A. Garcia, New regular black hole solution from nonlinear electrodynamics. Phys. Lett. B. **464**, 25 (1999)
25. E. Ayon-Beato, A. Garcia, Nonsingular charged black hole solution for nonlinear source. Gen. Rel. Grav. **31**, 629 (1999)
26. E. Ayon-Beato, A. Garcia, Regular black hole in general relativity coupled to nonlinear electrodynamics. Phys. Rev. Lett. **80**, 5056 (1998)
27. K.A. Bronnikov, Regular magnetic black holes and monopoles from nonlinear electrodynamics. Phys. Rev. D. **63**, 044005 (2001)
28. M.R. Mbonye, D. Kazanas, A Non-singular black hole model as a possible end-product of gravitational collapse. DPhys. Rev. D. **72**, 024016 (2005)
29. S.A. Hayward, Formation and evaporation of regular black holes. Phys. Rev. Lett. **96**, 031103 (2006)
30. S. Ansoldi, Spherical black holes with regular center: a review of existing models including a recent realization with Gaussian sources. arXiv:0802.0330 [gr-qc]
31. H. Nicolai, Quantum gravity: the view from particle physics. Fundam. Theor. Phys. **177**, 369 (2014)
32. J.M. Maldacena, The large N limit of superconformal field theories and supergravity. Int. J. Theor. Phys. **38**, 1113 (1999)
33. K.S. Stelle, BPS branes in supergravity. in *ICTP Summer School in High-energy Physics and Cosmology* (1998)
34. I. Antoniadis, N. Arkani-Hamed, S. Dimopoulos, G. Dvali, New dimensions at a millimeter to a Fermi and superstrings at a TeV. Phys. Lett. B. **436**, 257 (1998)
35. N. Arkani-Hamed, S. Dimopoulos, G. Dvali, The hierarchy problem and new dimensions at a millimeter. Phys. Lett. B. **429**, 263 (1998)
36. N. Arkani-Hamed, S. Dimopoulos, G. Dvali, Phenomenology, astrophysics and cosmology of theories with submillimeter dimensions and TeV scale quantum gravity. Phys. Rev. D. **59**, 086004 (1999)
37. L. Randall, R. Sundrum, A large mass hierarchy from a small extra dimension. Phys. Rev. Lett. **83**, 3370 (1999)
38. L. Randall, R. Sundrum, An alternative to compactification. Phys. Rev. Lett. **83**, 4690 (1999)
39. M. Gogberashvili, Our world as an expanding shell. Europhys. Lett. **49**, 396 (2000)
40. M. Gogberashvili, Hierarchy problem in the shell universe model. Int. J. Mod. Phys. D. **11**, 1635 (2002)
41. M. Gogberashvili, Four dimensionality in noncompact Kaluza-Klein model. Mod. Phys. Lett. A. **14**, 2025 (1999)
42. T. Banks, W. Fischler, A model for high-energy scattering in quantum gravity, (1999), unpublished paper
43. S.D. Mathur, The fuzzball proposal for black holes: an elementary review. Fortsch. Phys. **53**, 793 (2005)
44. C. Vafa, The string landscape and the swampland, (2005), Based on talks given at the Einstein Symposium in Alexandria, at the 2005 Simons Workshop in Mathematics and Physics, and the talk to have been presented at Strings 2005
45. D. Amati, M. Ciafaloni, G. Veneziano, Superstring collisions at Planckian energies. Phys. Lett. B. **197**, 81 (1987)
46. D. Amati, M. Ciafaloni, G. Veneziano, Classical and quantum gravity effects from Planckian energy superstring collisions. Int. J. Mod. Phys. A. **3**, 1615 (1988)
47. D. Amati, M. Ciafaloni, G. Veneziano, Can space-time be probed below the string size? Phys. Lett. B. **216**, 41 (1989)

48. G. Dvali, S. Folkerts, C. Germani, Physics of trans-Planckian gravity. Phys. Rev. D. **84**, 024039 (2011)
49. G. Dvali, G.F. Giudice, C. Gomez, A. Kehagias, UV-completion by classicalization. JHEP **08**, 108 (2011)
50. A. Aurilia, E. Spallucci, Why the length of a quantum string cannot be Lorentz contracted. Adv. High Energy Phys. **2013**, 531696 (2013)
51. L.J. Garay, Quantum gravity and minimum length. Int. J. Mod. Phys. A. **10**, 145 (1995)
52. A. Aurilia, E. Spallucci, Planck's uncertainty principle and the saturation of Lorentz boosts by Planckian black holes, (2013), unpublished essay submitted to the Gravity Research Foundation for the 2002–2003 competition
53. G. Dvali, C. Gomez, Self-completeness of Einstein gravity. (2010), unpublished paper
54. R.J. Adler, Six easy roads to the Planck scale. Am. J. Phys. **78**, 925 (2010)
55. B.J. Carr, The black hole uncertainty principle correspondence, in *1st Karl Schwarzschild Meeting on Gravitational Physics*, ed. by P. Nicolini, M. Kaminski, J. Mureika, M. Bleicher, Springer Proceedings in Physics, vol. 170 (Springer International Publishing, Switzerland, 2016), pp. 159–167
56. T. Padmanabhan, Planck length: Lost + found. Phys. Lett. B. **809**, 135774 (2020)
57. B. Carr, J. Mureika, P. Nicolini, Elementary Particles as Black Holes: Linking Experimental Tests in the Microscopic and Macroscopic Regimes, (2023), in preparation
58. B.J. Carr, J. Mureika, P. Nicolini, Sub-Planckian black holes and the generalized uncertainty principle. JHEP **07**, 052 (2015)
59. B. Carr, H. Mentzer, J. Mureika, P. Nicolini, Self-complete and GUP-modified charged and spinning black holes. Eur. Phys. J. C. **80**(12), 1166 (2020)
60. J. Mureika, P. Nicolini, Self-completeness and spontaneous dimensional reduction. Eur. Phys. J. Plus. **128**, 78 (2013)
61. M. Knipfer, S. Köppel, J. Mureika, P. Nicolini, Generalized uncertainty principle and black holes in higher dimensional self-complete gravity. JCAP **08**, 008 (2019)
62. P. Nicolini, Planckian charged black holes in ultraviolet self-complete quantum gravity. Phys. Lett. B. **778**, 88 (2018)
63. S.W. Hawking, Black hole explosions. Nat. **248**, 30 (1974)
64. S. Hawking, Gravitationally collapsed objects of very low mass. Mon. Not. Roy. Astron. Soc. **152**, 75 (1971)
65. B.J. Carr, S.W. Hawking, Black holes in the early Universe. Mon. Not. Roy. Astron. Soc. **168**, 399 (1974)
66. R. Bousso, S.W. Hawking, The Probability for primordial black holes. Phys. Rev. D. **52**, 5659 (1995)
67. P. Nicolini, Noncommutative black holes, the final appeal to quantum gravity: a review. Int. J. Mod. Phys. A. **24**, 1229 (2009)
68. G. Denardo, E. Spallucci, Switching off the black hole evaporation. Nuovo Cim. B. **44**, 381 (1978)
69. G.W. Gibbons, Vacuum polarization and the spontaneous loss of charge by black holes. Commun. Math. Phys. **44**, 245 (1975)
70. D.N. Page, Evidence against astrophysical dyadospheres. Astrophys. J. **653**, 1400 (2006)
71. P. Nicolini, E. Spallucci, Holographic screens in ultraviolet self-complete quantum gravity. Adv. High Energy Phys. **2014**, 805684 (2014)
72. G. Dvali, C. Gomez, Black hole macro-quantumness, (2012), unpublished paper
73. G. Dvali, C. Gomez, Black hole's quantum N-portrait. Fortsch. Phys. **61**, 742 (2013)
74. G. Dvali, C. Gomez, Black hole's 1/N hair. Phys. Lett. B. **719**, 419 (2013)
75. S. Ansoldi, A. Aurilia, E. Spallucci, Loop quantum mechanics and the fractal structure of quantum space-time. Chaos Solitons Fractals **10**, 197 (1999)
76. P. Nicolini, Quantum gravity and the zero point length. Gen. Rel. Grav. **54**, 106 (2022)
77. T. Padmanabhan, Duality and zero point length of space-time. Phys. Rev. Lett. **78**, 1854 (1997)
78. T. Padmanabhan, Hypothesis of path integral duality. 1. Quantum gravitational corrections to the propagator. Phys. Rev. D. **57**, 6206 (1998)

79. A. Smailagic, E. Spallucci, T. Padmanabhan, String theory T duality and the zero point length of space-time, (2003), unpublished paper
80. E. Spallucci, M. Fontanini, *Zero-point length, extra-dimensions and string T-duality*, (Nova Publishers, 2005)
81. M. Fontanini, E. Spallucci, T. Padmanabhan, Zero-point length from string fluctuations. Phys. Lett. B. **633**, 627 (2006)
82. N.V. Krasnikov, Nonlocal gauge theories. Theor. Math. Phys. **73**, 1184 (1987)
83. E.T. Tomboulis, Superrenormalizable gauge and gravitational theories, (1997), unpublished paper
84. L. Modesto, Super-renormalizable quantum gravity. Phys. Rev. D. **86**, 044005 (2012)
85. T. Biswas, E. Gerwick, T. Koivisto, A. Mazumdar, Towards singularity and ghost free theories of gravity. Phys. Rev. Lett. **108**, 031101 (2012)
86. A.O. Barvinsky, Nonlocal action for long distance modifications of gravity theory. Phys. Lett. B. **572**, 109 (2003)
87. H.W. Hamber, R.M. Williams, Nonlocal effective gravitational field equations and the running of Newton's G. Phys. Rev. D. **72**, 044026 (2005)
88. J.W. Moffat, Ultraviolet complete quantum gravity. Eur. Phys. J. Plus. **126**, 43 (2011)
89. L. Modesto, J.W. Moffat, P. Nicolini, Black holes in an ultraviolet complete quantum gravity. Phys. Lett. B. **695**, 397 (2011)
90. P. Gaete, P. Nicolini, Finite electrodynamics from T-duality. Phys. Lett. B. **829**, 137100 (2022)
91. P. Nicolini, E. Spallucci, M.F. Wondrak, Quantum corrected black holes from string T-duality. Phys. Lett. B. **797**, 134888 (2019)
92. J. Preskill, Cosmological production of superheavy magnetic monopoles. Phys. Rev. Lett. **43**, 1365 (1979)
93. J. Preskill, Magnetic monopoles. Ann. Rev. Nucl. Part. Sci. **34**, 461 (1984)
94. P. Gaete, K. Jusufi, P. Nicolini, Charged black holes from T-duality. Phys. Lett. B. **835**, 137546 (2022)
95. A. Connes, Noncommutative geometry and reality. J. Math. Phys. **36**, 6194 (1995)
96. A. Connes, Gravity coupled with matter and foundation of noncommutative geometry. Commun. Math. Phys. **182**, 155 (1996)
97. T. Schucker, Forces from Connes' geometry. Lect. Notes Phys. **659**, 285 (2005)
98. R.J. Szabo, Quantum field theory on noncommutative spaces. Phys. Rept. **378**, 207 (2003)
99. M.R. Douglas, N.A. Nekrasov, Noncommutative field theory. Rev. Mod. Phys. **73**, 977 (2001)
100. N. Seiberg, E. Witten, String theory and noncommutative geometry. JHEP **09**, 032 (1999)
101. S. Dimopoulos, G.L. Landsberg, Black holes at the LHC. Phys. Rev. Lett. **87**, 161602 (2001)
102. S.B. Giddings, S.D. Thomas, High-energy colliders as black hole factories: the end of short distance physics. Phys. Rev. D. **65**, 056010 (2002)
103. J. Mureika, P. Nicolini, E. Spallucci, Could any black holes be produced at the LHC? Phys. Rev. D. **85**, 106007 (2012)
104. P. Nicolini, J. Mureika, E. Spallucci, E. Winstanley, M. Bleicher, Production and evaporation of Planck scale black holes at the LHC, in *13th Marcel Grossmann Meeting on Recent Developments in Theoretical and Experimental General Relativity, Astrophysics, and Relativistic Field Theories*, pp. 2495–2497 (2015)
105. R.J. Glauber, Coherent and incoherent states of the radiation field. Phys. Rev. **131**, 2766 (1963)
106. A. Smailagic, E. Spallucci, UV divergence free QFT on noncommutative plane. J. Phys. A. **36**, L517 (2003)
107. A. Smailagic, E. Spallucci, Feynman path integral on the noncommutative plane. J. Phys. A. **36**, L467 (2003)
108. E. Spallucci, A. Smailagic, P. Nicolini, Trace anomaly in quantum spacetime manifold. Phys. Rev. D. **73**, 084004 (2006)
109. D.V. Vassilevich, Towards noncommutative gravity, in *Fundamental Interactions: A Memorial Volume for Wolfgang Kummer*. ed. by D. Grumiller, A. Rebhan, D. Vassilevich (World Scientific, Singapore, 2010), pp.293–302

110. S.M. Carroll, J.A. Harvey, V.A. Kostelecky, C.D. Lane, T. Okamoto, Noncommutative field theory and Lorentz violation. Phys. Rev. Lett. **87**, 141601 (2001)
111. C.E. Carlson, C.D. Carone, N. Zobin, Noncommutative gauge theory without Lorentz violation. Phys. Rev. D. **66**, 075001 (2002)
112. K. Morita, Lorentz invariant noncommutative QED. Prog. Theor. Phys. **108**, 1099 (2003)
113. P. Nicolini, A. Smailagic, E. Spallucci, Noncommutative geometry inspired Schwarzschild black hole. Phys. Lett. B. **632**, 547 (2006)
114. A.A. Tseytlin, On singularities of spherically symmetric backgrounds in string theory. Phys. Lett. B. **363**, 223 (1995)
115. M. Kober, P. Nicolini, Minimal scales from an extended Hilbert space. Class. Quant. Grav. **27**, 245024 (2010)
116. T.G. Rizzo, Noncommutative inspired black holes in extra dimensions. JHEP **09**, 021 (2006)
117. E. Spallucci, A. Smailagic, P. Nicolini, Non-commutative geometry inspired higher-dimensional charged black holes. Phys. Lett. B. **670**, 449 (2009)
118. S. Ansoldi, P. Nicolini, A. Smailagic, E. Spallucci, Noncommutative geometry inspired charged black holes. Phys. Lett. B. **645**, 261 (2007)
119. A. Smailagic, E. Spallucci, "Kerrr" black hole: the lord of the string. Phys. Lett. B. **688**, 82 (2010)
120. L. Modesto, P. Nicolini, Charged rotating noncommutative black holes. Phys. Rev. D. **82**, 104035 (2010)
121. P. Nicolini, E. Winstanley, Hawking emission from quantum gravity black holes. JHEP **11**, 075 (2011)
122. G.'t Hooft, The black hole interpretation of string theory. Nucl. Phys. B. **335**, 138 (1990)

Chapter 4
Regular Black Holes from Higher-Derivative Effective Delta Sources

Breno L. Giacchini and Tibério de Paula Netto

4.1 Introduction

It is widely expected that quantum effects would resolve the spacetime singularities present in classical solutions of general relativity. The process behind this regularization, nevertheless, is not entirely understood. It might ultimately rely on a consistent quantum description of gravity, which remains an open problem.

In many approaches to quantum gravity, the regularization mechanism can be described, usually approximately or effectively, via the replacement of pointlike localized structures with smeared objects. In this perspective, the effective source formalism is a useful framework for studying the resultant spacetime configurations. It has been applied, for example, in the context of noncommutative geometry, generalized uncertainty principle models, string theory, and higher-derivative gravity (see, e.g., [6, 14, 20, 37, 44, 50, 54, 56–58, 68, 69] and references therein, and also Chap. 3 in this book).

Our goal in this chapter is to introduce the effective source formalism and show some regularity properties of black holes constructed using such sources. We choose to work in the framework of higher-derivative gravity. The motivation for this choice is not only pedagogical but also because many interesting and general results about these models were obtained first using this formalism.

We emphasize, however, that the results presented here are very general and can be easily transposed to other frameworks. The generality obtained is a consequence of the wide scope in which we consider higher-derivative gravity. Even though this term

B. L. Giacchini (✉) · T. de Paula Netto (✉)
Department of Physics, Southern University of Science and Technology,
Shenzhen 518055, China
e-mail: breno@sustech.edu.cn

T. de Paula Netto
e-mail: tiberio@sustech.edu.cn

C. Bambi (ed.), *Regular Black Holes*, Springer Series in Astrophysics and Cosmology,
https://doi.org/10.1007/978-981-99-1596-5_4

has been more frequently associated with the Einstein-Hilbert action augmented by fourth-derivative structures, such as R^2 and $R^2_{\mu\nu}$, it can be applied to any extension of general relativity whose action contains more than two derivatives of the metric. In an even broader sense, it refers to gravitational theories whose propagator displays an improved behavior in the ultraviolet (UV) regime, as it happens, e.g., in some families of nonlocal gravity models defined by actions that are nonpolynomial in derivatives of the metric.

Solving the equations of motion in higher-derivative gravity is a highly complex task; in the case of static spherically symmetric metrics only in fourth-derivative gravity the space of solutions of the nonlinear field equations has been studied rigorously [12, 13, 21, 41–43, 47, 48, 60, 64, 65]. Regarding theories with six or more derivatives, the solutions obtained usually involve certain approximations of the field equations, such as the Newtonian or the small-curvature approximation. The effective source formalism proves to be a useful tool in both situations, as it enables the derivation of general results on the regularity of the solution without the need to solve all the equations for a specific model. As we show here, some features of the solution only depend on the behavior of the higher-derivative sector of the model, and this connection is made explicit via the effective source. Indeed, if the effect of the higher derivatives can be interpreted as the regularization of a singular source, one might study the effective source to investigate the regularity of the associated spacetime configuration.

Nevertheless, what is expected of a "regular" metric depends on the application under consideration. While the regularity of the metric components in a specific coordinate chart can suffice to define a bounded (modified) Newtonian potential, it might not be enough to avoid the divergence of components of the Riemann curvature tensor. Thinking of general covariance, a more useful definition of regular spacetime relies on the regularity of an invariant (or a particular set of invariants). In many cases, the Kretschmann scalar is chosen, but in other applications it might be necessary also to consider scalars involving covariant derivatives (see, e.g., [16, 38]). A simple example showing that the regularity of the Kretschmann scalar does not imply that all curvature-derivative invariants are bounded is provided by the Hayward metric [40],

$$
\mathrm{d}s^2 = -\left(1 - \frac{2Mr^2}{r^3 + 2L^3}\right)\mathrm{d}t^2 + \left(1 - \frac{2Mr^2}{r^3 + 2L^3}\right)^{-1}\mathrm{d}r^2 + r^2\mathrm{d}\Omega^2, \tag{4.1}
$$

where $\mathrm{d}\Omega^2$ is the metric of the unit two-sphere. It is straightforward to verify that while R, $R^2_{\mu\nu}$, $R^2_{\mu\nu\alpha\beta}$, and even the scalars cubic in curvatures are all bounded, invariants containing derivatives of curvatures might be singular; for instance,

$$
\Box^2 R \underset{r\to 0}{\sim} -\frac{504M}{L^6 r}. \tag{4.2}
$$

In this regard, the effective source formalism, as presented here, is helpful in the classification of the order of regularity of certain families of black hole solutions, according to the behavior of the effective source. Moreover, since field equations with similar effective sources are also obtained in different frameworks (even from other approaches to quantum gravity), these results can be applied in a broader context.

This chapter is organized as follows. In Sect. 4.2, we make a brief review about the importance of higher-derivative terms in the gravitational action, mainly motivated by quantum considerations. The effective source is introduced in Sect. 4.3, where we analyze the Newtonian limit of a generic higher-derivative gravity model. Important results on the effective sources are proved in Sect. 4.4, followed by a discussion on the definition of higher-order regularity of the metric, which is a crucial notion in the study of the regularity of curvature invariants with covariant derivatives. Then, in Sect. 4.5 we present some explicit examples of the general results obtained in the preceding sections by focusing on particular models, namely, polynomial and nonlocal gravity. In Sect. 4.6 we construct regular black hole solutions based on the results and examples derived in the other sections. Finally, our concluding remarks are given in Sect. 4.7.

Throughout this chapter, we use the mostly plus metric convention, with the Minkowski metric $\eta_{\mu\nu} = \operatorname{diag}(-1, +1, +1, +1)$. The Riemann curvature tensor is defined by

$$R^{\alpha}{}_{\beta\mu\nu} = \partial_\mu \Gamma^{\alpha}_{\beta\nu} - \partial_\nu \Gamma^{\alpha}_{\beta\mu} + \Gamma^{\alpha}_{\mu\tau}\,\Gamma^{\tau}_{\beta\nu} - \Gamma^{\alpha}_{\nu\tau}\,\Gamma^{\tau}_{\beta\mu}\,, \tag{4.3}$$

while the Ricci tensor and the scalar curvature are, respectively, $R_{\mu\nu} = R^{\alpha}{}_{\mu\alpha\nu}$ and $R = g^{\mu\nu} R_{\mu\nu}$. Also, we use the notation

$$G_{\mu\nu} = R_{\mu\nu} - \frac{1}{2} g_{\mu\nu} R \tag{4.4}$$

for the Einstein tensor, and we adopt the unit system such that $c = 1$ and $\hbar = 1$.

4.2 Higher-Derivative Gravity Models

The idea of generalizing the Einstein-Hilbert action by including higher-order terms can be traced back to the early years of general relativity (see, e.g., [29]). Nevertheless, only in the last decades these models have attracted more attention, and the main motivation for this comes from the quantum theory.

In fact, since the 1970s, it has been known that the quantum version of general relativity, based on the Einstein-Hilbert action (with or without the cosmological constant Λ),

$$S_{\text{EH}} = \frac{1}{16\pi G} \int \mathrm{d}^4 x \sqrt{-g}(R + 2\Lambda), \tag{4.5}$$

is not perturbatively renormalizable [23, 26, 39, 67]. This is mainly due to the negative mass dimension of the Newton constant G. However, if the gravitational action is enlarged by fourth-derivative terms,

$$S_{\text{grav}} = S_{\text{EH}} + \int \mathrm{d}^4x\sqrt{-g}\,\{\alpha_1 R^2_{\mu\nu\alpha\beta} + \alpha_2 R^2_{\mu\nu} + \alpha_3 R^2 + \alpha_4 \Box R\}\,, \tag{4.6}$$

where $\Box = g^{\mu\nu}\nabla_\mu\nabla_\nu$ is the d'Alembert operator, the corresponding quantum model turns out to be renormalizable [63]. It is interesting to notice that the structures in (4.6) also appear in the semiclassical theory in order to renormalize the vacuum fluctuations of quantum fields in a classical curved background (see, e.g., [10, 17] for an introduction).

One of the most severe drawbacks of higher-derivative models is the ghostlike particles that typically exist in the theory's spectrum. From the classical point of view, such modes are characterized by having negative kinetic energy, while at the quantum level, they usually cause violation of unitarity.[1] In recent years, however, considerable effort has been made to understand the ambiguous role of higher-derivative ghosts in perturbative quantum gravity. This led to a multiplicity of insights on how to tame (or avoid) the ghosts and restore unitarity[2] [3–5, 8, 9, 11, 28, 45, 46, 50, 51, 53, 66]. Among these proposals, we mention the cases of Lee-Wick gravity [51, 53], which is based on a formulation of polynomial-derivative gravity with ghosts with complex masses, and nonlocal ghost-free gravity [11, 45, 46, 50, 66], based on actions that are nonpolynomial in derivatives of the metric.

In the former case, one introduces in the action (4.6) operators that are polynomial in the d'Alembertian, namely,

$$\sum_{n=1}^{N-1} \int \mathrm{d}^4x\sqrt{-g}\left\{\alpha_{1,n}\, R_{\mu\nu\alpha\beta}\,\Box^n\, R^{\mu\nu\alpha\beta} + \alpha_{2,n}\, R_{\mu\nu}\,\Box^n R^{\mu\nu} + \alpha_{3,n}\, R\,\Box^n R\right\}. \tag{4.7}$$

Since each curvature contains two metric derivatives, the resulting action contains $2N+2$ derivatives. Compared with the fourth-derivative gravity (4.6), which can be at most strictly renormalizable, the presence of derivatives higher than four can render this theory superrenormalizable [5]. In the particular case of Lee-Wick gravity, the coefficients $\alpha_{\ell,n}$ are chosen so that all the ghostlike modes correspond to complex conjugate pairs of poles in the propagator.

In addition to the terms (4.7), it is possible to include structures of higher order in curvatures without spoiling the superrenormalizability of the model. This happens provided that such terms contain at most $2N+2$ derivatives of the metric, for example,

$$R^2\Box^{N-3}R^2 \quad \text{and} \quad R_{\mu\nu}R^{\mu\nu}\Box^{N-3}R_{\alpha\beta}R^{\alpha\beta}\,. \tag{4.8}$$

---

[1] Ghosts should not be mistaken for tachyons, which are instead characterized by negative mass squared.

[2] As the technical details regarding unitarity in these models lie beyond the scope of this text, we refer the interested reader to the original references mentioned above for further consideration.

Terms of this type can even be beneficial, as with a judicious choice of their front coefficients, they can be used to cancel the loop divergences and make the theory finite at the quantum level [5, 51, 52].

On the other hand, in the case of nonlocal ghost-free higher-derivative gravity models, nonlocalities are introduced at the classical level in such a way that the UV behavior of the graviton propagator gets modified without introducing new (ghost) degrees of freedom. The simplest model with these characteristics is described by the action

$$S_{\text{grav}} = S_{\text{EH}} + \frac{1}{16\pi G}\int \mathrm{d}^4x\sqrt{-g}\, G_{\mu\nu}\frac{e^{H(\Box)}-1}{\Box}R^{\mu\nu}\,, \tag{4.9}$$

where $H(z)$ is an entire function. Besides being free of ghosts, for some choices of the function $H(z)$ the model (4.9) can be (super)renormalizable. The reader can consult [45, 46, 50, 66] for the details.

4.3 Higher-Derivative Gravity in the Newtonian Limit

In the previous section we presented some families of higher-derivative gravity models whose main motivations come from the quantum field theory point of view. Here we start to consider these models in the classical domain to discuss whether (and how) they can lead to regular spacetime configurations. It is sound to begin with the simplest singularity in gravity, related to the Newtonian singularity proportional to $1/r$. In this spirit, in this section, we discuss in detail the Newtonian limit of a general higher-derivative gravity model.

4.3.1 Linearized Higher-Derivative Gravity

Instead of considering each model described in Sect. 4.2 separately, it is possible to develop a general formalism for linearized higher-derivative gravity. This comes from the observation that, in the linear approximation, any higher-derivative gravity can be expressed in the form of the action[3]

$$S_{\text{grav}} = \frac{1}{16\pi G}\int \mathrm{d}^4x\sqrt{-g}\,\left\{R + RF_1(\Box)R + R_{\mu\nu}F_2(\Box)R^{\mu\nu}\right\}, \tag{4.10}$$

where $F_{1,2}$ are arbitrary functions, called form factors, that depend on the specific model under consideration.

[3] In the action (4.10), we did not include total derivative terms since they do not contribute to the equations of motion, nor did we include the cosmological constant because it is irrelevant for metric perturbations around the Minkowski spacetime.

In fact, in the linear regime, we consider metric fluctuations $h_{\mu\nu}$ around the Minkowski spacetime,

$$g_{\mu\nu} = \eta_{\mu\nu} + h_{\mu\nu}, \qquad |h_{\mu\nu}| \ll 1, \tag{4.11}$$

and we are interested in equations of motion

$$\frac{\delta S_{\text{grav}}}{\delta h_{\mu\nu}} \tag{4.12}$$

that are in the first order in $h_{\mu\nu}$. Since the linear terms in (4.12) are originated from the second-order terms in the action, the only relevant higher-derivative structures in the linear regime are the ones capable of generating quadratic terms in $h_{\mu\nu}$. Expanding the curvatures in powers of the perturbation $h_{\mu\nu}$ (see the Appendix for the explicit formulas), we notice that the Riemann curvature tensor is already $O(h_{...})$ because the Minkowski background is flat. From this trivial observation, we can derive important conclusions about the types of terms that contribute to the linear approximation:

1. Terms of order higher than two in curvatures are irrelevant in this limit, since they are $O(h^3_{...})$. This is the case, e.g., of the terms in (4.8).
2. Every term quadratic in curvature is already $O(h^2_{...})$. Thus, if such terms contain covariant derivatives, only the zero-order term of the expansion of the derivative can contribute in the linear limit. So, in this case, we can simply trade $\nabla_\mu \mapsto \partial_\mu$ for all derivatives in curvature-squared terms. Therefore, by changing the order of the derivatives, applying integration by parts, ignoring surface terms, and using the Bianchi identities, it is possible to prove that any quadratic curvature term in the action can be reduced to three structures, namely,

$$RF_1(\Box)R\,, \quad R_{\mu\nu}F_2(\Box)R^{\mu\nu}\,, \quad \text{and} \quad R_{\mu\nu\alpha\beta}F_3(\Box)R^{\mu\nu\alpha\beta}\,. \tag{4.13}$$

3. Among these three structures, only two are linearly independent at order $h^2_{...}$. Indeed, using the formulas (4.150)–(4.152) of the Appendix, it is immediate to prove that, for any $F(\Box)$,

$$\big[R_{\mu\nu\alpha\beta}F(\Box)R^{\mu\nu\alpha\beta} - 4R_{\mu\nu}F(\Box)R^{\mu\nu} + RF(\Box)R\big]^{(2)} = 0, \tag{4.14}$$

where the superscript indicates that the equation is valid at order $h^2_{...}$. Hence, applying (4.14), it is possible to eliminate one of the curvature scalars in the basis (4.13). For example, if we start with the action

$$S_{\text{grav}} = \frac{1}{16\pi G}\int \mathrm{d}^4x\sqrt{-g}\,\big\{R + R\tilde{F}_1(\Box)R + R_{\mu\nu}\tilde{F}_2(\Box)R^{\mu\nu} + R_{\mu\nu\alpha\beta}\tilde{F}_3(\Box)R^{\mu\nu\alpha\beta}\big\}, \tag{4.15}$$

with the aid of (4.14) one can reduce it to (4.10) under the redefinition of the form factors

$$F_1(\Box) = \tilde{F}_1(\Box) - \tilde{F}_3(\Box), \qquad F_2(\Box) = \tilde{F}_2(\Box) + 4\tilde{F}_3(\Box). \tag{4.16}$$

With this procedure, we can eliminate the Riemann-squared term in the action. Note, however, that there is nothing special in removing the Riemann-squared term; according to the problem under consideration, one can use (4.14) to rewrite the action on the most convenient basis consisting of any two curvature-quadratic invariants (even including the square of the Weyl tensor).

Observation 4.1 In some situations, the identity (4.14) can be seen as the linearized version of the Gauss-Bonnet-like relations that hold in four-dimensional spacetime. Indeed, it is well known that the integrand of the topological Gauss-Bonnet term is a total derivative (see, e.g., [67]),

$$R^2_{\mu\nu\alpha\beta} - 4R^2_{\mu\nu} + R^2 = \text{total derivative}. \tag{4.17}$$

Although the same cannot be said if we insert any power of the d'Alembertian in between the curvatures in (4.17), it is possible to show that

$$R_{\mu\nu\alpha\beta}\Box^\ell R^{\mu\nu\alpha\beta} - 4R_{\mu\nu}\Box^\ell R^{\mu\nu} + R\Box^\ell R = O(R^3_{\dots}) \\ + \text{total derivative} \qquad \forall \ell \in \{1, 2, 3, \dots\}. \tag{4.18}$$

The proof of (4.18) involves commuting covariant derivatives and applying the Bianchi identities, see [5, 25] for the details.

Therefore, if $F(\Box)$ can be expressed as a power series, Eq. (4.18) implies in (4.14). On the other hand, identity (4.14) holds for any function F, even for nonanalytic form factors. This last observation is important because quantum corrections for the gravitational action typically have the form of $\log(-\Box/\mu^2)$. Further discussion on logarithmic quantum corrections to the Newtonian potential can be found in [19, 20, 24] and references therein.

Taking into account the above results, it follows that in the linear approximation, any higher-order term in the gravitational action is equivalent to the structures in (4.10). It is essential, however, to keep in mind that this equivalence is valid only at second order in the metric perturbation (i.e., at the level of linearized equations of motion); the same also applies to the freedom mentioned above for expressing the action in any preferred curvature basis. In fact, in applications where terms of higher order in the metric perturbation are relevant, different curvature invariants can give different contributions. For example, this happens at the classical level when one considers the solutions of the complete (nonlinear) field equations. At the quantum level, we mention that the quadratic part of the action is enough to determine the propagator of the model, while to define the structure of vertices, one has to examine the higher-order terms.

Using the formulas in the Appendix, it is possible to show that the $h_{\mu\nu}$-bilinear part of (4.10) is given by

$$S_{\text{grav}} = \frac{1}{32\pi G}\int \mathrm{d}^4x \Big[\frac{1}{2}h_{\mu\nu}a_1(\Box)\Box h^{\mu\nu} - \frac{1}{2}ha_2(\Box)\Box h - h^{\mu\nu}a_1(\Box)\partial_\mu\partial_\lambda h^\lambda_\nu + ha_2(\Box)\partial_\mu\partial_\nu h^{\mu\nu} + \frac{1}{2}h_{\alpha\beta}[a_1(\Box) - a_2(\Box)]\frac{\partial^\alpha\partial^\beta\partial^\mu\partial^\nu}{\Box}h_{\mu\nu}\Big], \tag{4.19}$$

where the functions $a_{1,2}(z)$ relate to the form factors $F_{1,2}(z)$ via

$$a_1(\Box) = 1 + F_2(\Box)\Box, \tag{4.20a}$$

$$a_2(\Box) = 1 - [4F_1(\Box) + F_2(\Box)]\Box. \tag{4.20b}$$

Finally, the coupling with matter is introduced through

$$S_{\text{m}} = \frac{1}{2}\int \mathrm{d}x\ T^{\mu\nu}h_{\mu\nu}, \tag{4.21}$$

where $T_{\mu\nu}$ is the energy-momentum tensor of matter, such that the total action is $S_{\text{total}} = S_{\text{grav}} + S_{\text{m}}$. The principle of least action,

$$\frac{\delta S_{\text{total}}}{\delta h_{\mu\nu}} = 0, \tag{4.22}$$

results in the linearized equations of motion,

$$\varepsilon^{\mu\nu} = -16\pi G\, T^{\mu\nu}, \tag{4.23}$$

where

$$\varepsilon^{\mu\nu} = a_1(\Box)\big(\Box h^{\mu\nu} - \partial^\mu\partial^\lambda h^\nu_\lambda - \partial^\nu\partial^\lambda h^\mu_\lambda\big) + a_2(\Box)\big[\eta^{\mu\nu}\big(\partial^\alpha\partial^\beta h_{\alpha\beta} - \Box h\big) + \partial^\mu\partial^\nu h\big] + [a_1(\Box) - a_2(\Box)]\,\frac{\partial^\mu\partial^\nu\partial^\alpha\partial^\beta}{\Box}h_{\alpha\beta}. \tag{4.24}$$

4.3.2 *Field Equations in the Newtonian Limit*

The Newtonian limit means that the gravitational field is weak and static. In this situation, we can adopt Cartesian coordinates $x^\mu = (t, x, y, z)$ and write the metric in a generic space-isotropic form

$$ds^2 = -(1+2\varphi)dt^2 + (1-2\psi)(dx^2+dy^2+dz^2)\,, \tag{4.25}$$

where φ and ψ ($|\varphi|, |\psi| \ll 1$) represent the two independent components of the metric, often called *Newtonian potentials*. Moreover, the Newtonian approximation also assumes that the matter is nonrelativistic, described by the energy-momentum tensor

$$T_{\mu\nu} = \delta^0_\mu \delta^0_\nu \rho\,, \tag{4.26}$$

where ρ is the mass density.

To obtain the equations for the potentials it is sufficient to consider the 00-component and the trace of Eq. (4.23). Since the metric is static, $\partial_0 h_{\mu\nu} = 0$; and we get, respectively,

$$\varepsilon_{00} = a_1(\Box)\Box h_{00} + \eta_{00} a_2(\Box)(\partial_\alpha\partial_\beta h^{\alpha\beta} - \Box h) = -16\pi G\,\rho\,, \tag{4.27}$$

$$\eta^{\mu\nu}\varepsilon_{\mu\nu} = [3a_2(\Box) - a_1(\Box)](\partial_\alpha\partial_\beta h^{\alpha\beta} - \Box h) = -16\pi G\eta^{00}\rho\,. \tag{4.28}$$

In our sign convention $\eta^{00} = -1$, $\eta^{ij} = \delta^{ij}$ and, consequently, $\Box = \Delta$ when applied to a static field. Inasmuch as $h_{00} = -2\varphi$ and $h_{ij} = -2\psi\delta_{ij}$, it follows

$$\Box h_{00} = -2\Delta\varphi \qquad \text{and} \qquad \partial_\alpha\partial_\beta h^{\alpha\beta} - \Box h = 2(2\Delta\psi - \Delta\varphi)\,. \tag{4.29}$$

Thus, we find for (4.27) and (4.28), respectively,

$$[a_1(\Delta) - a_2(\Delta)]\Delta\varphi + 2a_2(\Delta)\Delta\psi = 8\pi G\rho, \tag{4.30a}$$

$$[3a_2(\Delta) - a_1(\Delta)](2\Delta\psi - \Delta\varphi) = 8\pi G\rho. \tag{4.30b}$$

Proposition 4.1 *If* $a_1(\Box) = a_2(\Box) \equiv a(\Box)$, *the two Newtonian potentials* φ *and* ψ *are equal and determined by the modified Poisson's equation*

$$a(\Delta)\Delta\varphi = 4\pi G\rho\,. \tag{4.31}$$

Proof Putting $a_1(\Box) = a_2(\Box) \equiv a(\Box)$ in the system (4.30) we obtain

$$a(\Delta)\Delta\psi = 4\pi G\rho, \qquad a(\Delta)(2\Delta\psi - \Delta\varphi) = 4\pi G\rho\,. \tag{4.32}$$

Then, using the first equation in the second one, it follows

$$a(\Delta)\Delta\varphi = a(\Delta)\Delta\psi = 4\pi G\rho\,. \tag{4.33}$$

Since both potentials are defined by the same equations (and are subjected to the same boundary conditions), we get $\varphi = \psi$. □

Observation 4.2 The condition $a_1(\Box) = a_2(\Box) = a(\Box)$ means that [see Eq. (4.20)]

$$F_2(\Box) = -2F_1(\Box) = \frac{a(\Box) - 1}{\Box}. \tag{4.34}$$

So, substituting (4.34) into the gravitational action (4.10), we have

$$S_{\text{grav}} = \frac{1}{16\pi G} \int \mathrm{d}^4x \sqrt{-g} \Big\{ R + G_{\mu\nu} \frac{a(\Box) - 1}{\Box} R^{\mu\nu} \Big\}. \tag{4.35}$$

The action (4.35) represents the only family of higher-derivative models in which the two Newtonian potentials are equal and defined by the differential equation (4.31). Also, general relativity is recovered for $a_1(\Box) = a_2(\Box) = 1$ or, in an equivalent way, $F_1(\Box) = F_2(\Box) = 0$.

In the case of more general functions $a_1(\Box)$ and $a_2(\Box)$, it is convenient to rewrite the system (4.30) in terms of variables that depend only on the spin-0 or the spin-2 degrees of freedom. This can be done with the following definition.

Definition 4.1 The *spin-2 and spin-0 potentials* are defined, respectively, by

$$\chi_2 = \frac{\varphi + \psi}{2} \qquad \text{and} \qquad \chi_0 = 2\psi - \varphi. \tag{4.36}$$

Proposition 4.2 *In terms of the spin-2 and spin-0 potentials, the system of differential equations (4.30) decouples and can be expressed as a single-index equation,*

$$f_s(\Delta)\Delta\chi_s = 4\pi G\rho, \qquad s = 0, 2, \tag{4.37}$$

where

$$f_2(\Box) = a_1(\Box)\,, \tag{4.38a}$$

$$f_0(\Box) = \frac{3a_2(\Box) - a_1(\Box)}{2}\,. \tag{4.38b}$$

Proof The Eq. (4.30b) can be directly rewritten in terms of χ_0 and f_0 as (4.37). For the other equation, if we take three times the Eq. (4.30a) minus Eq. (4.30b) all the $a_2(\Delta)$-dependent terms cancel, and we get

$$a_1(\Delta)\Delta(\varphi + \psi) = 8\pi G\rho\,, \tag{4.39}$$

which is the same as (4.37) using the definitions for χ_2 and f_2. □

Once the solutions for $\chi_{0,2}$ are obtained, the original potentials φ and ψ can be recovered as a linear combination of them through the inverse transformation

$$\varphi = \frac{4}{3}\chi_2 - \frac{1}{3}\chi_0, \qquad \psi = \frac{2}{3}\chi_2 + \frac{1}{3}\chi_0. \tag{4.40}$$

In this sense, one can work with the spin-s potentials without loss of generality.

Observation 4.3 The spin-s potentials were introduced in [36] and applied in the subsequent works [18–20, 37]. From the technical point of view, they facilitate the considerations since the simplicity of Eq. (4.37) (in comparison to the system (4.30) of coupled equations) makes it possible to derive general results based on particular characteristics of the functions f_s. For instance, this decomposition is critical to applying the effective source formalism to a general higher-derivative gravity model in the Newtonian limit.

The spin-s potentials also have a clear physical interpretation. As the name suggests, these auxiliary potentials separate the contributions of the gauge-invariant scalar and spin-2 degrees of freedom of $h_{\mu\nu}$. To understand this, we need to remember the spin-2 and spin-0 projection operators for symmetric rank-2 tensors [7, 62], namely,

$$P^{(2)}_{\mu\nu,\alpha\beta} = \frac{1}{2}\left(\theta_{\mu\alpha}\theta_{\nu\beta} + \theta_{\mu\beta}\theta_{\nu\alpha}\right) - \frac{1}{3}\,\theta_{\mu\nu}\theta_{\alpha\beta}, \tag{4.41}$$

$$P^{(0-s)}_{\mu\nu,\alpha\beta} = \frac{1}{3}\,\theta_{\mu\nu}\theta_{\alpha\beta}, \tag{4.42}$$

where the transverse and longitudinal vector projection operators are, respectively,

$$\theta_{\mu\nu} = \eta_{\mu\nu} - \omega_{\mu\nu}, \qquad \omega_{\mu\nu} = \frac{\partial_\mu \partial_\nu}{\Box}. \tag{4.43}$$

The spin-s component of the field $h_{\mu\nu}$ is defined as

$$h^{(s)}_{\mu\nu} \equiv P^{(s)}_{\mu\nu,\alpha\beta}\, h^{\alpha\beta}. \tag{4.44}$$

Thus, using the above formulas and the Newtonian-limit metric (4.25), it is possible to show that the gauge-invariant spin-2 and scalar components of $h_{\mu\nu}$ are determined, respectively, by the spin-2 and spin-0 potentials,

$$h^{(2)}_{\mu\nu} = \left[\eta_{\mu 0}\eta_{\nu 0} - \frac{1}{3}\left(\eta_{\mu\nu} - \frac{\partial_\mu \partial_\nu}{\Box} \right) \right] \chi_2 \,, \tag{4.45}$$

$$h^{(0-s)}_{\mu\nu} = -\frac{2}{3}\left(\eta_{\mu\nu} - \frac{\partial_\mu \partial_\nu}{\Box} \right) \chi_0 \,. \tag{4.46}$$

It is also useful to recall that the gauge-independent part of the propagator associated to (4.10) is given by

$$G_{\mu\nu,\alpha\beta}(k) = \frac{P^{(2)}_{\mu\nu,\alpha\beta}}{k^2 f_2(-k^2)} - \frac{P^{(0-s)}_{\mu\nu,\alpha\beta}}{2k^2 f_0(-k^2)} . \tag{4.47}$$

Thus, the number of extra spin-s particles in the model depends on the number of roots of the equations $f_s(-k^2) = 0$. (For example, in local higher-derivative gravity, f_s is a polynomial, and the propagator can have many massive poles. On the other hand, for the nonlocal models of Eq. (4.9) the function f_s is the exponential of an entire function, and there is just the massless pole of the graviton, at $k^2 = 0$—like in the case of general relativity, for which $f_0(-k^2) = f_2(-k^2) = 1$.) In light of (4.37), this also means that the potential χ_s only depends on the spin-s sector of the theory.

4.3.3 Effective Delta Sources

With the previous proposition, we managed to reduce the system of linearized equations of motion to the simple form of decoupled equations for the potentials χ_s. Formally, we can rewrite Eq. (4.37) as a standard Poisson equation,

$$\Delta \chi_s = 4\pi G \rho_s , \tag{4.48}$$

with modified effective sources ρ_s defined through

$$\rho = f_s(\Delta)\, \rho_s . \tag{4.49}$$

This procedure involves the inversion of the operator $f_s(\Delta)$, which is not direct[4] in general and also depends on the shape of the original source ρ. Since in this section we investigate whether higher derivatives can resolve the $-1/r$ Newtonian singularity, here we consider the case of a static pointlike source with mass M, associated with a Dirac delta function,

[4] Since we only consider static solutions, the original d'Alembert operator $\Box$ is substituted by the Laplacian Δ. This avoids all the complications related to the choice of the appropriate Green function of the inverse operator in four-dimensional space with Lorentzian signature, especially for models with complex poles and nonlocalities (see, e.g., [22, 59] for further discussion).

$$\rho(\mathbf{r}) = M\delta(\mathbf{r}). \tag{4.50}$$

In the linear limit, there is no loss of generality in considering only pointlike sources, as more complicated matter distributions can be constructed using the *superposition principle*. For further examples, the reader can consult [31, 34, 36], where this procedure was applied to study the collapse of small-mass spherical shells of null fluid.

Definition 4.2 The *effective delta source* (or, simply, *effective source*) is given by

$$\rho_s(r) = \frac{M}{2\pi^2} \int_0^\infty \mathrm{d}k \, \frac{k \sin(kr)}{r f_s(-k^2)}, \tag{4.51}$$

where $r = |\mathbf{r}|$. It is important to notice that some conditions must be imposed on the function $f_s(x)$ for the integral (4.51) to be well defined. Namely, we assume that $f_s(-k^2) > 0$ for $k \in \mathbb{R}$, $f_s(0) = 1$, and that, if $f_s(-k^2)$ is not trivial, it diverges at least as fast as k^2 for $k \to \infty$. Although in some cases $f_s(x)$ can be a nonanalytic function (see [20] for an example), for the sake of simplicity of considerations *here we always assume that* $f_s(x)$ *is analytic*.

Under these assumptions the Fourier kernel associated with the function $1/f_s(-k^2)$ is well defined on the space of square-integrable functions and allows one to obtain (4.51) through the Fourier transform method applied to (4.49). In fact, starting from the integral representation of the Dirac delta function,

$$\delta(\mathbf{r}) = \int \frac{\mathrm{d}^3 k}{(2\pi)^3} e^{i\mathbf{k}\cdot\mathbf{r}}, \tag{4.52}$$

it follows

$$\rho_s = M\,[f_s(\Delta)]^{-1}\,\delta(\mathbf{r}) == M \int \frac{\mathrm{d}^3 k}{(2\pi)^3} \frac{e^{i\mathbf{k}\cdot\mathbf{r}}}{f_s(-k^2)}, \tag{4.53}$$

which can be written as (4.51) after integrating on the angular coordinates.

The conditions imposed by the Definition 4.2 to the functions $f_s(x)$ can be regarded as a restriction on the type of models we consider. Indeed, the requirement $f_s(0) = 1$ expresses that the theory recovers general relativity in the infrared limit. In the case of local models, $f_s(x)$ is a polynomial, and the condition $f_s(-k^2) > 0$ for all $k \in \mathbb{R}$ means, from the physical point of view, that the model does not have tachyons in the spectrum, while technically it ensures that the integrand in (4.51) does not have singularities on the integration interval. Besides that, in what concerns nonlocal models, the condition on the asymptotic behavior of $f_s(-k^2)$ on the real line acts as a constraint on the type of nonlocality of the theory. Namely, it requires that $f_s(-k^2) \sim k^2$ (or faster) for sufficiently large k, so that the propagator (4.47) has an improved behavior in the UV with respect to general relativity (for which $f_s(-k^2) = 1$). It is in this sense that the term "nonlocal higher-derivative gravity" should be understood throughout this chapter—in opposition to other nonlocal models that do not have this kind of improved propagator, e.g., the ones defined by form factors $F_s(\Box) \propto 1/\Box$ or $F_s(\Box) \propto 1/\Box^2$ [27, 49].

As a consequence of the definition of the effective sources ρ_s, formula (4.48) means that the effect of the higher-derivative operator $f_s(\Delta)$ on the Newtonian potentials can be treated equivalently as the smearing of the original delta source. In this spirit, we have the following definition.

Definition 4.3 The *mass function* $m_s(r)$ represents the effective mass inside a sphere of radius r,

$$m_s(r) = 4\pi \int_0^r \mathrm{d}x\, x^2 \rho_s(x). \tag{4.54}$$

Accordingly, it is expected that

$$\lim_{r\to\infty} m(r) = M. \tag{4.55}$$

This can be proved from the first equality in (4.53) by recalling that $f_s(0) = 1$, so that the first term in the series of $1/f(\Delta)$ gives the original delta function (whose integral in whole space is 1), while the other terms in the series produce derivatives of delta functions, which vanish upon integration.

With all these definitions, we can finally obtain the formal solution for the spin-s potentials by rewriting (4.48) in spherical coordinates,

$$\chi_s''(r) + \frac{2}{r}\,\chi_s'(r) = 4\pi G\, \rho_s(r)\,. \tag{4.56}$$

As it can be directly verified, we have the following result.

Theorem 4.1 *The solution of Eq. (4.56) for the effective delta source (4.51) reads*

$$\chi_s(r) = -\int_\infty^r \mathrm{d}x\, g_s(x), \tag{4.57}$$

where

$$g_s(r) = -\frac{G\, m_s(r)}{r^2} \tag{4.58}$$

and $m_s(r)$ is the mass function (4.54).

Before we consider explicit examples, in the next section we take a closer look at this general solution and present some results related to the regularity of the Newtonian potentials and curvature invariants.

Observation 4.4 In the case of the Proposition 4.1, i.e., if $a_1(\Box) = a_2(\Box) \equiv a(\Box)$, the Eq. (4.38) implies

$$f_2(\Box) = f_0(\Box) = a(\Box), \tag{4.59}$$

so that there is only one effective source

$$\rho_{\text{eff}}(r) \equiv \rho_2(r) = \rho_0(r), \tag{4.60}$$

where

$$\rho_{\text{eff}}(r) = \frac{M}{2\pi^2 r} \int_0^\infty \text{d}k \, \frac{k \sin(kr)}{a(-k^2)}, \tag{4.61}$$

and one mass function,

$$m(r) \equiv m_2(r) = m_0(r) = 4\pi \int_0^r \text{d}x \, x^2 \rho_{\text{eff}}(x). \tag{4.62}$$

Accordingly, all the potentials are equal in this situation, namely,

$$\varphi(r) = \psi(r) = \chi_2(r) = \chi_0(r). \tag{4.63}$$

Given the Theorem 4.1, the solution for the potential, in this case, is

$$\varphi(r) = -\int_\infty^r \text{d}x \, g(x), \tag{4.64}$$

where

$$g(r) = -\frac{G\,m(r)}{r^2}. \tag{4.65}$$

4.4 Properties of the Effective Sources and Newtonian Potentials

Important properties of the potentials $\chi_s(r)$ can be derived without actually specifying the functions $f_s(x)$ and solving the integrals involved in (4.57), but only by knowing some basic features of $f_s(x)$ translated into the effective source $\rho_s(r)$. In this section, we present necessary and sufficient conditions for the Newtonian potentials to be finite at $r = 0$. Then, after characterizing the models with a regular Newtonian-limit metric, we extend the discussion to the behavior of the derivatives of the potentials and the regularity of curvature invariants. The core elements in this consideration are the regularity properties of the effective delta sources, which will also be used to analyze the regular black holes constructed in Sect. 4.6.

4.4.1 Regularity and Higher-Order Regularity of the Effective Sources

Theorem 4.2 (Regularization of the source [37]) *If there exists a $k_0 > 0$ such that $k > k_0$ implies that $f_s(-k^2)$ grows at least as fast as k^4, then the effective source $\rho_s(r)$ is integrable and bounded. Moreover, $\rho_s(r)$ reaches its maximum at $r = 0$.*

Proof Let us define the function

$$G_s(r,k) = \frac{k \sin(kr)}{r f_s(-k^2)}, \tag{4.66}$$

which is the integrand in Eq. (4.51). Since it is assumed that $f_s(0) = 1$ and that $f_s(-k^2)$ does not change sign for $k > 0$, it follows that $G_s(r,k)$ is bounded on any compact of $\mathbb{R}^2$. The integrability of (4.66), thus, depends on its behavior as $k \to \infty$. Under the hypothesis of the theorem,

$$k > k_0 \quad \Longrightarrow \quad |G_s(r,k)| \leqslant \frac{c}{k^2} \tag{4.67}$$

for a constant c. Using the Weierstrass test it follows that $G_s(r,k)$ is integrable on k, even for $r = 0$, and the integral converges uniformly. Hence, $\rho_s(r)$ is continuous, integrable, and bounded, showing that the higher derivatives regularize the δ-singularity of the original source.

In what concerns the maximum, since $r, k > 0$ implies

$$|G_s(r,k)| = \frac{k|\sin(kr)|}{r f_s(-k^2)} \leqslant \frac{k^2}{f_s(-k^2)} = G_s(0,k)\,, \tag{4.68}$$

then

$$\int_0^\infty \mathrm{d}k\, G_s(r,k) \leqslant \int_0^\infty \mathrm{d}k\, |G_s(r,k)| \leqslant \int_0^\infty \mathrm{d}k\, G_s(0,k), \tag{4.69}$$

which means that $r = 0$ is the maximum of $\rho_s(r)$. In particular, $\rho_s(0) \neq 0$. □

As an application of this theorem, we note that if a local higher-derivative gravity model contains at least six derivatives of the metric in the spin-s sector, then the related effective source ρ_s is regular, regardless of whether the propagator has real or complex, simple or degenerate poles. On the other hand, fourth-derivative gravity does not satisfy the hypothesis of the theorem, as $f_s(-k^2)$ grows like k^2; indeed, this growth is not fast enough to result in a regular effective source, as we explicitly show in Sect. 4.5.1 (see also [37]).

For the considerations in the following sections, it is necessary to investigate the behavior of the odd-order derivatives of the effective sources. To this end, let us define the *order of regularity* of a regular function [20].

Definition 4.4 Given a bounded function $\xi : [0, +\infty) \longrightarrow \mathbb{R}$ and an integer $p \geqslant 0$, we shall say that p is the *order of regularity* of ξ if:

1. $\xi(r)$ is at least $2p$-times differentiable on $[0, +\infty)$ and $\xi^{(2p)}(r)$ is continuous.
2. If $p \geqslant 1$, the first p odd-order derivatives of $\xi(r)$ vanish as $r \to 0$, namely,

$$0 \leqslant n \leqslant p-1 \quad \Longrightarrow \quad \lim_{r\to 0} \xi^{(2n+1)}(r) = 0.$$

If these conditions are satisfied, we shall also say that the function $\xi(r)$ is *p-regular*.

Notice that 0-regularity only means that the function is bounded, or regular in the usual sense. Moreover, having Taylor's theorem in mind, one can say that a function $\xi(r)$ is p-regular if the first p odd-order coefficients of its Taylor polynomial around $r = 0$ are zero. In this sense, an analytic function $\xi(r)$ is ∞-regular if and only if it is an *even function*, i.e., such that $\xi(-r) = \xi(r)$. Also, the second condition of the definition means that $\xi^{(2n+1)}(r)$ vanishes at least linearly as $r \to 0$.

In terms of the definition of p-regularity, in the Theorem 4.2 we presented sufficient conditions for the effective sources to be 0-regular. In what follows, we extend this theorem for arbitrary order of regularity.

Theorem 4.3 (Higher-order regularity of the source [20]) *If there exists a $k_0 > 0$ such that $k > k_0$ implies that $f_s(-k^2)$ grows at least as fast as k^{4+2N} for an integer $N \geqslant 0$, then the effective source $\rho_s(r)$ is N-regular.*

Proof The case $N = 0$ was already proved in Theorem 4.2; so, hereafter we take $N \geqslant 1$. Notice that the function $G_s(r, k)$ in (4.66) is even in r, analytic, and has the series expansion

$$G_s(r,k) = \frac{k^2}{f_s(-k^2)} \sum_{\ell=0}^{\infty} \frac{(-1)^\ell}{(2\ell+1)!} (kr)^{2\ell}, \tag{4.70}$$

whence,

$$\lim_{r\to 0} \frac{\partial^n}{\partial r^n} G_s(r,k) = \begin{cases} 0, & \text{if } n \text{ is odd}, \\ \frac{(-1)^{n/2}}{(n+1)} \frac{k^{n+2}}{f_s(-k^2)}, & \text{if } n \text{ is even}. \end{cases} \tag{4.71}$$

Also, the derivatives with respect to r are bounded (for a fixed k). Indeed, since

$$\frac{\partial^{n+1}}{\partial r^{n+1}} G_s(r,k) = -\frac{n+1}{r}\frac{\partial^n}{\partial r^n} G_s(r,k) + \frac{k^{n+2}}{r f_s(-k^2)} \sin\left[kr + \frac{(n+1)\pi}{2}\right],$$

the extrema of $|\frac{\partial^n}{\partial r^n} G_s(r,k)|$ are limited by $\frac{k^{n+2}}{(n+1) f_s(-k^2)}$. Taking (4.71) and the analyticity of $G_s(r,k)$ in r into account, we have

$$\left|\frac{\partial^n}{\partial r^n} G_s(r,k)\right| \leqslant \frac{k^{n+2}}{(n+1) f_s(-k^2)}, \tag{4.72}$$

which generalizes (4.68).

Regarding the bound defined by (4.72) as a function of k it follows that if $f_s(-k^2)$ grows at least as fast as k^{n+4}, then the improper integral

$$\int_0^\infty \mathrm{d}k \, \frac{\partial^n}{\partial r^n} G_s(r,k) \tag{4.73}$$

converges uniformly for $r \geqslant 0$. In these circumstances, the source ρ_s is n-times continuously differentiable and we can apply differentiation under the integral sign in (4.51), namely,

$$\rho_s^{(n)}(r) = \frac{M}{2\pi^2} \int_0^\infty \mathrm{d}k \, \frac{\partial^n}{\partial r^n} G_s(r,k). \tag{4.74}$$

Furthermore, in the limit $r \to 0$ the function $\frac{\partial^n}{\partial r^n} G_s(r,k)$, for n odd, converges uniformly to zero on any compact, implying that the limit $r \to 0$ can be interchanged with the integral in (4.74) (see, e.g., [70]). Thus, the odd derivatives of the source vanish at $r = 0$ because of (4.71)—the condition is that $f_s(-k^2)$ is continuous and grows at least as fast as k^{n+4}. To rewrite this according to the hypothesis of the theorem we use the correspondence $n \mapsto 2N$. This means that if $f_s(-k^2)$ grows at least as fast as k^{4+2N} for an integer $N > 0$ then, up to the $2N$-th order, all the odd-order derivatives of the effective source vanish at $r = 0$, or

$$\lim_{r\to 0} \rho_s^{(2n+1)}(r) = 0, \qquad n = 0, 1, \ldots, N-1. \tag{4.75}$$

In other words, the effective source $\rho_s(r)$ is N-regular. □

Corollary 4.1 *If $f_s(-k^2)$ asymptotically grows faster than any polynomial, then the effective source $\rho_s(r)$ is ∞-regular, i.e., it is an even and analytic function of r.*

The previous theorems can be rewritten making explicit the relation between the behavior of the function $f_s(-k^2)$ for large k and the number of derivatives in a gravitational action:

Corollary 4.2 *If a local gravitational action has $2N + 6$ derivatives of the metric in the spin-s sector (for $N \geqslant 0$), then the effective source $\rho_s(r)$ is N-regular.*

Having established these results about the regularity of the effective source, we can study the behavior of the corresponding effective mass function, defined in (4.54).

Proposition 4.3 *If $f_s(-k^2)$ grows at least as fast as k^{4+2N} for an $N \geqslant 0$ then, near $r = 0$,*

$$m_s(r) \underset{r \to 0}{\sim} r^3, \tag{4.76}$$

Proof Under this hypothesis the effective source is N-regular, and from Theorem 4.2 it has maximum at $r = 0$, so $\rho_s(0) \neq 0$. Then, from Taylor's theorem and Eq. (4.54) we get $m(r) \approx 4\pi \rho_s(0) r^3/3$ for small enough r. □

However, in the case of $f_s(-k^2) \sim k^2$ asymptotically, it happens that $m(r) \sim r^2$ for small r, since $\rho_s(r) \sim 1/r$ (see Sect. 4.5.1 below and other examples in [20, 37]).

4.4.2 Regularity of Newtonian-Limit Solutions

Returning to the Newtonian-limit solutions in higher-derivative gravity in the context of the Theorem 4.1, we can use the regularity properties of the effective source ρ_s to deduce those of the potential χ_s. In fact, from the above results on the small-r behavior of the mass function $m_s(r)$ it follows that the function $g_s(r)$ defined in Eq. (4.58) is bounded in any model with higher derivatives in the spin-s sector,[5] thus the potential χ_s in Eq. (4.57) is also finite.

A stronger result can be obtained concerning the higher-order regularity of the potentials:

Theorem 4.4 (Higher-order regularity of the potential [20]) *If the effective source $\rho_s(r)$ is N-regular, then the potential $\chi_s(r)$ is $(N + 1)$-regular.*

Proof It follows from the relation between χ_s and ρ_s in terms of Theorem 4.1, the order of regularity of the effective source and Taylor's theorem. For instance, under the hypothesis of the theorem, the first odd-order term in the Taylor expansion of the source around $r = 0$ is at least of order r^{2N+1}, while for the potential $\chi_s(r)$ it is at least of order r^{2N+3}. In other words, $\chi_s(r)$ is $(N + 1)$-regular. □

The two previous results can be combined in a theorem relating the behavior of the function $f_s(-k^2)$ in the limit of large k and the order of regularity of $\chi_s(r)$.

Theorem 4.5 (Higher-order regularity of the potential [20]) *If there exists a $k_0 > 0$ such that $k > k_0$ implies that $f_s(-k^2)$ grows at least as fast as k^{2+2N} for an integer $N \geqslant 0$, then the potential $\chi_s(r)$ is N-regular.*

---

[5] Since the function $g_s(r)$ is related to the gravitational force exerted on a test particle, this can be interpreted in the following way: If $f_s(-k^2)$ grows at least as fast as k^4 for large k, the force vanishes linearly as $r \to 0$, for $g_s(r) = O(r)$; on the other hand, in the case of $f_s(-k^2) \sim k^2$ asymptotically, $g_s(0) \neq 0$ and the force is finite (but nonzero) at $r = 0$.

Proof The case $N = 0$ was proved at the beginning of this subsection, while the case $N \geqslant 1$ follows from the combination of Theorems 4.3 and 4.4. □

Corollary 4.3 *If $f_s(-k^2)$ asymptotically grows faster than any polynomial, then the potential $\chi_s(r)$ is ∞-regular, i.e., it is an even and analytic function.*

Taking into account the relation between the spin-s potentials and the Newtonian-limit metric through Eqs. (4.40) and (4.25), it follows that if $\chi_s(r)$ is N_s-regular the potentials $\varphi(r)$ and $\psi(r)$ (thus, the metric components) are N-regular, with $N \equiv \min\{N_0, N_2\}$. In particular, for a higher-derivative gravity model to have a regular Newtonian-limit metric, it must contain higher derivatives both in the spin-2 and spin-0 sectors. For example, in the incomplete polynomial-derivative model considered in [61], with $F_2(\Box) = 0$, the potentials $\varphi(r)$ and $\psi(r)$ are unbounded near the origin because only $\chi_0(r)$ is regular.

4.4.3 Curvature Invariants in the Newtonian Limit

As mentioned in the introduction of this chapter, the regularity of the metric components is not enough to avoid the occurrence of singularities in the curvature invariants. Consider, for instance, a generic Newtonian-limit metric in the form (4.25). The Kretschmann scalar associated with it reads

$$(R^2_{\mu\nu\alpha\beta})_{\text{lin}} = 4(\varphi''^2 + 2\psi''^2) + \frac{16}{r}\psi'\psi'' + \frac{8}{r^2}(\varphi'^2 + 3\psi'^2), \tag{4.77}$$

where we kept only the terms of the lowest order in the metric perturbation in consonance with the Newtonian approximation. Therefore, for the Kretschmann scalar to be bounded, it suffices that the potentials are twice continuously differentiable, and the limits

$$\lim_{r\to 0} \chi_s''(r) \quad \text{and} \quad \lim_{r\to 0} \frac{\chi_s'(r)}{r} \tag{4.78}$$

exist for both $s = 0, 2$—remember that φ and ψ are related to $\chi_{0,2}$ via (4.40).

These conditions are automatically satisfied if the potentials χ_s are at least 1-regular, but they do not hold if they are only 0-regular. It is not difficult to check that the situation of the other linearized curvature invariants, such as $(R)_{\text{lin}}$ and $(R^2_{\mu\nu})_{\text{lin}}$, is exactly the same.

Proposition 4.4 *All the Newtonian-limit scalars formed by the contraction of an arbitrary number of curvature tensors are bounded if the potentials χ_2 and χ_0 are at least 1-regular.*

Proof By definition, the Newtonian-limit scalars are evaluated at lowest order in the metric perturbation. Thus, since the curvatures are already $O(h_{\dots})$, any contraction of indices between curvatures is performed with the flat-space metric $\eta_{\mu\nu}$. Therefore, by dimensional arguments, such scalars are formed by combinations of products of χ_s'' and χ_s'/r. The desired result then comes from the fact that these functions are finite if the potential $\chi_s(r)$ is at least 1-regular. □

However, the potentials should be regular to a higher order for the regularity of scalars involving derivatives of the curvatures. This can be readily seen evaluating the invariant $\Box^n R$ (for an arbitrary n), which in the Newtonian limit reads

$$(\Box^n R)_{\text{lin}} = 2\Delta^{n+1}\chi_0 = 2\left[\chi_0^{(2n+2)} + \frac{2(n+1)}{r}\chi_0^{(2n+1)}\right]. \tag{4.79}$$

Hence, to regularize the scalar $\Box^n R$ it suffices to have a potential $\chi_0(r)$ that is $(n+1)$-regular. In addition, if $\chi_0(r)$ is analytic but $\chi_0^{(2n+1)}(0) \neq 0$, then $\Box^n R$ is not regular.

This simple example illustrates the relation between the higher-order regularity of the metric components and the cancellation of the singularity in scalars containing covariant derivatives of curvatures. A complete treatment of the problem at linear level was carried out in [20], where one can find the proof of the next theorem.

Definition 4.5 Given a certain $n \in \mathbb{N}$, we denote by $\mathscr{I}_{2n}$ the set of all the scalars that are polynomial in curvature tensors and their derivatives, with the restriction that the maximum number of covariant derivatives is $2n$. For example, $\mathscr{I}_0$ is the set of the scalars of type $R^N_{\dots}$ (contraction of an arbitrary number N of curvature tensors), while $\mathscr{I}_2$ also contains invariants like $R\Box R$ and $(\nabla_\lambda R_{\mu\nu\alpha\beta})^2$. Accordingly, $\mathscr{I}_{2n} \supset \mathscr{I}_{2(n-1)} \supset \cdots \supset \mathscr{I}_0$. We refer to a generic element of the set $\mathscr{I}_{2n} \setminus \mathscr{I}_0$ as a *curvature-derivative invariant.*

Theorem 4.6 (Reference [20]) *For each $n \in \mathbb{N}$, a sufficient condition for the Newtonian-limit regularity of all the elements in $\mathscr{I}_{2n}$ is that the potentials χ_0 and χ_2 are $(n+1)$-regular.*

Notice that Proposition 4.4 is the particular case of the theorem for $n = 0$. Taking into account the relation between the behavior of the function $f_s(-k^2)$ for large k and the order of regularity of the potential $\chi_s(r)$, given by Theorem 4.5, the previous theorem offers a characterization of the minimal higher-derivative gravity model with a regular Newtonian limit defined by curvature-derivative invariants:

Corollary 4.4 (Reference [20]) *If for k large enough $f_s(-k^2)$ grows at least as fast as k^{4+2N_s} for an integer $N_s \geqslant 0$, then all the linearized elements in $\mathscr{I}_{2N}$ (where $N \equiv \min\{N_0, N_2\}$) evaluated at the Newtonian-limit metric are regular. In other words, if a local gravitational action has $2N_s + 6$ derivatives of the metric in the spin-s sector, then all the Newtonian-limit scalars in the set $\mathscr{I}_{2N}$ are regular. In the nonlocal gravity models which $f_{0,2}(-k^2)$ asymptotically grows faster than any polynomial, all the curvature-derivative invariants are regular.*

4.5 Selected Examples

Here we show some explicit examples in which the general results presented in the previous sections can be verified. We focus on three interesting higher-derivative models, *viz.* fourth-derivative gravity, polynomial-derivative gravity, and exponential nonlocal gravity. Besides the theoretical importance of these models (as commented in Sect. 4.2), they serve well to illustrate the increasing order of regularity of the solutions as more derivatives are included in the action.

4.5.1 *Fourth-Derivative Gravity*

The solution for the Newtonian potential in fourth-derivative gravity was obtained by Stelle in 1977 [63, 64]. Let us explain how to reproduce this result in the formalism presented above.

The model is described by the action (4.6) with real dimensionless coefficients $\alpha_{1,2,3}$, thus the form factors $F_{1,2}$ in Eq. (4.10) are constants, namely,

$$F_1 = 16\pi G\,(\alpha_3 - \alpha_1)\ , \qquad F_2 = 16\pi G\,(4\alpha_1 + \alpha_2)\ . \tag{4.80}$$

These formulas follow from the comparison of Eqs. (4.6) and (4.15), and applying (4.16). Therefore, using (4.20) we find for the spin-s functions (4.38),

$$f_s(-k^2) = 1 + c_s k^2\,, \tag{4.81}$$

where

$$c_2 = -16\pi G(4\alpha_1 + \alpha_2)\,, \qquad c_0 = 32\pi G(\alpha_1 + \alpha_2 + 3\alpha_3)\,. \tag{4.82}$$

As discussed in the Observation 4.3, the roots of the equation $f_s(-k^2) = 0$ are related to the theory's massive spin-s degrees of freedom. Since we want to consider the general case with higher derivatives in both spin-0 and spin-2 sectors, we assume $c_{0,2} \neq 0$ and rewrite (4.81) in the more convenient form

$$f_s(-k^2) = \frac{k^2 + \mu_s^2}{\mu_s^2}\,, \tag{4.83}$$

where $k^2 = -\mu_s^2$ is the root of $f_s(-k^2) = 0$, i.e.,

$$\mu_s^2 \equiv \frac{1}{c_s}. \tag{4.84}$$

In order to avoid tachyons on the model and guarantee the existence of asymptotically flat solutions, we must assume $\mu_s^2 > 0$, or $c_s > 0$. Of course, in view of Eq. (4.82), this imposes some constraints on the parameters of the action. Note that this assumption is also in consonance with the requirement that the function $f_s(-k^2)$ does not change sign for $k \in \mathbb{R}$ (see Definition 4.2).

Using (4.51) and (4.83) we obtain the effective delta source,

$$\rho_s(r) = \frac{M\mu_s^2}{2\pi^2 r} \int_0^\infty \mathrm{d}k \, \frac{k \sin(kr)}{k^2 + \mu_s^2} = \frac{M\mu_s^2}{4\pi r} e^{-\mu_s r}. \tag{4.85}$$

The integral in (4.85) can be easily evaluated using Cauchy's residue theorem, noting that the integrand (as a complex function) has simple poles at $k = \pm i\mu_s$.

Now, inserting (4.85) into (4.54) gives the result for the effective mass function,

$$m_s(r) = M \left[1 - (1 + \mu_s r)e^{-\mu_s r}\right] \tag{4.86}$$

from which we obtain the function $g_s(r)$ in (4.58) and, finally, the spin-s potential (4.57),

$$\chi_s(r) = -\frac{GM}{r} \left(1 - e^{-\mu_s r}\right). \tag{4.87}$$

Substituting this expression into (4.40), the Newtonian potentials $\varphi(r)$ and $\psi(r)$ follow as a linear combination of $\chi_{0,2}$, namely,

$$\varphi(r) = -\frac{GM}{r}\left(1 - \frac{4}{3}e^{\mu_2 r} + \frac{1}{3}e^{\mu_0 r}\right), \tag{4.88}$$

$$\psi(r) = -\frac{GM}{r}\left(1 - \frac{2}{3}e^{\mu_2 r} - \frac{1}{3}e^{\mu_0 r}\right). \tag{4.89}$$

The small-r behavior of the functions $\rho_s(r), m_s(r), g_s(r)$, and $\chi_s(r)$ can be easily obtained from the previous formulas,

$$\rho_s(r) = \frac{M\mu_s^2}{4\pi}\left(\frac{1}{r} - \mu_s\right) + O(r), \qquad m_s(r) = \frac{M\mu_s^2}{2} r^2 + O(r^3),$$

$$g_s(r) = -\frac{GM\mu_s^2}{2} + O(r), \qquad \chi_s(r) = -GM\mu_s\left(1 - \frac{\mu_s}{2} r\right) + O(r^3).$$

Thus, we see that ρ_s diverges as $1/r$ and the mass function goes like r^2 for small r, the spin-s potential is finite at $r = 0$ and the force (proportional to g_s) tends to a nonzero constant. All these properties agree with the general results of Sect. 4.4 for a model with a function $f_s(-k^2)$ that grows like k^2.

Even though the Newtonian $1/r$ singularity in the potentials χ_s is regularized, they are not 1-regular, as $\chi_s'(0) = GM\mu_s^2/2 \neq 0$. Therefore, the metric has a curvature singularity at $r = 0$; indeed, using Eqs. (4.77), (4.88) and (4.89), we find that near the origin the Kretschmann scalar behaves like

$$(R^2_{\mu\nu\alpha\beta})_{\text{lin}} \underset{r\to 0}{\sim} \frac{8G^2M^2}{9r^2}(\mu_0^4 + \mu_0^2\mu_2^2 + 7\mu_2^4). \tag{4.90}$$

This divergence, though, is less strong than in general relativity, for which $(R^2_{\mu\nu\alpha\beta})_{\text{lin}} \sim r^{-6}$.

4.5.2 Polynomial Higher-Derivative Gravity

The case of polynomial-derivative gravity involves the action (4.6) enlarged by the higher-order terms (4.7), which contain $2N + 2$ derivatives of the metric. This is equivalent to have the action (4.10) with form factors $F_{1,2}(z)$ that are polynomials of maximum degree $N - 1$, for a certain $N \geqslant 2$. Note that we excluded the possibility of $N = 1$, for which the polynomials are trivial, and the model becomes the fourth-derivative gravity considered in the previous subsection. Under these conditions, let us consider that $f_s(z)$ is a real polynomial of degree $N_s \geqslant 2$ and, to simplify the analysis, that the equation $f_s(z) = 0$ has N_s distinct roots. (The more general case with degenerate roots can be found in [36, 37].) However, we do not restrict the occurrence of complex conjugate pairs of roots, as in the case of Lee-Wick gravity [51, 53].

Therefore, recalling that $f_s(0) = 1$, the function $f_s(z)$ can be factored as

$$f_s(z) = \prod_{i=1}^{N_s} \frac{\mu_{s,i}^2 - z}{\mu_{s,i}^2}. \tag{4.91}$$

To solve the integral in (4.51) we can apply the partial fraction decomposition,

$$\frac{1}{f_s(z)} = \prod_{i=1}^{N_s} \frac{\mu_{s,i}^2}{\mu_{s,i}^2 - z} = \sum_{i=1}^{N_s} C_{s,i} \frac{\mu_{s,i}^2}{\mu_{s,i}^2 - z}, \tag{4.92}$$

where

$$C_{s,i} = \prod_{\substack{j=1 \\ j\neq i}}^{N_s} \frac{\mu_{s,j}^2}{\mu_{s,j}^2 - \mu_{s,i}^2}. \tag{4.93}$$

Thus,

$$\rho_s(r) = \frac{M}{2\pi^2 r} \sum_{i=1}^{N_s} C_{s,i}\, \mu_{s,i}^2 \int_0^\infty \mathrm{d}k\, \frac{k \sin(kr)}{k^2 + \mu_{s,i}^2}. \tag{4.94}$$

Since each integral in (4.94) is the same as the one in (4.85), *mutatis mutandis*, the result is

$$\rho_s(r) = \frac{M}{4\pi r} \sum_{i=1}^{N_s} C_{s,i}\, \mu_{s,i}^2\, e^{-\mu_{s,i} r}. \tag{4.95}$$

As consequence,

$$m_s(r) = M \sum_{i=1}^{N_s} C_{s,i} [1 - (1 + \mu_{s,i} r) e^{-\mu_{s,i} r}] \tag{4.96}$$

and

$$\chi_s(r) = -\frac{GM}{r} \left(1 - \sum_{i=1}^{N_s} C_{s,i}\, e^{-\mu_{s,i} r} \right). \tag{4.97}$$

Some general comments about this solution are in order. Even if some of the quantities $\mu_{s,i}^2$ are complex, like in the case of Lee-Wick gravity, the potential (4.97) is a real function. The main reason for this is that the function $f_s(z)$ is a real polynomial with real coefficients, so the integral involved in the calculation of the effective source (4.51) cannot result in a complex function. Therefore, if the coefficients $C_{s,i}$ and some of the functions in the above formulas take complex values, their imaginary parts are canceled in the combinations present in (4.95), (4.96) and (4.97)—an explicit proof of this statement can be found in [35]. This cancellation happens because, from the fundamental theorem of algebra, the complex masses always occur in complex conjugate pairs. In particular, we notice that

$$\mathrm{Re}\left(C_{s,i}\, e^{-\mu_s r} \right) = [c_\mathrm{R} \cos(\mu_\mathrm{I} r) + c_\mathrm{I} \sin(\mu_\mathrm{I} r)]\, e^{-\mu_\mathrm{R} r}, \tag{4.98}$$

where we denote $\mu_s = \mu_\mathrm{R} + i\mu_\mathrm{I}$ and $C_{s,i} = c_\mathrm{R} + i c_\mathrm{I}$. Hence, the presence of complex poles in the propagator can result in oscillatory contributions to the effective source, the mass function, and the potentials [1, 6, 14, 35, 51].

The explicit verification of the order of regularity of the effective source and the potential is based on this lemma, whose proof can be found in [20] (see also [59]).

Lemma 4.1 *Given the polynomial (4.91) with N_s distinct roots $\mu_{s,i}^2$ ($i = 1, \ldots, N_s$), the following $N_s + 1$ identities are true,*

$$\sum_{i=1}^{N_s} C_{s,i}\,(\mu_{s,i})^{2n} = \begin{cases} 1, & \text{if } n = 0, \\ 0, & \text{if } n = 1, \ldots, N_s - 1, \\ (-1)^{N_s-1}\prod_{i=1}^{N_s}\mu_{s,i}^2, & \text{if } n = N_s, \end{cases} \quad (4.99)$$

with $C_{s,i}$ *defined in (4.93).*

Proposition 4.5 (Higher-order regularity in polynomial-derivative gravity [20]) *The effective source (4.95) is* $(N_s - 2)$*-regular and the potential (4.97) is* $(N_s - 1)$*-regular. Moreover, they cannot be regular to an order higher than that.*

Proof Once proving that the above functions $\rho_s(r)$ and $\chi_s(r)$ are analytic, we only have to show that the first odd-order coefficients of their Taylor series around $r = 0$ are null. Both results can be easily obtained by expanding the exponential functions in power series. Indeed, from Eq. (4.95) we get

$$\rho_s(r) = \frac{M}{4\pi}\sum_{\ell=0}^{\infty}\frac{(-1)^\ell}{\ell!}\left[\sum_{i=1}^{N_s} C_{s,i}\,(\mu_{s,i})^{\ell+2}\right] r^{\ell-1}. \quad (4.100)$$

The result now follows from the application of the above lemma, and for each $n = 1, \ldots, N_s - 1$ the formula (4.99) shows that $\rho_s(r)$ is $(n-1)$-regular. In particular, Eq. (4.99) with $n = 1$ shows that the effective source is bounded as r approaches 0 (and, thus, it is bounded everywhere). Whereas Eq. (4.99) with $n = N_s$ shows that the coefficient of the term of order r^{2N_s-3} is nonzero, being proportional to the product of all the quantities $\mu_{s,i}^2$. Thus, $\rho_s(r)$ is $(N_s - 2)$-regular, but it is not $(N_s - 1)$-regular.

The same reasoning can be applied to the potential $\chi_s(r)$ in (4.97),

$$\chi_s(r) = -GM\left[\left(1 - \sum_{i=1}^{N_s} C_{s,i}\right)\frac{1}{r} + \sum_{k=0}^{\infty}\frac{(-1)^{k+1}}{(k+1)!}\left(\sum_{i=1}^{N_s} C_{s,i}\,\mu_{s,i}^{k+1}\right) r^k\right] \quad (4.101)$$

The 0-regularity of the potential comes from Eq. (4.99) with $n = 0$, while the other N_s identities show that $\chi_s(r)$ is $(N_s - 1)$-regular, but it is not N_s-regular. □

The above result provides the explicit verification of the Theorems 4.3 and 4.5 in the context of polynomial-derivative gravity. Also, from the Taylor expansion of the effective mass function (4.96) around $r = 0$ it is straightforward to obtain

$$m_s(r) = -\frac{M}{3}\sum_{i=1}^{N_s} C_{s,i}\mu_{s,i}^3 r^3 + O(r^4), \quad (4.102)$$

in agreement with the Proposition 4.3. Finally, it is worth noticing that the statement of the last theorem can only hold if $N_s \geqslant 2$, a condition we assumed at the beginning of this subsection. This marks the critical difference between theories defined by actions with four and more than four metric derivatives in what concerns the regularization of the effective delta source, the 1-regularity of the potentials, and the cancellation of the curvature singularities.

The latter can be explicitly visualized by calculating some curvature invariants, which also serves as an example of Theorem 4.6. Let us consider the particular case of sixth-derivative gravity with $\mu_{0,i} = \mu_{2,i} \equiv \mu_i$ $(i = 1, 2)$. Since $N_2 = N_0 = 2$, from Proposition 4.5 we know that $\chi_{0,2}$ are 1-regular but they are not 2-regular. Using Eqs. (4.40), (4.77), (4.93) and (4.97) we can evaluate the linearized Kretschmann scalar near $r = 0$,

$$(R^2_{\mu\nu\alpha\beta})_{\text{lin}} = \frac{20(GM)^2\mu_1^4\mu_2^4}{3(\mu_1 + \mu_2)^2} + O(r), \tag{4.103}$$

which is finite, as the potentials $\chi_{0,2}(r)$ are 1-regular. However, since they are not 2-regular, we expect that there are singular scalars with two covariant derivatives; for instance, from (4.79) we have

$$(\Box R)_{\text{lin}} = -\frac{2GM\mu_1^2\mu_2^2}{r} + O(r^0) \tag{4.104}$$

in agreement with Theorem 4.6.

Observation 4.5 We close this example with a brief digression about the literature on Newtonian-limit solutions in polynomial higher-derivative gravity. In 1991, Quandt and Schmidt [61] published the result for the potential in the incomplete polynomial model, with higher derivatives only in the spin-0 sector, *i.e*, with $F_2(\Box) = 0$ in (4.10). Because the spin-2 sector of the model does not have higher derivatives, the potentials found still diverge like $-1/r$. The Newtonian potential for the complete polynomial model with real simple poles in the propagator was obtained in [55], where it was noted that the potential is finite at $r = 0$. In [35] it was shown that the cancellation of the singularity in the potential happens in any local model with higher derivatives in the spin-2 and spin-0 sectors, regardless of whether the number of derivatives is the same or the nature of the poles (real or complex, simple or degenerate). Explicit expressions for the potentials in the case of complex poles were first published in [1, 51]; some applications of solutions with complex and degenerate real poles were studied in [1, 2]. The complete expression for the potential in a general polynomial gravity theory, including complex poles and poles with any multiplicity, was obtained in [36]. Also, the decomposition of the potential in terms of its spin-parts was introduced, and it was proved that all models with at least six derivatives in both spin-2 and spin-0 sectors have a regular Newtonian limit without curvature singularities. The formulation of this problem in terms of effective delta sources was carried out in [37], aiming at extending the result to the case of nonlocal gravity models. Finally, in [20] the idea of higher-order regularity was introduced, and it was shown the relation between the number of derivatives in the action, the order of regularity of the potential, and the regularity of linearized curvature-derivative invariants.

4.5.3 Nonlocal Gravity

The last example we consider belongs to the class of nonlocal theories defined by the action (4.9). As discussed in the Observations 4.2 and 4.4, in this case, there is only one independent effective source $\rho_{\text{eff}}(r)$ and potential $\varphi(r)$, inasmuch as the spin-2 and spin-0 parts of the propagator have exactly the same behavior, related to the function [compare Eqs. (4.9) and (4.35)]

$$a(\Box) = e^{H(\Box)}. \tag{4.105}$$

There is a huge arbitrariness in the choice of the entire function $H(z)$, and the order of regularity of the effective delta source can be affected by this choice—in fact, as shown in Sect. 4.4, it depends on the behavior of $a(-k^2)$ for large k. Since the previous examples dealt with functions that behave like polynomials, here we focus on a model for which $a(-k^2)$ grows faster than any polynomial as $k \to \infty$. To this end, we choose the simplest entire function,[6] $H(z) = -z/\mu^2$, so that

$$a(-k^2) = e^{k^2/\mu^2}, \tag{4.106}$$

being $\mu > 0$ a parameter with mass dimension, known as the nonlocality scale.

In this case, the effective delta source obtained from (4.61) has a Gaussian profile,

$$\rho_{\text{eff}}(r) = \frac{M}{2\pi^2 r} \int_0^\infty \text{d}k \, \frac{k \sin(kr)}{e^{k^2/\mu^2}} = \frac{M\mu^3}{8\pi^{3/2}} \, e^{-\frac{1}{4}\mu^2 r^2}, \tag{4.107}$$

and the mass function (4.62) is given by

$$m(r) = M \left[\text{erf}\left(\frac{\mu r}{2}\right) - \frac{\mu r}{\sqrt{\pi}} \, e^{-\frac{1}{4}\mu^2 r^2} \right], \tag{4.108}$$

where

$$\text{erf}(x) = \frac{2}{\sqrt{\pi}} \int_0^x \text{d}t \, e^{-t^2} \tag{4.109}$$

is the error function. Finally, we have the potential

$$\varphi(r) = -\frac{GM}{r} \, \text{erf}\left(\frac{\mu r}{2}\right). \tag{4.110}$$

The solution (4.110) was first derived by Tseytlin [68] in the framework of string theory, using effective sources, and it was later obtained in the context of nonlocal gravity in [11, 50].

---

[6] For the evaluation of the potential with more complicated functions, see, e.g., [15, 30, 33] and, in particular, [14, 20, 37], for considerations regarding the effective source.

We observe that, in agreement with the results of Sect. 4.4, the source $\rho_{\text{eff}}(r)$ and the potential are even analytic functions of r. Indeed, using the series expansion of the error function, we get

$$\varphi(r) = -\frac{GM\mu}{\sqrt{\pi}} \sum_{n=0}^{\infty} \frac{(-1)^n}{(2n+1)n!} \left(\frac{\mu r}{2}\right)^{2n}. \tag{4.111}$$

Therefore, $\rho_{\text{eff}}(r)$ and $\varphi(r)$ are ∞-regular because $a(-k^2)$ grows faster than any polynomial. This ensures the regularity of *all* the curvature and curvature-derivative local invariants (see Corollary 4.4). For example, using (4.110) we get for the Kretschmann scalar (4.77),

$$(R^2_{\mu\nu\alpha\beta})_{\text{lin}} = \frac{5G^2M^2\mu^6}{3\pi} + O(r). \tag{4.112}$$

Moreover, it is easy to verify that $m(r)$ is an odd function, and its behavior is the one expected from Proposition 4.3.

Finally, we close the section of examples with a graph illustrating the increasing regularity of the effective delta source as the behavior of the function $f_s(-k^2)$ is improved. In Fig. 4.1 we compare the effective sources for models with four, six, and eight metric derivatives, and also the nonlocal model discussed above. From the graph it is clear that in the case of fourth-derivative gravity the source is still singular [see Eq. (4.85)], while it is regular in all the other cases. However, $\rho_s'(0) \neq 0$ for the sixth-derivative model, indicating that the source is not 1-regular. For the model with eight derivatives and the exponential nonlocal model, $\rho_s'(0) = 0$, confirming the 1-regularity of the sources. The qualitative behavior of the graphs of these two

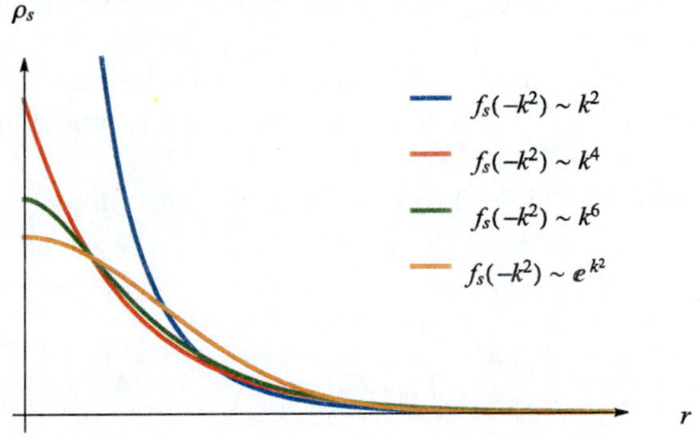

Fig. 4.1 Qualitative behavior of the effective sources for models with four (blue, $1/r$ divergence), six (red, 0-regular), and eight (green, 1-regular) derivatives, and for the nonlocal exponential model (yellow, ∞-regular). Notice the increasing order of regularity with the number of metric derivatives in the action. For the last two sources $\rho_s'(0) = 0$, but differences appear in the higher derivatives of ρ_s

sources is very similar; nonetheless, the differences appear when higher derivatives are evaluated, as ∞-regularity cannot be achieved in polynomial models.

4.6 Regular Black Holes from Effective Delta Sources

In the previous sections, we developed the formalism of effective sources. First, we showed how, in the Newtonian limit, the effect of the higher derivatives can be treated as the smearing of the original delta source. Subsequently, we derived the properties of the effective source depending on the behavior of the form factor and related them with the regularity of the curvature invariants. Now we shall apply the results about the effective sources to construct regular black hole metrics.

To this end, let us consider a generic static and spherically symmetric metric with line element written in the convenient form

$$\mathrm{d}s^2 = -A(r)e^{B(r)}\mathrm{d}t^2 + \frac{\mathrm{d}r^2}{A(r)} + r^2\mathrm{d}\Omega^2, \tag{4.113}$$

where $A(r)$ and $B(r)$ are two arbitrary functions to be determined. The main idea is to obtain solutions (in the above form) of the Einstein's equations

$$G^{\mu}{}_{\nu} = 8\pi G\, \tilde{T}^{\mu}{}_{\nu} \tag{4.114}$$

sourced by the effective energy-momentum tensor

$$\tilde{T}^{\mu}{}_{\nu} = \mathrm{diag}(-\rho_{\text{eff}}, p_r, p_\theta, p_\theta), \tag{4.115}$$

where ρ_{eff} is the effective source given by Eq. (4.61). Notice that $\tilde{T}^{\mu}{}_{\nu}$ includes not only the effective delta source, but also some nontrivial effective radial (p_r) and tangential (p_θ) pressures. These components are essential for the validity of the conservation equation $\nabla_\mu \tilde{T}^{\mu}{}_{\nu} = 0$ and, therefore, for the consistency of the system (4.114).

In fact, for the metric (4.113), the nonzero components of the Einstein tensor are

$$G^{t}{}_{t} = \frac{A'}{r} + \frac{A-1}{r^2} = G^{r}{}_{r} - \frac{AB'}{r}, \tag{4.116}$$

$$G^{\theta}{}_{\theta} = G^{\phi}{}_{\phi} = \frac{A'}{r} + \frac{1}{2}\frac{AB'}{r} + \frac{3}{4}A'B' + \frac{1}{4}AB'^2 + \frac{1}{2}A'' + \frac{1}{2}AB''. \tag{4.117}$$

Hence, Eq. (4.114) with (4.115) is a system of three equations to be solved for the two functions $A(r)$ and $B(r)$. A moment's reflection shows that the tt and rr components of (4.114) can be rearranged in the form

$$\frac{A'}{r} + \frac{A-1}{r^2} = -8\pi G \rho_{\text{eff}} \,, \tag{4.118a}$$

$$\frac{AB'}{r} = 8\pi G (\rho_{\text{eff}} + p_r) \,, \tag{4.118b}$$

in which the equation for $A(r)$ is decoupled. So, the solution for $B(r)$ can be obtained from Eq. (4.118b) after $A(r)$ is determined by Eq. (4.118a), and the remaining equation (for the component $\theta\theta$) is actually a consistency relation. The latter is equivalent to the conservation of $\tilde{T}^{\mu}{}_{\nu}$ and can be cast in the form

$$p_r' = -\frac{1}{2}\left(\frac{A'}{A} + B'\right)(p_r + \rho_{\text{eff}}) - \frac{2}{r}(p_r - p_\theta) \,. \tag{4.119}$$

Therefore, the field equations (4.114) are equivalent to the system formed by (4.118) and the conservation equation (4.119).

The solution for $A(r)$ can be directly obtained from Eq. (4.118a), as it only depends on the effective source. It is not difficult to verify that

$$A(r) = 1 - \frac{2Gm(r)}{r}, \tag{4.120}$$

where $m(r)$ is the same mass function defined in (4.62). Furthermore, taking into account the relation between $m(r)$ and the function $\varphi(r)$, given by Eq. (4.64), one can express Eq. (4.120) as

$$A(r) = 1 - 2r\varphi'(r). \tag{4.121}$$

Thus, the Newtonian-limit solutions obtained in the previous sections of this chapter can be used to construct solutions of Eq. (4.114).

In order to solve the remaining equations, it is necessary to specify the pressure components of the effective energy-momentum tensor. Here, we adopt the equation of state for the radial pressure,

$$p_r(r) = [A(r) - 1]\, \rho_{\text{eff}}(r), \tag{4.122}$$

as a means to generate regular black hole solutions. Now we can solve Eq. (4.118b) for $B(r)$, and determine p_θ through the conservation equation (4.119)—indeed, using (4.122) it reduces to

$$p_\theta = \frac{1}{4}\left[3A'r + (4 + B'r)A - 4\right]\rho_{\text{eff}} + \frac{A-1}{2}\, r\rho_{\text{eff}}' \,. \tag{4.123}$$

Let us mention that there are other possible choices for the pressure components that can lead to regular black hole solutions, but due to the space limitation, we shall only explicitly discuss the ones originated from the equation of state (4.122). A slightly more general result, though, is presented in the Theorem 4.9 below.

Having found the solution (4.120) [or, equivalently, (4.121)] for $A(r)$ and specified the pressure p_r, the function $B(r)$ follows from (4.118b), namely,

$$B(r) = 8\pi G \int_{\infty}^{r} \mathrm{d}x \, \frac{x\,[\rho_{\text{eff}}(x) + p_r(x)]}{A(x)} = 8\pi G \int_{\infty}^{r} \mathrm{d}x \, x \, \rho_{\text{eff}}(x), \qquad (4.124)$$

where we used (4.122). Therefore, we find for the metric (4.113),

$$\mathrm{d}s^2 = -\left(1 - \frac{2Gm(r)}{r}\right) e^{8\pi G \int_{\infty}^{r} \mathrm{d}x \, x \, \rho_{\text{eff}}(x)} \mathrm{d}t^2 + \frac{\mathrm{d}r^2}{\left(1 - \frac{2Gm(r)}{r}\right)} + r^2 \mathrm{d}\Omega^2. \quad (4.125)$$

Observation 4.6 Even though (4.114) is not the field equation of any higher-derivative model described previously in this chapter, sometimes it is used in an effective approach, as an approximation of the field equations of the models described by the action (4.35). The reason is the specific form of the higher-derivative structure in (4.35), which makes it possible to factor the operator $a(\Box)$ together with the Einstein tensor in the field equations. Indeed, taking the variational derivative of (4.35) with respect to the metric, we find

$$a(\Box) G^{\mu}{}_{\nu} + O(R^2_{\dots}) = 8\pi G \, T^{\mu}{}_{\nu}. \qquad (4.126)$$

Therefore, by the inversion of the form factor $a(\Box)$ one might regard (4.114) as an approximation of (4.126), sourced by a pointlike source, in spacetime regions where the curvature is small; more details about this procedure can be found in [6, 50]. It is worth mentioning that Eq. (4.114) was also used as effective equations in other frameworks—see, e.g., [44, 54, 56–58, 69] and Chap. 3 in this book.

Regardless of the physical interpretation of (4.114) described above, our point of view is that these equations are very interesting per se and deserve a detailed study. As we show below, by imposing some general requirements on the form factor $a(\Box)$, it is possible to obtain singularity-free black holes that are solutions of some equations of motion.

4.6.1 *General Properties of $A(r)$ and $B(r)$*

Similar to the case of the Newtonian limit, by just considering the asymptotic behavior of the form factor $a(z)$, translated into the effective source $\rho_{\text{eff}}(r)$, we can explain many important physical properties of the solutions.

The properties of $A(r)$, for instance, can be derived straightforwardly from the relation between it and the function $\varphi(r)$, via (4.121).

Theorem 4.7 *If the effective source $\rho_{\text{eff}}(r)$ is N-regular for an integer $N \geqslant 0$ (or, in an equivalent way, if the function $a(-k^2)$ asymptotically grows at least as fast as k^{4+2N}), then the function $A(r)$ is $(N+1)$-regular. Moreover,*

$$\lim_{r\to 0} A(r) = \lim_{r\to\infty} A(r) = 1. \tag{4.127}$$

Proof The first part is a direct consequence of Theorems 4.4 and 4.5, which imply that $r\varphi'(r)$ is $(N+1)$-regular. Also, since $\varphi(r)$ is at least 1-regular, near $r = 0$ we have $r\varphi'(r) = O(r^2)$ whence, using (4.121), $A(0) = 1$. The remaining limit follows from (4.55). □

As a corollary, if $a(-k^2)$ asymptotically grows faster than any polynomial, $A(r)$ is an even function.

Under the hypotheses of Theorem 4.7 for the effective source $\rho_{\text{eff}}(r)$, and using the result (4.127) for $A(r)$, it can be shown that the effective pressure components defined by Eqs. (4.122) and (4.123) have the following features:

1. They vanish asymptotically for large and small r.
2. They are finite at the horizons, i.e., in the spacetime regions where $A(r) = 0$.

Other choices of effective pressures can have this same qualitative behavior, or less stringent ones (e.g., being only finite, but nonvanishing, at $r = 0$), see the discussion after Theorem 4.9, below.

Another important consequence of Eq. (4.127), and the fact that $A(r)$ is bounded for all the effective delta sources considered above,[7] is the existence of a critical mass M_c such that $M < M_c$ implies $A(r) > 0$ for all r. This means that for this source there exists a *mass gap* for a solution to describe a black hole. Indeed, as $A(r)$ does not change sign if $M < M_c$, the metric does not have any horizon. On the other hand, if $M > M_c$, the solution has an even number of horizons because the function $A(r)$ changes sign an even number of times. This is in contrast with the case of the delta source in general relativity, for which black hole solutions exist regardless of the value of M. Further discussion on the mass gap for black hole solutions in higher-derivative gravity can be found in [31–34, 36].

In what concerns the function $B(r)$ in (4.124), its regularity properties can be directly deduced from the results on the order of regularity of the effective source $\rho_{\text{eff}}(r)$, stated in Theorem 4.3.

Theorem 4.8 *If the effective source $\rho_{\text{eff}}(r)$ is N-regular for an integer $N \geqslant 0$ (or, in an equivalent way, if the function $a(-k^2)$ asymptotically grows at least as fast as k^{4+2N}), then the function $B(r)$ in (4.124) is $(N+1)$-regular.*

---

[7] Notice that (4.127) is valid also for $a(-k^2) \sim k^2$, which is not covered by Theorem 4.7. Indeed, in this case $\varphi(r)$ and $A(r)$ are 0-regular, but not 1-regular; yet, $A(0) = 1$ because $r\varphi'(r) = O(r)$.

A more general theorem, valid for a broader family of equations of state for p_r can be formulated as follows.

Theorem 4.9 *Let $\sigma(x)$ be a continuous, N-regular function (for an integer $N \geqslant 0$), such that the function $\xi(x) = x\sigma(x)$ is integrable on $[0, +\infty)$. If we assume the equation of state*

$$p_r(r) = A(r)\sigma(r) - \rho_{\text{eff}}(r), \tag{4.128}$$

then $B(r)$ is $(N+1)$-regular and is given by

$$B(r) = 8\pi G \int_{\infty}^{r} \text{d}x \, x \, \sigma(x). \tag{4.129}$$

The proof is immediate after applying Taylor's theorem. Note that the equation of state (4.122) and the Theorem 4.8 correspond to the particular choice $\sigma(r) = \rho_{\text{eff}}(r)$.

Also, if $\sigma(r)$ is only 0-regular, then $B(r)$ is only 1-regular. This happens, e.g., for the choice $\sigma(r) \propto r\rho_{\text{eff}}(r)$ where $\rho_{\text{eff}}(r)$ is an effective delta source of any order of regularity (see [56] for an explicit example). In this case $B(r) - B(0) \propto m(r)$, which is not 2-regular, as it is immediate from Eq. (4.76). Therefore, for this choice of $\sigma(r)$, the Kretschmann scalar and the other curvature-squared invariants are bounded, but invariants containing at least two covariant derivatives, such as $\Box R$, might diverge (see Theorem 4.10 below).

4.6.2 *Curvature Regularity of the Solutions*

The Theorem 4.6 presented in Sect. 4.4.3 related the regularity order of the components of the metric (4.25) and the regularity of sets of linearized curvature-derivative invariants. In particular, it was shown that the occurrence of odd powers of r could make some curvature invariants singular at $r = 0$. A similar situation happens in the nonlinear regime. Since the complete analysis of the problem exceeds the scope of this text, we shall only present some results in this direction.

To this end, let us consider a generic static and spherically symmetric metric with line element (4.113), where $A(r)$ and $B(r)$ are two arbitrary regular analytic functions. Thus, they can be written in the form of power series,

$$A(r) = \sum_{\ell=0}^{\infty} a_\ell r^\ell, \qquad B(r) = \sum_{\ell=0}^{\infty} b_\ell r^\ell. \tag{4.130}$$

By direct calculation it is straightforward to verify that the conditions for the regularity of the invariants R, $R^2_{\mu\nu}$ and $R^2_{\mu\nu\alpha\beta}$ are

$$a_0 = 1, \tag{4.131a}$$

$$a_1 = b_1 = 0. \tag{4.131b}$$

For example, by substituting the series (4.130) into the expression for the Kretschmann scalar, we get

$$\begin{aligned} R^2_{\mu\nu\alpha\beta} &= 4\,\tfrac{1+a_0(a_0-2)}{r^4} + 8\,\tfrac{a_1(a_0-1)}{r^3} + 2\,\tfrac{2a_1(a_0b_1+2a_1)+4a_2(a_0-1)+a_0^2b_1^2}{r^2} \\ &+4\,\tfrac{a_1(a_0b_1^2+2a_0b_2+a_1b_1+6a_2)+2b_1(a_0^2b_2+a_0a_2)+2(a_0-1)a_3}{r} + O\left(r^0\right). \end{aligned} \tag{4.132}$$

Therefore, $R^2_{\mu\nu\alpha\beta}$ is finite if the conditions (4.131) are satisfied.

As proved in Theorems 4.7 and 4.8, if the effective source $\rho_{\text{eff}}(r)$ is regular, the functions $A(r)$ in (4.120) and $B(r)$ in (4.124) are at least 1-regular, thus $a_1 = b_1 = 0$; while the requirement $a_0 = A(0) = 1$ is automatically satisfied by the solution in (4.120), see (4.127). We conclude that for the curvature-squared invariants to be bounded, with the equation of state (4.122), it suffices that $\rho_{\text{eff}}(r)$ is regular.

The case of the invariants of the type $\Box^N R$ was considered in detail in Ref. [38], where the following result was proved.

Theorem 4.10 *Let $N_A \geqslant 1$ and $N_B \geqslant 1$ be, respectively, the order of regularity of the functions $A(r)$ and $B(r)$ in the metric (4.113), which also satisfy (4.131), and let $N \equiv \min\{N_A, N_B\}$. Then, for each $n \in \mathbb{N}$ such that $n \leqslant N-1$, the scalar $\Box^n R$ is finite at $r=0$, whereas the invariant $\Box^N R$ might be singular.*

The extension of this proposition to general curvature-derivative invariants is a more involved task, but it is expected that one similar to Theorem 4.6 also holds in the nonlinear regime. In particular, all the local curvature-derivative scalars are finite at $r=0$ if the metric (4.113) is an even and analytic function of r. Further considerations and examples can be found in Ref. [38].

4.6.3 Example of Regular Black Hole: The Case of Nonlocal Form Factor

As an explicit example of the procedure outlined in this chapter to construct regular black hole metrics, here we consider the nonlocal form factor $a(-k^2) = \exp(k^2/\mu^2)$. This case, in analogy with Sect. 4.5.3, displays an interesting feature: Since the form factor $a(-k^2)$ grows faster than any polynomial for large enough k, the metric is an even analytic function, and, as a consequence, all the curvature invariants are regular.

According to the discussion along this section, to obtain the explicit form of the metric (4.125) the only new ingredient is the function $B(r)$, as the functions $\rho_{\text{eff}}(r)$ and $m(r)$ for this form factor were already evaluated in (4.107) and (4.108), respectively. Thus, substituting the latter into (4.120) we find

$$A(r) = 1 - \frac{2GM}{r}\,\text{erf}\left(\frac{\mu r}{2}\right) + \frac{2GM\mu}{\sqrt{\pi}}\,e^{-\frac{1}{4}\mu^2 r^2}, \tag{4.133}$$

and using (4.107) and (4.124), we obtain

$$B(r) = \frac{GM\mu^3}{\sqrt{\pi}}\int_{\infty}^{r} \mathrm{d}x\, x\, e^{-\frac{1}{4}\mu^2 x^2} = -\frac{2G\mu M}{\sqrt{\pi}}\,e^{-\frac{1}{4}\mu^2 r^2}, \tag{4.134}$$

so that the metric (4.125), in this case, is given by

$$\begin{aligned} \mathrm{d}s^2 = &-\left[1 - \frac{2GM}{r}\,\text{erf}\left(\frac{\mu r}{2}\right) + \frac{2GM\mu}{\sqrt{\pi}}\,e^{-\frac{1}{4}\mu^2 r^2}\right]\exp\left(-\frac{2G\mu M}{\sqrt{\pi}}\,e^{-\frac{1}{4}\mu^2 r^2}\right)\mathrm{d}t^2 \\ &+\left[1 - \frac{2GM}{r}\,\text{erf}\left(\frac{\mu r}{2}\right) + \frac{2GM\mu}{\sqrt{\pi}}\,e^{-\frac{1}{4}\mu^2 r^2}\right]^{-1}\mathrm{d}r^2 + r^2\mathrm{d}\Omega^2. \end{aligned} \tag{4.135}$$

As mentioned before, this metric describes a black hole if the mass M is bigger than the critical mass M_c for the black hole formation. This is related to the minima of the function $A(r)$. In the specific case of (4.133) there is only one absolute minimum A_{min}. If $A_{\text{min}} < 0$ the function (4.133) flips sign twice and the equation $A(r) = 0$ has two roots, which we denote by $r_\pm$ (with $r_+ > r_-$). The values r_+ and r_- are, respectively, the positions of the event horizon and an inner horizon.[8] Otherwise, if $A_{\text{min}} > 0$ the function $A(r)$ does not change sign and the metric (4.135) does not describe a black hole. These features are shown in Fig. 4.2, where we plot the graph of $A(r)$ for the two distinct scenarios.

To determine the critical mass, it is useful to rewrite Eq. (4.133) in terms of a function $q(x)$ that only depends on the dimensionless combination μr, namely,

$$A(r) = 1 - 2GM\mu\, q(\mu r), \tag{4.136}$$

where

$$q(x) = \frac{1}{x}\,\text{erf}\left(\frac{x}{2}\right) - \frac{e^{-\frac{x^2}{4}}}{\sqrt{\pi}}. \tag{4.137}$$

As expected from Theorem 4.7, $q(x)$ is a continuously differentiable even function, it is bounded and satisfies $\lim_{x\to 0} q(x) = \lim_{x\to\infty} q(x) = 0$, in agreement with (4.127). In fact, the graph of (4.137) is depicted in Fig. 4.3, from which we see that it has an absolute maximum, $q_{\text{max}} = 0.263$ at the point $x = 3.02$. Therefore, the critical mass is obtained as the solution of the equation

$$A_{\text{min}} = 1 - 2GM_c\,\mu\, q_{\text{max}} = 0, \tag{4.138}$$

[8] In addition, the position of the event horizon is bounded by the Schwarzschild radius, $r_+ \leqslant r_s = 2GM$, because in this example $m(r) \leqslant M$.

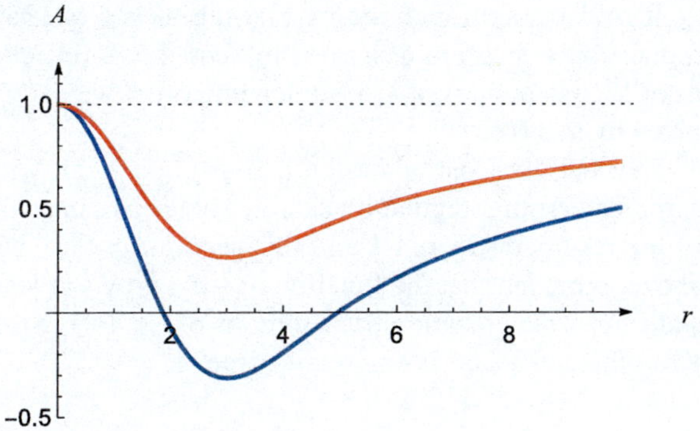

Fig. 4.2 Plot of $A(r)$ for $\mu = M_P = 1$ in two situations. In blue line, with $M = 2.5 > M_c = 1.9$, so that $A_{min} < 0$ and the equation $A(r) = 0$ has two solutions, *i.e*, we have two horizons at $r_- = 1.87$ and $r_+ = 4.67 < r_s = 5.0$. In red line, with $M = 1.4 < M_c$, so that $A_{min} > 0$ and the metric has no horizon

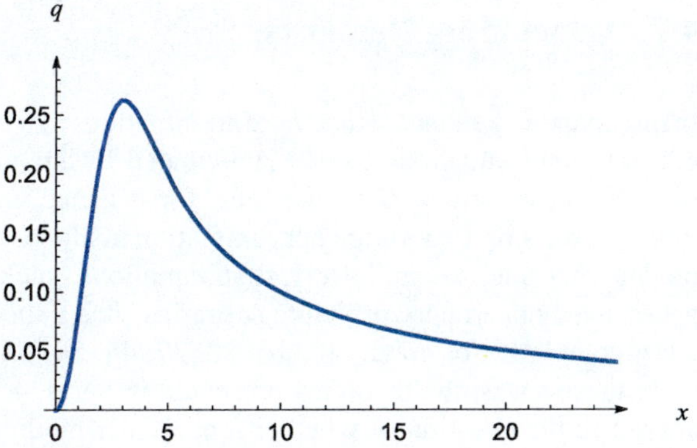

Fig. 4.3 Plot of (4.137). The function $q(x)$ has an absolute maximum $q_{max} = 0.263$ at $x = 3.02$ and asymptotically vanishes for small and large values of x

that is,

$$M_c = \frac{1}{2G\mu\, q_{max}} \approx 1.9\, \frac{M_P^2}{\mu}, \tag{4.139}$$

where we used $G = 1/M_P^2$ for the Planck mass.

Regarding the regularity of the solution, since the form factor (4.106) grows faster than any polynomial, the metric components $A(r)$ and $B(r)$ are even functions. Therefore, we expect all curvature invariants to be finite at the origin, even those constructed with derivatives of the curvature tensors. For example,

$$R(0) = \frac{G\mu^3 M}{\sqrt{\pi}}, \qquad \Box R(0) = -\frac{G\mu^5 M}{2\pi}\left(2G\mu M + 3\sqrt{\pi}\right), \tag{4.140}$$

$$\Box^2 R(0) = \frac{G\mu^7 M}{12\pi^{3/2}}(45\pi - 52G^2\mu^2 M^2 + 138\sqrt{\pi}G\mu M), \tag{4.141}$$

in agreement with the Theorem 4.10. Similar relations hold for invariants constructed using the Riemann and Ricci tensors.

It is also worth mentioning that the metric (4.135) is consistent with the Newtonian approximation described in Sect. 4.5.3. Indeed, expanding the temporal part of (4.135) in powers of GM, at leading order we get $g_{tt} = -[1 + 2\varphi(r)]$, with $\varphi(r)$ given by (4.110).

Last but not least, although this example dealt with the case of the exponential form factor, other regular black hole solutions can be obtained from the combination of the results of Sect. 4.4 and this section. In fact, within the procedure described above, provided that the function $a(-k^2)$ grows at least as fast as k^4 for large k, the static spherically symmetric solutions of Eq. (4.114) are regular in the sense that its Kretschmann scalar is singularity-free.

4.7 Concluding Remarks

In this chapter, we showed that there are situations in which the regularity of a solution of some field equations can be anticipated by knowing the regularity properties of an effective source. This happens, for instance, in the cases of the Newtonian limit and the small-curvature approximation in higher-derivative gravity and in other models that lead to similar effective equations, such as noncommutative gravity, generalized uncertainty principle scenarios, string theory, and other approaches for a UV completion of gravity [6, 14, 20, 37, 44, 50, 54, 56–58, 68, 69].

In those cases, the theorems presented in Sects. 4.4 and 4.6 offer an immediate answer to the question of whether a particular modification of the form factor (or, more generally, of the delta source) can lead to regular spacetime configurations and to which extent this regularity is maintained as one takes into account invariants formed by curvatures and their covariant derivatives. It should be stressed, however, that the solutions of the effective equations considered here might not describe all the properties of the original field equations of the models. This assessment can only be done case by case through the detailed analysis of the model and the assumptions underlying the approximations involved in the effective equations.

Appendix

Here we list the main formulas needed in Sect. 4.3, following the expansion (4.11). For the metric inverse and determinant, we have

$$g^{\mu\nu} = \eta^{\mu\nu} - h^{\mu\nu} + h^{\mu\lambda}h^{\nu}_{\lambda} + O(h^3_{...}), \tag{4.142}$$

$$\sqrt{-g} = 1 + \frac{1}{2}h - \frac{1}{4}h^2_{\mu\nu} + \frac{1}{8}h^2 + O(h^3_{...}), \tag{4.143}$$

where $h = \eta^{\mu\nu}h_{\mu\nu}$. As explained in Sect. 4.3, to obtain the bilinear part in $h_{\mu\nu}$ of the action (4.10) we need the Riemann and Ricci tensors only in the first order,

$$R^\alpha{}_{\beta\mu\nu} = \frac{1}{2}\,(\partial_\mu\partial_\beta h^\alpha_\nu - \partial_\nu\partial_\beta h^\alpha_\mu + \partial_\nu\partial^\alpha h_{\beta\mu} - \partial_\mu\partial^\alpha h_{\beta\nu}) + O(h^2_{...})\,, \quad (4.144)$$

$$R_{\mu\nu} = \frac{1}{2}\,(\partial_\lambda\partial_\mu h^\lambda_\nu + \partial_\lambda\partial_\nu h^\lambda_\mu - \Box h_{\mu\nu} - \partial_\mu\partial_\nu h) + O(h^2_{...})\,, \quad (4.145)$$

and the scalar curvature up to $O(h^2_{...})$, because of the term linear in R in (4.10),

$$R = R^{(1)} + R^{(2)} + O(h^3_{...}), \quad (4.146)$$

where

$$R^{(1)} = \partial_\alpha\partial_\beta h^{\alpha\beta} - \Box h, \quad (4.147)$$

$$R^{(2)} = h^{\alpha\beta}(\Box h_{\alpha\beta} + \partial_\alpha\partial_\beta h - 2\partial_\alpha\partial_\lambda h^\lambda_\beta) + \frac{3}{4}\,\partial_\lambda h_{\alpha\beta}\partial^\lambda h^{\alpha\beta}$$
$$-\frac{1}{2}\,\partial_\alpha h_{\lambda\beta}\partial^\beta h^{\alpha\lambda} - (\partial_\rho h^\rho_\lambda - \tfrac{1}{2}\partial_\lambda h)(\partial_\sigma h^{\sigma\lambda} - \tfrac{1}{2}\partial^\lambda h). \quad (4.148)$$

With these expressions one can derive the quadratic part of the terms in the action. Integrating by parts and ignoring unimportant surface terms, we get

$$\left[\sqrt{-g}\,R\right]^{(2)} = \frac{1}{4}\,h^{\mu\nu}\Box h_{\mu\nu} - \frac{1}{4}h\Box h + \frac{1}{2}h^{\mu\nu}\partial_\mu\partial_\nu h - \tfrac{1}{2}\,h^{\mu\nu}\partial_\mu\partial_\lambda h^\lambda_\nu, \quad (4.149)$$

$$\left[\sqrt{-g}\,RF(\Box)R\right]^{(2)} = hF(\Box)\Box^2 h - 2h^{\mu\nu}F(\Box)\Box\partial_\mu\partial_\nu h$$
$$+h^{\mu\nu}F(\Box)\partial_\mu\partial_\nu\partial_\alpha\partial_\beta h^{\alpha\beta}\,, \quad (4.150)$$

$$\left[\sqrt{-g}\,R_{\mu\nu}F(\Box)R^{\mu\nu}\right]^{(2)} = \frac{1}{4}h_{\mu\nu}F(\Box)\Box^2 h^{\mu\nu} + \frac{1}{4}hF(\Box)\Box^2 h$$
$$-\frac{1}{2}h^{\mu\nu}F(\Box)\Box\partial_\mu\partial_\nu h - \frac{1}{2}h^{\mu\nu}F(\Box)\Box\partial_\mu\partial_\lambda h^\lambda_\nu$$
$$+\frac{1}{2}h^{\mu\nu}F(\Box)\partial_\mu\partial_\nu\partial_\alpha\partial_\beta h^{\alpha\beta}\,, \quad (4.151)$$

$$\left[\sqrt{-g}\,R_{\mu\nu\alpha\beta}F(\Box)R^{\mu\nu\alpha\beta}\right]^{(2)} = h_{\mu\nu}F(\Box)\Box^2 h^{\mu\nu} - 2h^{\mu\nu}F(\Box)\Box\partial_\mu\partial_\lambda h^\lambda_\nu$$
$$+h^{\mu\nu}F(\Box)\partial_\mu\partial_\nu\partial_\alpha\partial_\beta h^{\alpha\beta}. \quad (4.152)$$

References

1. A. Accioly, B.L. Giacchini, I.L. Shapiro, Low-energy effects in a higher-derivative gravity model with real and complex massive poles. Phys. Rev. D **96**, 104004 (2017). arXiv:1610.05260
2. A. Accioly, B.L. Giacchini, I.L. Shapiro, On the gravitational seesaw in higher-derivative gravity. Eur. Phys. J. C **77**, 540 (2017). arXiv:1604.07348
3. D. Anselmi, On the quantum field theory of the gravitational interactions. J. High Energy Phys. **06**, 086 (2017). arXiv:1704.07728

4. D. Anselmi, M. Piva, Perturbative unitarity of Lee-Wick quantum field theory. Phys. Rev. D **96**, 045009 (2017). arXiv:1703.05563
5. M. Asorey, J.L. López, I.L. Shapiro, Some remarks on high derivative quantum gravity. Int. J. Mod. Phys. A **12**, 5711 (1997). arXiv:hep-th/9610006
6. C. Bambi, L. Modesto, Y. Wang, Lee-Wick black holes. Phys. Lett. B **764**, 306 (2017). arXiv:1611.03650
7. K.J. Barnes, Lagrangian theory for the second-rank tensor field. J. Math. Phys. **6**, 788 (1965)
8. C.M. Bender, P.D. Mannheim, No-ghost theorem for the fourth-order derivative Pais-Uhlenbeck oscillator model. Phys. Rev. Lett. **100**, 110402 (2008). arXiv:0706.0207
9. C.M. Bender, P.D. Mannheim, Exactly solvable PT-symmetric Hamiltonian having no Hermitian counterpart. Phys. Rev. D **78**, 025022 (2008). arXiv:0804.4190
10. N.D. Birrell, P.C.W. Davies, *Quantum Fields in Curved Space* (Cambridge University Press, Cambridge, England, 1982)
11. T. Biswas, E. Gerwick, T. Koivisto, A. Mazumdar, Towards singularity and ghost free theories of gravity. Phys. Rev. Lett. **108**, 031101 (2012). arXiv:1110.5249
12. A. Bonanno, S. Silveravalle, Characterizing black hole metrics in quadratic gravity. Phys. Rev. D **99**, 101501 (2019). arXiv:1903.08759
13. A. Bonanno, S. Silveravalle, The gravitational field of a star in quadratic gravity. J. Cosmol. Astropart. Phys. **08**, 050 (2021). arXiv:2106.00558
14. J. Boos, Gravitational Friedel oscillations in higher-derivative and infinite-derivative gravity? Int. J. Mod. Phys. D **27**, 1847022 (2018). arXiv:1804.00225
15. J. Boos, V.P. Frolov, A. Zelnikov, Gravitational field of static p-branes in linearized ghost-free gravity. Phys. Rev. D **97**, 084021 (2018). arXiv:1802.09573
16. J.N. Borissova, A. Eichhorn, Towards black-hole singularity-resolution in the Lorentzian gravitational path integral. Universe **7**, 48 (2021). arXiv:2012.08570
17. I.L. Buchbinder, I.L. Shapiro, *Introduction to Quantum Field Theory with Applications to Quantum Gravity* (Oxford University Press, Oxford, England, 2021)
18. L. Buoninfante, B.L. Giacchini, Light bending by a slowly rotating source in quadratic theories of gravity. Phys. Rev. D **102**, 024020 (2020). arXiv:2005.05355
19. N. Burzillà, B.L. Giacchini, T. de Paula Netto, L. Modesto, Newtonian potential in higher-derivative quantum gravity. Phys. Rev. D **103**, 064080 (2021). arXiv:2012.06254
20. N. Burzillà, B.L. Giacchini, T. de Paula Netto, L. Modesto, Higher-order regularity in local and nonlocal quantum gravity. Eur. Phys. J. C **81**, 462 (2021). arXiv:2012.11829
21. Y.F. Cai, G. Cheng, J. Liu, M. Wang, H. Zhang, Features and stability analysis of non-Schwarzschild black hole in quadratic gravity. J. High Energy Phys. **01**, 108 (2016). arXiv:1508.04776
22. G. Calcagni, Taming the beast: diffusion method in nonlocal gravity. Universe **4**, 95 (2018); [erratum: Universe **5**, 121 (2019)]. arXiv:1809.04292
23. S.M. Christensen, M.J. Duff, Quantizing gravity with a cosmological constant. Nucl. Phys. B **170**, 480 (1980)
24. T. de Paula Netto, L. Modesto, I.L. Shapiro, Universal leading quantum correction to the Newton potential. Eur. Phys. J. C **82**, 160 (2022). arXiv:2110.14263
25. S. Deser, A.N. Redlich, String induced gravity and ghost freedom. Phys. Lett. B **176**, 350 (1986)
26. S. Deser, P. van Nieuwenhuisen, One-loop divergences of quantized Einstein-Maxwell fields. Phys. Rev. D **10**, 401 (1974)
27. S. Deser, R.P. Woodard, Nonlocal cosmology. Phys. Rev. Lett. **99**, 111301 (2007). arXiv:0706.2151
28. J.F. Donoghue, G. Menezes, Unitarity, stability and loops of unstable ghosts. Phys. Rev. D **100**, 105006 (2019). arXiv:1908.02416
29. A. Eddington, *The Mathematical Theory of Relativity* (Cambridge University Press, Cambridge, England, 1924)
30. J. Edholm, A.S. Koshelev, A. Mazumdar, Behavior of the Newtonian potential for ghost-free gravity and singularity-free gravity. Phys. Rev. D **94**, 104033 (2016). arXiv:1604.01989

31. V.P. Frolov, Mass-gap for black hole formation in higher derivative and ghost free gravity. Phys. Rev. Lett. **115**, 051102 (2015). arXiv:1505.00492
32. V.P. Frolov, G.A. Vilkovisky, Spherically symmetric collapse in quantum gravity. Phys. Lett. B **106**, 307 (1981)
33. V.P. Frolov, A. Zelnikov, Head-on collision of ultrarelativistic particles in ghost-free theories of gravity. Phys. Rev. D **93**, 064048 (2016). arXiv:1509.03336
34. V.P. Frolov, A. Zelnikov, T. de Paula Netto, Spherical collapse of small masses in the ghost-free gravity. J. High Energy Phys. **06**, 107 (2015). arXiv:1504.00412
35. B.L. Giacchini, On the cancellation of Newtonian singularities in higher-derivative gravity. Phys. Lett. B **766**, 306 (2017). arXiv:1609.05432
36. B.L. Giacchini, T. de Paula Netto, Weak-field limit and regular solutions in polynomial higher-derivative gravities. Eur. Phys. J. C **79**, 217 (2019). arXiv:1806.05664
37. B.L. Giacchini, T. de Paula Netto, Effective delta sources and regularity in higher-derivative and ghost–free gravity. J. Cosmol. Astropart. Phys. **1907**, 013 (2019). arXiv:1809.05907
38. B.L. Giacchini, T. de Paula Netto, L. Modesto, Action principle selection of regular black holes. Phys. Rev. D **104**, 084072 (2021). arXiv:2105.00300
39. M.H. Goroff, A. Sagnotti, Quantum gravity at two loops. Phys. Lett. B **160**, 81 (1985)
40. S.A. Hayward, Formation and evaporation of regular black holes. Phys. Rev. Lett. **96**, 031103 (2006). arXiv:gr-qc/0506126
41. B. Holdom, On the fate of singularities and horizons in higher derivative gravity. Phys. Rev. D **66**, 084010 (2002). arXiv:hep-th/0206219
42. B. Holdom, 2-2-holes simplified. Phys. Lett. B **830**, 137142 (2022). arXiv:2202.08442
43. B. Holdom, J. Ren, Not quite a black hole. Phys. Rev. D **95**, 084034 (2017). arXiv:1612.04889
44. M. Isi, J. Mureika, P. Nicolini, Self-completeness and the generalized uncertainty principle. J. High Energy Phys. **11**, 139 (2013). arXiv:1310.8153
45. N.V. Krasnikov, Nonlocal Gauge theories. Theor. Math. Phys. **73**, 1184 (1987) [Teor. Mat. Fiz. **73**, 235 (1987)]
46. Yu.V. Kuz'min, Finite nonlocal gravity. Sov. J. Nucl. Phys. **50**, 1011 (1989) [Yad. Fiz. **50**, 1630 (1989)]
47. H. Lü, A. Perkins, C.N. Pope, K.S. Stelle, Black holes in higher-derivative gravity. Phys. Rev. Lett. **114**, 171601 (2015). arXiv:1502.01028
48. H. Lü, A. Perkins, C.N. Pope, K.S. Stelle, Spherically symmetric solutions in higher-derivative gravity. Phys. Rev. D **92**, 124019 (2015). arXiv:1508.00010
49. M. Maggiore, Nonlocal infrared modifications of gravity. A review. Fundam. Theor. Phys. **187**, 221 (2017). arXiv:1606.08784
50. L. Modesto, Super-renormalizable quantum gravity. Phys. Rev. D **86**, 044005 (2012). arXiv:1107.2403
51. L. Modesto, Super-renormalizable or finite Lee-Wick quantum gravity. Nucl. Phys. B **909**, 584 (2016). arXiv:1602.02421
52. L. Modesto, L. Rachwał, Super-renormalizable and finite gravitational theories. Nucl. Phys. B **889**, 228 (2014). arXiv:1407.8036
53. L. Modesto, I.L. Shapiro, Superrenormalizable quantum gravity with complex ghosts. Phys. Lett. B **755**, 279 (2016). arXiv:1512.07600
54. L. Modesto, J.W. Moffat, P. Nicolini, Black holes in an ultraviolet complete quantum gravity. Phys. Lett. B **695**, 397 (2011). (arXiv:1010.0680)
55. L. Modesto, T. de Paula Netto, I.L. Shapiro, On Newtonian singularities in higher derivative gravity models. J. High Energy Phys. **04**, 098 (2015). arXiv:1412.0740
56. P. Nicolini, E. Spallucci, Noncommutative geometry inspired wormholes and dirty black holes. Class. Quant. Grav. **27**, 015010 (2010). arXiv:0902.4654
57. P. Nicolini, A. Smailagic, E. Spallucci, Noncommutative geometry inspired Schwarzschild black hole. Phys. Lett. B **632**, 547 (2006). (arXiv:gr-qc/0510112)
58. P. Nicolini, E. Spallucci, M.F. Wondrak, Quantum corrected black holes from string t-duality. Phys. Lett. B **797**, 134888 (2019). arXiv:1902.11242
59. A. Pais, G.E. Uhlenbeck, On Field theories with nonlocalized action. Phys. Rev. **79**, 145 (1950)

60. J. Podolsky, R. Svarc, V. Pravda, A. Pravdova, Explicit black hole solutions in higher-derivative gravity. Phys. Rev. D **98**, 021502 (2018). arXiv:1806.08209
61. I. Quandt, H.J. Schmidt, The Newtonian limit of fourth and higher order gravity. Astron. Nachr. **312**, 97 (1991). arXiv:gr-qc/0109005
62. R.J. Rivers, Lagrangian theory for neutral massive spin-2 fields. Nuovo Cimento **34**, 386 (1964)
63. K.S. Stelle, Renormalization of higher derivative quantum gravity. Phys. Rev. D **16**, 953 (1977)
64. K.S. Stelle, Classical gravity with higher derivatives. Gen. Rel. Grav. **9**, 353 (1978)
65. R. Svarc, J. Podolsky, V. Pravda, A. Pravdova, Exact black holes in quadratic gravity with any cosmological constant. Phys. Rev. Lett. **121**, 231104 (2018). arXiv:1806.09516
66. E.T. Tomboulis, Superrenormalizable gauge and gravitational theories. arXiv:hep-th/9702146
67. G. 't Hooft, M. Veltman, One loop divergencies in the theory of gravitation. Ann. Inst. H. Poincaré Phys. Theor. A **20**, 69 (1974)
68. A.A. Tseytlin, On singularities of spherically symmetric backgrounds in string theory. Phys. Lett. B **363**, 223 (1995). (arXiv:hep-th/9509050)
69. Y. Zhang, Y. Zhu, L. Modesto, C. Bambi, Can static regular black holes form from gravitational collapse? Eur. Phys. J. C **75**, 96 (2015). (arXiv:1404.4770)
70. V.A. Zorich, *Mathematical Analysis II* (Springer, Berlin, Heidelberg, 2004)

Chapter 5
Black Holes in Asymptotically Safe Gravity and Beyond

Astrid Eichhorn and Aaron Held

5.1 Invitation: Black Holes, Quantum Scale Symmetry and Probes of Quantum Gravity

In General Relativity (GR), black holes harbor several unphysical properties: first, they contain a curvature singularity. The singularity renders black-hole spacetimes geodesically incomplete. Second, they contain a Cauchy horizon, unless their spin is set exactly to zero, which is not very likely to be astrophysically relevant. The Cauchy horizon leads to a breakdown of predictive evolution.

Therefore, it is not a question whether General Relativity breaks down for black holes. It is only a question, where and how it breaks down and what it is substituted by.

In this chapter, we explore the possibility that an asymptotically safe quantum theory of gravity provides a fundamental description of gravity and we focus on potential consequences for black holes. Asymptotically safe quantum gravity is an attractive contender for a quantum theory of gravity, because it stays as close as possible to GR in the sense that (i) the gravitational degrees of freedom are carried

A. Eichhorn (✉)
CP3-Origins, University of Southern Denmark, Campusvej 55, 5230 Odense M, Denmark
e-mail: eichhorn@cp3.sdu.dk

A. Held
Theoretisch-Physikalisches Institut, Friedrich-Schiller-Universität Jena,
Max-Wien-Platz 1, 07743 Jena, Germany
e-mail: aaron.held@uni-jena.de

The Princeton Gravity Initiative, Jadwin Hall, Princeton University,
Princeton, NJ 08544, United States

C. Bambi (ed.), *Regular Black Holes*, Springer Series in Astrophysics and Cosmology,
https://doi.org/10.1007/978-981-99-1596-5_5

by the metric,[1] (ii) the rules of standard (quantum) field theories still apply and (iii) as a consequence of (i) and (ii), a continuum spacetime picture continues to hold, where spacetime is a differentiable, Lorentzian manifold with a metric. Therefore, it is expected that black holes exist in asymptotically safe gravity, instead of being substituted by black-hole mimickers without event horizon. This immediately gives rise to the following questions: (i) What is the fate of the classical curvature singularity? (ii) What is the fate of the Cauchy horizon? (iii) Are all effects of quantum gravity hidden behind the event horizon and therefore in principle undetectable? In short, as we will review in this chapter, the answers are: (i) the singularity may be resolved completely or weakened and geodesic completeness may be achieved; (ii) the Cauchy horizon's fate is unknown, but current models contain a Cauchy horizon, raising questions of instability; (iii) in asymptotically safe quantum gravity, singularity resolution goes hand in hand with effects at all curvature scales, with their size determined by an undetermined parameter of the theory. If this parameter, and thus the onset of observable deviations, are *not* tied to the Planck scale, current astrophysical observations can start to place constraints on black-hole spacetimes motivated by asymptotically safe quantum gravity.

The main idea that this chapter relies on, is that quantum scale symmetry, which is realized in asymptotically safe gravity, "turns off" the gravitational interaction at high curvature scales, resolving spacetime singularities. In fact, asymptotic safety is a theory that fits well into a highly successful paradigm in fundamental physics, namely the use of symmetries. In asymptotic safety, quantum scale symmetry determines the high-curvature regime. Scale symmetry implies that there cannot be any distinct physical scales. Quantum scale symmetry is a bit more subtle, because it need not be realized at all scales: at low energy (or curvature) scales, quantum scale symmetry is absent, and distinct physical scales exist (e.g., masses of various elementary particles). Conversely, beyond a critical high energy scale, quantum scale symmetry is present, and no distinct physical scales can exist in that regime.

Classically, gravity comes with a scale, namely the Planck mass. In asymptotically safe gravity, the Planck mass is present at low curvature scales (where quantum scale symmetry is absent), but is absent at high curvature scales (where quantum scale symmetry is present). This translates into a dynamical weakening of gravity at high curvature scales, suggesting a resolution of spacetime singularities.

In Sect. 5.2, we open our discussion with a heuristic argument why quantum scale symmetry should lead to resolution or weakening of singularities. In Sect. 5.3 we then discuss, how asymptotic-safety inspired spacetimes are constructed through the method of Renormalization Group improvement. We also elaborate on ambiguities and potential pitfalls of the method. In Sect. 5.4 we discuss the resulting spherically symmetric spacetimes and in Sect. 5.5 we focus on axisymmetric, asymptotic-safety inspired spacetimes, highlighting universal results, such as singularity resolution,

[1] Attempts at formulating asymptotically safe gravity in terms of, e.g., the vielbein [64], the vielbein and the connection [34], a unimodular metric [44] or a generalized connection [62] also exist; however, metric gravity is by far the most explored of these options, see [97, 106] for textbooks and [8, 46, 49, 92, 98, 102] for reviews. The other options all start from theories which are classically equivalent to GR.

which are independent of some of the ambiguities of the Renormalization Group improvement. In Sect. 5.6, we go beyond stationary settings and discuss gravitational collapse, as well as the formation of black holes in high-energy scattering processes. Finally, in Sect. 5.7 we connect to observations, opening with a discussion on the scale of quantum gravity, on which detectability of quantum-gravity imprints depends. In Sect. 5.8, we broaden our viewpoint beyond asymptotic safety and discuss, how Renormalization Group improved black holes fit into the principled-parameterized approach to black holes beyond GR. In Sect. 5.9 we summarize the current state of the art and point out the open challenges and future perspectives of the topic.

We aim to be introductory and pedagogical, making this review suitable for non-experts. Thus, all sections also contain "further reading" subsections, where we briefly point out literature that goes beyond our discussion.

5.2 A Heuristic Argument for Singularity Resolution from Quantum Scale Symmetry

...where we discuss how quantum scale symmetry, because it implies the absence of physical scales, forces the Newton coupling to vanish or the Planck mass to diverge at asymptotically small distances. The resulting weakening of gravity is expected to resolve (or at least weaken) classical curvature singularities. This can be viewed as a consequence of the dynamical decoupling of transplanckian degrees of freedom, which is expected to occur if quantum scale symmetry is realized in gravity.

Scale symmetry and the absence of scales

Quantum scale symmetry is a form of scale symmetry. Scale symmetry says that no distinct physical scales can exist, i.e., the theory must be invariant under scale transformations, which is a form of self-similarity. Starting off with a theory that contains distinct physical mass scales, i.e., which is not scale-symmetric, one can arrive at scale symmetry in two distinct ways: one can either set all masses to zero or to infinity. The difference between classical and quantum scale symmetry is that in a theory with classical symmetry, all mass scales are set to zero. In contrast, in a theory with quantum scale symmetry, which does not feature classical scale symmetry, the dimensionless counterparts of mass scales are finite.[2] Therefore, the dimensionfull masses diverge in the small-distance regime of a quantum scale symmetric theory, cf. Fig. 5.1. Here, we already assume that the theory is not quantum scale symmetric at all scales, but that quantum scale symmetry only determines the short-distance

[2] In technical language, we state here that a theory with quantum scale symmetry but without classical scale symmetry has at least one non-vanishing canonically relevant or irrelevant coupling and not just canonically marginal ones. Due to its canonical dimension, the coupling implicitly defines a mass scale.

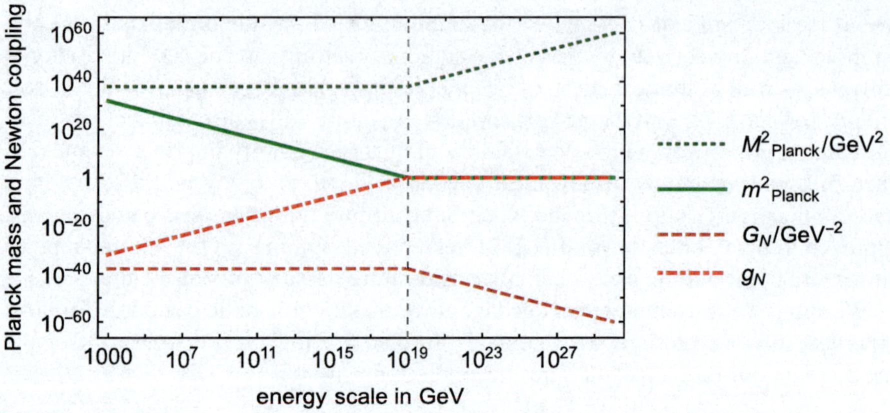

Fig. 5.1 We show the dimensionfull Planck mass M_{Planck}, which is constant in the classical gravity regime below 10^{19} GeV, and scales quadratically above. The dimensionless Planck mass m_{Planck} becomes constant there, as does the dimensionless Newton coupling g_N, whereas the dimensionfull Newton coupling G_N decreases quadratically. Thus, quantum scale symmetry translates into a dynamical decoupling of gravity, because the associated mass scale diverges and the corresponding interaction vanishes. As a consequence, gravitational modes beyond the classical Planck scale are dynamically removed from the theory. We argue that therefore curvature invariants are limited in asymptotically safe gravity and the limiting value is determined by the classical Planck scale, which is the transition scale to the quantum scale invariant regime

behavior. In contrast, above a critical distance scale (or below a critical mass scale), the theory leaves the quantum scale symmetric regime and distinct physical scales (e.g., masses for elementary particles) emerge.

Quantum scale symmetry has consequences for the degrees of freedom of the theory: simply put, because mass scales diverge, one may expect degrees of freedom to decouple. If a theory exhibits quantum scale symmetry only above a critical mass scale, then only those degrees of freedom at higher energies decouple.[3]

Interlude: Terminology

On a technical level, asymptotic safety is a statement about the scale-dependence of couplings in the action of a quantum field theory (QFT). Formally, this scale dependence is described by the flow with Renormalization Group (RG) scale k and affects not just the Newton coupling G_N, but all possible couplings C_i allowed by the fundamental symmetries of the action. The scale-dependence of their dimensionless counterparts $c_i = C_i\, k^{-d_i}$ is governed by the β-functions of the QFT, i.e., by

$$k\partial_k\, c_i(k) = \beta_{c_i}(\{u_j\}) \,. \tag{5.1}$$

[3] There are exceptions to this argument, namely when a theory has vanishing mass parameters and the diverging mass scale is related to a relevant interaction. In that case quantum scale symmetry corresponds to a strongly-coupled regime.

Asymptotic safety occurs at fixed points of the RG flow corresponds at which

$$\beta_{c_i}(\{u_{j*}\}) = 0\,, \tag{5.2}$$

hence the flow vanishes and quantum scale symmetry is achieved. If the respective QFT exhibits such an asymptotically safe scale-invariant regime at high energies, the limit $k \to \infty$ can be taken without divergences and the respective QFT is fundamental.

Of the couplings in the theory, only those that are RG relevant correspond to free parameters. Those are the couplings for which quantum fluctuations favor a departure from scale symmetry. If an asymptotically safe theory has only a finite number of such relevant parameters (for which there is a general argument), then the theory is UV finite with just a finite number of free parameters that determines the low-energy physics. This generalizes the perturbative notion of renormalizability (which is not available for Einstein gravity) to a nonperturbative form of renormalizability.

Quantum scale symmetry and weakening of gravity

In gravity, this is relevant for the Newton coupling. Building a classically scale symmetric theory of the gravitational field, one would put the Planck mass to zero. In a quantum scale symmetric theory, the dimensionless Newton coupling g_N becomes constant, as does the dimensionless Planck mass. The dimensionfull Newton coupling G_N thus scales quadratically with the inverse distance, and the dimensionfull Planck mass scales quadratically with the distance, see Fig. 5.1. The transition scale to the scale-symmetric regime is expected to be the classical Planck mass, see Sect. 5.7.1 for potential caveats to this expectation.

A diverging Planck mass means that the gravitational interaction decouples. The same can be inferred from the Newton coupling going to zero: the gravitational interaction becomes weaker at lower distances, such that the small-distance limit of quantum gravity is one where all degrees of freedom decouple. This ensures that divergences in the theory, be they in scattering cross-sections or in spacetime curvature invariants, are softened or even removed completely. For scattering cross-sections, the idea was originally put forward by Weinberg, who suggested that asymptotic safety ensures finiteness of cross-sections.

In turn, quantum scale symmetry, or the decoupling of the high-energy degrees of freedom, has consequences for the form of the solutions: because high-energy modes are absent from the theory, there is a limit to the maximum value of curvature one can reach: if we decompose the metric into eigenmodes of the Laplacian (with respect to some background metric), then all those eigenmodes with eigenfrequencies above a critical value decouple dynamically, and only the low-frequency modes are still present. Accordingly, if we consider a spacetime, which we write in terms of the basis of eigenmodes of the Laplacian, quantum scale symmetry implies that the high-frequency modes must be removed. Therefore, the curvature cannot exceed a limiting value, which is set by the transition scale to the scale-symmetric regime.

Singularity resolution from destructive interference in the path integral
There is a second, independent argument to support the expectation that asymptotically safe quantum gravity lifts curvature singularities. The argument was first made in a cosmological context [82] and then adapted to the black-hole context [16]. It is based on the Lorentzian path integral

$$Z = \int \mathscr{D}g_{\mu\nu}\, e^{i S_{\mathrm{AS}}[g_{\mu\nu}]}, \tag{5.3}$$

in which spacetime configurations interfere destructively with "nearby" configurations, if their action diverges. It is expected that a Riemann-squared (or, equivalently, Weyl-squared) term (as well as higher-order terms in the Riemann tensor) are present in S_{AS}, see, e.g., [4, 58, 61, 75]. These terms diverge when evaluated on classical, singular black-hole spacetimes. In turn, this results in destructive interference of such singular spacetimes with "neighboring" configurations. Therefore, one may expect that singular black-hole geometries are not part of the asymptotically safe path integral.[4] As a result, the spacetime structure of black holes in asymptotically safe quantum gravity may be expected to be regular.

In this argument, scale symmetry does not explicitly appear and one may thus wonder whether the two arguments we have given for singularity resolution are related. They are related, because the presence of higher-order curvature terms appear to be required by scale symmetry; a theory with just the Einstein-Hilbert action is not asymptotically safe, according to the state-of-the-art in the literature.

Further Reading

- **State of the art in asymptotically safe quantum gravity:**
  Recent reviews on asymptotically safe quantum gravity are [8, 48, 92, 98, 102]. [48, 102] are lecture notes that serve as an introduction to the topic. The books [97, 106] provide introductions to the topic and in-depth expositions of recent results. [98] reviews asymptotically safe quantum gravity and explains the tentative connection to tensor models, which, if they have a universal continuum limit, could be in the same universality class, i.e., have a continuum limit that yields the physics of asymptotically safe gravity. Reference [92] focuses on the distinction between the background metric and the fluctuation field, which is introduced by the gauge fixing as well as the infrared regulator term. Reference [8] addresses open challenges of asymptotically safe gravity, also responding to the questions raised in [41]. Asymptotically safe gravity-matter systems are reviewed in [47] and an update is provided in [49].

[4] Incidentally, this also implies that a "folk theorem" about the violation of global symmetries from quantum gravity may not be applicable to asymptotically safe quantum gravity.

5.3 Constructing Asymptotic-Safety Inspired Spacetimes

...where we explain how a classical theory or its solutions can be upgraded to asymptotic-safety inspired solutions by accounting for the scale-dependence of the theories' couplings. In this RG-improvement procedure, the key physics input is the identification of the RG scale – which determines up to which length scale quantum effects are present – with a scale of the system, e.g., a curvature scale.

Physically, RG improvement is based on a decoupling of modes in the path integral, that occurs, when physical scales act as a low-energy cutoff in the path integral.

Ideally, to derive the structure of black holes in asymptotic safety, the following program should be implemented: first, the asymptotically safe fixed point must be determined with sufficiently high precision and in a Lorentzian regime. Both of these points are open challenges, although progress towards precision is being made, see, e.g., [58, 59], as well as progress towards Lorentzian signature, see, e.g., [60].

Second, the effective action of the theory must be calculated, such that all quantum fluctuations are integrated out. The resulting effective action will contain various higher-order curvature operators, the coefficients of which are functions of the free parameters (the relevant couplings) of the theory.

Third, the effective equations of motion must be derived by applying the variational principle to the effective action. Finally, black-hole solutions to these effective equations of motion must be found.

This program is currently not (yet) feasible. Instead, a simpler program, based on the method of RG improvement, is implemented, which does not strictly derive the black-hole solutions in asymptotic safety. Instead, it gives rise to asymptotic-safety inspired black holes. Generally, it is expected that these asymptotic-safety inspired black holes may capture some of the salient features of black holes in full asymptotic safety.

5.3.1 Renormalization-Group Improvement and the Decoupling Mechanism

...where we review two classic flat-space examples of RG improvement.

RG improvement is a method in which a classical solution of the theory is taken; the coupling constants in that solution are replaced by their running counterparts and the Renormalization Group scale is replaced by a physical scale of the solution that is studied.

The expectation that RG improvement captures the effects of a scale-invariant regime in quantum gravity is based on the success of RG improvement in flat-space quantum field theories, which we briefly review here.

The method of RG improvement has been applied to many field theories. The case of massless quantum electrodynamics is particularly instructive and has, therefore, been used in [13] as an analogy to motivate the first application of RG-improvement in the context of black-hole spacetimes. In this case, the electromagnetic coupling e is the analogue of the Newton coupling G_N. The Coulomb potential

$$V_{\text{cl}} = e^2/(4\pi r) \tag{5.4}$$

serves as a simple flat-space analog of the gravitational potential. First, one replaces the electromagnetic coupling e by its RG-scale dependent counterpart (evaluated at one loop in perturbation theory)

$$e^2(k) = e^2(k_0)\frac{1}{1 - \frac{e^2(k_0)}{6\pi^2}\ln(k/k_0)}, \tag{5.5}$$

where $e^2(k_0)$ is the low-energy value and k_0 an infrared scale, corresponding to an inverse distance $k_0 = r_0^{-1}$. Next, the RG scale is identified with the only physical scale in the classical potential, which is the distance r between the two charged particles, i.e., $k \sim 1/r$. As a result, one recovers the well-known Uehling correction to the Coulomb potential, as obtained by a calculation of the one-loop effective potential [115],

$$V_{\text{RG-improved}}(r) = \frac{e^2(r_0^{-1})}{4\pi r}\left(1 + \frac{e^2(r_0^{-1})}{6\pi^2}\ln\left(\frac{r_0}{r}\right) + \mathcal{O}(e^4)\right), \tag{5.6}$$

where a series expansion in small $e^2(r_0^{-1})$ is done in the last step.

Beyond this simple example, an extensively studied case is the RG improvement of the scalar potential [32], e.g., for the Higgs scalar in the Standard Model, but also beyond. It is an instructive case, because the RG improvement is done not in terms of some external, fixed scale, as for the electrodynamic potential. Instead, the field itself is used as a scale. This is much closer in spirit to what we will do in gravity, where the physical scale necessarily depends on the field, i.e., the metric. Here, the classical scalar potential for the massless scalar ϕ

$$V_{\text{scalar cl}}(\phi) = \frac{\lambda}{4}\phi^4 \tag{5.7}$$

is RG-improved in the same two-step procedure. First, the classical coupling constant λ is replaced by its RG-scale dependent counterpart $\lambda(k)$, which at one loop reads

$$\lambda(k) = \frac{\lambda(k_0)}{1 - \frac{9}{8\pi^2}\ln\left(\frac{k}{k_0}\right)}. \tag{5.8}$$

Second, the RG-scale k is identified with the field value itself, i.e., $k \sim \phi$. The result captures the leading-order logarithmic quantum corrections to the classical potential. The scale identification $k \sim \phi$ is unambiguous in the massless case, because the field ϕ is the only dimensionful quantity in the classical potential.

In the above, we have used k for the RG scale. This notation is geared towards the functional Renormalization Group, where the scale k appears as an infrared cutoff in the path integral.[5] Field configurations with momenta below k are thereby suppressed. Thus, k initially serves as a "book-keeping" scale that sorts the field configurations in the path integral and enables a "step-wise" calculation of the path integral. As exposed in detail in [18], k acquires a physical meaning in settings with physical IR scales. For instance, in the above example of the Coulomb potential, field configurations with momenta lower than the inverse distance between the charges are not among the virtual quanta that are being exchanged. Thus, $r^{-1} = k$ acts as an IR cutoff. Similarly, in the case of the RG-improved Higgs potential, the propagator around any nonzero field value ϕ acquires an effective mass term $\sim \lambda\phi^2$, which suppresses quantum fluctuations with momenta lower than this mass.

More generally, the success of RG improvement is based on the decoupling of low-energy modes: if a physical scale s is present that acts as an IR cutoff, then one does not need to evaluate the path integral to $k < s$, but can stop at $k \approx s$. The effect of quantum fluctuations is then encoded in the scale-dependence of the couplings on k, which is equated to s. Thereby, if s is high, only UV modes contribute, whereas, if s is low, most modes in the path integral contribute.

To apply the method of RG improvement to gravitational solutions – black holes in particular – the starting point is no longer a scalar potential but a classical spacetime which is a solution to GR. This brings several added challenges, which are related to the fact that these solutions are typically expressed in a covariant form (i.e., as a spacetime metric) and not in an invariant form; and that the scales of the system usually involve the field itself, i.e., are constructed from the metric. Before turning to specific applications, we review (i) the scale dependence and (ii) the choice of scale identification. Finally, we also discuss pitfalls and, in particular, the altered role of coordinate-dependence.

5.3.2 *Scale Dependence of Gravitational Couplings*

...where we review the scale dependence of the gravitational coupling.

In analogy to the scale dependence of the electromagnetic coupling $e(k)$ in the previous section, RG improvement of gravitational systems starts by the replacement of classical gravitational couplings with their scale-dependent counterparts. In explicit

[5] In the above, one-loop examples, the scale-dependence of couplings on k agrees with that on μ, the RG scale that appears in perturbative renormalization.

calculations, we focus on the Newton coupling G_N and on the simplest possible interpolation between the quantum scale-invariant regime in the UV and the "classical" regime in the IR,[6] cf. Sect. 5.2 and Fig. 5.1, i.e.,

$$G_N(k) = \frac{G_0}{1 + (G_0/g_{N,*})\, k^2}. \tag{5.9}$$

Indeed, this scale dependence can formally be derived as an approximate solution to the Einstein-Hilbert truncation of the functional RG flow between asymptotic safety and a classical low-curvature regime [14]. Here, G_0 denotes the measured low-curvature value of the dimensionful Newton coupling. This value is approached for $k \to 0$. In the limit $k \to \infty$, the interpolation scales like $G_N(k) \sim g_{N,*}/k^2$. The coefficient $g_{N,*}$ is the fixed-point value of the dimensionless Newton coupling $g_N = G_N\, k^2$ in the asymptotically safe regime. It also determines the transition scale between classical scaling $G_N = \text{const}$ and asymptotically safe scaling $G_N \sim k^{-2}$. This transition scale is typically expected to coincide with the Planck scale. However, because asymptotic safety has further relevant parameters, the onset of quantum gravity effects may well be shifted away from the Planck scale, towards lower energies. We will discuss this possibility more extensively in Sect. 5.7, where it motivates a comparison of RG-improved black holes with observations, while treating $g_{N,*}$ as a free parameter, to be constrained by observations.[7]

One typically includes only the scale-dependence of the Newton coupling.[8] One may therefore expect that the RG improvement captures leading-order quantum effects in the semi-classical regime, but might no longer be fully adequate in the deep quantum regime. In other words, such asymptotic-safety inspired black-hole spacetimes may work well for comparisons with observations, which access horizon-scale physics. In contrast, the deep interior of such asymptotic-safety inspired black holes may still feature pathologies (Cauchy horizons, unresolved singularities, wormholes) which may only be resolved only once further couplings are included.

---

[6] We put "classical" in quotation marks, because it is not the $\hbar \to 0$ regime – in nature, $\hbar \to 0$ is at best an approximately observable limit – instead, the IR is the setting in which all quantum fluctuations in the path integral are present, and we are simply probing the effective action of the theory in its low-curvature regime.

[7] Note that such constraints on $g_{N,*}$ should not be misinterpreted as actual constraints on the fixed-point value. The latter is a non-universal quantity that is not directly accessible to measurements. It makes its way into observable quantities in the RG improvement procedure, because the procedure is an approximate one. Observations constrain a physical scale, namely the scale at which fixed-point scaling sets in. This scale depends on various couplings of the theory, but in our simple approximation it is only set by the fixed-point value of the Newton coupling.

[8] More complete studies with higher-order gravitational couplings would have to start from black-hole solutions in corresponding higher-order theories, which are typically not known at all or only for small or vanishing spin.

5.3.3 *Scale Identification for Gravitational Solutions*

...where we review the different choices for the scale identification in the RG improvement of gravitational solutions.

The RG scale dependence of the Newton coupling is a robust result of asymptotic safety. The other physical input for the RG-improvement – namely the scale identification – is less robust, because one may a priori identify the RG scale with quite distinct notions of scale. In the context of quantum-improved black-hole spacetimes, several different scale identifications have been put forward in the literature. In order to produce meaningful results, it is key that k^2 is identified with an invariant quantity (and not, e.g., a coordinate distance). We will discuss some possible choices below and compare them throughout the following applications.

- The arguably most obvious scale identification of the RG scale k is with a curvature scale. Such a scale identification follows the expectation that quantum-corrections to black holes are larger, the larger the local curvature scale. One thus identifies k^2 with the local value of a curvature scalar K_i such that

$$k^2 = \tilde{\xi} K_i^{n_i} \; . \tag{5.10}$$

Herein, $\tilde{\xi}$ determines a dimensionless number of order 1. The index i indicates that there can be multiple inequivalent curvature scalars. The respective exponent n_i is uniquely fixed by dimensional analysis, because no other scale should enter the identification Eq. (5.10). In highly symmetric situations, such as for Schwarzschild spacetime, all choices of curvature scalars $K_i \equiv K \; \forall i$ are equivalent, see also Sect. 5.4 below.
For cases with multiple distinct curvature scalars, one may identify the RG scale with the root mean square of all distinct curvature scalars, i.e.,

$$k^2 = \tilde{\xi} \left(\sum_i (K_i^2)^{n_i} \right)^{1/2} \; . \tag{5.11}$$

This also solves another potential issue of Eq. (5.10) which arises if curvature scalars change sign. Such a sign-change happens, for example, in the near-horizon region of the Kerr spacetime, cf. Sect. 5.5. A scale identification like Eq. (5.10) can thus result in complex-valued k^2, which is not possible for the RG scale. Taking an absolute value in Eq. (5.10) avoids this, but introduces non-smooth behavior in k and thus eventually in the RG-improved solution. The scale identification in Eq. (5.11) avoids both issues since the sum of all curvature scalars only vanishes in the approach to an asymptotically flat region of the spacetime.
By plugging in the explicit expression for $\left(\sum_i (K_i^2)^{n_i}\right)^{1/2}$ in terms of the coordinates x^μ, one obtains an expression $k^2 = k^2(x^\mu)$. For instance, for Schwarzschild spacetime, one obtains $k^2 \sim r^{-3}$. Because the spacetime may not be geodesically

complete for $r \geq 0$, one has to consider negative r as well. The correct identification (which ensures that $k^2 \geq 0$ always holds), is then in terms of $|r|$, not r.

- Another scale identification is with the (radial) geodesic distance d from the center of the black hole to an asymptotic observer, i.e.,

$$k^2 = \tilde{\xi} \frac{1}{d^2} . \tag{5.12}$$

For Schwarzschild spacetime, this assignment is unique, but for less symmetric settings such as Kerr spacetime it depends on the geodesic path connecting the center of the spacetime to an asymptotic observer. Another key difference to the scale identification with local curvature is that the geodesic distance d is a nonlocal quantity of the classical spacetime. This identification is closest in spirit to the RG-improvement of the Coulomb potential discussed in Sect. 5.3.1. However, it is less obvious that the geodesic distance of a (timelike) observer sets an IR cutoff in the path integral.
For this scale identification, similar considerations regarding $r < 0$ apply as for the first scale identification.

- Finally, in the presence of matter fields (e.g., to describe gravitational collapse), one can identify the RG scale with the trace of the stress-energy tensor [11, 12], i.e.,

$$k^4 = \tilde{\xi} g^{\mu\nu} T_{\mu\nu} . \tag{5.13}$$

For matter models with vanishing pressure, this scale identification reduces to $k^4 \sim \rho$, with $\rho = T^{00}$ the energy density. Similarly, in settings with a traceless energy-momentum tensor, one may choose T^{00} for the identification. In contrast to the trace $g^{\mu\nu} T_{\mu\nu}$, however, this is not a coordinate independent scale identification because T^{00} does not transform like a scalar.

Which scale identification is most suited may ultimately also depend on the questions one is interested in and which physical assumptions one wants to implement. For instance, a scale identification with temperature (as in [17]) appears most "natural" when one is interested in thermal properties of a black hole. In contrast, a scale identification with local curvature scalars matches the effective-field-theory assumption that all quantum-gravity modifications are tied to local curvature scales.
All scale identifications follow from dimensional analysis under the assumption that no second scale is relevant, i.e., one identifies the physical scale with an appropriate power of k, with only ξ allowed as the constant of proportionality. These dimensionless factors are either treated as free parameters, or determined by additional considerations: For instance, [14] argues to fix these dimensionless factors by matching to perturbative calculations in the context of effective field theory [39, 40]. Generically one may expect these dimensionless factors to be of order unity.

In spite of these differing choices, there are some results about RG improved black holes that are universal, whereas others depend on the scale identification. We will highlight both below.

Further Reading

- **RG improvement based on Hawking temperature:**
In the spirit of black-hole thermodynamics, it has been advocated [17] to use the temperature T of a black hole, i.e., to identify

$$k = \tilde{\xi}\, k_B\, T\ . \tag{5.14}$$

with k_B the Boltzmann constant. Such a scale identification follows the intuition from thermal field theory where the temperature can act akin to an infrared cutoff.
- **RG improvement of equations of motion or action:**
Departing from the more standard method of RG improvement at the level of the classical solution (i.e., the potential in electrodynamics and the black-hole geometry in the gravitational case), one may also RG improve at the level of the equations of motion [6, 78] or even the action [17, 104, 105]. There is an expectation that, if results from all three methods agree, the result is robust, see, e.g., [18]. However, we caution that there is no a priori reason to expect that RG improving the action gives a result that is closer to the full result than if one RG improves the solution.
- **Coordinate dependence of the RG improvement procedure:**
One potential pitfall in the RG-improvement of gravitational systems is coordinate dependence. This added difficulty can be understood by comparing the RG-improvement of the Coulomb potential in electromagnetism, cf. Sect. 5.3.1 and the RG-improvement of gravitational spacetimes: For the flat-space RG-improvement, the Coulomb potential as a starting point is itself a coordinate-invariant (scalar) quantity. In contrast, for the gravitational RG-improvement, the metric as a starting point does not transform as a scalar under coordinate transformations. This leads to a dependence of the RG improvement on the choice of coordinates in the classical metric [66]. For instance, using coordinates in which a spinning black-hole metric has coordinate singularities, RG improvement may lead to curvature singularities which are not present in other horizon-penetrating coordinate choices, cf. Sect. 5.5. One potential remedy is to perform the RG-improvement at the level of quantities which transform as scalars under coordinate transformations, see [66].

5.4 Spherically Symmetric Black Holes

...where we analyze the properties of asymptotic-safety inspired spherically symmetric spacetimes, focusing on the fate of the classical singularity and of the inner and outer event horizon. We also review results on black-hole thermodynamics.

5.4.1 *Construction of Asymptotic-Safety Inspired Black Holes*

...where we construct the RG improved spherically symmetric line element and discuss the universality of different scale identifications.

To obtain an asymptotic-safety inspired, spherically symmetric black hole, we start with the Schwarzschild metric in Schwarzschild coordinates t, r, θ, ϕ,

$$ds^2 = -f(r)dt^2 + f(r)^{-1}dr^2 + r^2 d\Omega^2, \tag{5.15}$$

with

$$f(r) = 1 - \frac{2G_N M}{r}. \tag{5.16}$$

The high degree of symmetry ensures that the choice of Schwarzschild coordinates is equivalent to the choice of horizon-penetrating Eddington-Finkelstein coordinates, cf. [66] for an explicit discussion.

We follow the two-step procedure of RG-improvement as outlined in Sect. 5.3: First, we upgrade the classical Newton coupling G_N to its RG-scale dependent counterpart $G_N(k)$, as given in Eq. (5.9). Second, we fix the scale identification. As we will see, the local scale identification with curvature (cf. Eq. 5.11 and [13]), and the non-local scale identification with the radial geodesic distance (cf. Eq. (5.12) and [14]) are equivalent. This equivalence is, once again, due to the high degree of symmetry of the Schwarzschild geometry. Therefore, there are universal results for RG improved spherically symmetric results.

To make the scale identification with local curvature explicit, we determine the curvature invariants. When evaluated in Schwarzschild spacetime, all curvature invariants built from the Ricci tensor vanish and all others are equivalent to the Kretschmann scalar

$$K = R_{\mu\nu\kappa\lambda}R^{\mu\nu\kappa\lambda} = \frac{48 G_N^2 M^2}{r^6} . \tag{5.17}$$

The Kretschmann scalar is manifestly positive. Hence, no average or absolute value needs to be taken and we find the scale identification

$$k^2 = \tilde{\xi}\sqrt{K} = \xi \frac{G_0 M}{\sqrt{r^6}} , \tag{5.18}$$

where we have included an a priori unknown constant of proportionality $\tilde{\xi}$.

Accordingly, the RG improved line-element is given by Eq. (5.15) with

$$f(r) = 1 - \frac{2G_N(k)\,M}{r} = 1 - \frac{2G_0\,M}{r\left(1 + \xi\frac{G_0^2 M}{\sqrt{r^6}}\right)} , \tag{5.19}$$

where we have absorbed the fixed-point value for the dimensionless Newton coupling $g_{N,*}$, cf. Eq. (5.9) and the constant of proportionality $\tilde{\xi}$ arising in the scale identification, cf. Eq. (5.18), into a single dimensionless constant

$$\xi = \sqrt{48\tilde{\xi}}/g_{N,*}. \tag{5.20}$$

Incidentally, this agrees with the mass function of a Hayward black hole [65] for $r > 0$.

The free parameter ξ determines the scale at which quantum-gravity effects become sizable: the larger ξ, the lower the curvature scale at which quantum-gravity effects are sizeable. For $\xi = 1$, quantum-gravity effects effectively set in at the Planck scale, because $f(r) \approx 1 - 2G_0M/r$ for any r greater than the Planck length. For astrophysical black holes, the curvature radius at the horizon is far above the Planck length. Thus, $\xi \approx 10^{95}$ would be required to achieve $\mathcal{O}(1)$ modifications at horizon scales [67]. (This estimate assumes a supermassive black holes with mass $M \sim 10^9 M_{\odot}$.) As we discuss in Sect. 5.7.1, such large choices of ξ may be relevant for black holes.

In the following, we analyze the above RG-improved Schwarzschild spacetime from the inside out: we start at the core and investigate the fate of the curvature singularity and of geodesic completeness; then we move on to the inner and outer horizon and the photon sphere.

5.4.2 *Singularity-Resolution*

...where we show that the asymptotic-safety inspired spherically symmetric black hole is regular with finite curvature invariants and we also discuss the status of geodesic completeness.

The asymptotic-safety inspired spherically symmetric black hole does not have a curvature singularity. This can be seen by calculating the curvature invariants[9] of the RG-improved metric. The RG-improved spacetime is no longer a vacuum solution of GR[10] and thus curvature invariants involving the Ricci tensor can be non-vanishing. In a general 4-dimensional spacetime, there are up to 14 curvature invariants which can be polynomially independent. To capture all possible (degenerate) cases at once, a complete set of 17 curvature invariants is required [25, 26, 122]. (At most) four of these are polynomially independent for the RG-improved spacetime at hand [66, App. D]. These four polynomially independent invariants may be chosen as R,

[9] We use the term 'curvature invariants' to refer to Riemann invariants, i.e., scalars built from contractions of any number of Riemann tensors and the metric. This does not include derivative invariants, i.e., those which involve additional covariant derivatives.

[10] Heuristically, one can think of the non-zero energy-momentum tensor as a contribution on the *left-hand-side* of the generalized Einstein equations, i.e., a contribution from higher-order curvature terms which arise in asymptotically safe gravity.

$R_{\mu\nu}R^{\mu\nu}$, $C_{\mu\nu\rho\sigma}C^{\mu\nu\rho\sigma}$, and $R_{\mu\nu}R_{\rho\sigma}C^{\mu\nu\rho\sigma}$, where R is the Ricci scalar, $R_{\mu\nu}$ is the Ricci tensor and $C_{\mu\nu\rho\sigma}$ is the Weyl tensor. They take the form

$$R = \frac{2M\left(r\,G_N'' + 2\,G_N'\right)}{r^2}, \tag{5.21}$$

$$R_{\mu\nu}R^{\mu\nu} = \frac{2M^2\left(r^2\left(G_N''\right)^2 + 4\left(G_N'\right)^2\right)}{r^4}, \tag{5.22}$$

$$C_{\mu\nu\rho\sigma}C^{\mu\nu\rho\sigma} = \left(\frac{2M\left(r\left(rG_N'' - 4G_N'\right) + 6G_N\right)}{\sqrt{3}r^3}\right)^2, \tag{5.23}$$

$$R_{\mu\nu}R^{\mu\nu}C^{\mu\nu\rho\sigma} = \frac{1}{3}\left(\frac{M\left(rG_N'' - 2G_N'\right)}{r^2}\right)^2\left(\frac{2M\left(r\left(rG_N'' - 4G_N'\right) + 6G_N\right)}{\sqrt{3}r^3}\right). \tag{5.24}$$

Primes denote radial derivatives, i.e., $G_N'(r) = \partial G_N(r)/\partial r$ and $G_N''(r) = \partial^2 G_N(r)/\partial r^2$. From these expressions, we can draw conclusions about the presence/absence of a curvature singularity, depending on the behavior of the scale-dependent Newton coupling near $r = 0$, i.e., depending on the leading exponent n in $G_N(r) \sim r^n + \mathcal{O}(r^{n+1})$:

- If $n = 3$, the curvature invariants take finite values at $r = 0$.
- If $n > 3$, all curvature invariants vanish and the spacetime is flat in the center.
- If $n < 3$, the curvature singularity remains, although it is weakened for $0 < n < 3$ compared to the Schwarzschild case.

From Eq. (5.18), we see that the RG-improved Newton coupling realizes the critical behavior, i.e., scales like $G_N(r) \sim r^3 + \mathcal{O}(r^4)$ for small r. In other words, RG-improvement suggests that asymptotically safe quantum gravity limits the maximal value that each curvature invariant can attain, and therefore these invariants tend to universal, i.e., mass-independent, limits,

$$R \overset{r\to 0}{\to} \frac{24}{G_0\xi}, \quad R_{\mu\nu}R^{\mu\nu} \overset{r\to 0}{\to} \frac{144}{G_0^2\xi^2}, \quad C_{\mu\nu\rho\sigma}C^{\mu\nu\rho\sigma} \overset{r\to 0}{\to} 0, \quad R_{\mu\nu}R^{\mu\nu}C^{\mu\nu\rho\sigma} \overset{r\to 0}{\to} 0, \tag{5.25}$$

see also Fig. 5.2. This is exactly what one expects from the dynamical decoupling of the high-energy modes of the metric. It implies that those modes with high eigenvalues of a (suitably chosen) Laplacian, which must be present in order to generate high values of the curvature, are absent. Therefore, the curvature radius is limited, and the limiting value is determined by $1/\sqrt{G_0\xi}$, i.e., by the quantum-gravity scale.

In contrast, the black-hole mass determines the radius at which effects become sizable: At fixed ξ, the departure of R from zero becomes of order one at smaller r/r_S, if M is made smaller, with r_S the classical Schwarzschild radius.

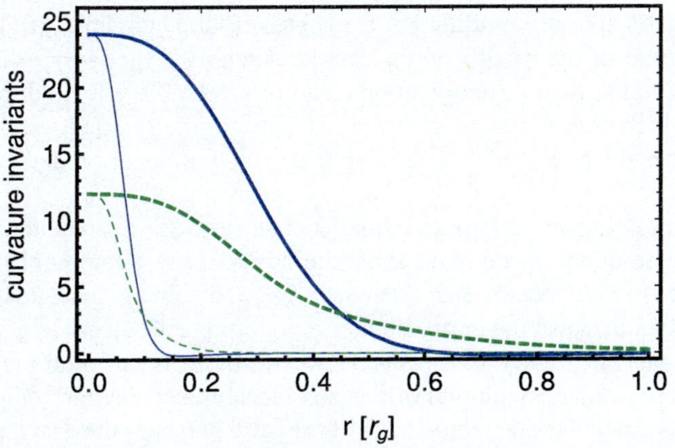

Fig. 5.2 We show the behavior of the curvature scalars R (blue continuous) and $\sqrt{R_{\mu\nu}R^{\mu\nu}}$ (green dashed) on the RG-improved Schwarzschild geomerty, cf. Eq. (5.18). The thick lines are for $M = 1$, the thin lines for $M = 10$; in all cases $G_0 = 1$ and $\xi = 1$

We conclude that RG-improvement suggests that an asymptotically safe scaling regime resolves the curvature singularity at the center of Schwarzschild spacetime.

Resolution of curvature singularities does not imply geodesic completeness, because the two are independent requirements. It nevertheless turns out that the RG improved black hole is geodesically complete as well. This can be seen by following [125], where it is pointed out that the Hayward black hole is geodesically incomplete for $r \geq 0$ and therefore needs to be extended to $r < 0$. This then also applies to RG-improved spherically symmetric black holes. In the $r < 0$ region, geodesics terminate at a curvature singularity for the Hayward black hole. However, in that region, the RG-improved black hole differs from the Hayward black hole: As we pointed out in Sect. 5.3, a sensible scale identification, which ensures that $k^2 > 0$, forces us to use $|r|$ instead of r. This, as has first been pointed out by [110], is sufficient to remedy the curvature singularity at $r < 0$ and also avoids a pole in the geodesic equation that was found in [125]. Thus, the maximal extension of RG improved spherically symmetric black holes is geodesically complete.

Another independent requirement is completeness of the spacetime for accelerating observers, which, to the best of our knowledge, has not yet been checked for these black holes.

5.4.3 *Spacetime Structure*

...where we explain how the regularization by quantum gravity affects the spacetime at all scales, such that the black hole is more compact than its classical counterpart and also has a more compact photon sphere, i.e., casts a smaller shadow.

The spacetime is not only modified in the center of the black hole, but everywhere, although the size of the modifications decreases with the (geodesic) distance from the center. Whether or not modifications lead to detectable effects depends on the value of ξ, which determines at what scale quantum-gravity effects are sizable. For now we discuss the effects at a qualitative level, and comment on detectability later on.

Because the Newton coupling is smaller than its classical counterpart, gravity is weaker in the quantum regime. Thus, the horizon, the photon sphere (and any special surface in the spacetime) are more compact. At a first glance, this may appear counterintuitive, because the converse is true for compact objects (e.g., stars): they are *less* compact, if gravity is weaker. However, there is an important difference between compact objects without horizon and black holes: in a horizonless compact object, such as a neutron star, some form of pressure balances the gravitational force and this balance determines the location of the surface. In a black hole, the horizon is not a material surface and its location is determined by gravity alone: the expansions of in- and outgoing null geodesic congruences are negative (i.e., the lightcone tilts "inwards"), once gravity is locally strong enough. Thus, a weakening of gravity results in a more compact horizon – in the limit of vanishing gravitational force, the size of the horizon would collapse to zero.

To determine the location of the horizon, we proceed as in the Schwarzschild case: The horizon respects the Killing symmetries of the spacetime, i.e., it is a function of r and θ only and can thus be described by the condition $h(r,\theta) = 0$. Because the horizon is a null surface, its normal is a null vector, i.e., $n^\mu = \partial^\mu h(r,\theta)$ is null:

$$0 = g^{\mu\nu} n_\mu n_\nu = g^{\mu\nu} \left(\partial_\mu h(r,\theta)\right) \left(\partial_\nu h(r,\theta)\right). \tag{5.26}$$

For the Schwarzschild spacetime, $g^{r\theta} = 0$, which continues to hold at finite ξ. Additionally, the horizon may not violate spherical symmetry, and thus $\partial_\theta h(r,\theta) = 0$. This leaves us with

$$g^{rr} = 0. \tag{5.27}$$

For the RG improved spacetime, the condition reads

$$0 = 1 - \frac{2G_0 M}{r\left(1 + \xi \frac{G_0^2 M}{r^3}\right)}. \tag{5.28}$$

This condition has two or no real and positive solutions. The black hole therefore has an outer and inner horizon. At large values of the parameter ξ, the two horizons can merge and leave behind a horizonless object, cf. Fig. 5.3. The larger the mass of the black hole, the larger the value of ξ_{crit} for which the horizons merge,

$$\xi_{\text{crit}} = \frac{32}{27} G_0 M^2. \tag{5.29}$$

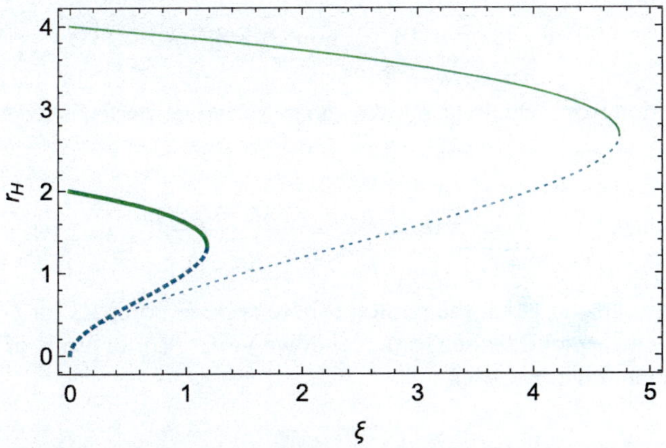

Fig. 5.3 We show the location of the outer (green continuous) and inner (cyan dotted) horizon as a function of ξ for $M = 1$ (thick lines) and $M = 2$ (thin lines) and $G_0 = 1$

The inner horizon is a Cauchy horizon. This is not only problematic when setting up initial-value-problems in the spacetime, it even suggests the presence of an instability. This instability has been discussed extensively in the literature [9, 10, 20–23, 37], and is the subject of Chap. 10 in this book, to which we refer the interested reader.

To bridge the gap between theoretical studies and observations, we investigate the photon sphere, which determines the size of the black-hole shadow. We explicitly prove that the photon sphere is more compact for $\xi > 0$ than it is for $\xi = 0$. This serves as one example of a general result, namely that all special hypersurfaces in a black-hole spacetime (event horizon, photon sphere, (ergosphere for finite spin)) are more compact [54].

The photon sphere is the border of the spacetime region within which infalling photons inevitably fall into the event horizon, i.e., it corresponds to the innermost circular photon orbit. The radius of this orbit can be determined by using four constants of motion, namely the energy E and all three components of the angular momentum L. We thereby obtain three equations for generic null geodesics parameterized by the affine parameter λ:

$$\left(\frac{dr}{d\lambda}\right)^2 = -V_r(r), \quad \frac{d\phi}{d\lambda} = \frac{L}{r^2}, \quad \frac{dt}{d\lambda} = \frac{E}{f(r)}, \tag{5.30}$$

where the effective radial potential is

$$V_r(r) = -\left(E^2 - f(r)\frac{L^2}{r^2}\right). \tag{5.31}$$

For circular geodesics, $dr/d\lambda = 0$ for all λ, which entails that

$$V_r(r) = 0, \qquad \frac{d}{dr} V_r(r) = 0. \tag{5.32}$$

The final equation from which the photon sphere r_γ is determined is the explicit form of Eq. (5.32):

$$0 = f'(r = r_\gamma) - \frac{2f(r = r_\gamma)}{r_\gamma}. \tag{5.33}$$

From here, one can show that the photon sphere is more compact for $\xi > 0$ than in the classical Schwarzschild case. To that end, we write $f(r)$ in terms of its classical part and the quantum correction,

$$f(r) = 1 - \frac{2G_0 M}{r} f_{\text{qm}}(r), \tag{5.34}$$

where

$$f_{\text{qm}}(r) = \frac{1}{1 + \xi \frac{G_0^2 M}{r^3}}. \tag{5.35}$$

For our derivation, the critical properties of $f_{\text{qm}}(r)$ are i) that it is everywhere smaller than one, because it weakens gravity everywhere and ii) that its derivative is positive, because the quantum effects fall off with the radial distance. Then, Eq. (5.33) becomes

$$0 = -\frac{G_0 M}{r_\gamma} f'_{\text{qm}}(r = r_\gamma) - \frac{1}{r_\gamma} + 3\frac{G_0 M}{r_\gamma^2} f_{\text{qm}}(r = r_\gamma), \tag{5.36}$$

from where $r_\gamma = 3G_0 M$ follows for $\xi = 0$. For $\xi > 0$, one can rewrite Eq. (5.36) by using the two conditions on $f_{\text{qm}}(r)$:

$$r_\gamma = \frac{3G_0 M\, f_{\text{qm}}(r = r_\gamma)}{1 + G_0 M\, f'_{\text{qm}}(r = r_\gamma)} < \frac{3G_0 M}{1 + G_0 M\, f'_{\text{qm}}(r = r_\gamma)} < 3G_0 M. \tag{5.37}$$

Therefore, the photon sphere of an asymptotic-safety inspired, spherically symmetric black hole is smaller than for its Schwarzschild counterpart with the same mass.

In addition, Eq. (5.36) has more than one solution: in concert with an inner horizon, an inner circular orbit appears. In contrast to the outer one, it has different stability properties and is therefore not relevant for images of black holes. The two photon spheres also annihilate at a critical value of $\xi_{\text{crit},2}$, which, however, is larger than the critical value for horizon-annihilation, ξ_{crit}. Therefore, within the interval $\xi \in [1.185, 1.808]G_0M^2$ there is a horizonless object that generates photon rings in its image [54].

5.4.4 *Thermodynamics*

...where we review calculations of black-hole entropy and black-hole evaporation in asymptotic-safety inspired black holes. We highlight that the evaporation process may end in a remnant.

We have seen that regularity implies a second inner horizon and thus a critical mass M_{crit} (for a fixed quantum-gravity scale ξ) at which the inner and outer horizon coincide. As we will see below, this modifies the Bekenstein-Hawking temperature of the black-hole and thus the evaporation process. While the original discussion was given in the context of RG-improved black holes [14], the following applies to spherically-symmetric regular black holes more generally.

The Bekenstein-Hawking temperature arises generically from quantum field theory on a curved background spacetime which contains a (black-hole) horizon. It thus applies to any asymptotically flat, spherically symmetric line element

$$ds = -f(r)dt^2 + \frac{1}{f(r)}dr^2 + r^2 d\Omega \;, \tag{5.38}$$

for which the largest root of $f(r = r_h) = 0$ implies a horizon at $r = r_h$. In a setting which does not account for the backreaction of the quantum fields, and is thus independent of the gravitational dynamics, the presence of an event horizon implies Hawking radiation with an associated temperature

$$T_{\text{BH}} = \frac{f'(r_h)}{4\pi} \;. \tag{5.39}$$

For very large asymptotic-safety inspired black holes with

$$M \gg M_{\text{crit}} = \frac{3^{\frac{3}{2}}}{\sqrt{32 G_0}}\sqrt{\xi} \tag{5.40}$$

the exterior spacetime approximates the classical Schwarzschild spacetime very well and thus recovers the classical temperature $T_{\text{BH}} = 1/(8\pi G M)$. As the regular black hole evaporates, its temperature first grows, as in the case without RG-improvement. However, as the evaporation continues, the black hole approaches the critical mass M_{crit} at which the inner and outer horizon merge, cf. Fig. 5.3. At this point, $f'(r_h)|_{M_{\text{crit}}} = 0$, and thus the temperature of the RG-improved black hole must vanish. Inbetween the temperature thus peaks and then decreases to zero.

This has a profound implication, namely, that a vanishing temperature at finite mass M_{crit} suggests a stable end state of Hawking evaporation of an asymptotic-safety inspired black hole [14]. An application of the Boltzmann law suggests that it takes infinite proper time for the asymptotic-safety inspired black hole to cool down to its finite remnant state. It has not been investigated, whether these remnants pose problems, as is generically expected for remnants [28], such as the overproduction problem. It has to be stressed, though, that the pair-production of black hole remnants (usually argued to lead to instabilities), depends on the dynamics of the theory, which is not yet fully understood in the case of asymptotically safe gravity. As in the classical-spacetime case, one can also assign an entropy to evaporating black holes [14, 18, 56, 57].

Both the classical and the RG-improved scenario for black-hole evaporation make use of a calculation for Hawking radiation on a fixed background spacetime and thus explicitly neglect backreaction of the radiation onto the black-hole spacetime. In the classical case, where the late-time temperature diverges, this approximation must necessarily break down when the black hole evaporates to sufficiently sub-Planckian mass. In the RG-improved case, the issue of backreaction could be less severe, because the temperature remains bounded throughout the entire evaporation process. However, the asymptotically safe dynamics for quantum fields on curved backgrounds contain non-standard non-minimal interactions (schematically of the form "kinetic term × curvature"), cf. [49, 55] for recent reviews. Their effect on the Hawking evaporation system has not been taken into account yet.

Further Reading

- **Schwarzschild (Anti-) deSitter black holes:**

  (i) In [77], the authors perform an RG improvement of Schwarzschild-de Sitter black holes, where they also include the running of the cosmological constant. Using the same scale-identification as in the Schwarzschild case then results in a singular spacetime, because the cosmological constant scales with k^2. Given that this system features several distinct scales, such a scale identification may no longer be appropriate. Whether a different scale identification should be performed, or whether the RG improvement procedure is too simple to properly deal with the situation with several distinct scales is an open question.

  (ii) In [1], the authors include the scaling of the cosmological constant away from the fixed-point regime, obtaining a non-singular black-hole spacetime when conditions on the scaling exponents hold.

  (iii) In [113] the author instead considers unimodular gravity, where the cosmological constant arises as a constant of integration at the level of the equations of motion and which may also be asymptotically safe [44, 45]. Therefore, it is not part of the RG-improvement, but remains classical. Accordingly, the resulting black holes are regular.

- **Leading-order quantum-gravity correction:**
  In [14], the authors use a result for $G_N(k)$ which includes additional scale-dependence around the transition scale and thereby reproduces the leading-order quantum-gravity correction at large r.
- **Iterated RG improvement:**
  In [99], the author iterates the RG improvement, starting with Eq. (5.19) as the first step of an iteration procedure. The final result of this procedure is the Dymnikova line element [42]. The same iteration procedure was used to describe the formation and evaporation of black holes in [18].

- **Evaporation of a Schwarzschild black hole:**
  The evaporation of a Schwarzschild black hole due to Hawking radiation has been effectively described by a Vaidya metric in which the mass function decreases with time [70]. In [7], the authors RG-improve said Vaidya metric as a model to incorporate the quantum fluctuations in the late-time evolution of Hawking evaporation. In agreement with the discussion above, they find that the final state of the evaporation process is a Planck-sized cold remnant.
- **Black holes beyond four dimensional spacetime:**
  Going beyond the four-dimensional spacetime setting, black solutions have been RG-improved in [86] see also [126]. The authors consider Myers-Perry-black holes, which are higher-dimensional rotating solutions of higher-dimensional GR. Their study finds that the ring singularity is softened enough to achieve geodesic completeness. They also find a minimum black-hole mass, related to the weakening of gravity through asymptotic safety: below a critical mass, gravity is no longer strong enough to form an event horizon.

5.5 Spinning Asymptotic-Safety Inspired Black Holes

...where we consider spinning asymptotic-safety inspired black holes and explore the fate of the classical curvature singularity, the fate of the Cauchy horizon, the event horizon, the ergosphere and the photon spheres.

Spinning black holes have a much more intricate structure than their spherically symmetric counterparts. This leads to several ambiguities in the RG improvement procedure, which are mostly irrelevant in the spherically symmetric case and that we discuss below.

Despite its complexity, the Kerr spacetime can be uniquely characterized by a single – remarkably simple – complex invariant

$$\mathscr{C} = \frac{G_0\, M}{(r - i\, a\, \cos(\theta))^3}\,, \tag{5.41}$$

where r is the radial coordinate and θ the polar angle with respect to the black hole's spin axis.[11] The more standard Weyl invariants follow directly from $\mathscr{C}$ as

$$C_{\mu\nu\rho\sigma}C^{\mu\nu\rho\sigma} = \mathrm{Re}\left[48\,\mathscr{C}^2\right]\,, \tag{5.42}$$

$$C_{\mu\nu\rho\sigma}\overline{C}^{\mu\nu\rho\sigma} = \mathrm{Im}\left[48\,\mathscr{C}^2\right]\,, \tag{5.43}$$

[11] The definition of the coordinates $(r,\ \theta)$ is equivalent in many of the standard coordinate systems for Kerr spacetime and, in particular, agrees with the $(r,\ \theta)$ as defined in ingoing Kerr and Boyer-Lindquist coordinates.

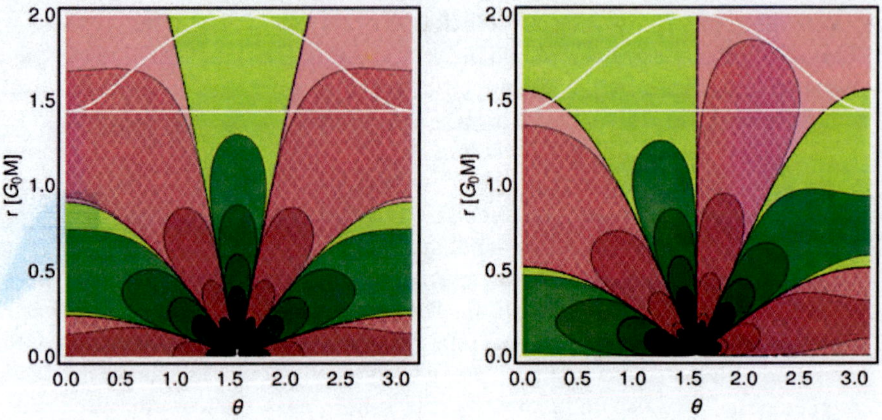

Fig. 5.4 We show contour plots of the two polynomially independent real invariants $C_{\mu\nu\rho\sigma}C^{\mu\nu\rho\sigma}$ (left-hand panel) and $C_{\mu\nu\rho\sigma}\overline{C}^{\mu\nu\rho\sigma}$ (right-hand panel) of Kerr spacetime. Increasingly positive values are shaded in increasingly dark shades of green (non-hatched). Increasingly negative values are shaded in increasingly dark shades of red (hatched). The two thick white lines indicate the location of the event horizon (lower line) and the ergosphere (upper line)

where $C_{\mu\nu\rho\sigma}$ and $\overline{C}_{\mu\nu\rho\sigma} = 1/2\,\epsilon_{\mu\nu\kappa\lambda}C^{\kappa\lambda}{}_{\rho\sigma}$ denote the Weyl tensor and the dual Weyl tensor. All other Riemann invariants are either polynomially dependent or vanish.

The key distinction to Schwarzschild spacetime is the angular dependence on θ. This angular dependence has consequences for the construction of spinning RG improved black holes.

First, in contrast to Schwarzschild spacetime, Kerr spacetime is characterized by two, not one, real invariants.

Second, because of their angular dependence, these two real invariants can change sign – even in spacetime regions external to the event horizon, cf. Fig. 5.4.

Third, this angular dependence is also mirrored in many of the key physical aspects that distinguish Kerr spacetime from Schwarzschild spacetime: for instance, the ergosphere and thus frame dragging is most prominent close to the equatorial plane (cf. outer white line in Fig. 5.4), and the singularity is a ring singularity, confined to $\theta = \pi/2$ (cf. dark-shaded region in Fig. 5.4).

All of these effects vanish in the continuous non-spinning limit for which Kerr spacetime reduces to Schwarzschild spacetime. This is made explicit by noting that the complex invariant $\mathscr{C}$ turns real and its square converges to the Kretschmann scalar, i.e., $\mathscr{C}^2 \overset{a\to 0}{\longrightarrow} R_{\mu\nu\rho\sigma}R^{\mu\nu\rho\sigma}$, in the Schwarzschild limit, cf. Eqs. (5.41) and (5.17).

The above understanding of Kerr spacetime indicates that any RG-improvement that captures the key difference between spinning and non-spinning black-hole spacetimes needs to incorporate angular dependence. In the following, we will first discuss the key difficulties that come with this added complexity: In Sect. 5.5.1, we show that the choice of scale identification now impacts physical conclusions. In particular, we summarize which conclusions appear robust and which other ones do not. We also argue why we expect a scale identification with local curvature to give the most physical results.

We also point out that horizon-penetrating coordinates are crucial to avoid naked singularities which arise from coordinate singularities through the RG improvement. We will then specify to an RG-improvement based on local curvature invariants and horizon-penetrating coordinates and examine the resulting physical aspects of the resulting RG-improved Kerr spacetime.

5.5.1 *Universal and Non-universal Aspects of RG Improvement*

...where we demonstrate that the RG improvement of axisymmetric spacetimes is less universal than in spherically symmetric spacetimes. Starting from Kerr spacetime in horizon-penetrating coordinates, we discuss which conclusions about the RG-improved spacetime depend on the choice of scale identification and, in contrast, which conclusions remain universal.

We start out from Kerr spacetime in horizon-penetrating (ingoing Kerr) coordinates $(u, r, \chi = \cos(\theta), \phi)$:

$$\begin{aligned} ds^2 = &-\frac{r^2 - 2G_N Mr + a^2\chi^2}{r^2 + a^2\chi^2} du^2 + 2\,du\,dr - 4\frac{G_N Mar}{r^2 + a^2\chi^2}\left(1 - \chi^2\right) du\,d\phi \\ &+ \frac{1 - \chi^2}{r^2 + a^2\chi^2}\left(\left(a^2 + r^2\right)^2 - a^2\left(r^2 - 2G_N Mr + a^2\right)\cdot\left(1 - \chi^2\right)\right) d\phi^2 \\ &-2a\left(1 - \chi^2\right) dr\,d\phi + \frac{r^2 + a^2\chi^2}{1 - \chi^2} d\chi^2 \,. \end{aligned} \tag{5.44}$$

For the classical case, i.e., for constant Newton coupling $G_N \equiv G_0$, this metric is equivalent to the above representation of Kerr spacetime in terms of curvature invariants, cf. (5.41). At the level of the metric, RG-improvement is implemented in the same way as for the non-spinning case, i.e., by the two-step replacement (cf. Sect. 5.3)

$$G_N \overset{\text{scale dependence}}{\longrightarrow} G_N(k) \overset{\text{scale identification}}{\longrightarrow} G_N(k(r,\,\chi))\,. \tag{5.45}$$

For now, we need not specify to any particular choice of scale dependence and scale identification, because one can determine all of the curvature invariants of the RG-improved spacetime for general $G_N(r,\,\chi)$ [66]. Under such an RG-improvement, the complex invariant of Kerr spacetime reads

$$\mathscr{C} = \frac{(r - ia\chi)\left[r(r - ia\chi)MG_N'' - 2(2r + ia\chi)MG_N'\right] + 6(r + ia\chi)MG_N}{6\,(r - ia\chi)^3(r + ia\chi)}\,. \tag{5.46}$$

Here and in the following, primes are used to denote derivatives with respect to the radial coordinate. We remind the reader that $C_{\mu\nu\rho\sigma}C^{\mu\nu\rho\sigma} = \mathrm{Re}\left[48\,\mathscr{C}^2\right]$ and $C_{\mu\nu\rho\sigma}\overline{C}^{\mu\nu\rho\sigma} = \mathrm{Im}\left[48\,\mathscr{C}^2\right]$. Since the RG-improved spinning spacetime is no longer Ricci flat, it is characterized by a second complex invariant

$$\mathscr{C}_2 = \mathscr{C} \times \left[\frac{M\left(r\left(r^2 + a^2\chi^2\right)G_N'' - 2(r^2 - a^2\chi^2)G_N'\right)}{\left(r^2 + a^2\chi^2\right)^2}\right]^2, \tag{5.47}$$

which in turn is related to the mixed curvature invariants

$$R^{\mu\nu}R^{\rho\sigma}C_{\mu\nu\rho\sigma} = \mathrm{Re}\,[4\,\mathscr{C}_2]\,, \tag{5.48}$$
$$R^{\mu\nu}R^{\rho\sigma}\overline{C}_{\mu\nu\rho\sigma} = \mathrm{Im}\,[4\,\mathscr{C}_2]\,. \tag{5.49}$$

In addition, there are two polynomially independent[12] Ricci invariants, i.e.,

$$g^{\mu\nu}R_{\mu\nu} = \frac{2\left(r\left(r^2 + a^2\chi^2\right)MG_N'' + 2a^2\chi^2MG_N' + \left(2r^2MG_N'\right)\right)}{\left(r^2 + a^2\chi^2\right)^2}\,, \tag{5.50}$$

$$R_{\mu\nu}R^{\mu\nu} = \frac{2\left(\left(r\left(r^2 + a^2\chi^2\right)MG_N'' + 2a^2\chi^2MG_N'\right)^2 + \left(2r^2MG_N'\right)^2\right)}{\left(r^2 + a^2\chi^2\right)^4}\,. \tag{5.51}$$

All other curvature invariants of the general RG-improved spacetime can be expressed as polynomials of the above invariants [66, App. D]. It is quite remarkable that no curvature invariants depend on angular χ-derivatives of the Newton coupling $G_N(r, \chi)$.

With this complete polynomially independent set of curvature invariants at hand, we are in a position to derive generic conditions on $G_N(r, \chi)$ which are required to remove the curvature singularity of Kerr spacetime. We observe that the only potential curvature singularity occurs when $\chi = 0$ and $r = 0$. Just like for the spherically-symmetric case in Sect. 5.4, the removal of this singularity depends on the leading exponent n in the expansion of $G_N(r, \chi)$ close to $\chi = 0$ and $r = 0$, i.e., $G_N(r, \chi) \sim r^n + \mathcal{O}(r^{n+1}\chi)$. Once again, we find that:

- If $n = 3$, the curvature invariants are finite at $r = 0$. (There is a subtlety about uniqueness of the limit under exchange of $r \to 0$ and $\chi \to 0$ which we will discuss below.)

[12] It has not been proven that the displayed invariants are polynomially independent, hence there could, in principle, be further polynomial relations among them.

- If $n > 3$, all curvature invariants vanish and the spacetime is locally flat in the center.
- If $n < 3$, the curvature singularity remains, although it is weakened for $0 < n < 3$.

We can compare this to the outcome of the different scale identification procedures introduced in Sect. 5.3. In contrast to the spherically-symmetric case, we will see that the choice of scale identification makes a difference when applied to the case of Kerr spacetime.

We start with the scale-identification with radial geodesic distance which is closest in spirit to the flat-spacetime RG-improvement of the Coulomb potential, cf. Eq. (5.12).

Reference [103] uses a radial path in the spacetime,

$$k = k(r, \theta) = \left(\int_0^r d\bar{r} \sqrt{\left| \frac{\bar{r}^2 + a^2 \cos^2\theta}{\bar{r}^2 + a^2 - 2m\bar{r}} \right|} \right). \tag{5.52}$$

The integral can be performed, e.g., in the equatorial plane, $\theta = \pi/2$, which is in fact the choice the authors work with. It is thereby assumed that quantum-gravity effects have the same size at all values of θ for a given r. The resulting scaling of the Newton coupling is

$$G_N(r) \sim \frac{G_0}{4\xi} r^4 + \mathcal{O}(r^5) \,. \tag{5.53}$$

It is non-local in a sense that it uses a quantity defined at $\theta = \pi/2$ also away from $\theta = \pi/2$.

A similar non-local approximation has been made for the scale-identification with curvature invariants. Reference [93] uses the Kretschmann scalar in the equatorial plane, i.e.,

$$k = \xi \, (K(r, \theta = \pi/2)))^{\frac{1}{4}} = \left(\frac{48 G_0^2 M^2}{r^6} \right)^{\frac{1}{4}}. \tag{5.54}$$

Also here the assumption is that quantum-gravity effects have the same size at all angles. In fact, one can see that the scale identification effectively neglects all effects due to the breaking of spherical symmetry, i.e., the Kretschmann scalar in the equatorial plane is proportional to the one in a non-spinning Schwarzschild spacetime. The resulting scaling of the Newton coupling is

$$G_N(r) \sim \frac{1}{4\sqrt{3}\xi \, M} r^3 + \mathcal{O}(r^4). \tag{5.55}$$

The scale-identification with the curvature scales of an equivalent Schwarzschild black hole thus results in the critical scaling that is necessary to remove the ring singularity of Kerr spacetime.

In contrast to the above, local physics in the Kerr spacetime depends on the angle with respect to the spin axis, cf. discussion around Fig. 5.4. Thus, any scale identification that means to capture this local physics will necessarily need to take into account θ-dependence.

This was first discussed in [67], where the θ-dependence of the Kretschmann scalar was (partially) taken into account. The reasoning starts from the observation that the singularity in Kerr spacetime is a ring singularity, such that curvature invariants only diverge in the equatorial plane. Accordingly, curvature invariants such as the Kretschmann scalar show an angular dependence, cf. Eq. (5.41). Accordingly, the curvature is larger, i.e., closer to the Planck scale, in the equatorial plane than at other values of θ. One may thus argue that quantum-gravity effects are overestimated by the choices in [93, 103]. Instead, the angular dependence of Eq. (5.41) should be taken into account. The angular dependence of curvature invariants is more comprehensively accounted for in [51], see also Sect. 5.5.2 below. This results in the same scaling as Eq. (5.55) in the equatorial plane.

In contrast to all choices above, at larger distances, relevant, e.g., for black-hole thermodynamics, [86] does not use any information on the geometry at high curvature, neither local nor non-local, but identifies $k \sim r$ based on dimensional grounds, where r is the radial coordinate in Boyer-Lindquist coordinates. The resulting scaling is

$$G_N(r) \sim \frac{G_0}{\xi} r^2 + \mathcal{O}(r^4). \tag{5.56}$$

As one might expect from such a large-distance approximation, this scaling is insufficient to remove the curvature singularity.

Irrespective of the choice of scale identification, or at least among the choices discussed above, we find the following universal features.

- A resolution of the curvature singularity, cf. Table 5.1.
- a more compact event horizon than in the classical case.
- the possibility of a horizonless spacetime, obtained through a merging of outer and inner horizon, when quantum effects are made large.

In the following, we focus on the local scale identification. This brings us close in spirit to effective field theory, where quantum effects lead to higher-order curvature

Table 5.1 We list the behavior of two curvature invariants for the various scale identifications. The singularity is resolved if a diffeomorphism invariant quantity is used in the scale identification, such as a radial path or a curvature invariant

Weyl^2 at $r, \chi \to 0$	Weyl^2 at $\chi, r \to 0$	Ricci at $r, \chi \to 0$	Ricci at $\chi, r \to 0$	Scale-identification	References
0	0	0	0	Eq. (5.52)	[103]
0	0	$2\sqrt{3}/(\xi)$	0	$k \sim \sqrt{K(\theta = \pi/2)}$	[93]
0	$-1/(\xi)^2$	$2\sqrt{3}/(\xi)$	0	Eq. (5.57)	[51]

terms. These become important at a given spacetime point, when the local curvature exceeds a critical value. In the same vein, we expect higher-order curvature terms in the effective action for asymptotic safety to become important, when the local curvature exceeds a critical value. To mimick this, we choose a local curvature scale for the scale identification in Sect. 5.5.2 below.

To conclude this section, we point out that Boyer-Lindquist type coordinates with coordinate singularities can be used in the non-local scale identifications, where $G_N = G_N(r)$. In contrast, the local scale identification, in which $G_N = G_N(r, \theta)$ is not compatible with Boyer-Lindquist type coordinates. The reason is that the coordinate singularity of the Kerr spacetime in Boyer-Lindquist coordinates can be upgraded to a curvature singularity in the presence of $G_N = G_N(r, \theta)$. In [66], it is discussed more generally, that RG improvement can result in such pathologies, if coordinates are not chosen with due care.

5.5.2 *The Line Element for the Locally Renormalization-Group Improved Kerr Spacetime*

...where we discuss the scale identification with local curvature in detail and derive the resulting axisymmetric spacetime.

We will work in horizon-penetrating coordinates, to avoid turning coordinate singularities into unphysical curvature singularities [66]. Our starting point is the Kerr metric in ingoing Kerr coordinates, cf. Eq. (5.44).

To capture the angular dependence of the spinning black-hole spacetime, we perform a scale identification based on the local curvature invariants of the Kerr spacetime. The Kerr spacetime is fully characterized by the single complex invariant $\mathscr{C}$, cf. Eq. (5.41). Thus, we identify

$$k^2 = \sqrt{48}\,\tilde{\xi}\,|\mathscr{C}| \equiv \tilde{\xi}\left(\left(C_{\mu\nu\rho\sigma}C^{\mu\nu\rho\sigma}\right)^2 + \left(C_{\mu\nu\rho\sigma}\overline{C}^{\mu\nu\rho\sigma}\right)^2\right)^{1/4}. \tag{5.57}$$

This implements the more general scale identification with the root-mean-square of all polynomially independent curvature invariants discussed in Sect. 5.3. Moreover, it smoothly connects to the spherically symmetric case in the limit of $a \to 0$, cf. Sect. 5.4.

With the approximate scale-dependence as given in Eq. (5.9), we find the RG-improved Newton coupling

$$G_N(r,\theta) = \frac{G_0}{1 + \frac{G_0^2 M\,\xi}{\left(r^2+a^2\cos(\theta)^2\right)^{3/2}}}, \tag{5.58}$$

where we have used the re-definition $\xi = \sqrt{48\tilde{\xi}}/g_{N,*}$, as in the spherically symmetric case. This functional dependence characterizes the respective RG-improved Kerr black hole which can be represented either by using Eq. (5.58) by the metric (Eq. (5.44)) or equivalently by the curvature invariants discussed in Sect. 5.3.

5.5.3 *Spacetime Structure and Symmetries*

...where we discuss that the event horizon of a spinning, asymptotic-safety inspired black hole is more compact than its classical counterpart. Further, we highlight that for such a black hole, the Killing and the event horizon do not agree for the local scale-identification. Moreover, its symmetry properties differ from Kerr: circularity, which is an additional isometry of Kerr spacetime, is not realized for this black hole.

Working our way from asymptotic infinity to the regular core[13] of the asymptotic-safety inspired black hole, we encounter the photon shell,[14] the ergosphere, and then the event horizon. While all these surfaces/regions persist, they are more compact than for the respective Kerr black hole. The reason is the same as in the spherically symmetric case: the quantum effects weaken gravity and therefore a black hole of a larger mass would be needed to exert the same effects on timelike and null geodesics; first, introducing a rotation for timelike geodesics at the ergosphere, then, preventing stable photon orbits at the photon sphere and finally, redirecting outgoing null geodesics towards smaller radii. Thus, if the asymptotic mass M is held fixed, these surfaces are all located at smaller radii than in the classical case.

The persistence of the photon shell and the ergosphere is relevant for phenomenology: Supermassive black holes may launch jets through the Blandford-Znajek process, which requires the ergosphere to be present. The photon shell comprises the region in spacetime in which unstable closed photon orbits are possible and hence results in photon rings in black-hole images, which may be detectable with very-large-baseline interferometry [19]. As we discuss below, the geodesic motion in the RG-improved black-hole spacetime is no longer separable. Thus, we cannot present analytical results for closed photon orbits in the photon shell. A similar proof of increased compactness goes through in the equatorial plane as it does for the spherically symmetric case, cf. Sect. 5.4.3; at arbitrary θ, increased compactness can be shown analytically. The increase in compactness is not uniform; but, due to frame-dragging, larger on the prograde side; and also larger in the equatorial plane than away from it. In Sect. 5.7, we briefly discuss the qualitative effects resulting from numerical studies.

---

[13] Curvature invariants are finite, but not single-valued at $r \to 0$, $\chi \to 0$. For $r > 0$, the spacetime is not geodesically complete and needs an extension to $r < 0$. There, the dependence on r^2 (instead of r) in Eq. (5.58) ensures the absence of curvature singularities, see [110].

[14] In the spherically-symmetric case, closed photon orbits can only arise at a 2-dimensional surface at fixed radius, cf. Sect. 5.4.3 – hence the name photon sphere. For stationary axisymmetric spacetimes and, in particular, for Kerr spacetime [2, 108], several classes of closed photon orbits can occur which cover an extended 3-dimensional region – hence, the name photon shell.

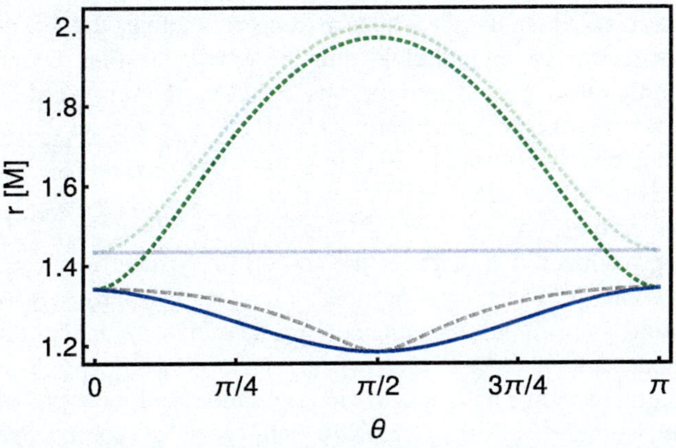

Fig. 5.5 We show the ergoregion (green dotted), the Killing horizon (gray dashed), and the event horizon (blue continuous) of the RG-improved black hole, cf. Sect. 5.5.2 for $a = 0.9\,M$ and near-critical deviation parameter $\xi = 0.13$. In lighter shading and for reference, we also show the Killing/event horizon (blue continuous) and the ergosphere (dotted green) of the corresponding Kerr black hole

The overall increase in radial compactness depends on $\chi = \cos(\theta)$. This is because the quantum-gravity effects, i.e., the weakening of gravity, is always strongest in the equatorial plane. Therefore all surfaces are most compact at $\chi = 0$, cf. Fig. 5.5. For instance, the horizon exhibits a "dent", i.e,. the deviation between the classical and the quantum horizon is largest at $\chi = 0$, cf. Fig. 5.5.

There is an additional effect that relates to the event horizon: in the Kerr spacetime, the event horizon is also a Killing horizon, so that a constant surface gravity can be defined. This is no longer the case for the asymptotic-safety inspired black hole, see [51].[15] The event horizon, described by Eq. (5.26), which results in $g^{rr} = 0$ for the Kerr spacetime, features an angular dependence, necessitating a numerical solution of the horizon equation. Specifically, the horizon equation for the spinning asymptotic-safety inspired black hole is

$$g^{rr}(r = H(\theta), \theta) + g^{\theta\theta}(r = H(\theta), \theta) \left(\frac{dH}{d\theta}\right)^2 = 0. \tag{5.59}$$

The Killing horizon in turn is described by the condition

$$g_{u\phi}^2 - g_{uu}g_{\phi\phi} = 0. \tag{5.60}$$

One can confirm that these are the same condition in the Kerr spacetime, but not in the RG improved spacetime based on the local improvement procedure.

---

[15] Consequences for black hole thermodynamics have not been explored yet.

The inner horizon which is a Cauchy horizon persists; thus, the RG improvement can only remove some of the pathologies of the Kerr spacetime. Because the RG-improvement only takes the quantum-gravity effect on the most relevant curvature coupling, i.e., the Newton coupling, into account, one may expect that a more complete treatment is more accurate at large curvatures; thus, the line element may be a fair description of a black hole in asymptotically safe gravity at large enough r, but miss effects which may, e.g., resolve the Cauchy horizon (and the $r < 0$ region).

The RG improvement also modifies the spacetime symmetries. The two Killing symmetries, stationarity and axisymmetry are left intact. The only way to break them would be to build an artificial dependence on time and/or azimuthal angle into the RG improvement.

However, Kerr spacetime has a less obvious symmetry, which is called circularity. Circularity is an isometry of Kerr spacetime which is easiest to see in Boyer-Lindquist coordinates (t, r, θ, ϕ): under a simultaneous mapping of $t \to -t$ and $\phi \to -\phi$, the spacetime is invariant. This is no longer the case for the RG improved black hole (unless one works with a non-local scale identification as in Eq. (5.54)).[16] The more general way of testing circularity is by testing the following conditions on the Ricci tensor, see [91]:

$$\begin{aligned}
\xi_1^{[\mu}\xi_2^{\nu}\nabla^{\kappa}\xi_1^{\lambda]} &= 0 \text{ at at least one point},\\
\xi_2^{[\mu}\xi_1^{\nu}\nabla^{\kappa}\xi_2^{\lambda]} &= 0 \text{ at at least one point},\\
\xi_1^{\mu} R_{\mu}^{\;[\nu}\xi_2^{\kappa}\xi_1^{\lambda]} &= 0 \text{ everywhere},\\
\xi_2^{\mu} R_{\mu}^{\;[\nu}\xi_1^{\kappa}\xi_2^{\lambda]} &= 0 \text{ everywhere}.
\end{aligned} \tag{5.61}$$

Because the Killing vector for axisymmetry vanishes on the axis of rotation, the first two conditions are always satisfied. The latter two conditions are nontrivial and impose a condition on the Ricci tensor. Because spinning black holes are vacuum spacetimes in GR, all four conditions hold there. In contrast, one should not expect the Ricci tensor to be special (in the sense of respecting the above symmetry) in a quantum theory any more – at least not if effective field theory reasoning, which determines the size of deviations from the Kerr spacetime through local curvature invariants, holds.[17]

Finally, there is also no generalized Carter constant, i.e., geodesic motion is not separable, and energy and angular momentum are the only conserved quantities. This is because the spacetime no longer features a Killing tensor, as it does in GR, where this tensor gives rise to the Carter constant.

---

[16] Because Killing symmetries can be made manifest in the spacetime metric, RG improvement can always be made to respect Killing symmetries. However, any symmetry that is not a Killing symmetry, but instead only expressible as a condition on the Riemann tensor, need not be respected by the RG improvement.

[17] See, however, [120] for a proof that circularity holds for black holes which are perturbatively connected to the Kerr solution.

As a consequence, RG improved spinning black holes are not included in the parameterizations of black-hole spacetimes in [72, 73], which assumes a generalized Carter constant; nor in [80], which assumes circularity, but requires a more general parameterization [36].

5.5.4 Horizonless Spacetimes

...where we discuss how increasing ξ leads to a loss of horizon at $\xi_{\rm crit}(a)$. For near-critical spin, $\xi_{\rm crit}(a)$ can be made very small, such that Planck-scale modifications of the black-hole spacetime suffice to dissolve the horizon.

The inner and outer horizon approach each other, when ξ increases. Intuitively, this follows from the fact that the limit $\xi \to \infty$ is a gravity-free limit, i.e., the line element becomes that of Minkowski spacetime. The transition between $\xi = 0$ (a Kerr black hole) and $\xi \to \infty$ (Minkowski spacetime) must therefore include a mechanism in which the event horizon of the Kerr spacetime is resolved. Mathematically, this can occur when the solutions to the horizon equation become complex, which they can only do as a pair. This requires them to become equal just before they become complex, i.e., the inner and outer horizon annihilate.

The critical value of ξ, at which the annihilation occurs depends on the spin parameter a. For a classical Kerr black hole, the horizons annihilate at $|a| = M$. Close to $|a| = M$, where the horizons are close to each other (i.e., at values of r which are close), the tiny increase in compactness that is caused by a Planck-scale ξ is enough to trigger an annihilation of the horizons. More specifically, if we set $|a| = M\,(1 - \delta a)$, we obtain

$$\delta a = \frac{\xi}{2}\frac{G_0}{M^2} + \mathcal{O}(\xi^2). \tag{5.62}$$

For Planck-scale effects, i.e., $\xi = 1$, the black hole therefore has to be very close to the critical spin value. It is an open question whether or not this is achievable in settings beyond GR and specifically asymptotically safe gravity, see [71, 85, 121] for general studies for regular black holes.

Further Reading

- **Impact of the cosmological constant:**
  Reference [93] also considers Kerr-(A)dS black holes and includes the cosmological constant in the RG improvement. Just like in the spherically symmetric case, the scaling of the cosmological constant $\Lambda \sim k^2$ dominates the large-curvature behavior and reintroduces a curvature singularity. Again, as in the case of spherical symmetry, performing the RG improvement in this way appears to be insufficient to capture the expected quantum effects. This again highlights the limitations of the RG improvement procedure in settings with several physical scales, where

one needs to distinguish carefully which scales matter for the quantum effects and which do not.

- **Change of coordinates after RG improvement:**
For the nonlocal scale-identifications, a coordinate transformation from Boyer-Lindquist coordinates to horizon-penetrating coordinates is straightforward and mimicks the classical coordinate transformation, see, e.g., [103]. In contrast, the coordinate transformation is more involved for the local scale identification, see [36], and leads to a line element which has additional non-vanishing components in comparison to the classical case. This is directly linked to the breaking of a spacetime symmetry, circularity, by the quantum-gravity effects, see the discussion above.

5.6 Formation of Asymptotic-Safety Inspired Black Holes

...where we discuss the formation of asymptotic-safety inspired black holes. First, we analyze the spacetime structure of RG-improved black holes formed from gravitational collapse. Second, we discuss whether the formation of black holes in high-energy scattering processes is to be expected in asymptotically safe gravity.

5.6.1 Spacetime Structure of Gravitational Collapse

...where we review the spacetime structure and the fate of the classical singularity in asymptotic-safety inspired gravitational collapse.

Up to here, we have looked at black holes as "eternal" objects, i.e., as idealized stationary solutions to gravitational theories. As such, they are merely of academic interest, or, at best constitute approximations to time-dependent systems. However, black holes gain physical importance because we expect them to form as the generic final state of gravitational collapse – at least in GR.

In GR, singularity theorems [95] imply that gravitational collapse ends in the formation of geodesic singularities. At the same time, gravitational collapse generically results in the formation of black-hole horizons, which has led to the weak cosmic censorship conjecture in GR [94]. In colloquial terms, weak cosmic censorship states that, given physical and generic[18] initial conditions, the dynamics of GR ensures that all singularities are hidden behind horizons with respect to asymptotic observers, cf. [31] for a mathematical formulation.

[18] The word 'physical' refers to physically realistic matter models [96, 119], see also the discussion below Eq. (5.65). The word 'generic' is important since non-generic initial conditions result in counter-examples to weak cosmic censorship [29], even in restricted sectors (spherically symmetric GR with a real massless scalar field) for which the mathematical conjecture has been proven [30].

Weak cosmic censorship in GR may suggest a form of "quantum gravity censorship": if regions of very high/diverging curvature are generically hidden behind horizons, then so are the effects of quantum gravity (at least, where they are large). Models of gravitational collapse which lead to naked singularities within GR are thus of particular interest for quantum gravity, because they may lead to regions of spacetime with large quantum-gravity effects, and not shielded from asymptotic observers.

A simple model for the exterior spacetime of gravitational collapse is given by the ingoing Vaidya metric [116–118]

$$ds^2 = -f(r,v)\,dv^2 + 2\,dv\,dr + r^2\,d\Omega^2 \quad \text{with} \quad f(r,v) = 1 - \frac{2\,m(v)}{r}\,. \tag{5.63}$$

For $m(v) = M$, the Vaidya spacetime reduces to Schwarzschild spacetime in ingoing Eddington-Finkelstein coordinates. For general $m(v)$, a calculation of the Einstein tensor confirms that this metric is a solution to the Einstein field equations with stress-energy tensor

$$T_{\mu\nu} = \frac{\dot{m}(v)}{4\pi\,r^2} l_\mu\,l_\nu\,, \tag{5.64}$$

where $l_\mu = -\partial_\mu\,v$ is tangent to ingoing null geodesics. This stress-energy tensor describes so-called null dust, i.e., a pressure-less fluid with energy density $\rho = \dot{m}/(4\pi\,r^2)$ and ingoing fluid 4-velocity l_μ. For null dust which satisfies the weak (or null) energy condition, the energy density $\rho \geq 0$ must not be negative, and we see that the mass function $m(v)$ must not decrease.

The Vaidya spacetime – as well as other null-dust collapse models – admit initial conditions which form spacelike singularities that always remain hidden behind horizons. However, they also permit initial conditions which form naked singularities, visible to (at least some) asymptotic observers, cf. [74] for review. For instance, the simple choice of a linearly increasing mass function

$$m(v) = \begin{cases} 0 & v < 0 \\ \lambda\,v & 0 \leqslant v \leqslant \overline{v} \\ \overline{m} & v > \overline{v} \end{cases} \tag{5.65}$$

results in a naked curvature singularity if $\lambda \geqslant 1/(16\,G_0)$ and has been one of the earliest models discussed in the context of cosmic censorship in GR. At this point, it is important that pressureless fluids may not be considered as describing physically relevant initial data since they can form density singularities even in flat spacetime, cf. [96, 119] for a more complete discussion.

Irrespective of whether or not the above null-dust collapse-models are considered as physically relevant counterexamples to the cosmic-censorship conjecture, they may be used as toy models for potential classical cosmic-censorship violations. Thus, we ask whether quantum effects can remove the naked singularity. We highlight, that

(i) if, in GR, physically relevant cosmic censorship violations exist, and (ii) if the naked singularity is lifted by quantum gravity effects, then the result is a quantum-gravity region not shielded from asymptotic observers by a horizon. Depending on the properties of such a region, its effects may be detectable, e.g., in electromagnetic radiation emitted from or traversing the region.[19]

Spherically-symmetric RG-improved gravitational collapse of null-dust models has been investigated in [11, 12, 27, 111, 112, 114]. The specific example of the Vaidya spacetime has been RG-improved in [11, 12]. In this work, the running Newton coupling has been approximated by Eq. (5.9). Moreover, the energy density ρ of the null dust in the Vaidya metric was used to set the RG scale k: On dimensional grounds, this implies

$$k^4 \sim \rho \tag{5.66}$$

for the scale identification. This results in a radially and advanced-time dependent cutoff,

$$k^2 = k(r, v) = \tilde{\xi} \left(\frac{\dot{m}(v)}{4\pi r^2} \right)^{\frac{1}{4}}, \tag{5.67}$$

leading to the RG improved lapse function

$$f(r, v) = 1 - \frac{2G_0\, m(v)}{r + \frac{G_0}{\sqrt{4\pi}} \xi^2 \sqrt{\dot{m}(v)}}. \tag{5.68}$$

For these choices, the antiscreening effect of RG improvement is not strong enough to fully remove the curvature singularity. The resulting RG-improved Vaidya metric still contains a naked, although now integrable, curvature singularity [11]. The authors also highlight that quantum effects even seem to broaden the range of initial conditions for which the formed singularity is naked.

A different null-dust collapse-model (Lemaitre-Tolman-Bondi [15, 83, 109]) has been RG-improved in [112], while also matching the interior solution to an exterior RG-improved Schwarzschild spacetime. Notably, [112] uses the scale identification with geodesic distance (originally proposed in [14]). In that case, the central curvature singularity (naked or hidden) is removed by the RG improvement.

While the two null-dust collapse-models are distinct and an explicit comparative study remains to be performed, we expect that a key difference is the use of distinct scale identifications. It is a universal result that RG-improvement weakens the curvature singularities which arise in simple null-dust models of gravitational collapse, cf. also [27]. Whether or not the curvature singularities are fully lifted, seems to depend on the scale identification, just as it does for stationary black holes.

---

[19] Note that there are spacetimes which satisfy weak cosmic censorship in GR, but their RG-improved counterparts do not satisfy "quantum gravity censorship": For instance, a Kerr-black hole, with sub-, but near-critical spin is an example, since its RG-improved counterpart can be horizonless, cf. [53] and Sect. 5.5.4.

Further Reading

- **RG improvement of gravitational collapse based on decoupling:**
  The RG-improvement of spherically-symmetric gravitational collapse has recently been investigated [18] in the context of the decoupling mechanism [105] and with the iterative RG-improvement developed in [99], to find a self-consistent RG improved model that describes a black hole from its formation through gravitational collapse to its evaporation through Hawking radiation. The study finds that the classical curvature singularity of a Vaidya-Kuroda-Papapetrou spacetime is weakened, but not fully lifted.

5.6.2 *Formation of Black Holes in High-Energy Scattering*

...where we discuss the expectation that scattering at transplanckian center-of-mass energies must necessarily result in a black hole and point out that it may not be realized in asymptotically safe gravity.

In GR, confining an energy density or mass to a spacetime region smaller than the associated Schwarzschild radius is expected to lead to black-hole formation. Thorne's hoop conjecture accordingly says that if one can localize particles to within their Schwarzschild radius, a black hole will form. If one uses a particle's de Broglie wavelength as the radius of a region within which the particle is localized, black-hole formation is expected to set in at the Planck scale. This argument, which has also been studied numerically in [43], appears to be at odds with the idea of asymptotic safety: if indeed transplanckian scattering inevitably leads to black holes, then black holes dominate the high-energy spectrum of a theory. In turn, because more massive black holes have larger area, there is a UV-IR-duality in the theory: the deeper one tries to probe in the UV, the further in the IR one ends up. If such a picture is correct, there does not appear to be any room for asymptotic safety at transplanckian scales.

We propose, see also [3, 8] that this apparent problem can be resolved as follows: to decide whether or not a black hole forms, it is not enough to state that the energy is transplanckian, because transplanckian energies are reached both in classical and in quantum processes. That black holes form in transplanckian, classical processes is not in doubt – after all, astrophysical black holes all have masses corresponding to highly transplanckian energies. Such processes are characterized by an impact parameter much larger than the Planck length. In contrast, we are interested in a regime where energies are transplanckian and impact parameters subplanckian, because this is the quantum regime. However, because it is the quantum regime, it is incorrect to apply GR, and therefore it is a priori unclear whether or not a black hole forms.

The outcome of scattering processes at transplanckian center-of-mass energy and subplanckian impact parameter depends on the dynamics of quantum gravity. To obtain some intuition for what the outcome could be in asymptotically safe quantum gravity, we perform an RG improvement of the hoop conjecture. To that end, we simply take the classical Schwarzschild radius, $R_{S,\,\mathrm{cl}} = \sqrt{2G_N M}$ (with $c = 1$), and upgrade G_N to its scale-dependent counterpart, $G_N = G_N(k)$. We subsequently identify $k = \xi/b$, where b is the impact parameter and ξ is a number of order one. We thereby obtain an RG-improved R_S that we can compare to its classical counterpart. If $R_S(b) \geq R_{S,\,\mathrm{cl}}$, then the classical argument underestimates the impact parameter, at which black holes form. Conversely, if $R_S(b) < R_{S,\,\mathrm{cl}}$, then the classical argument overestimates the impact parameter, at which black holes form. Because $G_N(k < M_{\mathrm{Planck}}) = \mathrm{const}$, the classical and the quantum estimate agree for superplanckian impact parameter. For subplanckian impact parameter, $G_N(k > M_{\mathrm{Planck}}) \sim k^{-2} \sim b^2$ implies that the RG-improved R_S shrinks linearly with decreasing impact parameter:

$$R_S(b) = \sqrt{2M}\sqrt{\frac{G_0}{1+\xi\frac{G_0}{b^2}}} \approx \sqrt{\frac{2M}{\xi}}b \tag{5.69}$$

In this regime, the critical radius at which black-hole formation occurs is therefore smaller than suggested by the classical estimate. Therefore, the classical hoop conjecture does not generically apply in this simple RG-improved setup. Whether this simple argument captures the relevant gravitational dynamics in asymptotic safety is of course an open question.

Whether or not black holes form is therefore currently an unanswered question. The weakening of gravity associated with asymptotic safety implies that the answer may be negative, because, simply put, the Schwarzschild radius decreases faster (as a function of impact parameter) than the impact parameter itself.

5.7 Towards Observational Constraints

...where we first discuss theoretical expectations on ξ and argue that, irrespective of theoretical considerations, any observational avenue to put constrains on deviations from GR, should be explored. In this spirit, we discuss how different types of observations, both gravitational and electromagnetic, can be used to constrain asymptotic-safety inspired black holes.

Observational insight into quantum gravity is scarce. Nevertheless, it is crucial that theoretical progress should always be confronted with observation. Therefore, in Sects. 5.7.2 and 5.7.3, we discuss how the constructed asymptotic-safety inspired black holes can be confronted with recent electromagnetic and gravitational-wave observations of astrophysical black holes and which part of the parameter space may be accessed by current observations.

By construction, the previously discussed asymptotic-safety inspired black holes derived from an RG-improvement of the Newton coupling introduce a single new physics scale which is tied to ξ. All observational constraints will thus effectively constrain the scale ξM_{Planck}. In Sect. 5.7.1, we discuss how ξ relates to theoretical expectations about the scale of quantum gravity.

In the context of observational constraints, it is important to keep in mind that beyond GR, black-hole uniqueness theorems need not hold. A simple example is Stelle gravity, in which there is more than one spherically-symmetric black-hole solution [88, 89, 100, 101] (although in this case thermodynamics [90] and linear instabilities [69] suggest that only one solution is stable at a given set of parameters). In particular, beyond GR, if there is no black-hole uniqueness, it may well depend on the formation history, which metric describes a black hole, and supermassive and solar-mass-black holes may be described by different metrics.[20] Constrains from different populations of black holes can therefore only be merged under the added assumption of black hole uniqueness.

5.7.1 The Scale of Quantum Gravity

...where we discuss theoretical expectations on the scale of quantum gravity relevant for black hole physics. We argue that a twofold strategy is called for, in which (i) we remain agnostic with respect to theoretical considerations and explore how far observations can probe and (ii) we remain conservative with respect to theoretical considerations and investigate whether Planck-scale modifications can be enhanced to make quantum-gravity effects significant.

It is often assumed that quantum-gravity effects are negligible in astrophysical black holes. The argument is based on the low value of the curvature at the horizon, compared to the Planck scale. However, the Planck scale is actually based on a simple dimensional analysis that uses no information whatsoever on the dynamics of quantum gravity. Therefore, it is conceivable that in a given quantum-gravity theory, effects are present at other scales; and, indeed, examples exist in quantum gravity theory.

Within asymptotic safety, the relevant scale is the transition scale, at which the scaling of G_N (and/or further gravitational couplings) changes. In pure-gravity calculations which include only the Newton coupling and cosmological constant, this scale roughly agrees with the Planck scale. This implies $\xi \approx 1$, which in turn means that modifications on horizon scales are tiny, unless the spin is close to criticality, see [53]. Given this initial expectation that $\xi \approx 1$, we argue that a twofold strategy is called for: First, we explore whether effects tied to $\xi \approx 1$ can be enhanced to impact scales that can be probed. We find that spin can act as a lever arm in this context. Second, we challenge this initial theoretical expectation; on the one hand by reflecting on freedom within the theory and on the other hand by remaining agnostic and exploring which range of ξ can be constrained with current observations.

[20] Our simple RG-improvement procedure can of course not account for this possibility, because it starts from a unique starting point, namely a Kerr black hole.

Within the theory, there may be scope for $\xi > 1$. First, the transition scale to the fixed-point regime is known to depend on the matter content of the theory; and can be significantly lower, if a large number of matter fields is present, see [35, 38].

Further, asymptotic safety has more than one relevant parameter. Each relevant parameter can be translated into a scale. Specifically, there is evidence for three relevant parameters [58]: the Newton coupling (which sets the Planck scale), the cosmological constant (which sets the corresponding low-energy scale) and a superposition of two curvature-squared couplings. The last one can set a scale significantly different from the Planck scale; e.g., in [63], the free parameter is used to obtain Starobinsky inflation, which translates into a scale four orders of magnitude below the Planck mass. Instead of accommodating Starobinsky inflation, that same free parameter can be used to lower the scale at which a departure from Einstein gravity is significant even further.[21]

There is also an argument based solely on GR, independent of the specifics of a quantum gravity theory, that suggests that quantum-gravity effects are not confined to the core of black holes, and may indeed become important at scales where the local curvature of the Kerr spacetime is still low: this argument considers perturbations on top of the Kerr spacetime, which experience an infinite blueshift at the Cauchy horizon and thereby destabilize the Cauchy horizon so that its curvature increases and a singularity forms. This suggests that quantum-gravity effects are important at the Cauchy horizon – a location where the spacetime curvature of the Kerr spacetime is still relatively low.[22]

The above reasoning, and in fact, any other determination of the scale at which quantum gravity is significant, is so far purely theoretical. Thus, it is critical to place the determination of this scale on a different footing, i.e., an observational one. In the last few years, new observational opportunities have become available, including most importantly LIGO-Virgo and the EHT. This enables us to place observational constraints on ξ without relying purely on theoretical considerations.

5.7.2 Constraints from Electromagnetic Signatures

... where we explain how data from electromagnetic observations could now and in the future be used to constrain the scale of quantum gravity in asymptotic safety. We also highlight the importance of nonlocal observables such as the presence/absence of an event horizon.

[21] An inclusion of such higher-order couplings for black holes would require the construction of spinning black-hole solutions in higher-curvature gravity, followed by an RG improvement. Due to the technical complexity of this task, it has not been attempted.

[22] This reasoning does not invalidate effective field theory, because not all terms in the effective action remain low. Because perturbations become blueshifted, the kinetic terms describing perturbations of scalars/fermions/vectors/metric fluctuations become large, signalling the breakdown of effective field theory and the need for a quantum theory of gravity.

The effects introduced by the RG improvement can be recast as an effective mass function $M(r, \chi)$, much like for other classes of regular black holes in the literature. Because the angular dependence is subleading, we discuss only the radial dependence here. A reconstruction of this mass function requires observations that probe the spacetime at several distinct distance scales: for instance, at large distances $r \gg r_g$, where r_g is the gravitational radius, M approaches a constant; at shorter distances, M is always smaller than this asymptotic constant. For two supermassive black holes, M 87* and Sag A^*, measurements at two different distances are available: for each of these supermassive black holes, M can be inferred from stellar orbits at $r \gg r_g$, and M can also be inferred from the diameter of the black-hole shadow imaged by the Event Horizon Telescope. If ξ is very large, then these two measurements are expected to be significantly different from each other. Given that for both supermassive black holes, they agree within the errors of the respective measurements, constraints on ξ can be derived. In practise, these constraints are of the order $\xi \simeq 10^{95}$ for M 87*, [67], see also [81].

Going beyond spherical symmetry, additional potential signatures in the shadow image are expected if ξ is large. This expectation arises because idealized calculations of the shadow boundary show a deviation of its shape from the corresponding shape in the Kerr spacetime, see Fig. 5.6. This deviation is due to the same physics which cause a "dent" in the horizon: in the equatorial plane, all special surfaces of the black hole (horizon, ergosphere, photon shell) experience the highest increase in compactness in the equatorial plane. There is therefore a "dent" in all these surfaces.

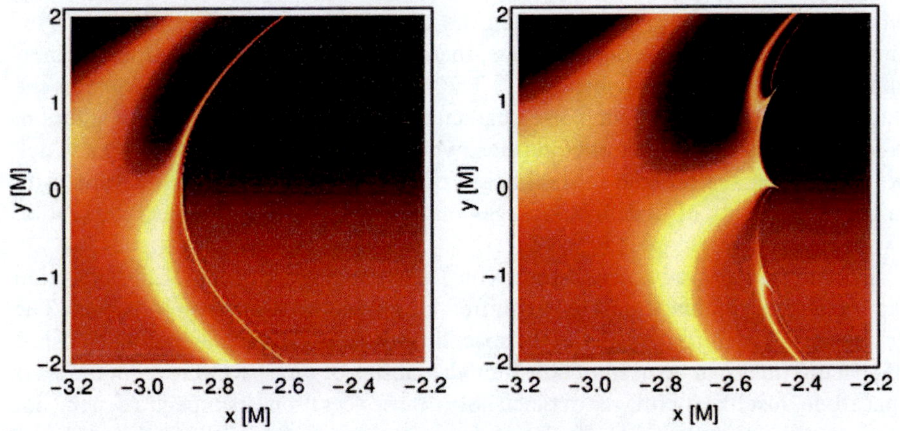

Fig. 5.6 We compare a detailed view of the prograde shadow boundary of a Kerr black hole (left-hand panel) with that of the corresponding RG-improved black hole, cf. Sect. 5.5.2. We choose $a = 0.9\,M$ and compare the Kerr case ($\xi = 0$) with a near-critical deviation parameter $\xi = 0.131$. The latter value is chosen such that the differences between both images are maximized. In both cases, the black hole is viewed at near-edge-on inclination $\theta_{\mathrm{obs}} = 9\pi/20$, where θ_{obs} is the angle between the black-hole spin axis and the vector pointing towards the observer, and the emission arises from an a geometrically thin disk model, cf. [53, 'slow model' in Table 1]

In turn, a "dent" in the photon shell is most visible on the prograde image side, where frame dragging causes geodesics to approach the event horizon more closely than on the retrograde side. In addition, such a "dent" implies that, if the inclination is not edge on (i.e., orthogonal to the spin axis), the shadow boundary is not reflection symmetric about the horizontal axis through the image. These features are tied to the angular dependence in $G_N(r, \theta)$ and thus to the local choice of scale identification, cf. Sect. 5.5.2, which violates the circularity conditions. They are not present for nonlocal RG-improvements which result in $G_N(r)$.

A final effect relates to the photon rings, i.e., the higher-order lensed images of the accretion disk. These higher-order images approach the shadow boundary, with the more highly lensed images found further inwards. The distance between these photon rings depends on the spacetime and on the astrophysics, i.e., characteristics of the accretion disk. However, for a given accretion disk, the photon rings are more separated, the larger ξ is. This can be inferred, because higher-order rings are generated by null geodesics which experience a larger effective mass. It can be confirmed by explicitly calculating simulated images for a given model of an accretion disk, as, e.g., in [52]. This effect is present for both the local and the nonlocal RG-improvement since it arises from the overall increase in compactness.

Spinning black holes also give access to much lower values of ξ: For the spin parameter close to criticality, $|a| \lesssim M$, the two horizons of the Kerr spacetime are very close to each other. In GR, they disappear at $|a| = M$, upon which a naked singularity is left behind. Thus, no physical process can spin up a black hole in GR to $|a| = M$, unless cosmic censorship in GR is violated.

In the RG-improved black hole, the two horizons disappear at $|a| = M\,(1 - \delta a)$, where $\delta a = \frac{\xi}{2}\frac{G_0}{M^2} + \mathcal{O}(\xi^2)$ [53]. Upon the disappearance of the horizons, a non-singular spacetime is left behind. Thus, there is no restriction on $|a|$ from a generalized cosmic censorship conjecture.[23] The image of such a spacetime contains the "standard" photon rings which also characterize images of black holes. In addition, it contains a series of inner photon rings, which are constituted by geodesics which are blocked by the horizon in the images of black holes. This additional intensity in the inner region of black hole images may be detectable by the next-generation Event Horizon Telescope, [50].

Additional observational constraints on ξ arise from spectroscopy. Emission from the vicinity of the black hole is redshifted on its way to a far-away observer. The gravitational redshifts depends on the spacetime. Therefore, the gravitational redshift of spectral lines can be used to constrain deviations of the spacetime from the Kerr spacetime. Just like for images of black holes, these accretion-disk spectra depend not just on spacetime-properties but also on astrophysics, i.e., properties of the accretion disk. Therefore, constraints on deviations from GR also depend on assumptions about the accretion disks.

[23] Whether or not a regular black hole can be overspun is not settled; see the discussion and references in [53].

For RG-improved black holes, this has been studied in [123, 124]. X-ray spectroscopy is possible, e.g., for known X-ray binaries. Because their masses are lower than that of the supermassive black holes observed by the EHT, the resulting constraints on ξ are stronger and [124] obtains $\xi \simeq 10^{77}$.

5.7.3 Towards Constraints from Gravitational-Wave Signatures

...where we discuss what it would take to constrain asymptotic-safety inspired black holes from gravitational-wave observations.

In contrast to the previous Sect. 5.7.2, constraining gravity theories from gravitational waves, e.g., through LIGO-Virgo observations, or future gravitational-wave interferometers as well as pulsar-timing arrays, requires knowledge of a dynamics, not just of a background spacetime. For asymptotically safe gravity, the full dynamics is currently unknown. Nevertheless, one can make contact with LIGO-Virgo observations, specifically the ringdown phase of a merger signal, by making a series of assumptions. First, one assumes that quasinormal mode (QNM) frequencies deviate from GR as a function of ξ, which controls the scale of quantum gravity. Second, one assumes that QNM frequencies for tensor modes and the eigenfrequencies of scalar (or Dirac or vector) field oscillations have a similar dependence on ξ. Third, one assumes that the relevant equation to determine the eigenfrequencies for a scalar field is the Klein-Gordon-equation on the background of an asymptotic-safety inspired black hole. The second and third assumption place strong constraints on the underlying dynamics; e.g., it is assumed that scalar fields do not have sizable nonminimal couplings. One may well question whether these strong assumptions are justified. We leave it as an important future line of research to calculate QNM frequencies from the full dynamics in asymptotically safe gravity in order to check whether the assumptions were justified.

Under these assumptions, the dependence of the QNM spectrum on ξ was studied in [79, 84, 87, 107]. The RG improvement of the spacetime results in higher real part of the fundamental mode, but lower imaginary part. The oscillations are therefore less dampened, i.e., the RG improved black hole is a better oscillator than its classical counterpart. The size of the deviations depends on ξ and on the mass, and increases with decreasing mass. For the smallest black holes,[24] for which $M \sim 3.5 M_{\text{Planck}}$, the deviation is found to be about 20%. In [79], it was further found that while the fundamental mode only deviates very slightly from its value in Schwarzschild spacetimes for astrophysical black holes, larger deviations may occur for overtones.

Going beyond the study of just scalar-field perturbations, one could take the corresponding equations for tensor perturbations, the Regge-Wheeler-equation and the Zerilli equation, and solve for the quasinormal-mode frequencies on the background

[24] For smaller mass, there is no longer an event horizon.

spacetime of an asymptotic-safety inspired black hole. However, proceeding in this way relies on an assumption that one already knows to be false, namely that the perturbation equations are the same as in GR. This is known to be incorrect, because higher-order curvature terms are known to exist. In the presence of higher-order curvature terms, the equations of motion may still be recast in the form of the Einstein equations, with an effective energy-momentum tensor that contains the higher-order terms. However, for quasinormal modes, the second variation of the action is required, i.e., perturbations on top of the equations of motion. Thus, it is no longer equivalent to work with the full equations of motion or to work with the Einstein equations plus an effective energy-momentum tensor. The lack of knowledge about the full dynamics is therefore a major obstacle to make contact with gravitational wave observations.

5.8 Generalization Beyond Asymptotic Safety: The Principled-Parameterized Approach to Black Holes

...where we generalize from RG improved black-hole spacetimes to families of black holes which incorporate a set of fundamental principles. They arise within the principled-parameterized approach to black holes which aims to connect fundamental principles to properties of black-hole spacetimes, while also being general enough to cover whole classes of theories beyond GR.

The method of RG improvement can be generalized, resulting in the principled-parameterized approach to black holes, introduced in [51, 52]. In this approach, fundamental principles are built into phenomenological models of black-hole spacetimes. For instance, in the case of asymptotically safe gravity, these fundamental principles are quantum scale symmetry for the couplings and a locality principle in the RG improvement. By generalizing the approach beyond asymptotic safety, we obtain families of black-hole spacetimes, with free parameters or even free functions.

This principled-parameterized approach lies inbetween two other approaches to black-hole spacetimes beyond GR, and unites their strengths.

First, a principled approach starts from a specific theory beyond GR, in which black-hole solutions are computed. Their properties, e.g., with regards to their images, are then calculated and can ultimately be compared with data. As an advantage, this approach connects the fundamental principles of a specific theory directly with properties of the black-hole spacetimes. As a disadvantage, a comprehensive exploration of black-hole spacetimes beyond GR is in practise impossible, given how difficult it is to find rotating solutions to theories beyond GR.

Second, a parameterized approach starts from the most general parameterization of black-hole spacetimes. Their properties can then be computed, in some cases with the general functions that parameterize the spacetime, in others, for specific choices. As an advantage, this approach is comprehensive and can in principle cover any theory beyond GR that has a black-hole spacetime as a solution. As a disadvantage, the connection to fundamental principles is lost.

The principled-parameterized approach, proposed in [51, 52] and further developed in [36, 54], starts from fundamental principles which are built into black-hole spacetimes at a heuristic level, such that families of spacetimes, parameterized by free functions and/or free parameters, satisfy these principles. Thereby, a connection between properties of spacetimes and fundamental principles is kept. At the same time, the approach is more comprehensive, in principle covering whole classes of theories at once.

As an example, a useful set of principles may be:

(i) *Regularity*: Gravity is weakened at high curvatures, such that curvature singularities are resolved.
(ii) *Locality*: Spacetimes are modified locally, such that the size of deviations from the Kerr spacetime depends on the size of curvature invariants in the Kerr spacetime.
(iii) *Simplicity*: The modifications of the spacetime accommodate (i) and (ii) in the simplest possible way, such that no more than a single new-physics scale is introduced.

The first principle is expected to cover quantum gravity theories, but may also be demanded of classical modifications of gravity. The second principle arises from an effective-field-theory approach to beyond-GR theories, where modifications to the dynamics are given by higher powers in curvature and therefore the Kerr spacetime is expected to remain an (approximate) solution at sufficiently low curvature scales, but large deviations are expected, whenever the curvature scale in the Kerr spacetime becomes large. Finally, the third principle is a requirement on the simplicity of the underlying theory, which should come with a single new-physics scale which determines the size of all effects.

Asymptotically safe gravity may be expected to respect all three principles, which is why the RG improvement procedure results in a specific example of a black-hole metric respecting these principles. The piece of information that is specific to asymptotic safety is the function $G(k)$, which becomes $G(r, \chi)$ by the RG improvement procedure. Going beyond asymptotic safety, the three principles require a modification of the strength of gravity and can thereby be incorporated by functions $G(r, \chi)$. Alternatively, in order to stay in a system of units where $G = 1$, one can upgrade the mass parameter to a mass function, $M(r, \chi)$. These two descriptions are equivalent, because it is the product $G \cdot M$ that enters the classical Kerr metric.

To properly implement the locality principle, the use of horizon-penetrating coordinates is best. The reason is that in Boyer-Lindquist coordinates, which are otherwise a popular choice for modifications of the Kerr spacetime, the upgrade $M \to M(r)$ is usually viable, but the upgrade $M \to M(r, \chi)$ generically leads to divergences in curvature invariants at the location of the classical event horizon. The reason for these curvature divergences is the coordinate divergence at this location in the classical spacetime. To disappear at the level of curvature invariants, a delicate cancellation between the various divergent terms in the metric and its derivates must occur. This delicate cancellation is disturbed by the upgrade $M \to M(r, \chi)$.

The locality principle also generically leads to non-circularity of the spacetime, simply because the circularity conditions Eq. (5.61) do not generically hold if $M = M(r, \chi)$.When one gives up the locality principle, one can, e.g., use the value of the Kretschmann scalar in the equatorial plane to determine the size of deviations from the Kerr spacetime at all angles. Then, the spacetime is generically circular and can therefore be transformed into Boyer-Lindquist coordinates by the standard transformation. Thus, such spacetimes are captured by the parameterization in [80]. Such non-local modifications are generically what arises when the Janis-Newman algorithm is used to construct axisymmetric counterparts of spherically symmetric black holes. Black-hole spacetimes constructed through the Janis-Newman algorithm[25] therefore generically appear to be in conflict with the locality principle.

In turn, non-circularity means that popular parameterizations of spacetimes beyond GR [5, 24, 72, 73, 80] do not describe all such black holes, because they all make the assumption of circularity.[26] Instead, the use of Boyer-Lindquist-type coordinates (which the above parameterizations typically use) requires the introduction of additional metric coefficients [36].

Finally, in [51, 52] it was observed that the angular dependence of the mass function results in characteristic features of shadows of black holes, cf. also Fig. 5.6. While it is an open question whether these features can also be produced through spacetimes that do not obey the locality principle, it is clear that local modifications that also obey regularity and simplicity, generically lead to such features. This is an example how the principled-parameterized approach establishes a (not necessarily one-to-one) connection between principles of fundamental physics and (in principle observable) image features of black-hole spacetimes. Asymptotically safe gravity is one example of a specific theory that is expected to obey a particular set of principles and therefore constitutes one example of a theory, which may be informed through observational constraints placed on black-hole spacetimes in the principled-parameterized approach.

5.9 Summary, Challenges and Open Questions

Asymptotic safety realizes a weakening of gravity at high scales, because it realizes scale symmetry in the UV. In turn, this implies that the Planck scale in the sense of the onset of a strong-coupling gravity becomes a low-energy “mirage”: the Planck scale is in fact the transition scale to the scaling regime, in which dimensionful quantities scale according to the canonical dimension. Thereby the Newton coupling scales quadratically towards zero, as the energy scale is increased.

---

[25] For which we are not aware of a reason why it should work beyond GR in the sense of upgrading a spherically symmetric *solution* of the theory to an axisymmetric *solution*; instead, beyond GR, it is just one particular way of reducing spherical symmetry to axisymmetry.

[26] It is typically not made explicit.

This behavior is the basis for RG improved black holes. The RG improvement procedure posits that one can capture the leading quantum effects in a classical system by (i) substituting coupling constants by their scale dependent counterparts and (ii) identifying a suitable physical scale they depend on.

The procedure has been implemented for the Schwarzschild spacetime, the Kerr spacetime, and several spacetimes that model gravitational collapse. Ambiguities in the procedure concern mainly step (ii), because a suitable scale to describe the onset of quantum effects in black holes may be either a curvature scale or a geodesic distance in vacuum spacetimes, or the matter density in non-vacuum spacetimes. Despite these ambiguities, there is universality, namely in the resolution of curvature singularities for vacuum black holes and their weakening or resolution for gravitational-collapse spacetimes. Further, all physically distinct surfaces that characterize black-hole spacetimes (e.g., horizon, photon sphere, ergosphere) are more compact than in the classical case.

We argue that to most accurately model quantum effects in the Kerr spacetime, the angular dependence of curvature invariants has to be accounted for. This leads to spacetimes which fall outside popular parameterizations of spacetimes beyond GR [36], because they do not feature a generalized Carter constant and break circularity.

Next, we address the – critical! – link to observations. We challenge the established wisdom that quantum-gravity effects are necessarily confined to the Planck length $\ell_{\text{Planck}} \approx 10^{-35}$ m. Our challenge is based on the fact that the Planck length is the result of a simple dimensional estimate, which uses no dynamical information on quantum gravity whatsoever. It is further based on the assumption that quantum gravity only has one scale, and that this scale is "natural", i.e., differs from ℓ_{Planck} only by factors of order 1.

In particular, in asymptotically safe quantum gravity, there are several free parameters, one of which is linked to higher-curvature terms. The free parameter can be used to set the coupling for these terms and might make them important already at significantly sub-planckian curvature scales.

Given these considerations, the following strategy to probe quantum gravity is a promising one:

(a) We remain agnostic and conscious that the expectation of quantum-gravity effects not above $\ell_{\text{Planck}} \approx 10^{-35}$ m is a purely theoretical one. We therefore keep the quantum-gravity scale as a free parameter and constrain it from observations as best possible. Generically, these constraints concern length scales much above the Planck scale. One may therefore also regard these constrains as constraints on classical modifications of GR.
(b) We remain conservative in our theoretical assumptions and search for settings in which quantum-gravity effects at $\ell_{\text{Planck}} \approx 10^{-35}$ m results in observable effects. We find that the spin can serve as a lever arm to increase the imprint of quantum gravity.

For part (a) of the strategy, the use of electromagnetic signatures is currently most advanced for asymptotic-safety inspired black holes. Both EHT observations as well as X-ray reflection spectroscopy of accretion disks can set limits on the quantum-gravity scale. These limits are, as is to be expected, very high above ℓ_{Planck}. The study of quasinormal modes still faces significant challenges, most importantly because it requires knowledge of the gravitational dynamics beyond GR, and the use of the perturbation equations from GR with an asymptotic-safety inspired black hole as a background is at best an approximation, but not guaranteed to be viable.

For part (b) of the strategy, the next-generation Event Horizon Telescope may be in a position to discover whether a horizonless spacetime, achieved through a combination of Planck-scale effects with high-spin-effects, is a viable description of M87* and/or Sgr A^*.

The key outstanding question in this line of research is surely the derivation of a black-hole solution from the dynamical equations of asymptotically safe quantum gravity. These are currently not available, because the full effective action in asymptotically safe gravity has not yet been calculated.

A first step in this direction has recently been undertaken in [76], where possible effective actions are constrained by demanding that certain regular black holes constitute a solution. Simultaneously, one can start by including the leading-order terms in the effective action, which are curvature-squared terms. While those are universally expected in quantum gravity, the values of the couplings are expected to satisfy a relation in asymptotic safety [8, 46, 49, 92, 98, 102]. Therefore, studying black holes in curvature-squared gravity [33, 68, 69, 88–90, 100, 101] may provide information on asymptotically safe gravity.

RG-improved black holes also constitute one example within a broader line of research, namely the *principled-parameterized* approach to black holes. In this approach, phenomenological models are constructed from classical spacetimes, by including modifications based on fundamental principles. For instance, RG-improved black holes satisfy regularity (i.e., no curvature singularities), locality (i.e., deviations from GR are parameterized by the size of the local curvature), simplicity (i.e., there is a single scale which sets deviations from GR). More broadly, modifications of GR may satisfy similar or different sets of principles, which can be incorporated in phenomenological models. In turn, one can constrain many such phenomenological models by observations.

Because the current status of black holes in many theories of quantum gravity is unsettled, with very few rigorous derivations of black-hole spacetimes from theories of quantum gravity (or, in the spinning case, even classical modifications of gravity), such a principled-parameterized approach is a promising strategy to make contact with observations. This can also guide observational searches, because one can extract where observational signatures may be present. Conversely, observational constraints may provide information on whether the fundamental principles we base our quantum-gravity theories on are indeed viable. Phenomenological models such as RG-improved black holes or more broadly black holes in the principled-parameterized approach therefore play a critical role, given the current state-of-the-art of the field.

Acknowledgements AE is supported by a research grant (29405) from VILLUM FONDEN. The work leading to this publication was supported by the PRIME programme of the German Academic Exchange Service (DAAD) with funds from the German Federal Ministry of Education and Research (BMBF). A. Held acknowledges support by the Deutsche Forschungsgemeinschaft (DFG) under Grant no 406116891 within the Research Training Group RTG 2522/1.

References

1. A. Adeifeoba, A. Eichhorn, A. Platania, Towards conditions for black-hole singularity-resolution in asymptotically safe quantum gravity. Class. Quant. Grav. **35**(22), 225007 (2018). arXiv:1808.03472 [gr-qc]
2. J.M. Bardeen, Timelike and null geodesics in the Kerr metric, in *Black Holes (Les Astres Occlus)* (1973), pp. 215–239
3. S. Basu, D. Mattingly, Asymptotic safety, asymptotic darkness, and the hoop conjecture in the extreme UV. Phys. Rev. D **82**, 124017 (2010). arXiv:1006.0718 [hep-th]
4. D. Benedetti, P.F. Machado, F. Saueressig, Asymptotic safety in higher-derivative gravity. Mod. Phys. Lett. A **24**, 2233–2241 (2009). arXiv:0901.2984 [hep-th]
5. S. Benenti, M. Francaviglia, Remarks on certain separability structures and their applications to general relativity. Gen. Relativ. Gravit **10**(1), 79–92 (1979)
6. A. Bonanno, M. Reuter, Cosmology with selfadjusting vacuum energy density from a renormalization group fixed point. Phys. Lett. B **527**, 9–17 (2002). arXiv:astro-ph/0106468
7. A. Bonanno, M. Reuter, Spacetime structure of an evaporating black hole in quantum gravity. Phys. Rev. D **73**, 083005 (2006). arXiv:hep-th/0602159
8. A. Bonanno, A. Eichhorn, H. Gies, J.M. Pawlowski, R. Percacci, M. Reuter, F. Saueressig, G.P. Vacca, Critical reflections on asymptotically safe gravity. Front. Phys. **8**, 269 (2020). arXiv:2004.06810 [gr-qc]
9. A. Bonanno, A.-P. Khosravi, F. Saueressig, Regular black holes with stable cores. Phys. Rev. D **103**(12), 124027 (2021). arXiv:2010.04226 [gr-qc]
10. A. Bonanno, A.-P. Khosravi, F. Saueressig, Regular evaporating black holes with stable cores. arXiv:2209.10612 [gr-qc]
11. A. Bonanno, B. Koch, A. Platania, Cosmic censorship in quantum Einstein gravity. Class. Quant. Grav. **34**(9), 095012 (2017). arXiv:1610.05299 [gr-qc]
12. A. Bonanno, B. Koch, A. Platania, Gravitational collapse in quantum Einstein gravity. Found. Phys. **48**(10), 1393–1406 (2018). arXiv:1710.10845 [gr-qc]
13. A. Bonanno, M. Reuter, Quantum gravity effects near the null black hole singularity. Phys. Rev. D **60**, 084011 (1999). arXiv:gr-qc/9811026
14. A. Bonanno, M. Reuter, Renormalization group improved black hole space-times. Phys. Rev. D **62**, 043008 (2000). arXiv:hep-th/0002196
15. H. Bondi, Spherically symmetrical models in general relativity. Mon. Not. R. Astron. Soc. **107**(5-6), 410–425 (1947)
16. J.N. Borissova, A. Eichhorn, Towards black-hole singularity-resolution in the Lorentzian gravitational path integral. Universe **7**(3), 48 (2021). arXiv:2012.08570 [gr-qc]
17. J.N. Borissova, A. Held, N. Afshordi, Scale-invariance at the core of quantum black holes. arXiv:2203.02559 [gr-qc]
18. J.N. Borissova, A. Platania, Formation and evaporation of quantum black holes from the decoupling mechanism in quantum gravity. arXiv:2210.01138 [gr-qc]
19. A.E. Broderick et al., The photon ring in M87*. Astrophys. J. **935**, 61 (2022). arXiv:2208.09004 [astro-ph.HE]
20. R. Carballo-Rubio, F. Di Filippo, S. Liberati, C. Pacilio, M. Visser, On the viability of regular black holes. JHEP **07**, 023 (2018). arXiv:1805.02675 [gr-qc]

21. R. Carballo-Rubio, F. Di Filippo, S. Liberati, C. Pacilio, M. Visser, Inner horizon instability and the unstable cores of regular black holes. JHEP **05**, 132 (2021). arXiv:2101.05006 [gr-qc]
22. R. Carballo-Rubio, F. Di Filippo, S. Liberati, C. Pacilio, M. Visser, Regular black holes without mass inflation instability. JHEP **09**, 118 (2022). arXiv:2205.13556 [gr-qc]
23. R. Carballo-Rubio, F. Di Filippo, S. Liberati, M. Visser, Opening the Pandora's box at the core of black holes. Class. Quant. Grav. **37**(14), 14 (2020). arXiv:1908.03261 [gr-qc]
24. V. Cardoso, P. Pani, J. Rico, On generic parametrizations of spinning black-hole geometries. Phys. Rev. D **89**, 064007 (2014). arXiv:1401.0528 [gr-qc]
25. J. Carminati, R.G. McLenaghan, Algebraic invariants of the Riemann tensor in a four-dimensional Lorentzian space. J. Math. Phys. **32**(11), 3135–3140 (1991)
26. J. Carminati, E. Zakhary, Algebraic completeness for the invariants of the Riemann tensor, in *The Ninth Marcel Grossmann Meeting*. ed. by V.G. Gurzadyan, R.T. Jantzen, R. Ruffini (2002), pp.831–834
27. R. Casadio, S.D.H. Hsu, B. Mirza, Asymptotic safety, singularities, and gravitational collapse. Phys. Lett. B **695**, 317–319 (2011). arXiv:1008.2768 [gr-qc]
28. P. Chen, Y.C. Ong, D.-H. Yeom, Black hole remnants and the information loss paradox. Phys. Rept. **603**, 1–45 (2015). arXiv:1412.8366 [gr-qc]
29. M.W. Choptuik, Universality and scaling in gravitational collapse of a massless scalar field. Phys. Rev. Lett. **70**, 9–12 (1993)
30. D. Christodoulou, The instability of naked singularities in the gravitational collapse of a scalar field. Ann. Math. **149**(1), 183–217 (1999). http://www.jstor.org/stable/121023
31. D. Christodoulou, On the global initial value problem and the issue of singularities. Class. Quantum Gravity **16**(12A), A23 (1999). https://doi.org/10.1088/0264-9381/16/12A/302
32. S.R. Coleman, E.J. Weinberg, Radiative corrections as the origin of spontaneous symmetry breaking. Phys. Rev. D **7**, 1888–1910 (1973)
33. J. Daas, K. Kuijpers, F. Saueressig, M.F. Wondrak, H. Falcke, Probing quadratic gravity with the event horizon telescope. arXiv:2204.08480 [gr-qc]
34. J.E. Daum, M. Reuter, Renormalization group flow of the holst action. Phys. Lett. B **710**, 215–218 (2012). arXiv:1012.4280 [hep-th]
35. S. de Alwis, A. Eichhorn, A. Held, J.M. Pawlowski, M. Schiffer, F. Versteegen, Asymptotic safety, string theory and the weak gravity conjecture. Phys. Lett. B **798**, 134991 (2019). arXiv:1907.07894 [hep-th]
36. H. Delaporte, A. Eichhorn, A. Held, Parameterizations of black-hole spacetimes beyond circularity. Class. Quant. Grav. **39**(13), 134002 (2022). arXiv:2203.00105 [gr-qc]
37. F. Di Filippo, R. Carballo-Rubio, S. Liberati, C. Pacilio, M. Visser, On the inner horizon instability of non-singular black holes. Universe **8**(4), 204 (2022). arXiv:2203.14516 [gr-qc]
38. P. Donà, A. Eichhorn, R. Percacci, Matter matters in asymptotically safe quantum gravity. Phys. Rev. D **89**(8), 084035 (2014). arXiv:1311.2898 [hep-th]
39. J.F. Donoghue, General relativity as an effective field theory: the leading quantum corrections. Phys. Rev. D **50**, 3874–3888 (1994). arXiv:gr-qc/9405057
40. J.F. Donoghue, Leading quantum correction to the Newtonian potential. Phys. Rev. Lett. **72**, 2996–2999 (1994). arXiv:gr-qc/9310024
41. J.F. Donoghue, A critique of the asymptotic safety program. Front. Phys. **8**, 56 (2020). arXiv:1911.02967 [hep-th]
42. I. Dymnikova, Vacuum nonsingular black hole. Gen. Rel. Grav. **24**, 235–242 (1992)
43. W.E. East, F. Pretorius, Ultrarelativistic black hole formation. Phys. Rev. Lett. **110**(10), 101101 (2013). arXiv:1210.0443 [gr-qc]
44. A. Eichhorn, On unimodular quantum gravity. Class. Quant. Grav. **30**, 115016 (2013). arXiv:1301.0879 [gr-qc]
45. A. Eichhorn, The renormalization group flow of unimodular f(R) gravity. JHEP **04**, 096 (2015). arXiv:1501.05848 [gr-qc]
46. A. Eichhorn, Status of the asymptotic safety paradigm for quantum gravity and matter. Found. Phys. **48**(10), 1407–1429 (2018). arXiv:1709.03696 [gr-qc]

47. A. Eichhorn, An asymptotically safe guide to quantum gravity and matter. Front. Astron. Space Sci. **5**, 47 (2019). arXiv:1810.07615 [hep-th]
48. A. Eichhorn, Asymptotically safe gravity, in *57th International School of Subnuclear Physics: In Search for the Unexpected* (2020). arXiv:2003.00044 [gr-qc]
49. A. Eichhorn, Status update: asymptotically safe gravity-matter systems. Nuovo Cim. C **45**(2), 29 (2022). arXiv:2201.11543 [gr-qc]
50. A. Eichhorn, R. Gold, A. Held, Horizonless spacetimes as seen by present and next-generation event horizon telescope arrays. arXiv:2205.14883 [astro-ph.HE]
51. A. Eichhorn, A. Held, From a locality-principle for new physics to image features of regular spinning black holes with disks. JCAP **05**, 073 (2021). arXiv:2103.13163 [gr-qc]
52. A. Eichhorn, A. Held, Image features of spinning regular black holes based on a locality principle. Eur. Phys. J. C **81**(10), 933 (2021). arXiv:2103.07473 [gr-qc]
53. A. Eichhorn, A. Held, Quantum gravity lights up spinning black holes. arXiv:2206.11152 [gr-qc]
54. A. Eichhorn, A. Held, P.-V. Johannsen, Universal signatures of singularity-resolving physics in photon rings of black holes and horizonless objects. arXiv:2204.02429 [gr-qc]
55. A. Eichhorn, M. Schiffer, Asymptotic safety of gravity with matter. arXiv:2212.07456 [hep-th]
56. K. Falls, D.F. Litim, Black hole thermodynamics under the microscope. Phys. Rev. D **89**, 084002 (2014). arXiv:1212.1821 [gr-qc]
57. K. Falls, D.F. Litim, A. Raghuraman, Black holes and asymptotically safe gravity. Int. J. Mod. Phys. A **27**, 1250019 (2012). arXiv:1002.0260 [hep-th]
58. K. Falls, N. Ohta, R. Percacci, Towards the determination of the dimension of the critical surface in asymptotically safe gravity. Phys. Lett. B **810**, 135773 (2020). arXiv:2004.04126 [hep-th]
59. K.G. Falls, D.F. Litim, J. Schröder, Aspects of asymptotic safety for quantum gravity. Phys. Rev. D **99**(12), 126015 (2019). arXiv:1810.08550 [gr-qc]
60. J. Fehre, D.F. Litim, J.M. Pawlowski, M. Reichert, Lorentzian quantum gravity and the graviton spectral function. arXiv:2111.13232 [hep-th]
61. H. Gies, B. Knorr, S. Lippoldt, F. Saueressig, Gravitational two-loop counterterm is asymptotically safe. Phys. Rev. Lett. **116**(21), 211302 (2016). arXiv:1601.01800 [hep-th]
62. H. Gies, A.S. Salek, Asymptotically safe Hilbert-Palatini gravity in an on-shell reduction scheme. arXiv:2209.10435 [hep-th]
63. G. Gubitosi, R. Ooijer, C. Ripken, F. Saueressig, Consistent early and late time cosmology from the RG flow of gravity. JCAP **12**, 004 (2018). arXiv:1806.10147 [hep-th]
64. U. Harst, M. Reuter, The 'Tetrad only' theory space: nonperturbative renormalization flow and Asymptotic Safety. JHEP **05**, 005 (2012). arXiv:1203.2158 [hep-th]
65. S.A. Hayward, Formation and evaporation of regular black holes. Phys. Rev. Lett. **96**, 031103 (2006). arXiv:gr-qc/0506126
66. A. Held, Invariant renormalization-group improvement. arXiv:2105.11458 [gr-qc]
67. A. Held, R. Gold, A. Eichhorn, Asymptotic safety casts its shadow. JCAP **06**, 029 (2019). arXiv:1904.07133 [gr-qc]
68. A. Held, H. Lim, Nonlinear dynamics of quadratic gravity in spherical symmetry. Phys. Rev. D **104**(8), 084075 (2021). arXiv:2104.04010 [gr-qc]
69. A. Held, J. Zhang, Instability of spherically-symmetric black holes in quadratic gravity. arXiv:2209.01867 [gr-qc]
70. W.A. Hiscock, Models of Evaporating Black Holes. Phys. Rev. D **23**, 2813 (1981)
71. J. Jiang, Y. Gao, Investigating the gedanken experiment to destroy the event horizon of a regular black hole. Phys. Rev. D **101**(8), 084005 (2020). arXiv:2003.07501 [hep-th]
72. T. Johannsen, Regular black hole metric with three constants of motion. Phys. Rev. D **88**(4), 044002 (2013). arXiv:1501.02809 [gr-qc]
73. T. Johannsen, D. Psaltis, A metric for rapidly spinning black holes suitable for strong-field tests of the no-hair theorem. Phys. Rev. D **83**, 124015 (2011). arXiv:1105.3191 [gr-qc]
74. P.S. Joshi, D. Malafarina, Recent developments in gravitational collapse and spacetime singularities. Int. J. Mod. Phys. D **20**, 2641–2729 (2011). arXiv:1201.3660 [gr-qc]

75. B. Knorr, The derivative expansion in asymptotically safe quantum gravity: general setup and quartic order. SciPost Phys. Core **4**, 020 (2021). arXiv:2104.11336 [hep-th]
76. B. Knorr, A. Platania, Sifting quantum black holes through the principle of least action. Phys. Rev. D **106**(2), L021901 (2022). arXiv:2202.01216 [hep-th]
77. B. Koch, F. Saueressig, Structural aspects of asymptotically safe black holes. Class. Quant. Grav. **31**, 015006 (2014). arXiv:1306.1546 [hep-th]
78. G. Kofinas, V. Zarikas, Avoidance of singularities in asymptotically safe quantum Einstein gravity. JCAP **10**, 069 (2015). arXiv:1506.02965 [hep-th]
79. R.A. Konoplya, A.F. Zinhailo, J. Kunz, Z. Stuchlik, A. Zhidenko, Quasinormal ringing of regular black holes in asymptotically safe gravity: the importance of overtones. JCAP **10**, 091 (2022). arXiv:2206.14714 [gr-qc]
80. R. Konoplya, L. Rezzolla, A. Zhidenko, General parametrization of axisymmetric black holes in metric theories of gravity. Phys. Rev. D **93**(6), 064015 (2016). arXiv:1602.02378 [gr-qc]
81. R. Kumar, B.P. Singh, S.G. Ghosh, Shadow and deflection angle of rotating black hole in asymptotically safe gravity. Ann. Phys. **420**, 168252 (2020). arXiv:1904.07652 [gr-qc]
82. J.-L. Lehners, K.S. Stelle, A Safe Beginning for the Universe? Phys. Rev. D **100**(8), 083540 (2019). arXiv:1909.01169 [hep-th]
83. G. Lemaître, L'Univers en expansion. Annales de la Société Scientifique de Bruxelles **53**, 51 (1933)
84. J. Li, Y. Zhong, Quasinormal modes for electromagnetic field perturbation of the asymptotic safe black hole. Int. J. Theor. Phys. **52**, 1583–1587 (2013)
85. Z. Li, C. Bambi, Destroying the event horizon of regular black holes. Phys. Rev. D **87**(12), 124022 (2013). arXiv:1304.6592 [gr-qc]
86. D.F. Litim, K. Nikolakopoulos, Quantum gravity effects in Myers-Perry space-times. JHEP **04**, 021 (2014). arXiv:1308.5630 [hep-th]
87. D.-J. Liu, B. Yang, Y.-J. Zhai, X.-Z. Li, Quasinormal modes for asymptotic safe black holes. Class. Quant. Grav. **29**, 145009 (2012). arXiv:1205.4792 [gr-qc]
88. H. Lu, A. Perkins, C.N. Pope, K.S. Stelle, Black holes in higher-derivative gravity. Phys. Rev. Lett. **114**(17), 171601 (2015). arXiv:1502.01028 [hep-th]
89. H. Lü, A. Perkins, C.N. Pope, K.S. Stelle, Spherically symmetric solutions in higher-derivative gravity. Phys. Rev. D **92**(12), 124019 (2015). arXiv:1508.00010 [hep-th]
90. H. Lü, A. Perkins, C.N. Pope, K.S. Stelle, Lichnerowicz modes and black hole families in ricci quadratic gravity. Phys. Rev. D **96**(4), 046006 (2017). arXiv:1704.05493 [hep-th]
91. A. Papapetrou, Champs gravitationnels stationnaires a symetrie axiale. Ann. Inst. H. Poincare Phys. Theor. **4**, 83–105 (1966)
92. J.M. Pawlowski, M. Reichert, Quantum gravity: a fluctuating point of view. Front. Phys. **8**, 551848 (2021). arXiv:2007.10353 [hep-th]
93. J.M. Pawlowski, D. Stock, Quantum-improved Schwarzschild-(A)dS and Kerr-(A)dS spacetimes. Phys. Rev. D **98**(10), 106008 (2018). arXiv:1807.10512 [hep-th]
94. R. Penrose, Gravitational collapse: the role of general relativity. Riv. Nuovo Cim. **1**, 252–276 (1969)
95. R. Penrose, Gravitational collapse and space-time singularities. Phys. Rev. Lett. **14**, 57–59 (1965)
96. R. Penrose, The question of cosmic censorship. J. Astrophys. Astron. **20**, 233–248 (1999)
97. R. Percacci, An introduction to covariant quantum gravity and asymptotic safety, in *100 Years of General Relativity*, vol. 3, (World Scientific, 2017)
98. A.D. Pereira, Quantum spacetime and the renormalization group: progress and visions, in *Progress and Visions in Quantum Theory in View of Gravity: Bridging Foundations of Physics and Mathematics* (2019). arXiv:1904.07042 [gr-qc]
99. A. Platania, Dynamical renormalization of black-hole spacetimes. Eur. Phys. J. C **79**(6), 470 (2019). arXiv:1903.10411 [gr-qc]
100. J. Podolský, R. Švarc, V. Pravda, A. Pravdova, Black holes and other exact spherical solutions in quadratic gravity. Phys. Rev. D **101**(2), 024027 (2020). arXiv:1907.00046 [gr-qc]

101. V. Pravda, A. Pravdova, J. Podolsky, R. Svarc, Exact solutions to quadratic gravity. Phys. Rev. D **95**(8), 084025 (2017). arXiv:1606.02646 [gr-qc]
102. M. Reichert, Lecture notes: functional renormalisation group and asymptotically safe quantum gravity. PoS **384**, 005 (2020)
103. M. Reuter, E. Tuiran, Quantum gravity effects in the Kerr spacetime. Phys. Rev. D **83**, 044041 (2011). arXiv:1009.3528 [hep-th]
104. M. Reuter, H. Weyer, Quantum gravity at astrophysical distances? JCAP **12**, 001 (2004). arXiv:hep-th/0410119
105. M. Reuter, H. Weyer, Renormalization group improved gravitational actions: a Brans-Dicke approach. Phys. Rev. D **69**, 104022 (2004). arXiv:hep-th/0311196
106. M. Reuter, F. Saueressig, *Quantum Gravity and the Functional Renormalization Group: The Road towards Asymptotic Safety* (Cambridge University Press, 2019)
107. A. Rincón, G. Panotopoulos, Quasinormal modes of an improved Schwarzschild black hole. Phys. Dark Univ. **30**, 100639 (2020). arXiv:2006.11889 [gr-qc]
108. E. Teo, Spherical photon orbits around a Kerr black hole. Gen Relativ Gravit **35**(11), 1909–1926 (2003)
109. R.C. Tolman, Effect of inhomogeneity on cosmological models. Proc. Natl. Acad. Sci. **20**(3), 169–176 (1934)
110. R. Torres, Non-singular quantum improved rotating black holes and their maximal extension. Gen. Rel. Grav. **49**(6), 74 (2017). arXiv:1702.03567 [gr-qc]
111. R. Torres, F. Fayos, Singularity free gravitational collapse in an effective dynamical quantum spacetime. Phys. Lett. B **733**, 169–175 (2014). arXiv:1405.7922 [gr-qc]
112. R. Torres, Singularity-free gravitational collapse and asymptotic safety. Phys. Lett. B **733**, 21–24 (2014). arXiv:1404.7655 [gr-qc]
113. R. Torres, Nonsingular black holes, the cosmological constant, and asymptotic safety. Phys. Rev. D **95**(12), 124004 (2017). arXiv:1703.09997 [gr-qc]
114. R. Torres, F. Fayos, On the quantum corrected gravitational collapse. Phys. Lett. B **747**, 245–250 (2015). arXiv:1503.07407 [gr-qc]
115. E.A. Uehling, Polarization effects in the positron theory. Phys. Rev. **48**, 55–63 (1935)
116. P. Vaidya, The gravitational field of a radiating star. Proc. Natl. Inst. Sci. India A **33**, 264 (1951)
117. P.C. Vaidya, Nonstatic solutions of Einstein's field equations for spheres of fluids radiating energy. Phys. Rev. **83**, 10–17 (1951)
118. P.C. Vaidya, An analytical solution for gravitational collapse with radiation. Astrophys. J. **144**, 943 (1966)
119. R.M. Wald, Gravitational collapse and cosmic censorship (1997), pp. 69–85. arXiv:gr-qc/9710068
120. Y. Xie, J. Zhang, H.O. Silva, C. de Rham, H. Witek, N. Yunes, Square peg in a circular hole: choosing the right ansatz for isolated black holes in generic gravitational theories. Phys. Rev. Lett. **126**(24), 241104 (2021). arXiv:2103.03925 [gr-qc]
121. S.-J. Yang, Y.-P. Zhang, S.-W. Wei, Y.-X. Liu, Destroying the event horizon of a nonsingular rotating quantum-corrected black hole. JHEP **04**, 066 (2022). arXiv:2201.03381 [gr-qc]
122. E. Zakhary, C.B.G. McIntosh, A complete set of Riemann invariants. Gen. Relat. Gravit. **29**(5), 539–581 (1997)
123. Y. Zhang, M. Zhou, C. Bambi, Iron line spectroscopy of black holes in asymptotically safe gravity. Eur. Phys. J. C **78**(5), 376 (2018). arXiv:1804.07955 [gr-qc]
124. B. Zhou, A.B. Abdikamalov, D. Ayzenberg, C. Bambi, S. Nampalliwar, A. Tripathi, Shining X-rays on asymptotically safe quantum gravity. JCAP **01**, 047 (2021). arXiv:2005.12958 [astro-ph.HE]
125. T. Zhou, L. Modesto, Geodesic incompleteness of some popular regular black holes. arXiv:2208.02557 [gr-qc]
126. C.M. Chen, Y. Chen, A, Ishibashi, N, Ohta, D, Yamaguchi, Running Newton coupling, scale identification, and black hole thermodynamics. Phys. Rev. D **105** (10), 106026 (2022) arXiv:2204.09892 [hep-th], https://doi.org/10.1103/PhysRevD.105.106026

Chapter 6
Regular Black Holes in Palatini Gravity

Gonzalo J. Olmo and Diego Rubiera-Garcia

6.1 Introduction to Metric-Affine Gravity and Resolution of Space-Time Singularities

In the construction of the theories for the gravitational field considered in this chapter, the classical paradigm views gravity as the manifestation of a dynamical space-time, i.e., a differentiable manifold $\mathscr{M}$ endowed with geometrical entities acting not only as the canvas in which events happen, but which also react in a non-trivial way to the motion of energy and observers living on it. The first element in this view - the geometrical entities - is tightly attached to the roles attributed to the main characters of the geometry, namely, the metric g and the affine connection Γ. While the metric is the responsible of the causal structure of space-time and, as such, it is associated to local measurements of distances, angles, areas, volumes, etc, the affine connection is associated to free-fall, defining the notion of parallelism and entering into the covariant derivatives. As such, these two objects are conceptually and operationally independent from each other. The second aspect - the dynamical behavior - is given by the particular way the geometrical elements react (and the other way round) to the motion of the matter fields, and is given by the field equations derived from the action defining the theory. The original formulation by Einstein of his General Theory of Relativity (GR) amounts to associate the affine connection to the metric via the Christoffel symbols of the latter, thus rendering a privileged role to a particular connection - the Levi-Civita one - in order for its covariant derivative to preserve the

G. J. Olmo
Departamento de Física Teórica and IFIC, Centro Mixto Universidad de Valencia - CSIC. Universidad de Valencia, Burjassot, 46100 Valencia, Spain
e-mail: gonzalo.olmo@uv.es

D. Rubiera-Garcia (✉)
Departamento de Física Teórica and IPARCOS, Universidad Complutense de Madrid, E-28040 Madrid, Spain
e-mail: drubiera@ucm.es

C. Bambi (ed.), *Regular Black Holes*, Springer Series in Astrophysics and Cosmology,
https://doi.org/10.1007/978-981-99-1596-5_6

metric, and selects the lowest-order scalar object in the action - the curvature scalar - able to endow the theory with dynamics. This way, using the duo $\{\mathcal{M}, g\}$ one finds the canonical (*curvature-based*) formulation of GR, whose Einstein-Hilbert action yields the dynamical (Einstein) equations of the theory.

Since these events took place, theoreticians and geometers alike have not been idle. Among their many findings, for the sake of this chapter we underline the (nowadays) well understood fact that the affine connection can be split, in general, into three pieces, namely, curvature, torsion, and non-metricity, each with its own physical interpretation: the rotation of a vector transported along a closed curve, the non-closure of parallelograms when two vectors are parallely transported along each other, and the variation of a vector's length when parallel-transported. It turns out that, when the action of the gravitational theory is built upon the lowest scalar of each of these pieces (thus switching off the other two), the resulting dynamical equations are precisely the Einstein ones [21]. Diffeomorphism and Lorentz invariances are preserved, and the background solutions of each formulation of GR turn out to be the same in such a way that it is only at the level of their respective boundary terms (relevant for some applications) that they can be distinguished from each other. Such three versions are called (canonical) GR, the teleparallel equivalent of GR [8, 49], and the (symmetric) teleparallel GR [52, 54], respectively. Fully enlarging GR to include non-vanishing contributions of every such piece of the affine connection yields a theory called metric-affine gravity, given by a triplet $\{\mathcal{M}, g, \Gamma\}$ with a priori unspecified relations between g and Γ. Such a theory contains not only the three equivalent formulations of GR on its corresponding limits, but also other gravitational theories explored in the literature such as Einstein-Cartan, Weyl gravity, etc. Such a picture can be very well dubbed as *affinesia*.[1]

In view of the discussion above, in order to look for a way out of the unavoidable existence of space-time singularities within GR as guaranteed by the singularity theorems [73], one is faced at a crossroads with (at least) two important decisions to make: (i) what are the underlying geometrical elements of the theory as well as the a priori relation (if any) among them, and (ii) what is the action of the theory yielding (via its field equations) dynamics to them and allowing observers to "gravitate". In metric-affine (Palatini) theories of gravity the path taken is defined by the following choices: (i) restore metric and affine connection to their original roles as independent entities, taking the latter to have both non-vanishing curvature, torsion, and non-metricity pieces, and (ii) enlarge the Einstein-Hilbert action towards a new scalar action in such a way that the target theory has a number of minimum properties: diffeomorphism plus Lorentz invariances, compatibility with weak-field experiments (which means it reduces to GR in the suitable limit), and absence of extra propagating degrees of freedom (removing new scalar-like fields or additional tensorial polarizations). These constraints are conservative enough so as not to enter into conflict with those experiments that GR passes [77], but at the same time furnishing the resulting theory with a larger flexibility of the geometrical architecture which can also be put

[1] Credit to Jose Beltrán Jiménez who, as far as we know, is the person to be blamed for coining this term.

to observational test [10]. This chapter amounts to a substantive discussion of this framework and the insights it has brought to our knowledge regarding the nature of space-time singularities and how to resolve them in physically reasonable enough settings. For simplicity, and since we will be dealing with bosonic fields, which are oblivious to the torsional part of the affine connection [2], we shall neglect it from our considerations, though some comments and references will be provided when necessary.

There are many theories that can accommodate these two requirements. In order to start our discussion, we shall start with the essentials, and progress from there to the most general family of such theories studied in detail so far, dubbed as Ricci-based gravities. Let us begin.

6.2 Gravitational Models

6.2.1 $f(\mathcal{R})$ Gravity

The simplest extension of GR that we will consider is given by $f(\mathcal{R})$ gravity [57], where f is an arbitrary function of the curvature scalar $\mathcal{R} = g^{\mu\nu}R_{\mu\nu}(\Gamma)$, where the Ricci tensor $R_{\mu\nu} \equiv R^{\alpha}{}_{\mu\alpha\nu}$ is built from the Riemann tensor

$$R^{\rho}{}_{\sigma\mu\nu} = \partial_{\mu}\Gamma^{\rho}_{\nu\sigma} - \partial_{\nu}\Gamma^{\rho}_{\mu\sigma} + \Gamma^{\rho}_{\mu\lambda}\Gamma^{\lambda}_{\nu\sigma} - \Gamma^{\rho}_{\nu\lambda}\Gamma^{\lambda}_{\mu\sigma} \tag{6.1}$$

made up of an independent affine (torsionless) connection, $\Gamma \equiv \Gamma^{\lambda}_{\mu\nu}$, which is a priori unrelated to any metric. It is worth pointing out that this Palatini curvature scalar is different from the metric curvature scalar, $R = g^{\mu\nu}R_{\mu\nu}(g)$, where in this case the Ricci tensor $R_{\mu\nu}(g)$ is the one derived from the Christoffel symbols of the space-time metric $g_{\mu\nu}$ entering in the definition of the action. This fact makes the dynamics of Palatini $f(\mathcal{R})$ gravity to dramatically depart from its metric counterpart [32].

Projective invariance: The Einstein-Hilbert action is invariant under a class of (projective) transformations of the form $\tilde{\Gamma}^{\lambda}_{\mu\nu} = \Gamma^{\lambda}_{\mu\nu} + \xi_{\mu}\delta^{\lambda}_{\nu}$, where ξ_{μ} is a 1-form vector field. However, under this transformation the Ricci tensor is not invariant, transforming as $\tilde{R}_{\mu\nu} = R_{\mu\nu} + F_{\mu\nu}$, where the tensor $F_{\mu\nu} = \partial_{\mu}\xi_{\nu} - \partial_{\nu}\xi_{\mu}$. Since according to this, only the symmetric part of the Ricci tensor in $R_{\mu\nu} = R_{(\mu\nu)} + R_{[\mu\nu]}$ (parenthesis and brackets denote symmetrization and anti-symmetrization, respectively) is invariant under projective transformations, gravity theories based on the full Ricci tensor are prone to developing ghost-like instabilities associated to the anti-symmetric piece [20]. In order to avoid this problem, in the construction of the most general class of theories consider in this Chapter, namely, Ricci-Based Gravity theories (RBGs), we will just consider the symmetric part of the Ricci tensor $R_{(\mu\nu)}$, but remove the parenthesis for notational simplicity. This subtlety is irrelevant in the $f(\mathcal{R})$ case, because the contraction with the metric filters out the antisymmetric part.

Let us define its action as

$$S = \frac{1}{2\kappa^2}\int d^4x\sqrt{-g}f(\mathcal{R}) + S_m(g_{\mu\nu}, \psi_m) \tag{6.2}$$

where $\kappa^2 = 8\pi$ in $G = c = 1$ units, g is the determinant of the space-time metric $g_{\mu\nu}$, and $S_m = \int d^4x\sqrt{-g}\mathcal{L}_m(g_{\mu\nu}, \psi_m)$ is the matter action of a set of matter fields denoted collectively by ψ_m. Note that the independent connection Γ does not enter in the construction of the matter sector, which will have relevant implications when discussing geodesic behaviour later.

The field equations of this theory are obtained by independent variation of the action (6.2) with respect to metric and connection, which yields the two systems of equations (see [57] for a detailed derivation including torsion)

$$f_{\mathcal{R}} R_{\mu\nu} - \frac{1}{2}g_{\mu\nu}f = \kappa^2 T_{\mu\nu} \tag{6.3}$$

$$\nabla^{\Gamma}_{\beta}(\sqrt{-g}f_{\mathcal{R}}g^{\mu\nu}) = 0 \tag{6.4}$$

where $f_{\mathcal{R}} \equiv \frac{df}{d\mathcal{R}}$ and $T_{\mu\nu} \equiv \frac{-2}{\sqrt{-g}}\frac{\delta S_m}{\delta g^{\mu\nu}}$ is the stress-energy tensor of the matter fields. It is thus immediately seen the deviance with respect to the metric formulation of $f(R)$ gravity (calligraphic letter removed to highlight its metric character): in such a case, the fact that Γ is Levi-Civita of $g_{\mu\nu}$, replaces the two sets of equations above by the single equation

$$f_R R_{\mu\nu} - \frac{1}{2}g_{\mu\nu}f - [\nabla_\mu\nabla_\nu f_R - g_{\mu\nu}\Box R] = \kappa^2 T_{\mu\nu} \tag{6.5}$$

The presence of two derivative operators acting upon the scalar function $f(R)$, which itself contains two derivatives, has two consequences: first the theory generally contains fourth-order equations of motion, and second, the object $\phi = f_R$ can effectively be seen as a new (scalar) propagating degree of freedom, which introduces some difficulties in order to make the theory compatible with weak-field limit observations [27, 55, 56]. As opposed to this, in the Palatini formulation, such derivative operators are missing, which turns the field equations as second-order and there are no extra degrees of freedom to worry about. The price to be paid is that in this formulation one needs to solve the system of equations (6.4) in order to find the affine connection. In the present case, this is easily done by just noting that, after contraction of (6.3) with the metric $g_{\mu\nu}$, one finds an algebraic equation $\mathcal{R}f_{\mathcal{R}} - 2f(\mathcal{R}) = \kappa^2 T$, where $T = g^{\mu\nu}T_{\mu\nu}$ is the trace of the stress-energy tensor. This result tells us that $\mathcal{R} = \mathcal{R}(T)$, i.e., the curvature scalar can be written as a function of the trace T of the matter fields. This implies that in the Palatini version of the theory, the scalar field carries no dynamics and its role will be reduced to introducing a deformation in the way the matter fields source the gravitational dynamics.

In turn, this result allows to solve the connection equations (6.4) by introducing a new rank-two tensor $q_{\mu\nu}$ related to the space-time metric and the matter fields via the (conformal) relation

$$q_{\mu\nu} = \frac{1}{f_{\mathcal{R}}(T)} g_{\mu\nu} \tag{6.6}$$

where we recall that $f_{\mathcal{R}} \equiv df/d\mathcal{R}$ is a function of the trace T. In terms of $q_{\mu\nu}$, Eq. (6.4) reads as $\nabla_\beta^\Gamma(\sqrt{-q}q^{\mu\nu}) = 0$, i.e., Γ can be solved as the Christoffel symbols of $q_{\mu\nu}$. Equipped with the relation above, contracting again in (6.3) with $g^{\alpha\mu}$, and suitably rearranging terms, one arrives at

$$R^\mu{}_\nu(q) = \frac{\kappa^2}{f_{\mathcal{R}}^2}\left(\frac{f}{2}\delta_\nu^\mu + \kappa^2 T^\mu{}_\nu\right) \tag{6.7}$$

where $T^\mu{}_\nu \equiv T^{\mu\alpha}g_{\alpha\nu}$. Since Γ is Levi-Civita of q, then $R^\mu{}_\nu(q) \equiv q^{\mu\alpha}R_{\alpha\nu}$ is the usual Ricci tensor computed with the Christoffel symbols of $q_{\mu\nu}$. Moreover, since both f and $f_{\mathcal{R}}$ are functions of the matter sources, Eq. (6.7) is actually nothing more than Einstein equations (for q) coupled to a modified stress-energy tensor on its right-hand side. Therefore, they can be solved by resorting to the usual analytical and numerical methods developed within GR in order to find an expression for $q_{\mu\nu}$. Subsequently, using (6.6) it is trivial to find the corresponding solution for the space-time metric $g_{\mu\nu}$. The bottleneck of this procedure is to actually be able to solve the equation $\mathcal{R} = \mathcal{R}(T)$ under a workable enough form: for instance, in the quadratic case, $f(\mathcal{R}) = \mathcal{R} + \alpha\mathcal{R}^2$, with α a constant with dimensions of length squared, one finds $\mathcal{R} = -\kappa^2 T$, which is actually the same result as in GR. For more involved functional dependencies, this would introduce additional difficulties in solving the field equations, so one would be forced to introduce numerical methods.

This simple extension of GR illustrates the benefits and drawbacks of working in the Palatini approach: on the former we find a system of second-order field equations without extra propagating degrees of freedom that can be solved with standard methods, while for the latter one needs to find first a suitable algebraic way to solve the connection equations feeding the metric one. This procedure can be generalized to other cases of interest, as we shall do next.

6.2.2 *Quadratic Gravity*

The theory of quantized fields in a curved space-time tells us that an ultraviolet completion of GR must come under the form of higher-order contributions in the scalar objects out of curvature contractions, suppressed by inverse powers of the fundamental quantum (Planck) scale [25, 70]. This yields the natural generalization of $f(\mathcal{R})$ gravity by including another contraction of the Ricci tensor, $\mathcal{Q} = g^{\mu\alpha} g^{\nu\beta} R_{\mu\nu} R_{\alpha\beta}$, via an arbitrary function of these two invariants, i.e. $f(\mathcal{R}, \mathcal{Q})$. Therefore, the theory (6.8) is generalized to

$$S = \frac{1}{2\kappa^2} \int d^4x \sqrt{-g} f(\mathcal{R}, \mathcal{Q}) + S_m(g_{\mu\nu}, \psi_m) \tag{6.8}$$

with the same considerations and notations as before. The (Palatini) field equations associated to this action are again found by independent variations with respect to metric and connection, which in this case amount to

$$f_{\mathcal{R}} R_{\mu\nu} - \frac{1}{2} g_{\mu\nu} f + 2 f_{\mathcal{Q}} R_{\mu\alpha} R^{\alpha}{}_{\nu} = \kappa^2 T_{\mu\nu} \tag{6.9}$$

$$\nabla_{\beta}^{\Gamma} (\sqrt{-g} (f_{\mathcal{R}} g^{\mu\nu} + 2 f_{\mathcal{Q}} R^{\mu\nu})) = 0 \tag{6.10}$$

where a new contribution in $f_{\mathcal{Q}} \equiv \frac{df}{d\mathcal{Q}}$ is apparent. In the attempt to implement the same strategy as in the $f(\mathcal{R})$ case to solve the connection equations, one first notes that by defining a tensor $M^{\mu}{}_{\nu} \equiv g^{\mu\alpha} R_{\alpha\nu}$, then the objects $\mathcal{R}$ and $\mathcal{Q}$ are simply given in terms of traces of this object as $\mathcal{R} = \text{Tr}(M^{\mu}{}_{\nu})$, $\mathcal{Q} = \text{Tr}(M^{\mu}{}_{\alpha} M^{\alpha}{}_{\nu})$, respectively. The shape of each of these matrices is obtained from (6.9) rewritten as $2 f_{\mathcal{Q}} \hat{M}^2 + f_{\mathcal{R}} \hat{M} - \frac{f}{2} \hat{I} = \kappa^2 T$, where in all cases hats represent matrices. This implies that, similarly as in the $f(\mathcal{R})$ case above, one can solve this equation as $\hat{M} \equiv \hat{M}(\hat{T})$, i.e., both $\mathcal{R}$ and $\mathcal{Q}$ can be solved as functions of the stress-energy sources. In turn, this gives again consistency to introducing a new rank-two tensor as an attempt to solve the connection equations (6.10) via $\sqrt{-q} q^{\mu\nu} = \sqrt{-g} g^{\mu\alpha} \Sigma_{\alpha}{}^{\nu}$, where the matricial object $\hat{\Sigma} = f_{\mathcal{R}} \hat{I} + 2 f_{\mathcal{Q}} \hat{M}$. Operating these relations in (6.10) one arrives at

$$q_{\mu\nu} = |\hat{\Sigma}|^{1/2} \Sigma_{\mu}{}^{\alpha} g_{\alpha\nu} \tag{6.11}$$

where vertical bars denote a determinant. Furthermore, by contracting the metric field equations (6.9) with $M_{\mu}{}^{\alpha} \Sigma_{\alpha}{}^{\nu}$ one arrives at

$$R_{\mu}{}^{\nu}(q) = \frac{1}{|\hat{\Sigma}|^{1/2}} \left(\frac{f}{2} \delta_{\mu}^{\nu} + \kappa^2 T_{\mu}{}^{\nu} \right) \tag{6.12}$$

Since again Γ is Levi-Civita of q, and both $\mathcal{R}$ and $\mathcal{Q}$ are functions of the matter sources, Eq. (6.12) can again be read off as a system of Einstein-like equations sourced with a modified stress-energy tensor on its right-hand side, now including those extra contributions from the $\mathcal{Q}$-piece. Similarly as in the $f(\mathcal{R})$ case, application of standard analytical/numerical methods to solve these equations for the metric $q_{\mu\nu}$, and subsequent use of the (matter-mediated) transformation (6.11) allows to put the corresponding results into a solution for the space-time metric $g_{\mu\nu}$. The bottleneck of this procedure is again to be able to find a workable solution for the matrix $\hat{\Sigma}$, which is more involved than in the $f(\mathscr{R})$ case given the disformal rather than conformal transformation (6.11) between q and g. This resolution can be achieved for some simple enough examples, such as the quadratic gravity theory $f(\mathscr{R}, \mathscr{Q}) = \mathscr{R} + a\mathscr{R}^2 + b\mathscr{Q}$, with (a, b) some parameters with dimensions of length squared.

It is worth pointing out that one could add to this quadratic gravity theory further contractions of the Riemann tensor such as the Kretschmann scalar $K = R_{\alpha\beta\gamma\delta}R^{\alpha\beta\gamma\delta}$. However, we do not currently possess the necessary geometrical methods to deal with the resolution of the corresponding connection field equations. Instead, we shall head in the next section in a different direction.

6.2.3 *Eddington-Inspired Born-Infeld Gravity*

Adding further powers of the curvature scalars is not the only possible way to keep generalizing these theories. Indeed, every scalar Lagrangian of weight −4, appearing in the integral of the action, could make a candidate to a gravitational theory. A proposal whose relevance in the community has quite grown in the last few years was originally considered by Vollick [76] and then popularized by Bañados and Ferreira [13], being generally known as Eddington-inspired Born-Infeld gravity (EiBI), and is given by the action:

Born-Infeld-type theories: There is quite a long tradition of invoking square-root modifications of classical actions: from the relativistic Lagrangian of point particles or the Born-Infeld modification of classical electrodynamics to cure the electron's self-energy problem [26], to the results of Fradkin and Tseytlin [37] showing that actions of the Born-Infeld type arise in different scenarios related to M-theory, to finally arrive to the gravitational arena. Indeed, Born-Infeld type formulations of the gravitational field have been considered according to several frameworks and to address many astrophysical phenomena (see [24] for a review).

$$S_{EiBI} = \frac{1}{\kappa^2 \varepsilon} \int d^4x \left[\sqrt{-|g_{\mu\nu} + \varepsilon R_{\mu\nu}|} - \lambda\sqrt{-g} \right] + S_m(g_{\mu\nu}, \psi_m) \tag{6.13}$$

where ε is a smallness scale with dimensions of length squared, encoding the deviations with respect to GR. In this sense, for $|R_{\mu\nu}| \ll \varepsilon^{-1}$, this theory boils down to

$$S_{EiBI} \approx \int d^4x\sqrt{-g}\left(\frac{\mathcal{R}}{2\kappa^2} - \Lambda_{eff}\right) - \frac{\varepsilon}{4\kappa^2}\int d^4x\sqrt{-g}\left(-\frac{\mathcal{R}^2}{2} + \mathcal{Q}\right) + \mathcal{O}(\varepsilon^2) \tag{6.14}$$

which is nothing but GR with an effective cosmological constant $\Lambda_{eff} = \frac{\lambda-1}{\varepsilon}$ supplemented with higher-order curvature corrections, which at the quadratic level it is actually an example of a $f(\mathcal{R}, \mathcal{Q})$ theory. A Palatini formulation of this theory can be achieved if one introduces the new-rank two tensor as $q_{\mu\nu} = g_{\mu\nu} + \varepsilon R_{\mu\nu}$, in such a way that a variation of the action (6.13) with respect to the metric yields

$$\frac{\sqrt{-q}}{\sqrt{-g}} q^{\mu\nu} - g^{\mu\nu} = \kappa^2 \varepsilon T^{\mu\nu} \tag{6.15}$$

with (once again) Γ being Levi-Civita of q. The above equations can be written in a more convenient (and workable) way by writing formally the relation between the two metrics of the theory under the algebraic relation

$$q_{\mu\nu} = g_{\mu\alpha}\Omega^{\alpha}{}_{\nu} \tag{6.16}$$

To obtain the shape of the matrix $\hat{\Omega}$ we just need to contract the metric field equations (6.15) with the metric $g_{\nu\alpha}$ to find that its components are given by the algebraic equation

$$|\Omega|^{1/2}(\Omega^{-1})^{\mu}{}_{\nu} = \lambda\delta^{\mu}_{\nu} - \kappa^2\varepsilon T^{\mu}{}_{\nu} \tag{6.17}$$

which therefore can be determined once a matter sector is specified in the theory. Moreover, this matrix can also be used to rewrite the EiBI Lagrangian in (6.13) in terms of it as

$$\mathcal{L}_G = \frac{|\Omega|^{1/2}}{\varepsilon\kappa^2} - \lambda \tag{6.18}$$

while suitably working upon the metric field equations (6.13) one can rewrite them, with the help of $\hat{\Omega}$ and the expression above, as

$$R^{\mu}{}_{\nu}(q) = \frac{1}{|\Omega|^{1/2}}\left(\mathcal{L}_G\delta^{\mu}_{\nu} + \kappa^2 T^{\mu}{}_{\nu}\right) \tag{6.19}$$

This way we arrive (once again) to a set of Einstein-like equations sourced with a modified stress-energy tensor. The above three examples manifest a general trend: by introducing a sort of "auxiliary" metric $q_{\mu\nu}$ such that the independent connection is Levi-Civita of it, one can recast the field equations of the theory in Einstein-like form with its right-hand side emerging as the result of a sort of non-minimal coupling of the matter fields, while an algebraic *deformation* allows to transform the solution for $q_{\mu\nu}$ into a solution for the space-time metric $g_{\mu\nu}$. One can thus wonder what is the most general theory sharing all these features, which we tackle next.

6.2.4 The Ricci-Based Gravity Family

We wish to build a (projectively-invariant) theory of gravity based on every possible contractions of the (symmetric part of the) Ricci tensor with the metric. We can thus tentatively write

$$S_{RBG} = \frac{1}{2\kappa^2} \int d^4x \sqrt{-g} \mathscr{L}_G(g_{\mu\nu}, R_{(\mu\nu)}) + S_m(g_{\mu\nu}, \psi_m) \tag{6.20}$$

In order to guarantee that the function $\mathscr{L}_G$ is an scalar object, its functional dependence on its argument must be via traces of the object $M^\mu{}_\nu \equiv g^{\mu\alpha} R_{(\alpha\nu)}$. These theories are dubbed as Ricci-based gravities (or RBGs for short) and includes, in particular, the three cases above, since $\mathscr{R} = \text{Tr}(M^\mu{}_\nu)$, $\mathscr{Q} = \text{Tr}(M^\mu{}_\alpha M^\alpha{}_\nu)$, while the EiBI Lagrangian comes from a combination of the polynomial invariants associated to $M^\mu{}_\nu$ (for a general description on how to build actions from such invariants see [22]). To verify the validity of this general construction, we need to obtain the field equations associated to the action (6.20). By performing independent variations with respect to the metric and the connection, and assuming an algebraic relation between them of the form (6.16), one arrives at the metric field equations (see [2] for a detailed derivation including torsion)

$$G^\mu{}_\nu(q) = \frac{1}{|\Omega|^{1/2}} \left(T^\mu{}_\nu - \left(\mathscr{L}_G + \frac{T}{2} \right) \delta^\mu_\nu \right) . \tag{6.21}$$

This equation confirms our expectations: that the whole RBG family admits an Einstein-like representation of its field equations (for the metric q) with its right-hand side being that of a modified stress-energy tensor, since $\mathscr{L}_G \equiv \mathscr{L}_G(T_\mu{}^\nu, g_{\mu\nu})$. The space-time metric $g_{\mu\nu}$ is obtained as an algebraic deformation of $q_{\mu\nu}$ via the fundamental relation (6.16) and, consequently, we shall dub $\hat{\Omega}$ as the *deformation matrix*. In physically sensible scenarios, this matrix will typically follow the same algebraic structure as the one of the stress-energy tensor $T^\mu{}_\nu$, though other solutions can also be found [19]. Examples of this are the conformal transformation (6.6) of the $f(\mathscr{R})$ case, or the disformal transformations (6.11) and (6.17) of the $f(\mathscr{R}, \mathscr{Q})$ and EiBI cases. Furthermore, the relations obtained so far imply that the deformation matrix is, in general, a function of both the matter fields and the space-time metric, thus the extra terms present in the right-hand side of the RBG field equations (6.21) will also be functions of the matter. These features have important consequences. Indeed, the RBG field equations are always (as opposed to their metric counterparts) second-order and, moreover, in absence of matter, $T_\mu{}^\nu = 0$, its solutions (for q) are those of GR plus (possibly) a cosmological constant term, while as for the space-time metric itself one has $g_{\mu\nu} = q_{\mu\nu}$ (modulo a trivial re-scaling). This implies that RBG theories do not propagate other degrees of freedom beyond the two polarizations of the gravitational field (gravitational waves) travelling (in vacuum) at the speed of light (see [44] for the EiBI case), allowing them to naturally pass constraints from the observations of gravitational waves and their electromagnetic counterpart from binary neutron stars mergers [1]. Modifications with respect to GR predictions occur only inside matter sources, where in addition to effects related to the total mass-energy of the sources the new dynamics is also fed by energy-density effects, which is a feature absent within GR. This is yet another trademark of these theories with important consequences for the structure of their solutions.

There is, however, a difficulty to solve the field equations (6.21) due to the fact that, while their left-hand side is written in terms of $q_{\mu\nu}$, its right-hand side will be typically be written in terms of $g_{\mu\nu}$, whose inversion in terms of $q_{\mu\nu}$ using the fundamental relation (6.16) is not always easy (or even possible) depending of the algebraic properties of the deformation matrix in relation to the matter fields feeding it. In sufficiently symmetric scenarios, it is actually possible to invert this relation and write the field equations entirely in terms of $q_{\mu\nu}$, but for more complex (and interesting) scenarios, advanced techniques are required.

6.2.5 Einstein Frame and the Mapping Method

The large resemblance of (6.21) with the canonical Einstein equations strongly suggests that a full Einstein representation should be possible, i.e.:

$$G^{\mu}{}_{\nu}(q) = \kappa^2 \tilde{T}^{\mu}{}_{\nu}(q) \tag{6.22}$$

for a new stress-energy tensor $\tilde{T}^{\mu}{}_{\nu}(q)$. To this end, let us introduce a set of auxiliary fields $\Sigma^{\mu}{}_{\nu}$ to rewrite the action (6.20) as

$$S_{RBG}(g_{\mu\nu}, \Gamma^{\lambda}_{\mu\nu}, \Sigma^{\mu}{}_{\nu}, \psi_m) = \frac{1}{2\kappa^2}\int d^4x\sqrt{-g}\left[\mathcal{L}_G(\Sigma^{\mu}{}_{\nu}) + \left(g^{\mu\alpha}R_{\alpha\nu} - \Sigma^{\mu}{}_{\nu}\frac{\partial \mathcal{L}_G}{\partial \Sigma^{\mu}{}_{\nu}}\right)\right] + S_m(g_{\mu\nu}, \psi_m) \tag{6.23}$$

It is easy to see that under a variation with respect to $\Sigma^{\mu}{}_{\nu}$ one finds that $\Sigma^{\mu}{}_{\nu} = g^{\mu\alpha}R_{\alpha\nu}$, provided that $\frac{\partial^2 F}{\partial \Sigma^{\mu}{}_{\nu}\partial \Sigma^{\rho}{}_{\lambda}} \neq 0$. Next, by introducing the auxiliary metric $q_{\mu\nu}$ via the definition

$$\sqrt{-q}q^{\mu\nu} \equiv \sqrt{-g}g^{\mu\alpha}\frac{\partial \mathcal{L}_G}{\partial \Sigma^{\alpha}{}_{\nu}} \tag{6.24}$$

one finds that the action above can be rewritten as

$$S_{RBG}(g_{\mu\nu}, \Gamma^{\lambda}_{\mu\nu}, \Sigma^{\mu}{}_{\nu}, \psi_m) = \frac{1}{2\kappa^2}\int d^4x\sqrt{-q}q^{\mu\nu}R_{\mu\nu}(\Gamma) + \tilde{S}_m(g_{\mu\nu}, \Sigma^{\mu}{}_{\nu}, \psi_m) \tag{6.25}$$

where we identify this action as nothing but the Einstein-Hilbert action of GR (for q) with a non-minimally coupled matter field sector. Therefore, full consistence of this action with the Einstein-like field equations (6.21) has been achieved, but we are not nearer of solving our original riddle, namely, whether it is possible to write an stress-energy tensor satisfying the correspondence

$$\tilde{T}^{\mu}{}_{\nu}(q) \equiv -\frac{2}{\sqrt{-q}}\frac{\delta \tilde{S}_m}{\delta q^{\mu\nu}} = \frac{1}{|\Omega|^{1/2}}\left(T^{\mu}{}_{\nu} - \left(\mathcal{L}_G + \frac{T}{2}\right)\delta^{\nu}_{\mu}\right) \tag{6.26}$$

removing systematically all the dependencies out in $g_{\mu\nu}$ on the right in terms of $q_{\mu\nu}$-contributions. As current research can tell us, this can only be done on a case-by-case basis of matter sources. To this end, let us consider quite a general scenario of anisotropic fluids given by the stress-energy tensor (in the RBG frame)

$$T^{\mu}{}_{\nu} = (\rho + P_{\perp})u^{\mu}u_{\nu} + (P_r - P_{\perp})\xi^{\mu}\xi_{\nu} + P_{\perp}\delta^{\mu}_{\nu} \tag{6.27}$$

where we have introduced the unit (time-like and space-like vectors, respectively) $g_{\mu\nu}u^{\mu}u^{\nu} = -1$ and $g_{\mu\nu}\xi^{\mu}\xi^{\nu} = +1$, while ρ is the energy density of the fluid, P_r its pressure in the direction of ξ^{μ} and $P_{\perp}$ its pressure in the direction orthogonal to ξ^{μ}. Note also that, in a comoving system, this stress-energy tensor can be simply cast as $T^{\mu}{}_{\nu} = \text{diag}(-\rho, P_r, P_{\perp}, P_{\perp})$. Due to the orthogonality of these vectors, the deformation matrix must inherit the same algebraic structure, that is,

$$\Omega^{\mu}{}_{\nu} = \alpha\delta^{\mu}_{\nu} + \beta u^{\mu}u_{\nu} + \gamma\xi^{\mu}\xi_{\nu} \tag{6.28}$$

where the expressions for the three functions $\{\alpha, \beta, \gamma\}$ are (both in RBG and matter sector) model-dependent. Inserting these expressions into the one for the stress-energy tensor in the GR frame, Eq. (6.26), yields

$$\tilde{T}^{\mu}{}_{\nu}(q) = \frac{\kappa^2}{|\Omega|^{1/2}}\left[\left(\frac{\rho - P_r}{2} - F\right)\delta^{\mu}_{\nu} + (\rho + P_{\perp})u^{\mu}u_{\nu} + (P_r - P_{\perp})\xi^{\mu}\xi_{\nu}\right] \tag{6.29}$$

so if we propose a formally similar expression as that of (6.27) but with new unit time-like $q_{\mu\nu}v^{\mu}v^{\nu} = -1$ and space-like $q_{\mu\nu}\chi^{\mu}\chi^{\nu} = +1$ vectors, and new functions characterizing the fluid, $\{\rho^q, P_r^q, P_{\perp}^q\}$, one finds the identifications

$$P_{\perp}^q = \frac{1}{|\Omega|^{1/2}}\left[\frac{\rho - P_r}{2} - \mathcal{L}_G\right] \tag{6.30}$$

$$\rho^q + P_r^q = \frac{\rho + P_r}{|\Omega|^{1/2}} \tag{6.31}$$

$$\rho^q - P_{\perp}^q = \frac{\rho - P_{\perp}}{|\Omega|^{1/2}} \tag{6.32}$$

These relations allow to find a correspondence between the functions characterizing the fluid in the two frames once an RBG Lagrangian, given by a function $\mathcal{L}_G$, is specified. Furthermore, supplied with the relations $u^\mu u_\nu = v^\mu v_\nu$ and $\xi^\mu \xi_\nu = \chi^\mu \chi_\nu$, this also allows to write the deformation matrix in terms of those functions defining the fluid in the RBG frame. In turn, this allows one to solve the fundamental relation (6.15) in order to yield a solution for $g_{\mu\nu}$ once an expression for $q_{\mu\nu}$ is given. Since the latter corresponds to the solution of the GR problem coupled to a set of matter fields $\tilde{\psi}_m$ whose stress-energy tensor is given by $\tilde{T}^\mu{}_\nu(q, \tilde{\psi}_m)$, the above relations *map* the seed solution $q_{\mu\nu}$ into a new solution $g_{\mu\nu}$ corresponding to a given RBG theory coupled to a set of matter fields ψ_m with a stress-energy tensor $T^\mu{}_\nu(g, \psi_m)$. Moreover, working upon (6.29) this also allows (through a straightforward but tedious algebraic procedure) to find the correspondence between the gravitational Lagrangians on each side of the correspondence. It is worth pointing out that this result holds true irrespective of any symmetries involved in any problem under consideration, since we did not assume anything on the background solutions, but just worked at the level of the action and the general field equations.

The bottom line of this discussion is the power of the mapping method as a solution-generator machine. Indeed, it allows to generate solutions of a given pair $\{\text{RBG}(g), \psi_m\}$ starting from any known solution of the pair $\{\text{GR}(q), \tilde{\psi}_m\}$, with the correspondences above allowing one to find the relations between both the matter fields and the gravity Lagrangians. For the sake of this chapter we shall be interested in two particular cases of interest: scalar fields, which satisfy $P_\perp = P_r$, and (non-linear) electromagnetic fields, for which $P_r = -\rho$.

The poltergeist nature of the mapping: The mapping makes curious alchemy on the functional dependence of the action: for instance, quadratic $f(\mathcal{R})$ gravity coupled to a standard scalar/Maxwell field maps into GR coupled to quadratic scalar/Maxwell field, while EiBI gravity coupled to a Maxwell/Born-Infeld electrodynamics maps into GR coupled to a Born-Infeld/Maxwell electrodynamics!. This has side-effects in the way the structure of the corresponding solutions inherit properties from the original side of the mapping they came from.

This concludes our theoretical considerations on the theories and methods to be considered in this chapter. Next we shall head to the description of the most relevant black hole solutions found in the literature, before moving on to discuss their regularity.

6.3 Spherically Symmetric Black Hole Solutions

6.3.1 A Case-Sample on the Direct Attack to Solve the Palatini Equations

To start with our recollection of the black hole solutions of interest found in Palatini theories of gravity, let us first analyze in detail a case-sample of how to solve their field equations via direct attack. This involves *a tale of two frames*, since we shall be solving the field equations for $q_{\mu\nu}$, i.e., Eq. (6.21), while at the same time working out its relation with the space-time metric $g_{\mu\nu}$, i.e., Eq. (6.16). In order to simplify this discussion, let us consider the Palatini theory in which the relation between these two metrics is the simplest, as given by the conformal relation (6.6) of the $f(\mathscr{R})$ case. In this scenario, the fact that the new dynamics enters just via the trace T of the stress-energy tensor prevents one from using any source of matter that yields a traceless stress-energy tensor, since the corresponding solutions would reduce to those of GR. This, unfortunately, leaves aside the interesting case of Maxwell electrodynamics, which would allow us to find the counterpart of the Reissner-Nordström black hole in these theories. Therefore, we shall consider non-linear electromagnetic (NED) fields, which are described by a function $\varphi(X, Y)$ of the field invariants

$$X = -\frac{1}{2} F_{\mu\nu} F^{\mu\nu} \; ; \; Y = -\frac{1}{2} F_{\mu\nu} F^{\star\mu\nu} \, , \tag{6.33}$$

where $F_{\mu\nu} = \partial_\mu A_\nu - \partial_\nu A_\mu$ is the field strength tensor of the vector potential A_μ and $F^{\star\mu\nu} = \frac{1}{2}\varepsilon^{\mu\nu\alpha\beta} F_{\alpha\beta}$ its dual. Since we are interested in static, spherically symmetric solutions, the magnetic part of the electrostatic field, built from the vector potential $A_\mu = (A_t, 0, 0, 0)$, can be neglected, which entails $Y = 0$. This way, the stress-energy associated to these fields, which reads

$$T^\mu{}_\nu = -\frac{1}{4\pi} \left(\varphi_X F^\mu{}_\alpha F^\alpha{}_\nu - \frac{\delta^\mu_\nu}{2} \varphi(X, 0) \right) , \tag{6.34}$$

where $\varphi_X \equiv d\varphi/dX$, becomes in the present case

$$T^\mu{}_\nu = \frac{1}{8\pi} \mathrm{diag}(\varphi(X) - 2X\varphi_X, \varphi(X) - 2X\varphi_X, \varphi(X), \varphi(X)) \tag{6.35}$$

On the other hand, we have the electromagnetic field equations

$$\nabla_\mu (\varphi_X F^{\mu\nu} + \varphi_Y F^{\star\mu\nu}) = 0 \, . \tag{6.36}$$

Since these equations are coupled to the space-time metric $g_{\mu\nu}$, which in static, spherically symmetric configurations reads as

$$ds^2 = g_{tt} dt^2 + g_{rr} dr^2 + r^2 d\Omega^2 \, , \tag{6.37}$$

where $d\Omega^2 = d\theta^2 + \sin^2\theta d\phi^2$ is the volume element in the unit two-sphere, for electrostatic configurations these are solved as

$$\varphi_X F^{tr} = \frac{Q}{r^2\sqrt{-g_{tt}g_{rr}}}, \tag{6.38}$$

where F^{tr} is the single non-vanishing component of the field strength tensor in this setting, while Q is an integration constant identified as the electric charge for a given configuration. Moreover, using the fact that $X = -g_{tt}g_{rr}(F^{tr})^2$ we can rewrite the above equation as

$$X\varphi_X^2 = \frac{Q^2}{r^4} \tag{6.39}$$

which is an algebraic equation allowing to find the expression of the invariant X (via a quadrature) when a NED Lagrangian is given. For further reference, we find it convenient to introduce at this stage the well known Born-Infeld (BI) electrodynamics, which reads

$$\varphi_{BI}(X, Y) = 2\beta^2\left(1 - \sqrt{1 - \frac{X}{\beta^2} + \frac{Y}{4\beta^2}}\right), \tag{6.40}$$

where β is the Born-Infeld parameter. For this theory, Eq. (6.39) tells us that $E(r) \equiv F^{tr} = \frac{\beta Q}{\sqrt{\beta^2 r^4 + Q^2}}$. At large distances, $r \to \infty$, one finds that $E(r)$ reduces to the Maxwell field, $E(r) \approx Q/r^2$, while near the center, $r \to 0$, the finiteness of $E(r)$ makes the total energy associated to the electrostatic field to be finite. However, within GR this fact does not solve the singularity problem [33], since neither the completeness of geodesics nor the finiteness of curvature scalars is achieved. Further on this topic later.

Now that we have all the electromagnetic sector under control we can proceed with the resolution of the field equations (6.7). Given the symmetry of the stress-energy tensor in 2 × 2 blocks, such equations can be conveniently written as (here $\hat{I}_{2\times2}$ and $\hat{0}_{2\times2}$ are the identity and zero matrices, respectively)

$$R^\mu{}_\nu(q) = \frac{1}{2f_R^2}\begin{pmatrix} (f + 2\kappa^2 T^t{}_t)\hat{I}_{2\times2} & \hat{0}_{2\times2} \\ \hat{0}_{2\times2} & (f + 2\kappa^2 T^\theta{}_\theta)\hat{I}_{2\times2} \end{pmatrix} \tag{6.41}$$

therefore retaining the symmetry of the matter source. To solve them we propose a static, spherically symmetric line element on the $q_{\mu\nu}$ frame as given by the convenient form

$$ds_q^2 = q_{\mu\nu}dx^\mu dx^\nu = -e^{2\psi(x)}A(x)dt^2 + \frac{dx^2}{A(x)} + x^2 d\Omega^2 \tag{6.42}$$

for the metric functions $\{\psi(x), A(x)\}$ parameterized in terms of a new radial coordinate x. To handle the left-hand side of the equations (6.7), one uses canonical methods to compute the components of the Ricci tensor (since it is built in terms of the Christoffel symbols of the line element (6.42)), with the result

$$R^t{}_t = -\frac{A}{2}\left[\frac{A_{xx}}{A} - \left(\frac{A_x}{A}\right)^2 + 2\psi_{xx} + \left(\frac{A_x}{A} + 2\psi_x\right)\left(\frac{A_x}{A} + \psi_x + \frac{2}{x}\right)\right] \tag{6.43}$$

$$R^r{}_r = -\frac{A}{2}\left[\frac{A_{xx}}{A} - \left(\frac{A_x}{A}\right)^2 + 2\psi_{xx} + \left(\frac{A_x}{A} + 2\psi_x\right)\left(\frac{A_x}{A} + \psi_x\right) + \frac{2}{x}\frac{A_x}{A}\right] \tag{6.44}$$

$$R^\theta{}_\theta = (\sin^2\theta)R^\phi{}_\phi = \frac{1}{x^2}\left[1 - A(1 + x\psi_x) - xA_x\right] \tag{6.45}$$

Now, considering the subtraction $R^t{}_t - R^r{}_r = 0$, the last equality being a trivial consequence of the 2×2 block symmetry of the right-hand side of (6.41), one finds that ψ =constant, which can be set to zero by a redefinition of the temporal coordinate without loss of generality. Plugging this result into (6.45), and working out the right-hand side of (6.41), one finds that

$$\frac{1}{x^2}(1 - A(x) - xA_x) = \frac{1}{2f_{\mathcal{R}}^2}\left(f + \frac{\kappa^2}{4\pi}\varphi\right) \tag{6.46}$$

Next, proposing the usual mass ansatz

$$A(x) = 1 - \frac{2M(x)}{x} \tag{6.47}$$

the above equation is solved as

$$M_x \equiv \frac{dM}{dx} = \frac{x^2}{4f_{\mathcal{R}}^2}\left(f + \frac{\kappa^2}{4\pi}\varphi\right) \tag{6.48}$$

To keep progressing we need to rewrite this last expression in terms of the variables of the space-time metric (6.37). This is easily done in view of the conformal transformation (6.6), which implies the relation between the radial coordinates in each frame as

$$x^2 = r^2 f_{\mathcal{R}} \tag{6.49}$$

The non-trivial radial function: The tale of two frames usually ends up (at least in some branches of the solutions of the theory) in a non-trivial and non-monotonic behaviour of the radial function in the space-time metric in relation to the auxiliary one, i.e., $r^2(x)$. The consequences of this result (when it happens) for the regularity of the corresponding space-times must be discussed on a case-by-case basis, as shall be seen below.

Taking a derivative here one arrives at

$$\frac{dr}{dx} = \frac{1}{f_{\mathcal{R}}^{1/2}\left[1 + \frac{r f_{\mathcal{R},r}}{2 f_{\mathcal{R}}}\right]} \tag{6.50}$$

and replacing it in (6.48) we get

$$M_r \equiv \frac{dM}{dr} = \frac{r^2}{4 f_{\mathcal{R}}^{3/2}}\left(f + \frac{\kappa^2}{4\pi}\varphi\right)\left(f_{\mathcal{R}} + r f_{\mathcal{R},r}\right) \tag{6.51}$$

Typically, this function can be generically integrated as $M(r) = r_S(1 + \delta_1 G(r))/2$, where $r_S = 2M_0$ is the Schwarzschild radius, δ_1 encodes the relevant constants in the problem, and $G(r)$ is the primitive of M_r, thus a model-dependent function. Therefore, the space-time line element can be written as

$$ds_g^2 = \frac{1}{f_{\mathcal{R}}}\left(-A(x)dt^2 + \frac{dx^2}{A(x)} + r^2(x)d\Omega^2\right) \tag{6.52}$$

where the radial function is implicitly given by (6.49) while the metric function $A(r(x))$ is written as

$$A(r) = 1 - \frac{1 + \delta_1 G(r)}{\delta_2 r f_{\mathcal{R}}^{1/2}} \tag{6.53}$$

where δ_2 is another parameter collecting any additional constant that have emerged from our manipulations. This is how far we can go without further specifying our setting. Specific cases of interest can be now obtained by setting both a NED function $\varphi(X(r))$ and a shape for the $f(\mathcal{R})$ gravity function, which would allow to uniquely determine the right-hand side of (6.51) and, therefore, to provide a unique solution to this problem.

Let us thus set, for the gravity sector, the quadratic model

$$f(\mathcal{R}) = \mathcal{R} - \sigma\mathcal{R}^2 \tag{6.54}$$

and for the matter sector the BI electrodynamics of Eq. (6.40). In order to work with dimensionless variables, it is convenient to introduce the following definitions: $r_q^2 = \kappa^2 Q^2/4\pi$, $l_\beta^2 = 1/(\kappa^2\beta^2)$ and the dimensionless radial function $z = r/r_c$, with $r_c^2 = (4\pi)^{1/2} r_q l_\beta$. This way, Eqs. (6.51) and (6.53) still hold [with the radial function $r(x)$ replaced by its dimensionless version $z(x)$, where an r_c factor has also been reabsorbed in the x-coordinate], while the relevant gravity and matter functions become [59]

$$\varphi(z) = 2\left(1 - \frac{z^4}{z^4+1}\right) \; ; \; f(z) = \frac{\eta(z)}{2\pi}\left(1 - \frac{\alpha}{2}\eta(z)\right) \; ; \; f_{\mathcal{R}} = 1 - \alpha\eta(z) \tag{6.55}$$

where $\alpha \equiv \sigma/(2\pi l_\beta^2)$ and we have introduced the function

$$\eta(z) = \frac{(z^2 - \sqrt{z^4+1})^2}{z^2\sqrt{z^4+1}} \tag{6.56}$$

while the constants δ_1 and δ_2 appearing in (6.53) read

$$\delta_1 = 2(4\pi)^{3/4}\frac{r_q}{r_S}\sqrt{\frac{r_q}{l_\beta}} \quad ; \quad \delta_2 = \frac{r_c}{r_S} \tag{6.57}$$

which completes our construction. The line element (6.52) with the definitions above represent a generalization of the Reissner-Nordström (RN) solution characterized by mass, charge, and gravity parameter. The RN solution is recovered in the limit $r \to \infty$, as can be verified by expansion of the metric functions in such a limit, i.e.,

$$-g_{tt} = g_{rr}^{-1} \approx 1 - \frac{1}{\delta_2 z} + \frac{\delta_1}{16\pi\delta_2 z^2} + \mathcal{O}(z^{-4}) \tag{6.58}$$

which after restoring back the r-notation is easily recognized as the actual Reissner-Nordström solution of GR, $-g_{tt} = g_{rr}^{-1} = 1 - \frac{2M}{r} + \frac{Q^2}{8\pi r^2} + \mathcal{O}(r^{-4})$. Deviations with respect to such a solution will occur as the new energy density contributions driven by the interplay between the quadratic $f(\mathcal{R})$ gravity and BI effects, and encoded in the single parameter α, become relevant. This has important consequences for the global structure of the corresponding solutions in terms of the number and type of horizons, though for the sake of this part of the chapter we are more interested in describing the consequences for the regularity of these space-times, a topic that will be fully addressed in Sect. 6.4. In this sense, the expression of the radial function $z(x)$, obtained from Eq. (6.49) becomes also a critical aspect in the characterization of the regularity of these solutions, though an explicit expression is not always possible.

6.3.2 *Other Spherically Symmetric Solutions*

Having concluded with our detailed analysis of the procedure to solve the field equations by direct attack, we shall continue elaborating our space of solutions, relevant for the discussion of regular black holes. We shall first add other matter sources to our discussion of quadratic $f(\mathcal{R})$ gravity, and afterwards move to more complex gravitational theories.

Quadratic $f(\mathcal{R})$ gravity with anisotropic fluids [62]. We next consider a version of the anisotropic fluid (6.27) given by

$$T^{\mu}{}_{\nu} = \text{diag}(-\rho, -\rho, \alpha\rho, \alpha\rho) \tag{6.59}$$

where the constant α is constrained within the range $0 < \alpha \leq 1$ to satisfy the classical energy conditions, with the upper limit $\alpha = 1$ corresponding to the usual Maxwell electrodynamics. We consider the coupling of this matter source to the quadratic $f(\mathscr{R})$ model (6.54). The energy density of the fluid can be found via integration of the conservation equation of the stress-energy tensor (this conservation being a consequence of the fact that the independent connection does not enter into the matter sector), $\nabla_\mu T^{\mu\nu} = 0$, as $\rho = \rho_0/r(x)^{2(1+\alpha)}$, where ρ_0 is a (dimensionful) integration constant. The resolution of this problem amounts to the line element (6.52), where the new dimensionless scale is defined as $r_c^{2(1+\alpha)} = (4\sigma)\kappa^2\rho_0(1-\alpha)$ so that the constant $\delta_1 = r_c^3/(4\sigma r_S)$ and the relevant function characterizing the metric reads (for $\alpha \neq 1/2$):

Classical energy conditions: Classical matter fields must satisfy several conditions to fulfil fundamental facts on our understanding about the geometrical and physical components of our Universe. These conditions are typically given by the following [50] (we do not consider here averaged energy conditions):

- Null energy condition (NEC): $T_{\mu\nu}n^\mu n^\nu \geq 0$, with $n^\mu n_\mu = 0$ a null vector. NEC implies $\rho + P_i \geq 0 \; \forall i = 1, 2, 3$.
- Weak energy condition (WEC): $T_{\mu\nu}u^\mu u^\nu \geq 0$, with $u^\nu u_\nu = -1$ a time-like vector. WEC implies $\rho \geq 0$ and $\rho + P_i \geq 0 \; \forall i = 1, 2, 3$.
- Strong energy condition (SEC): $T_{\mu\nu}u^\mu u^\nu \geq -T/2$. SEC implies $\rho + \sum_{i=1}^3 P_i \geq 0$ and $\rho + P_i \geq 0 \; \forall i = 1, 2, 3$.
- Dominant energy condition (DEC): $-T^\mu{}_\nu u^\mu$ is a future-oriented null or time-like vector. DEC implies $\rho \geq 0$ and $\rho \geq |P_i| \; \forall i = 1, 2, 3$.

The reach of such conditions is inherently limited by the fundamental quantum nature of the matter fields; for instance, the Casimir effect violates them [72].

$$G(z) = \frac{z^{-4\alpha-1}\left(\frac{z^{2\alpha+2}\sqrt{1-z^{-2(\alpha+1)}}\left(2\alpha^2+\alpha+2z^{2\alpha+2}-3\right)}{z^{2\alpha+2}-1} - \frac{8\alpha\left(\alpha^2-1\right){}_2F_1\left(\frac{1}{2},\frac{4\alpha+1}{2\alpha+2};\frac{6\alpha+3}{2\alpha+2};z^{-2(\alpha+1)}\right)}{4\alpha+1}\right)}{2(\alpha-1)(2\alpha-1)}, \tag{6.60}$$

where ${}_2F_1\,(a, b, c; y)$ is a hypergeometric function. As for the behaviour of the radial function in these solutions, it is implicitly given by the expression

$$x = \frac{\sqrt{z^{2(1+\alpha)} - 1}}{z^\alpha} \tag{6.61}$$

Likewise in the previous case, this solution can be seen as a deformation of the RN solution of GR inside the matter sources, modifying the structure of horizons and having relevant implications for the regularity of these space-times.

Quadratic $f(\mathcal{R})$ gravity with more general anisotropic fluids [18]: Let us consider now the more general anisotropic fluid

$$T^{\mu}{}_{\nu} = \text{diag}(-\rho, -\rho, K(\rho), K(\rho)) \tag{6.62}$$

where the free function $K(\rho)$ characterizes the fluid, with $K(\rho) = \alpha\rho$ for the previous case. NEDs are naturally included within this class of fluids via the identifications $-8\pi\rho = \varphi - 2(X\varphi_X + Y\varphi_Y)$, $8\pi K(\rho) = \varphi$. In order to find explicit solutions, let us consider the ansatz for the free function as $K(\rho) = \rho + \beta\rho^2$ and introduce the definitions $\tilde{\beta} = s_\beta|\beta|\rho_0$ (with $s_\beta = \pm 1$ the sign of β) and $r_c = (\beta\rho_0)^{1/4} r_0$. Then the density of the fluid becomes $\rho(z) = \frac{\rho_m}{z^4 - s_\beta}$, with $\rho_m = 2/|\beta|$, and one also finds the constant $\delta_1 = \kappa^2 \rho_m (r_0|\tilde{\beta}|^{1/4})^3 / r_S$. Now there are four different solutions for the function G_z depending on the signs of β and γ as (here we have defined $\gamma \equiv 8\kappa^2\rho_m|\sigma|$)

$$G_z = \frac{z^2\left(1 - \frac{\gamma(1-3z^4)}{(z^4+1)^3}\right)\left(1 - \frac{\gamma}{(z^4+1)^3}\right)}{(z^4+1)\left(1 - \frac{\gamma}{(z^4+1)^2}\right)^{3/2}} \quad \text{for } \sigma > 0, \beta < 0 \tag{6.63}$$

$$G_z = \frac{z^2\left(1 - \frac{\gamma(3z^4+1)}{(z^4-1)^3}\right)\left(1 - \frac{\gamma}{(z^4-1)^3}\right)}{(z^4-1)\left(\frac{\gamma}{(z^4-1)^2} + 1\right)^{3/2}} \quad \text{for } \sigma > 0, \beta > 0 \tag{6.64}$$

$$G_z = \frac{z^2\left(1 + \frac{\gamma(1-3z^4)}{(z^4+1)^3}\right)\left(1 + \frac{\gamma}{(z^4+1)^3}\right)}{(z^4+1)\left(1 + \frac{\gamma}{(z^4+1)^2}\right)^{3/2}} \quad \text{for } \sigma < 0, \beta < 0 \tag{6.65}$$

$$G_z = \frac{z^2\left(1 + \frac{\gamma(1+3z^4)}{(z^4+1)^3}\right)\left(1 + \frac{\gamma}{(z^4-1)^3}\right)}{(z^4-1)\left(1 - \frac{\gamma}{(z^4-1)^2}\right)^{3/2}} \quad \text{for } \sigma < 0, \beta > 0\,. \tag{6.66}$$

and that we shall call them Type I, Type II, Type III, and Type IV, respectively. Accordingly, the behaviour of the radial function $z(x)$ is different in each case. In particular, only when σ and β have opposite signs, do the radial function $z(x)$ implements a bouncing behaviour.

Quadratic gravity with Maxwell electrodynamics [47, 60, 61]. Let us now upgrade our setting to consider the case of the quadratic model

$$f(\mathscr{R},\mathscr{Q})=\mathscr{R}+l^2(a\mathscr{R}^2+\mathscr{Q}) \tag{6.67}$$

where l is a new parameter with dimensions of length characterizing the scale of the corrections, and a is a dimensionless constant. As the matter source, the fact that the new gravitational dynamics associated to these theories has access to the full structure of the $T^\mu{}_\nu$, and not just to its trace, like in the $f(\mathscr{R})$ case, allows us to consider a standard Maxwell field as the matter source. Therefore, we set $\varphi(X)=X$ and proceed to solve the corresponding field equations. To this end, a straightforward but tedious algebraic exercise acting upon the trace of the metric field equations (6.9), allows one to find $\mathscr{Q}=\frac{\tilde{\kappa}^2 Q^2}{r^8}$, with $\tilde{\kappa}^2\equiv\kappa^2/4\pi$. It is illustrative now to write the full expression of the RBG field equations (6.41) in this case as

$$R^\mu{}_\nu(q)=\frac{\tilde{\kappa}^2Q^2}{2r^4}\begin{pmatrix}-\frac{1}{\Omega_+}\hat{I} & \hat{0}\\ \hat{0} & \frac{1}{\Omega_-}\hat{I}\end{pmatrix}. \tag{6.68}$$

where we have defined the objects $\Omega_\pm=1-1/z^4$ and in this case $r_c^4=\kappa^2Q^2l^2$. Note that the conformal factor $f_\mathscr{R}$ of the previous cases degenerates into the objects $\Omega_\pm$, which manifests itself at several levels in the structure of the corresponding solution. To find the latter, one follows the same tale of the two frames as in the quadratic $f_\mathscr{R}$ case above, which yields the result

$$ds^2=-\frac{A(z)}{\Omega_+}dt^2+\frac{1}{\Omega_+A(z)}dx^2+z^2(x)d\Omega^2\ ;\ A(z)=1-\frac{1+\delta_1G(z)}{\delta_2 z\Omega_-^{1/2}} \tag{6.69}$$

$$\delta_1=\frac{1}{2r_S}\sqrt{\frac{r_q^3}{l}};\ \delta_2=\frac{r_c}{r_S}\ ;\ G_z=\frac{\Omega_+}{z^2\Omega_-^{1/2}} \tag{6.70}$$

where in this case the radial function satisfies $x^2=z^2\Omega_-$, which can be inverted to find the explicit expression

$$z^2(x)=\frac{x^2+\sqrt{x^2+4}}{2} \tag{6.71}$$

We have yet another generalization of the RN solution of GR. However, this also introduces interesting modifications to the global structure of the solutions, in the sense that if $\delta_1>\delta_c$, with $\delta_c\approx-0.572$, one finds the presence of two (Cauchy and event) horizons, likewise in the RN solution of GR, while if $\delta_1<\delta_c$ then a single horizon is present, thus resembling more the Schwarzschild black hole of GR. The case $\delta_1=\delta_c$ has peculiar properties beyond its horizon structure (having a single horizon or not, depending on the absolute value of the electric charge), which will be very relevant in our discussion of the regularity of black hole solutions. We point out that similar comments on the horizons and critical configurations apply to solutions out of fluids discussed previously in this section.

EiBI gravity with Maxwell electrodynamics [63]. The EiBI gravity given by the action (6.13) is particularly amicable for this sort of computations. This is so thanks to the simpler structure of the equation for the deformation matrix, Eq. (6.17). Indeed, once the stress-energy tensor of the matter fields is specified (in particular, its symmetry in blocks), finding a solution to (6.17) is as simple a mathematical exercise as in their quadratic gravity cousins. Indeed, for a Maxwell field, $\varphi(X) = X$, the solution to this equation reads simply

$$\Omega^{\mu}{}_{\nu} = \begin{pmatrix} \Omega_{+}^{(\varepsilon)}\hat{I} & \hat{0} \\ \hat{0} & \sigma_{-}^{(\varepsilon)}\hat{I} \end{pmatrix}, \quad \Omega_{\pm}^{(\varepsilon)} = \lambda \pm \frac{|s_{\varepsilon}|}{z^4} \tag{6.72}$$

where now $r_c^4 = r_q^2 l_\varepsilon^2$ with the new scale $\varepsilon = -2l_\varepsilon^2$, and $s_\varepsilon = \pm 1$ is the sign of ε. The field equations in this case become

$$R^{\mu}{}_{\nu}(q) = \frac{1}{\varepsilon} \begin{pmatrix} \frac{(\Omega_{-}^{(\varepsilon)}-1)}{\Omega_{-}^{(\varepsilon)}}\hat{I} & \hat{0} \\ \hat{0} & \frac{(\Omega_{+}^{(\varepsilon)}-1)}{\Omega_{+}^{(\varepsilon)}}\hat{I} \end{pmatrix} \tag{6.73}$$

At this state one can note that, in all the solutions studied so far, the deformation matrix has a similar algebraic structure in blocks as the stress-energy tensor, and this feature is inherited by the corresponding field equations. To solve them, one follows the same strategy as in the previous case, with the surprising result that one gets exactly to the same Eqs. (6.69), (6.70) and (6.71) as in the quadratic case. This coincidence is not accidental, as shall be clear in the discussion of the mapped solutions below.

$f(\mathscr{R})$ gravity and EiBI gravity with Euler-Heisenberg electrodynamics [40]. In order to compare quadratic $f(\mathscr{R})$ and EiBI gravity solutions on equal footing, we can consider, for instance, the coupling of both of them to the same NED, which we take to be given by the Euler-Heisenberg (EH) electrodynamics. The action of the latter is written as (again, we restrict ourselves to purely electrostatic solutions, so $Y = 0$)

$$\varphi(X) = X + \beta X^2 \tag{6.74}$$

The corresponding equations of motion can be solved via a quadrature, with the resulting expression for the field invariant [43]:

$$X(z) = \frac{1}{6\beta\, z^{4/3}} \left[\left(1 + \sqrt{1+z^4}\right)^{1/3} + \left(1 - \sqrt{1+z^4}\right)^{1/3} \right]^2 . \tag{6.75}$$

where $r_c^4 = 54\pi r_q^2 l_\beta^2$ and $l_\beta^2 = \beta/\kappa^2$. Coupling this field to the quadratic $f(\mathcal{R})$ gravity model (6.54) yields the solution (6.52) and (6.53) with the following definitions

$$f = \frac{2}{9\pi}\tau^4(z)\left(1 - \frac{\tilde{\sigma}}{2}\tau^4(z)\right) \ ; \ f_R = 1 - \tilde{\alpha}\,\tau^4(z) \tag{6.76}$$

$$\varphi = \frac{\tau^2(z)}{6\pi}\left(1 + \frac{2}{3}\tau(z)\right) \quad ; \ \tau(z) = \text{Sinh}\left[\frac{1}{3}\ln\left[\frac{1}{z^2}\left(1 + \sqrt{z^4+1}\right)\right]\right] \tag{6.77}$$

$$\delta_1 = \frac{(54\pi)^{3/4}}{2r_S}\sqrt{\frac{r_q^3}{l_\beta}} \tag{6.78}$$

and $\tilde{\sigma} \equiv 4\sigma/(9\pi l_\beta^2)$. A bounce in the radial function is present in the branch $\tilde{\sigma} > 0$ and its minimum is located at $z_c = \sqrt{2a/(a^2-1)}$ with $a = \exp[3\text{ArcSinh}(|\tilde{\sigma}|^{-1/4}]$.

Now, considering the same problem now with EiBI gravity (6.13), one arrives to the line element (6.69), where the relevant functions are now defined as (here $l_\varepsilon^2 = \varepsilon/(12\pi l_\beta^2)$)

$$G_z = z^2(\Omega_- - 1)\,\Omega_-^{1/2}\left(1 + \frac{z\Omega_{-,z}}{2\,\Omega_-}\right) \tag{6.79}$$

$$\Omega_+ = \lambda - l_\varepsilon^2\,\tau^2(z)\left(1 + \frac{2\,\tau^2(z)}{3}\right) \ ; \ \Omega_- = \lambda + l_\varepsilon^2\,\tau^2(z)\,(1 + 2\,\tau^2(z))\,, \tag{6.80}$$

$$\delta_1 = \frac{r_c^3}{r_S\,\varepsilon} \ ; \ \delta_2 = \frac{r_c}{r_S}\,, \tag{6.81}$$

In this case, only for $l_\varepsilon^2 > 0$ does the radial function $z(x)$ attain a minimum at a certain z_0.

EiBI gravity with anisotropic fluids [51]. The anisotropic fluid configuration (6.62) with $K(\rho) = \rho + \beta\rho^2$ solved in the quadratic $f(\mathcal{R})$ case above also admits an exact solution in the EiBI gravity case. The solution is again given by the line element (6.69) and similar definitions as in the $f(\mathcal{R})$ case while the relevant functions characterizing the metric are now given by the compact expressions

$$G_z = \frac{z^2\Omega_1}{(z^4 - s_\beta)\Omega_2^{1/2}} \ ; \ \Omega_1 = 1 - s_\varepsilon\xi^2\left(\frac{z^4 + s_\beta}{(z^4 - s_\beta)^2}\right) \ ; \ \Omega_2 = 1 + \frac{s_\varepsilon\xi^2}{z^4 - s_\beta} \tag{6.82}$$

where now $\delta_1 = \frac{r_c^3}{r_S l_m^2}$, $l_m = \beta/(2\kappa^2)$ and $\xi = l_\varepsilon^2/l_\beta^2$. Likewise in the quadratic $f(\mathcal{R})$ case, the corresponding solutions split into four different types depending on the combinations of the signs of $s_\beta = \beta/|\beta|$ and $s_\varepsilon = \varepsilon/|\varepsilon|$. In Type-I $\{s_\varepsilon = -1, s_\beta = -1\}$ a bounce is present at $z_c = (\xi^2 - 1)^{1/4}$ provided that $\xi^2 > 1$, in Type-II $\{s_\varepsilon = -1, s_\beta = +1\}$ a bounce is always present at $z_c = (1 + \xi^2)^{1/4}$, while in Type-III $\{s_\varepsilon = +1, s_\beta = -1\}$ and in Type-IV $\{s_\varepsilon = +1, s_\beta = +1\}$ no bounces are present.

Functional extensions of EiBI gravity with Maxwell electrodynamics [12]: By regarding the object $|\Omega|$ as the main building block in constructing a metric-affine action, one can consider a family of functional extensions of EiBI gravity defined by the action

$$S = \frac{1}{\kappa^2 \varepsilon} \int d^4x \sqrt{-g}[f(|\Omega|) - \lambda] + S_m(g_{\mu\nu}, \psi_m) \quad (6.83)$$

where the choice $f(|\Omega|) = |\Omega|^{1/2}$ corresponds to the usual EiBI gravity Lagrangian. Similar methods as in the other members of the RBG family can be applied in this case and, in particular, the tale of two frames still holds. Using such methods, and considering again a Maxwell field as the matter source, one can obtain an exact solution for the family $f(|\Omega|) = |\Omega|^{n/2}$ under (once again) the form (6.69), now with the definitions

$$G_z = -z^2 \frac{(|\hat{\Omega}|^{\frac{n}{2}} - \sigma_+)\Omega_-^{1/2}}{n^2 |\hat{\Omega}|^{\frac{2n-1}{2}}} \left[(n-1)|\hat{\Omega}|^{\frac{n}{2}} \left(1 + \frac{n}{4} \frac{z|\hat{\Omega}|_z}{|\hat{\Omega}|} \right) + \sigma_+ \right] . \quad (6.84)$$

where now we have to upgrade our definitions of the $\Omega_\pm$ matrices to

$$\Omega_\pm = (n-1)|\hat{\Omega}|^{\frac{n}{2}} + \sigma_\pm \ ; \ \sigma_\pm = \lambda \mp X \ . \quad (6.85)$$

and similar definitions for the z variable and constant δ_1 as in the quadratic gravity and EiBI cases apply. Again, for $\varepsilon < 0$ and in the cases $1/2 < n \leq 1$ a bounce in the coordinate $z(x)$ driven by the relation $x = z\Omega_-^{1/2}$ is found.

EiBI gravity with scalar fields [3]. A scalar field with Lagrangian density $\mathcal{L}_m = X - V(\phi)$, where in this case $X = \partial_\mu \phi \partial^\mu \phi$, can be seen, in the static, spherically symmetric setting, $ds_g^2 = -A(x)dt^2 + B^{-1}(x)dx^2 + r^2(x)d\Omega^2$, as a sort of anisotropic fluid with stress-energy tensor

$$T^\mu{}_\nu = \text{diag}\left(-\frac{L_m}{2}, B\phi_x^2 - \frac{L_m}{2}, B\phi_x^2 - \frac{L_m}{2}, B\phi_x^2 - \frac{L_m}{2} \right) \quad (6.86)$$

where $\phi_x \equiv d\phi/dx$. Therefore we see that the main novelty of this case as compared to the electromagnetic and fluid ones studied before is that the 2×2-block structure of the stress-energy of the former is replaced by a 1×3-block one in the latter. However, the methods of the previous cases and the tell of two frames work equally fine (though with some technical adjustments), and one proposes two suitable line elements of the form (here C_0 is a constant)

$$ds_g^2 = -e^\nu dt^2 + \frac{1}{C_0^2 W^4 e^{-\nu}} dx^2 + \frac{1}{W^2} d\Omega^2 \ , \quad (6.87)$$

$$ds_q^2 = -e^{\tilde{\nu}} dt^2 + \frac{1}{C_0^2 \tilde{W}^4 e^{-\tilde{\nu}}} dy^2 + \frac{1}{\tilde{W}^2} d\Omega^2 \ , \quad (6.88)$$

subject to the relations $e^{\tilde{\nu}} = \Omega_+ e^{\nu}$; $\tilde{W}^2 = W^2/\Omega_+$; $dy = \frac{\Omega_-}{|\Omega|^{1/2}} dx = |\lambda - X_\varepsilon|^{-1} dx$, where $X_\varepsilon = \frac{\varepsilon \kappa^2 C^2}{2r^4 A}$ and the functions $\Omega_+ = A_+^{1/2} A_-^{1/2}$, $\Omega_- = A_+^{3/2} A_-^{1/2}$ with $A_\pm = (\lambda + \varepsilon\kappa^2 V \pm \frac{\varepsilon\kappa^2}{2} B\phi_x^2)$. The corresponding field equations can be conveniently written (in the q-frame) as

$$\tilde{\nu}_{yy} = -\frac{\kappa_0^2 \Omega_+^3}{X_\varepsilon} \left(1 - \tfrac{\lambda + X_\varepsilon}{\Omega^{1/2}}\right) \tag{6.89}$$

$$0 = \tilde{W}_{yy} - \tilde{\nu}_y \tilde{W}_y - \frac{\kappa_0^2 \Omega_+^3}{2\Omega^{1/2}} \tilde{W} \tag{6.90}$$

$$\tilde{W}\tilde{W}_{yy} - \tilde{W}_y^2 = -\frac{e^{\tilde{\nu}}}{C_0^2} + \frac{\kappa_0^2 \Omega_+^3}{2X_\varepsilon} \left(1 - \tfrac{\lambda + X_\varepsilon}{\Omega^{1/2}}\right) \tilde{W}^2 \,, \tag{6.91}$$

where $\kappa_0^2 \equiv \kappa^2 \nu_0^2$ (ν_0 another integration constant) has dimensions of length^{-2}, while C_0^2 has dimensions of length^4. These equations are impermeable to analytical resolution, so one has to resort to numerical methods. The latter reveal the presence of solitonic-like configurations in the spectrum of solutions, with non-trivial consequences for the analysis of the regularity of these space-times.

Higher-dimensional solutions. For the sake of completeness of this section, let us mention that higher-dimensional solutions can be found for the RBG family using the flexibility of the framework developed above. Indeed, the RBG field equations (6.21) are naturally generalized to this case as

$$G^{\mu}{}_{\nu}(q) = \frac{\kappa^2}{|\Omega|^{\frac{1}{(d-2)}}} \left(T^{\mu}{}_{\nu} - \left(\mathcal{L}_G + \frac{T}{2}\right) \delta^{\nu}_{\mu}\right) \tag{6.92}$$

where d is the number of space-time dimensions and $\mathcal{L}$ the gravity Lagrangian. Assuming a Maxwell field, now with $X = Q^2/(r^{2(d-2)}$, and the EiBI action generalized to higher-dimensions, remarkably the line element (6.69) still holds [with the extensions of the volume element part of the unit sphere to the $(D-2)$-dimensional space-time], while the relevant functions become now [16]

$$A(z) = 1 - \left(\frac{1 + \delta_1 G(z)}{\delta_2 \Omega_-^{\frac{d-3}{2}} z^{d-3}}\right) \; ; \; G_z = -z^{d-2} \left(\frac{\Omega_- - 1}{\Omega_-^{1/2}}\right) \left(\lambda + \frac{1}{z^{2(d-2)}}\right) \tag{6.93}$$

$$\Omega_- = \left(\lambda + \frac{1}{z^{2(d-2)}}\right)^{\frac{2}{d-2}} \; ; \; \Omega_+ = \frac{\left(\lambda - \frac{1}{z^{2(d-2)}}\right)}{\left(\lambda + \frac{1}{z^{2(d-2)}}\right)^{\frac{d-4}{d-2}}} \tag{6.94}$$

where now the relevant constants read

$$\delta_1 \equiv \frac{(d-3)r_c^{d-1}}{r_S l_\varepsilon^2} \; ; \; \delta_2 \equiv \frac{(d-3)r_c^{d-3}}{r_S}. \tag{6.95}$$

with $\varepsilon = -l_\varepsilon^2$ and $r_c^{2(d-2)} \equiv l_\varepsilon^2 \kappa^2 Q^2/(4\pi)$. Bouncing behaviours in the radial function $z(x)$ are again found in the branch $\varepsilon < 0$. Similarly, solutions can be found for certain $f(\mathscr{R})$ gravity models [15]; for instance, in the model $f(\mathscr{R}) = \mathscr{R} - \sigma\mathscr{R}^{5/2}$ the corresponding trace equations (recall that in higher dimensions the trace of the stress-energy tensor of a Maxwell field is non-vanishing) yields a simple solution of the curvature scalar, $\mathscr{R} = -\frac{2r_q^4}{3r^6}$, which in turn allows to find analytical solutions. Indeed, for every $f(\mathscr{R})$ theory, bouncing behaviours will be found according to the zeroes of the equation $x^2 = z^2 f_{\mathscr{R}}^{(2-n)/2}$, which will be present in the branch $\sigma > 0$.

6.3.3 Mapping-Generated Solutions

Having discussed a bunch of solutions obtained from direct attack of the field equations, we now turn to those generated via the mapping. Now, the focus of our discussion comes to working out the mapping equations (6.30), (6.31) and (6.32), finding the correspondences between the stress-energy tensors on both GR/RBG sides, and starting from a seed solution of the former reconstructing the corresponding action plus matter field on the latter. Let us discuss some examples.

Mastering the mapping: The mapping is a powerful but difficult-to-tame tool. It generically maps the same kind of matter fields into each other, though described by different Lagrangians and coupled to different RBGs. After the identification of such a combination of gravity + matter is worked out (which does *not* depend on any particular symmetry of the problem under consideration), its application to finding new exact solutions from a known *seed* solution (the latter typically within GR) is outrageously simple. See [5] for a conceptual discussion of this procedure.

Mapping electromagnetic fields [4]. As mentioned above, (non-linear) electromagnetic fields can be seen as anisotropic fluids, which we parameterize, in the RBG frame, via a function $\varphi(X)$ with stress-energy tensor

$$T^\mu{}_\nu = \text{diag}(-\rho, -\rho, K_\varphi(\rho), K_\varphi(\rho)) \tag{6.96}$$

and in the GR frame by a $\Phi(Z)$ with

$$\tilde{T}^\mu{}_\nu = \text{diag}(-\tilde{\rho}, -\tilde{\rho}, \tilde{K}_\Phi(\rho), \tilde{K}_\Phi(\rho)) \tag{6.97}$$

The mapping is particularly helpful when relating GR and EiBI gravity, where the resulting expressions are remarkably simple, being given by (tildes indicate an implicit factor $\varepsilon\kappa^2$)

$$\tilde{\rho}_{\text{EiBI}} = \frac{\lambda\tilde{\rho}_{\text{GR}} - (\lambda - 1)}{1 - \tilde{\rho}_{\text{GR}}} \; ; \; \tilde{K}_{\text{EiBI}} = \frac{\lambda\tilde{K}_{\text{GR}} + (\lambda - 1)}{1 + \tilde{K}_{\text{GR}}} \tag{6.98}$$

where the labels "GR" and "EiBI" refer to the respective sides of the correspondence. Taking $\lambda = 1$, one can reconstruct the corresponding Lagrangians on each side as

$$\text{EiBI} + X \leftrightarrow \text{GR} + \hat{\Phi}(Z) = \frac{1}{2}\left(-1 + \sqrt{1 + \frac{\varepsilon\kappa^2 Z}{2\pi}}\right) \tag{6.99}$$

On the left-hand side of this correspondence we have the same setting of EiBI gravity with a Maxwell field discussed in the previous section, a problem that the reader may remember we solved via direct attack of the field equations, while on the right-hand (GR) side we recognize the BI electrodynamics upon the identification $\beta^2 = -2\pi/(\varepsilon\kappa^2)$, which only holds true in the branch $\varepsilon < 0$. The latter is a well known problem that admits a general solution for every NED as

$$M(x) = \frac{r_S}{2} + \frac{\kappa^2}{2}\int_r^\infty x^2 T^t{}_t(x)dx \tag{6.100}$$

where $T^t{}_t$ is the stress-energy tensor of the (in the present case) BI electrodynamics with the identification made above. This expression feeds the mapping equations (6.98) together with the fluid's conservation equation discussed in previous section, which amounts to the equation $\rho(1 - \varepsilon\kappa^2\rho) = \frac{Q^2}{8\pi x^4}$, and which upon integration yields exactly the expression (6.70) of the quadratic gravity/EiBI gravity case.

The main lesson of this quick exercise is to show how hardly-won solutions can be found via a much direct procedure. Note that the correspondence of theories works in both ways, i.e., if in the mapping equation (6.98) we set Maxwell in the GR side and assume EiBI as the target RBG theory on the other side of the correspondence, then one can reconstruct the corresponding Lagrangians as

$$\text{EiBI} + \Phi(X) = \frac{4\pi}{\kappa^2\varepsilon}\left(1 - \sqrt{1 - \frac{\kappa^2\varepsilon X}{2\pi}}\right) \leftrightarrow \text{GR} + Z \tag{6.101}$$

where in the EiBI side of the correspondence we see again the appearance of a BI-type electrodynamics, with the right signs of the constants under the integral if we assume the identification $\beta^2 = 2\pi/(\kappa^2\varepsilon)$, where one could decide to impose $\varepsilon > 0$ in order to match the expectation of β^2 being a positive quantity in the GR scenario (although one could also explore the branch with $\varepsilon < 0$ oblivious to where this solution came originally from). Via this identification one can use once again the fluid's mapping equations (6.98) to arrive to the line element (6.70), in this case with the mass function

$$M(z) = \frac{r_S}{2} - \frac{\kappa^2 Q^2}{8\pi\sqrt{2}r_c} \frac{1}{\sqrt{z^2 + \sqrt{z^4 + 4s}}} \tag{6.102}$$

where now $r_c = |\varepsilon|\kappa^2 Q^2/(8\pi)$ and $s \equiv \varepsilon/|\varepsilon|$ is the sign of ε. As for the radial function, it behaves in this case as $z^2(x) = \frac{x^4 - s}{x^2}$, which implies that a bounce is present in the branch $s = -1$ only.

Since the structure of the mapping above works for any electromagnetic field configuration, there is no limit to what we can achieve now in these scenarios starting from as many known seed solutions as desired. For instance, the Majumdar-Papapetrou configurations [48, 69] are a subset of solutions of the Einstein-Maxwell-dust system in which the mass is exactly tuned to the electric charge, and which can be interpreted as a collection of extreme black holes in static equilibrium given by the line element [42]

$$ds^2 = -U^{-2}(R)dt^2 + U^2(dR^2 + R^2(d\theta^2 + \sin^2\theta d\phi^2)) , \tag{6.103}$$

where the function $U(r)$ satisfies the equation $\nabla^2 U = -4\pi\rho U^3$ thus granting some freedom to choose either the energy density ρ or the $U(r)$ field itself. For instance, the so-called Bonnor stars [45] are described by an external solution $U(r) = 1 + \frac{m}{r}$ and an internal one $U(r) = 1 + \frac{m}{r_0} + \frac{m(r_0^2 - r^2)}{2r_0^3}$ for a certain $r \geq r_0$. This scenario corresponds to the identification (6.101) under the map, so working again the relation between fluids (6.98) in the present case, we get to the general solution of the corresponding problem in the EiBI+ BI side, which in the spherically symmetric case reads [58]

$$ds^2 = -\frac{1}{U^2}\left(1 + \frac{\varepsilon\kappa^2}{16\pi}\frac{(\nabla U)^2}{U^4}\right)dt^2 + \frac{\varepsilon\kappa^2}{8\pi}\frac{(dU)^2}{U^2} + U^2\left(1 - \frac{\varepsilon\kappa^2}{16\pi}\frac{(\nabla U)^2}{U^4}\right)d\mathbf{x}^2 , \tag{6.104}$$

where the EiBI+BI corrections in ε to GR solutions are evident. This actually describes a family of wormholes in static equilibrium, with also relevant consequences for the regularity of the corresponding space-times.

For the sake of completeness, it is worth pointing out that the mapping also works in higher and lower-dimensional versions of the RBG theories, which therefore allows one to consider seed solutions of interest from GR within such cases. For instance, the BTZ solution [14] describes a family of rotating configurations in the $2 + 1$ dimensional Einstein-Maxwell systems with cosmological constant, which interpolates between black holes and a pure AdS space-time, including the case of a regular horizonless solution with mass $M = -1$ disconnected from the space of black hole solutions by a mass gap. Since the correspondence (6.101) works equally fine in this case, a solution to the equivalent problem in the EiBI+BI side of the correspondence is found as [41]

$$\frac{ds^2}{\eta} = \left(M + \frac{Q^2}{2}\log\left(\frac{x}{x_0}\right) - \frac{x^2}{l^2} - \frac{\varepsilon\kappa^2 J^2 Q^2}{64\pi\eta x^4}\right)dt^2 + H^{-1}dx^2 \quad (6.105)$$
$$+ x^2\left(1 - \frac{\varepsilon\kappa^2 Q^2}{16\pi\eta x^2}\right)d\phi^2 - J\left(1 - \frac{\varepsilon\kappa^2 Q^2}{16\pi\eta x^2}\right)dtd\phi\,,$$

where x_0 is some reference scale, $\eta \equiv (1 - 2\varepsilon\Lambda)$, $H(x) = -M - \frac{Q^2}{2}\log(x/x_0) + \frac{x^2}{l^2} + \frac{J^2}{4x^2}$, and we have taken $\kappa^2 = 8\pi$ and $\Lambda = -1/l^2$. As usual, the sign of ε controls whether in the new ε-deformed configurations there is ($s_\varepsilon = -1$) or not ($s_\varepsilon = +1$) a bounce in the radial function.

Mapping scalar fields. Similar correspondences to those of (6.99) and (6.101) also hold when one considers scalar fields as the matter source, though now the algebraic structure of the corresponding stress-energy tensor makes quite more difficult to prove it so [67]. Specifically, assuming a scalar Lagrangian in the RBG frame given by $\mathscr{L}_m = -\frac{1}{2}P(X, \phi)$, with $X = g^{\mu\nu}\partial_\mu\phi\partial_\mu\phi$, and another scalar Lagrangian in the GR frame as $\tilde{\mathscr{L}}_m = -\frac{1}{2}K(Z, \phi)$, with $Z = q^{\mu\nu}\partial_\mu\phi\partial_\nu\phi$, one can show the explicit shape of such correspondences as (in the quadratic $f(\mathscr{R})$ and EiBI cases) [6]

$$f(\mathscr{R}) = \mathscr{R} - \sigma\mathscr{R}^2 + (P = X - 2V) \leftrightarrow \text{GR} + K = \frac{Z + \sigma\kappa^2 Z^2}{1 + 8\sigma\kappa^2 V} - \frac{2V}{1 + 8\sigma\kappa^2 V} \quad (6.106)$$

$$f(\mathscr{R}) = \mathscr{R} - \sigma\mathscr{R}^2 + P = \frac{X - \sigma\kappa^2 X^2}{1 - 8\sigma\kappa^2 V} - \frac{2V}{1 - 8\sigma\kappa^2 V} \leftrightarrow \text{GR} + K = Z - 2V \quad (6.107)$$

$$\text{GR} + K = Z - 2V \leftrightarrow \text{EiBI} + P = \frac{2\left(\sqrt{1 + \varepsilon\kappa^2 X} - \lambda(1 + \varepsilon\kappa^2 V)\right)}{\varepsilon\kappa^2\left(1 + \varepsilon\kappa^2 V)\right)} \quad (6.108)$$

Let us recall that the power of the mapping reaches its apex when a seed solution is known under analytical exact form in the GR side. In the present case, this is achieved for the free scalar case, where the corresponding solution was found by Wyman in [78]. Therefore, setting $V = 0$ in Eqs. (6.107) and (6.108) one gets a quadratic-like scalar field Lagrangian in the quadratic $f(\mathscr{R})$ gravity case, and a square-root Born-Infeld-type in the EiBI case, yet another manifestation of the transfer of the functional dependencies between the gravity and matter Lagrangians when moving from one side of the mapping to the other. With these ingredients, finding the corresponding Wyman-like solutions in the quadratic $f(\mathscr{R})$ and EiBI frames is immediate and read [7]

$$ds^2_{f(R)} = \frac{1}{f_R}ds^2_{GR} = (1 + 2\alpha\kappa^2 Z)\Big[-e^\nu dt^2 + \frac{e^\nu}{W^4}dy^2 + \frac{1}{W^2}(d\theta^2 + \sin\theta^2 d\varphi^2)\Big], \quad (6.109)$$

where $Z \equiv W^4 e^{-\nu}$ the field invariant of the Wyman case, and

$$ds^2_{EiBI} = -e^{\nu}dt^2 + \left(\frac{e^{\nu}}{W^4} - \varepsilon\kappa^2\right)dy^2 + \frac{1}{W^2}(d\theta^2 + \sin\theta^2 d\varphi^2)\,, \qquad (6.110)$$

respectively.

This concludes our presentation of the space of solutions found within several Palatini theories of gravity, obtained either via direct attack of the field equations or via the shortcut provided by the mapping method. It is now time to turn to the analysis of the regularity of all these configurations.

6.4 Regularity Criteria

6.4.1 Curvature Divergences

Since (most) modern gravitational theories (including GR itself) see gravitation as a manifestation of a geometrical effect, then when a "singularity" is present in a given configuration our physical intuition tells us that this must be caused by something going ill with the underlying geometrical structure of the space-time [31]. In order to avoid considering artificial singularities caused out of a bad choice of basis on our tensorial quantities, in characterizing singularities it seems natural to resort to objects that are invariant under coordinate transformations, i.e., scalar geometrical objects. Natural among them are curvature scalars, namely, different contractions of the Riemann tensor with the metric. This intuitive view amounts to build objects like $g_{\mu\nu}R^{\mu\nu}$, $R_{\mu\nu}R^{\mu\nu}$, $R^{\alpha}{}_{\beta\gamma\delta}R_{\alpha}{}^{\beta\gamma\delta}$, ..., and consider a space-time as singular whenever *any* of such scalar objects is divergent. For instance, Schwarzschild space-time would be singular because in this case $g_{\mu\nu}R^{\mu\nu} = R_{\mu\nu}R^{\mu\nu} = 0$ but $R^{\alpha}{}_{\beta\gamma\delta}R_{\alpha}{}^{\beta\gamma\delta} = 48M^2/r^6$, and therefore the geometry is singular at $r = 0$ due to the divergence of the latter invariant there. *Mutatis mutandis*, a space-time has a chance of being regular if all possible curvature scalars one can think of are finite. This procedure is so popular, besides its physical intuitiveness, thanks to its easy implementation: it can be actually programmed in any computation software such as Mathematica with a few lines of code (at least in the spherically symmetric case), and applied to any configuration of mathematical/physical interest.

Since most of our solutions are deformations of the Reissner-Nordström solution of GR inside the matter sources, it is instructive to begin our analysis by considering the behaviour of its curvature scalars. The three more "popular" ones are given by

$$R_{RN} = 0\,,\ Q_{RN} = \frac{r_q^4}{r^8}\,,\ K_{RN} = \frac{12r_S^2}{r^6} - \frac{24r_S r_q^2}{r^7} + \frac{14r_q^4}{r^8}\,. \qquad (6.111)$$

Projective invariance and curvature scalars: The fact that neither the Ricci nor the Riemann tensor are invariant under projective transformations, makes the scalar object K to be non-projectively invariant as well ($\mathscr{Q}$ is projectively invariant since it is built from the symmetric part of the Ricci tensor alone, which is invariant), i.e., $K(\tilde{\Gamma}) = K(\Gamma) + 4F_{\mu\nu}F^{\mu\nu}$. This means that one can always find a suitable gauge for ξ_μ compensating the divergences in K so as to render it finite everywhere [17]. However, there is yet another scalar object, $P \equiv R^{\alpha\beta}{}_{\mu\nu}R_{\alpha\beta}{}^{\mu\nu}$, which is invariant both under coordinate and projective transformations. This invariant typically diverges as strongly as K itself, so that such divergences cannot be so simply gauged away. For the sake of this section we shall take the gauge $\xi_\mu = 0$, where the scalar invariants take their canonical expressions.

How can one then find regular space-times via the behaviour of curvature scalars? Instead of writing off different matter sectors and solve their corresponding field equations to see whether they give rise to regular solutions or not, one can use the backward procedure: to *engineer* space-times with any desired form of its line element, and then drive the Einstein equations back in order to find the matter sources threading such a geometry. In particular, one could seek space-times having a chosen set of their curvature scalars finite.[2] Focusing on the spherically symmetric case, and using the fact that the conflictive location of such scalars corresponds typically to $r = 0$, this analysis finds that in order to acquire the finiteness of such invariants, the metric must behave there as [34]

$$-g_{tt} = g_{rr}^{-1} \approx 1 - \frac{\Lambda}{3}r^2 + \mathscr{O}(r^3) \tag{6.112}$$

where Λ is a constant. This is such a strong departure from the behaviour of the RN metric there, $-g_{tt} \sim +Q^2/r^2$, that one may wonder what type of matter fields could generate it. Actually, NEDs are able to do the job [35], though the corresponding models do not come free of theoretical difficulties as every other model does in trying to overcome the constraints of the singularity theorems (at least within GR). This kind of configurations go collectively under the name of *de Sitter* cores [9, 46], and are a well known mechanism invoked in the literature for the removal of space-time singularities.

Is it possible to find curvature singularity-free solutions within Palatini theories of gravity without resorting to engineering constructions or de Sitter cores. To illustrate this issue, we consider our already familiar quadratic gravity/EiBI geometry of Eq. (6.69). For convenience we introduce Eddington-Finkelstein coordinates $dv = dt + \frac{dx}{A}$, which turns such a line element into

[2] Note, however, that one should establish a criterion to define a fundamental set of scalar invariants because the number of possible scalar functions is infinite.

$$ds^2 = -\frac{A}{\Omega_+}dv^2 + \frac{2dvdx}{\Omega_+} + z^2(x)d\Omega^2 \tag{6.113}$$

Since the object Ω_+ is positive and well-behaved everywhere, it cannot be a source of singularities anywhere. Troubles can thus only come out of the behaviour of the function $A(x)$, similarly as in GR. Therefore, expanding the function $B(x) = A(x)/\Omega_+$ in the line element above in series around the minimum value of the radial function $z = 1$ ($r = r_c$ in dimensionful coordinates), one finds

$$B(z) \approx \frac{(\delta_1 - \delta_c)}{4\delta_c\delta_2}\left(\frac{1}{\sqrt{z-1}} + \frac{9}{4}\sqrt{z-1} - \ldots\right) + \\ + \frac{1}{2}\left(1 - \frac{\delta_1}{\delta_2}\right) + \left(1 - \frac{2\delta_1}{3\delta_2}\right)(z-1) + \mathcal{O}(z-1)^2 \tag{6.114}$$

where we recall that δ_c is a constant coming out of the resolution of the field equations. Therefore we see that in this case the leading-order divergence is quite mild, $\sim (z-1)^{-1/2}$, and significantly improves the RN behaviour. Moreover, this expressions also hints that the class of solutions with $\delta_1 = \delta_c$ may have even improved properties. Therefore, replacing this condition first in the general expression of the $B(z)$ function, and expanding the result again in power series, we get the result

$$B(z) \approx \frac{1}{2}\left(1 - \frac{\delta_c}{\delta_2}\right) + \left(1 - \frac{2\delta_c}{3\delta_2}\right)(z-1) - \frac{1}{2}\left(1 - \frac{8\delta_c}{5\delta_2}\right)(z-1)^2 + \mathcal{O}(z-1)^3, \tag{6.115}$$

which is finite as $z \to 1$ and singles out this case as a special class of configurations. Note that these expressions also allow to classify the structure of horizons of the corresponding solutions, which can be two (Cauchy and event), a single one (degenerate or not) or none, depending on the combination of parameters. Therefore, we may have either black holes with different horizon structures, or horizonless compact objects.

With the expressions above for $B(z)$ one can now compute the behaviour of the corresponding curvature scalars. In the general case, these read

$$r_c^2 R(g) \approx \left(-4 + \frac{16\delta_c}{3\delta_2}\right) + O(z-1) + \ldots \\ - \frac{1}{2\delta_2}\left(1 - \frac{\delta_c}{\delta_1}\right)\left[\frac{1}{(z-1)^{3/2}} - O\left(\frac{1}{\sqrt{z-1}}\right)\right], \tag{6.116}$$

$$r_c^4 Q(g) \approx \left(10 + \frac{86\delta_1^2}{9\delta_2^2} - \frac{52\delta_1}{3\delta_2}\right) + O(z-1) + \ldots \\ + \left(1 - \frac{\delta_c}{\delta_1}\right)\left[\frac{6\delta_2 - 5\delta_1}{3\delta_2^2(z-1)^{3/2}} + O\left(\frac{1}{\sqrt{z-1}}\right)\right] \\ + \left(1 - \frac{\delta_c}{\delta_1}\right)^2\left[\frac{1}{8\delta_2^2(z-1)^3} - O\left(\frac{1}{(z-1)^2}\right)\right] \tag{6.117}$$

$$r_c^4 K(g) \approx \left(16 + \frac{88\delta_1^2}{9\delta_2^2} - \frac{64\delta_1}{3\delta_2}\right) + O\,(z-1) + \ldots \tag{6.118}$$
$$+ \left(1 - \frac{\delta_c}{\delta_1}\right)\left[\frac{2\,(2\delta_1 - 3\delta_2)}{3\delta_2^2 (z-1)^{3/2}} + O\left(\frac{1}{\sqrt{z-1}}\right)\right] +$$
$$+ \left(1 - \frac{\delta_c}{\delta_1}\right)^2\left[\frac{1}{4\delta_2^2 (z-1)^3} + O\left(\frac{1}{(z-1)^2}\right)\right].$$

From these expressions we observe that curvature divergences are softened from $\sim 1/r^8$ of RN to $\sim 1/(z-1)^3$ in the present case. Moreover, in the special case $\delta_1 = \delta_c$, these scalars become all finite and given by the constant terms appearing in this expressions [note that again one arrives to this result by first substituting $\delta_1 = \delta_c$ in the expression of $B(z)$ and then finding the corresponding shapes of them]. This points out to the striking nature of these $\delta_1 = \delta_c$ configurations: they correspond to regular (in the sense of free of curvature divergences) black holes with a single non-degenerated horizon if $\delta_1 > \delta_2$, and to horizonless regular objects otherwise, thus covering both scenarios of interest.

Two questions come about naturally. Does this mean that those configurations with $\delta_1 = \delta_c$ are free of singularities, and those with $\delta_1 \neq 0$ are "singular"? Is this a general property of all configurations in Palatini theories of gravity? The answer to the latter is negative, as is proven by the case of quadratic $f(\mathscr{R})$ gravity coupled either to a Born-Infeld [59] or to Euler-Heisenberg electrodynamics [40], where the invariant K is divergent at the center of the solutions no matter the choice of their parameters. However, other matter configurations show different features. Indeed, for curvature scalars to be finite, we already saw before that the metric at the center must be finite, and this can be achieved for certain fluid configurations and in certain combination of parameters of the gravity+matter models. This is the case, for instance, of the configurations of Eq. (6.64). Here there is no bounce in the radial function (as opposed to the case of a quadratic/EiBI gravity with a Maxwell field discussed above), but the energy density becomes divergent at $z = 1$. Because of this feature, it is natural to see the latter surface as the boundary of the space-time, and consider the expansion of the metric functions there, which yields the result

$$-g_{tt} \approx \frac{r_S\delta_1}{3r_c\gamma} + \mathscr{O}(z-1)\ ;\ g_{rr} \approx \frac{48r_c}{r_S\delta_1} + \mathscr{O}(z-1)^2 \tag{6.119}$$

As a consequence of this latter behaviour, all curvature scalars for these configurations are also finite. One could of course argue that the divergence of the energy density at $z = 1$ in these configurations would be a strong argument against the regularity and/or physical plausibility of these space-times. There are yet more examples of this kind within the setting above though now in a different branch of the theory as defined by Eq. (6.65), where one finds that while there is no bounce in the radial function, in this case the configurations can reach the center $z = 0$, where the metric functions become

$$-g_{tt} \approx \frac{1}{1+\gamma}\left(1 - \frac{r_S \delta_c^\gamma}{3r_c} z^2\right) \; ; \; g_{rr} \approx 1 + \frac{r_S \delta_c^\gamma}{3r_c} z^2 \tag{6.120}$$

where δ_c^γ is a constant playing the same role as its namesake in the quadratic/EiBI gravity case discussed above. In such a case, the geometry becomes de Sitter at the center, $R_{\mu\nu} = \frac{\delta_c^\gamma}{\delta_2} g_{\mu\nu}$, which yields finite curvature scalars with finite energy density everywhere, thus removing the objections above on the plausibility of these configurations. Indeed we note that, as opposed to GR-based attempts, this curvature-regularity via de Sitter cores can be naturally achieved with physically sensible matter fields satisfying basic energy conditions.

It turns out to be a generic property of $f(\mathcal{R})$ theories that solutions implementing a bounce in the radial function are prone to develop curvature divergences at the bounce location no matter the choice of the (gravity+matter) model parameters, while those with a monotonic behaviour of the radial function contain some sub-cases where the presence of curvature divergences can be removed everywhere. On the opposite side, we have quadratic/EiBI gravity, in which most configurations have curvature divergences at the bounce (whenever it exists), but at the same time some special configurations (as given by the choice of model parameters) can be found for which curvature scalars are finite everywhere.

However, the characterization of the degree of singularity of a space-time via divergences in the curvature scalars as pathology markers has been long contested in the literature. Arguments against such a line of reasoning come in several types. Firstly, given the many geometrical scalars that can be built beyond the $\{R, Q, K\}$ trio, in this approach one should be systematic in computing all possible such objects from every possible geometrical entity in order to make sure that they are all finite. Second, it could happen that every possible such scalar is regular and still pathological behaviours are present. But even more importantly, one may question this privileged role attributed to the behaviour of some geometrical quantities rather than to the behaviour of observers instead, driven by the implicit idea that an ill-behaviour of the former will necessarily be translated on something pathological occurring to the latter. Indeed, this view on the prominence of curvature scalars can be discredited by presenting explicit counter-examples in which the blow-up of (some) curvature scalars do not necessarily translate into any pathologies of observers propagating through the space-times holding them [17]. This idea connects with a more powerful concept in order to pin down the potential presence of pathologies in a given space-time, namely, the completeness of geodesic paths on them.

6.4.2 Geodesic Completeness and Mechanisms for Its Restoration

Placing the focus upon the idea of geodesic completeness over that of curvature divergences is simply a recognition of the motto that *the very existence of observers is more important that their potential suffering*. From a physical point of view, the requirement of geodesic completeness makes utter sense. Geodesics correspond to the universal free-falling motion of idealized (point-like) time-like observers, and are a natural consequence of the weak equivalence principle imbued in GR and in most extensions of it, which demands the trajectories of all such particles to be independent of their internal composition and structure. This way, the viewpoint above simply hovers the idea that in a physically reasonable space-time all time-like (physical observers) and null (light rays) should not be allowed to cease to exist suddenly or to pop up out of nothingness. Otherwise, the principles of causality, predictability and determinism would be seriously jeopardized. Note, however, that a geodesic completeness check of all trajectories within a given space-time does not guarantee its regularity: it could still harbour other types of pathologies affecting the structure of observers propagating on it. In other words, geodesic completeness may be regarded as a *minimum* necessary condition for a (black hole) space-time to have a chance of being regular, working on a separated and an unrelated logic as that of curvature scalars.

Working with geodesic completeness has yet another advantage: the fact that it is the core concept built in the theorems on singularities. These may be formulated under different combinations of hypothesis, but all of them resort to geodesic completeness as the marker of something going ill in the underlying structure of space-time [31, 73]. For the sake of this chapter, we shall adhere to Penrose's formulation of these theorems.

Penrose's 1965 singularity theorem (null version) [71]: If a given maximally extendible space-time $(M, g_{\mu\nu})$ satisfies the following conditions:

1. M contains a closed future-trapped null surface, i.e., a two-dimensional compact embedded sub-manifold surface S such that the two families of light rays emerging orthogonally from S to future infinite have a convergent expansion.
2. M admits a non-compact, connected Cauchy hypersurface, i.e., an achronal hypersurface which is met once and only once by all causal geodesics.
3. The null congruence condition, $R_{\mu\nu}n^{\mu}n^{\nu} \geq 0$, with $n^{\mu}n_{\mu} = 0$ a null vector, holds.

then such a space-time contains at least one incomplete null geodesic at a certain value u_0 of its affine parameter u. The theorem can be trivially upgraded to its time-like version by considering a time-like vector $u^{\mu}u_{\mu} = -1$ in the third condition, and a minor (technical) refinement of its hypothesis.

(Very) roughly speaking, the first condition of the theorem states that a black hole is present (we shall not deal here with cosmological singularities), the second that we have a well posed initial data problem, while the third is a technical condition that guarantees the focusing of geodesics at certain value of the affine parameter $u = u_0$ and the development of a singularity. The latter condition is equivalent, via the

Einstein equations, to the (in the null case) NEC, therefore linking the singular/regular character of the space-time to fundamental lab-verified features of the matter fields threading it. Central to the proof of the theorems is the idea of *maximally extensible* space-time, i.e., a space-time that cannot be further extended beyond the focusing point u_0, this assumptions designed to avoid cheating the theorems out by considering singular space-times that have made them so by artificially removing a given region of them to provoke the incompleteness of geodesic paths (e.g., Minkowski space with a single point removed). In other words, a geodesic path is non-defined beyond u_0 because *there is literally no further space it could be occupied by it*. In this part of the chapter, we shall stick to this idea and check the geodesic (in)completeness of (some of) the black hole solutions introduced in the previous sections.

In facing the issue of building the geodesic equations of our Palatini theories, one could question which one of the two metrics, the auxiliary one, $q_{\mu\nu}$, or the space-time one, $g_{\mu\nu}$, should be considered? We point out that in our construction of RBGs, as given by the action (6.20), we did not allow the auxiliary metric to couple to the matter sector of the theories. This condition guarantees that the matter fields will follow the laws of motion dictated by the space-time metric, while the auxiliary metric's role is reduced to induce a deformation of the geometrical background in which a given observer carries out its geodesic motion, which is consistent with the fact that it can be integrated out in favour of additional matter fields (seeing the theory as GR with non-linear couplings). Under these conditions, the geodesic equations retains its usual form

$$\frac{d^2x^\mu}{du^2} + \Gamma^\mu_{\alpha\beta}\frac{dx^\alpha}{du}\frac{dx^\beta}{du} = 0 \tag{6.121}$$

where $\Gamma^\mu_{\alpha\beta}$ are the components of the connection computed with the usual Christoffel symbols of the space-time metric $g_{\mu\nu}$. The Hamiltonian formulation of a point-particle action in a spherically symmetric space-time of the form $ds_g^2 = -C(x)dt^2 + B(x)dx^2 + r^2(x)d\Omega^2$ yields two conserved quantities, namely, the energy per unit mass, $E = C(x)\dot{t}$, and the angular momentum per unit mass, $L = \dot{\varphi}r^2(x)\sin^2\theta$, where dots denote derivatives with respect to the affine parameter u. By the freedom granted by spherical symmetry one can take the motion to be confined to the plane $\theta = \pi/2$ without loss of generality, in such a way that the geodesic equation (6.121) can be written in this background as

$$\frac{C(x)}{B(x)}\left(\frac{dx}{du}\right)^2 = E^2 - V(x) \tag{6.122}$$

which is somewhat akin to the equation of motion of a single particle moving in a one-dimensional effective potential of the form

$$V(x) = C(x)\left(\frac{L^2}{r^2(x)} - k\right) \tag{6.123}$$

where $k = \{-1, 0, +1\}$ corresponds to the sign of the Hamiltonian itself (properly re-scaled) and indicate the nature of the geodesics: time-like $k = -1$ (physical observers), null $k = 0$ (photons) and space-like $k = +1$ (tachyons - hypothetical particles moving faster than the speed of light).

Our goal now in verifying the regularity of the space-times considered so far is to inspect every time-like and null geodesic (disregarding space-like trajectories as unphysical) using the equation above, and to see whether each of them can be extended to arbitrarily large values of its affine parameter. Since in our analysis above we found that $f(\mathcal{R})$ theories manifest a qualitatively different behaviour in their metric functions as compared to quadratic/EiBI gravity, we shall consider separately the behaviour of their corresponding geodesics.

For $f(\mathcal{R})$ gravity the line element is given by Eq. (6.52), so that the geodesic equation (6.122) is particularized to this case as

$$\frac{1}{f_{\mathcal{R}}^2}\left(\frac{dx}{du}\right)^2 = E^2 - V(x)\ ;\ V(x) = \frac{A(x)}{f_{\mathcal{R}}}\left(\frac{L^2}{r^2(x)} - k\right) \tag{6.124}$$

In those cases in which a bounce is present, i.e., when $x^2 = z^2/f_{\mathcal{R}}$ implements a non-monotonic behaviour in $z(x)$ (recall that this is only possible when $f_{\mathcal{R}}$ has a zero), it is more convenient to rewrite the equation above in terms of z itself as

$$\frac{du}{dz} = \pm \frac{\left(1 + \frac{z f_{\mathcal{R},z}}{2 f_{\mathcal{R}}}\right)}{f_{\mathcal{R}}^{1/2}\sqrt{E^2 + g_{tt}\left(k + \frac{L^2}{r_c^2 z^2}\right)}} \tag{6.125}$$

where a factor r_c has also been re-scaled in the affine parameter u, while the $\pm$ signs correspond to outgoing/ingoing geodesics. The potentially problematic region in these cases correspond to the location of the bounce itself, $z = z_c$ (as opposed to the RN case, where no bounce is present and one can reach the point $z = 0$ instead), so we have to integrate this equation for both time-like and null geodesics, and spanning all possible values of the energy and the angular momentum. Starting with null $k = 0$ radial $L = 0$ geodesics is our first natural choice, since these are incomplete in the RN geometry of GR. In this case, the above equation reads simply

$$\pm E\frac{du}{dz} = \frac{\left(1 + \frac{z f_{\mathcal{R},z}}{2 f_{\mathcal{R}}}\right)}{f_{\mathcal{R}}} \tag{6.126}$$

so all comes down to the behaviour of the function $f_{\mathcal{R}}$. At large distances, $z \gg 1$, where $f_{\mathcal{R}} \to 1$ (i.e., GR), one finds the expected behaviour there, $Eu \approx \pm z$. On the other hand, at the bounce location $z \to 1$, since $f_{\mathcal{R}} = 1 - 2\sigma\kappa^2 T$, one expects the incompleteness of geodesics as far as this factor goes to a constant by the same reasons as above. This implies that only when the gravity constant σ and the trace of the stress-energy tensor have opposite signs will this factor go to zero and one can expect new scenarios of regularity. Indeed, this is exactly the case when the bounce in the radial function is present, since it comes from the relation $x = z f_{\mathcal{R}}^{1/2}$, which requires the presence of a zero in $f_{\mathcal{R}}$. Therefore, bounce scenarios are precisely those in which one can hope to achieve completeness of this kind of geodesics. Take for instance the quadratic $f(\mathcal{R})$ case coupled to BI electrodynamics studied in Sect. 6.3.1. A bounce is present at a minimum value of the radial function given by [11]

$$z_c^4(\alpha) = \frac{1+2\alpha-\sqrt{1+4\alpha}}{2\sqrt{1+4\alpha}} \tag{6.127}$$

in such a way that the expansion of the function $f_{\mathcal{R}}$ there becomes of the form $f_{\mathcal{R}} \approx c(\alpha)(z-z_c) + \mathcal{O}(z-z_c)^2$ (with c a constant whose shape is irrelevant for our purposes here) and the geodesic equation (6.126) thus becomes

$$\pm E(u(z)-u_0) \approx \mp \frac{z_c}{c^{1/2}(z-z_c)^{1/2}} \tag{6.128}$$

where u_0 is an integration constant. This expression implies that, as the bounce location $z = z_c$ is approached, the affine parameter diverges to $\pm\infty$. This can be interpreted in terms of ingoing (outgoing) null radial geodesics taking an infinite affine time to get to (depart from) the bounce radius z_c; in other words, $z = z_c$ represents the actual boundary of the space-time, which is pushed in these theories to infinite affine distance. Therefore, since the affine parameter can be indefinitely extended in both directions (to asymptotic infinity on one side, and to the infinitely-displaced bounce location on the other), these null radial geodesics are complete in these geometries. This is a strong departure from the GR (RN) result, where the fact that the center of the solution, $r = 0$, is reached in finite affine time by these geodesics, without possibility of further extension beyond this point, makes it to be (null) geodesically incomplete, hence singular from this point of view.

What about the remaining geodesics? From (6.125) one can see that either when k or L^2 are non-vanishing, the term g_{tt} under the square-root will contribute non-trivially to the geodesic behaviour. Indeed, in the case of BI electrodynamics considered above this term behaves as $g_{tt} \sim -1/(z-z_c)^2$ as $z \to z_c$. This means that before the surface $z = z_c$ can be reached, the term inside the square root will vanish, and a turning point is reached: in other words, the particle finds an infinite potential barrier preventing any such trajectory from reaching the region $z = z_c$ (similarly as in the usual RN solutions). Therefore, geodesic completeness is achieved for all time-like and null trajectories by the "trick" of effectively pushing the future (past) boundary of the space-time to unreachable distance to any kind of (causal) observers or information. This same conclusion will be met by any other geometry within the $f(\mathcal{R})$ class coupled to suitable matter fields having qualitative similar behaviours in their metric functions, namely, having a bounce in the radial function and having its metric function $A(z)$ with a suitable divergent behaviour. For instance, both the EH electrodynamics [40], the type-I anisotropic fluid (6.63) with $\gamma > 1$ (in order to support a bounce in the radial function) [18], and the type-IV anisotropic fluid (6.66) of the same fluid, have a similar functional dependence on the $f_{\mathcal{R}}$ and g_{tt} variables as the BI case, and therefore are also null and time-like geodesic complete.

A different kind of behaviour is met by the type-III anisotropic fluid (6.63), corresponding to the de Sitter cores discussed in the previous sub-section. Indeed, in this case nothing prevents all geodesics to reach the center $r = 0$ and, being the curvature perfectly regular, one could tentatively conclude that nothing should prevent these geodesics to continue their path through these cores. However, should we really grant to (regular) curvature the authority to determine when the continuation of geodesic paths is allowed? To analyze this question we turn now to the quadratic/EiBI gravity solutions coupled to Maxwell electrodynamics (in the branch $\varepsilon < 0$). Particularizing the geodesic equation (6.122) to the line element of these theories, Eq. (6.69), yields the result

$$\frac{1}{\Omega_+^2}\left(\frac{dx}{du}\right)^2 = E^2 - V(x)\,;\;\; V(x) = \frac{A(x)}{\Omega_+}\left(\frac{L^2}{r^2(x)} - k\right) \tag{6.129}$$

Doing the same gymnastics as before, in the null radial case the above equation becomes

$$\frac{1}{\Omega_+^2}\left(\frac{dx}{du}\right)^2 = E^2 \tag{6.130}$$

For physical insight, it is again more useful to rewrite this equation in terms of the radial function $z(x)$ using the relation of this case $dx/dz = \pm\Omega_+/\Omega_-^{1/2}$ which yields

$$\pm\, E du = \frac{dz}{\Omega_-^{1/2}} \tag{6.131}$$

and, therefore, in this case all comes down to the behaviour of the function Ω_-. Again, difficulties may arise at the bounce location, $z = 1$, where the function $\Omega_- = 1 - 1/z^4$ has a zero. In any case, the above equation admits an exact integral of the form

$$\pm\, E \cdot u(x) = \begin{cases} {}_2F_1[-\frac{1}{4}, \frac{1}{2}, \frac{3}{4}; \frac{r_c^4}{r^4}]z & \text{if } x \geq 0 \\ 2x_0 - {}_2F_1[-\frac{1}{4}, \frac{1}{2}, \frac{3}{4}; \frac{r_c^4}{r^4}]z & \text{if } x \leq 0 \end{cases}, \tag{6.132}$$

where $\pm$ corresponds to ingoing/outgoing geodesics in the $x > 0$ ($x < 0$) regions, while ${}_2F_1[a, b, c; y]$ is a hypergeometric function and the integration constant $x_0 = {}_2F_1[-\frac{1}{4}, \frac{1}{2}, \frac{3}{4}; 1] = \frac{\sqrt{\pi}\Gamma[3/4]}{\Gamma[1/4]} \approx 0.59907$ comes from matching the expressions of the asymptotic regions $x \gtrless 0$ across the surface $x = 0$. This expression clearly shows that the two regions on both sides of the bounce, $x > 0$ and $x < 0$, can be continuously connected across it, therefore guaranteeing the extensibility of these geodesics to arbitrarily large values of their affine parameter. This behaviour also allows to introduce the canonical interpretation for the nature of the solution, corresponding to the presence of a wormhole structure [75] consisting of two asymptotically flat space-times $x^+ \in (0, +\infty)$ and $x^- \in (-\infty, 0)$, and connected by a throat of non-vanishing areal radius $r^2(x) = r_c^2$ located at $x = 0$.

For every other geodesic, the effective potential in (6.129) is non-vanishing. Moreover, given the finiteness of the factor Ω_+ everywhere (which takes a value two at the bounce location itself), this problem again corresponds effectively to a classical-mechanical problem of the scattering in a central potential $V(r)$. In turn, this potential is governed by the metric function g_{tt} as given by the expansion (6.114) near $z = 1$, which flips sign depending on whether $\delta_1 \gtrless \delta_c$. This is relevant in order to determine whether a turning point will be present (i.e., an infinity in the potential) or if geodesics will be able to approach the central region of the solutions (at least for some values of their energy). Indeed, for $\delta_1 > \delta_c$ (Reissner-Norström-like solutions) the potential becomes infinitely repulsive at $x = 0$, which makes all these geodesics to be repelled back to asymptotic infinity before being able to reach to $x = 0$, while for $\delta_1 < \delta_c$ (Schwarzschild-like solutions) one finds an infinitely attractive potential near $x = 0$ instead, making all geodesics with enough energy to overcome the maximum of the potential barrier to unavoidably intersect this point. However, likewise for null radial geodesics above, nothing prevents such geodesics to cross the wormhole throat $x = 0$ and propagate through the other asymptotically flat space-time to arbitrarily large values of their affine parameter. And finally, in the case $\delta_1 = \delta_c$ (Minkowski-like solutions, since the curvature is everywhere regular in this case), the potential has a certain maximum near $x = 0$ and, therefore, depending on its energy every geodesic will be either deflected at a certain distance from $x = 0$ or cross it to freely propagate to new regions. Similar features can be found for other matter sources within this theory. For instance, the anisotropic fluid configuration of the EiBI case given by Eq. (6.82) has similar solutions as those above for the choices of Type-II $\{s_\varepsilon = -1, s_\beta = +1\}$ and Type-I $\{s_\varepsilon = -1, s_\beta = +1\}$ (with $l_\varepsilon^2/l_\beta^2 > 1$), with a bounce $z(x)$ and a similar description of the type of horizons as well as the behaviour of geodesics and of curvature scalars, i.e., finite when $\delta_1 = \delta_c^{(\gamma)}$ and divergent otherwise. It is worth pointing out that in such a model there are also de Sitter cores within the Type-II $\{s_\varepsilon = -1, s_\beta = +1\}$ provided that $l_\varepsilon^2/l_\beta^2 < 1$ and which arise when $\delta_1 = \delta_c^{(\gamma)}$, with similar comments as those made in the quadratic $f(\mathscr{R})$ case.

One could naturally wonder whether the presence of a bounce in the radial function is a *sufficient* condition for the singularity-removal, given the possibility they allow to extend the geodesics to new regions of the manifold. The answer to this question is negative. Indeed, counter-examples of this are typically given by scalar fields, which are prone to yield geodesically incomplete space-times no matter what their radial functions are doing [7, 41]. Moreover, the analysis of the conditions for having geodesically complete solutions are much more obscure than in the cases studied above.

From our discussion above, we found that geodesic completeness is achieved via (at least) three different mechanisms (see [28, 29] for an extended discussion on this point):

- The innermost region of the solutions, $x = 0$ ($r = r_c$ or $z = 1$) is pushed to infinite affine distance, thus being an effective (past/future) boundary of the space-time. Null radial geodesics take an infinite time to get/depart from there, while any other geodesic is repelled back at some distance. Another space-time region may lie beyond $x < 0$ ($x > 0$) but this is not accessible from the $x > 0$ ($x < 0$) side. This is the typical scenario in (quadratic) $f(\mathcal{R})$ gravity.
- The bounce in the radial function allows for the extensibility of (null and time-like) geodesics beyond $x = 0$ ($r = r_c$), provided that they are able first to overcome the (maximum of the) potential barrier. This is typically ascribed to quadratic/EiBI gravity.
- A de Sitter core is formed at $r = 0$, where curvature is everywhere regular and geodesics can be extended. They may arise in $f(\mathcal{R})$/quadratic/EiBI gravity.

The bottom line of the discussion above is that the completeness of all null and time-like geodesics can be achieved in (some of) these space-times, regardless of what curvature is doing. It should be pointed out that the singularity-removal must necessarily come at the expense of the violations of any of the conditions imposed by the singularity theorems. In most approaches to this issue within GR, energy conditions are those paying the prize in order to achieve the de-focusing of geodesics, typically summoning upon quantum-mechanical arguments (either from the matter or the gravitational fields) to save the day. In the Palatini theories of gravity considered in this work, the fact that the field equations can be written in purely Einstein form with a modified stress-energy tensor given by (6.26) implies that one may have scenarios in which the matter field stress-energy tensor derived from the space-time metric satisfies standard conditions while the effective one (sourcing the q-formulated Einstein equations in these theories, Eq. (6.26)) violates them, so that the null (time-like) congruence condition is bypassed and one can achieve geodesic completeness restoration. This is exactly what happens in most of the geometries considered here, since we chose matter fields satisfying energy conditions from the onset.

6.4.3 The Non-equivalence Between Geodesic Completeness and Curvature Divergences

Our discussion on the extensibility of geodesics (mostly) forsook the issue with curvature divergences. In the $f(\mathcal{R})$ case curvature divergences are typically present, and certainly so in those solutions whose inner region has been pushed to asymptotic infinity. In such a case, the wormhole has only one accessible side, since the other side cannot be reached by observers living in the former at any value of their (endless) affine parameters. Therefore, these curvature divergences should be devoid of any physical meaning since no observer will ever be able to interact with them and experience any type of pathological behaviour. On the contrary, in those models in which a bounce in the radial function is present and accessible to some subsets of geodesics, the presence of curvature divergences at the bounce must be taken seriously. Remarkably, we have seen that there are geodesically complete solutions with divergent curvatures at the bounce (the quadratic/EiBI gravity coupled to some classes of electromagnetic/fluids with $\delta \neq \delta_c$) but also a discrete sub-family of solutions with finite curvature everywhere provided that $\delta_1 = \delta_c$. This is quite an unpleasant situation, since any small perturbation to the latter (via e.g. addition of mass or charge) will take the constants of the solution out of this constraint and, therefore, to fall back into the curvature-divergent family. Moreover, since radial null geodesics in those solutions are insensitive to the specific values of the charge and mass, discriminating between them as complete or incomplete based on the behavior of curvature scalars seems rather arbitrary. Should curvature (divergences) have an actual physical meaning *in all cases* it would make little sense that a regular solution could be turned into a singular one by means of any small perturbation. We also point out that in our framework geodesically incomplete solutions with finite curvature (like in the quadratic $f(\mathscr{R})$ case coupled to the Type-II fluid of Eq. (6.64)) are also possible and, of course, so geodesically incomplete solutions with divergent curvature (e.g. the Type-I fluids of Eq. (6.63) with $\gamma < 1$) are.

The bottom line of this discussion is the lack of correlation between the (in) completeness of geodesics and the behaviour of curvature scalars. This goes against the initial intuition expressed at the beginning of this section, but agreed with our concern expressed at the conceptual troubles of using curvature scalars as a proxy for supposedly ill-behaviours of the geometry. We next want to take a closer and more physical look at the consequences of having unbound curvature when extended bodies, as opposed to the idealized scenario of geodesic point-like ones, cross regions with such an unbound curvature.

6.4.4 Tidal Forces and Congruences of Geodesics

After all, what is the problem with unbounded curvatures unless they induce any kind of uttermost destructive consequences upon extended bodies? It turns out that addressing this question can be made using methods developed decades ago, which we now bring forward. We are thus dealing with tidal forces upon extended observers.

Strong versus weak singularities. A strong (curvature) singularity occurs when all bodies falling into it are crushed to zero volume, no matter what their physical features may be [30, 36, 74]. This statement, therefore, captures the intuitive idea of space-time singularities as geometric abhorrent phenomena not related to the properties of specific matter configurations. Conversely, a weak singularity occurs when a body has a chance of surviving to an encounter with it. The strong singularity criterion can be enhanced to include some cases where the reference volume remains finite but the body itself undergoes unacceptably large deformations [53, 68].

Let us start by modelling any such observer by a set of points in geodesic motion but cohesioned by any kind of physical or chemical forces, i.e., a *congruence of geodesics*, where local differences in the gravitational field have an impact on the separation between every pair of geodesics as the time goes by. The idea is to define an infinitesimal volume associated to such a congruence and follow its evolution as the potentially problematic region is approached. If such a volume goes to zero, this would entail the crushing of the congruence, i.e., of the observer, hence the presence of a (strong) curvature singularity. Let us label the congruence by a set $x^\mu = x^\mu(u, \xi)$, where u is the affine parameter for every geodesic in the congruence and ξ identifies the specific geodesic under consideration. Thus, the tangent vector for a given geodesic is $u^\mu = \partial x^\mu/\partial u$ while the Jacobi vector field connecting infinitesimally close geodesics in the congruence is $Z^\mu = \partial x^\mu/\partial\xi$ and satisfies the geodesic deviation equation

$$\frac{D^2 Z^\alpha}{d\lambda^2} + R^\alpha{}_{\beta\mu\nu} u^\beta Z^\mu u^\nu = 0 , \tag{6.133}$$

where $DZ^\alpha/d\lambda \equiv u^\kappa \nabla_\kappa Z^\alpha = Z^\beta \nabla_\beta u^\alpha$. The above equation yields six independent Jacobi vectors which are orthogonal to the basis of vectors associated to an adapted orthonormal tetrad parallel-transported along the congruence, e_a, as $Z = Z^a e_a$ and DZ^a/du. Note that in such a case Eq. (6.133) allows to find the components of the Jacobi field at any time starting from a reference time u_i as $Z^a(u) = A^a{}_b(u) Z^b(u_i)$ (if the three Jacobi fields are zero at u_i then one can define this relation with its derivatives instead). Moreover, every three linearly independent solutions of (6.133) as $Z_i = Z_i^a e_a$, $i = 1, 2, 3$ allow to define the volume invariant as

$$V(u) = \det(Z_1^a, Z_2^b, Z_3^c) \tag{6.134}$$

Therefore one just simply needs to keep track of this volume element after finding a suitable ansatz for the Jacobi fields under the (spherical) symmetry of the system, allowing to solve the equation (6.133). For instance, in the Schwarzschild case one finds $V(u) \sim u^{1/3}$, which implies its vanishing as $u \to 0$, therefore implying the existence of a crushing-type (strong) singularity alongside the incompleteness of geodesics. On the contrary, in the quadratic/EiBI gravity system coupled to electromagnetic fields (i.e., the geodesically complete solutions with $\delta_1 \leq \delta_c$, since those are the only time-like trajectories able to reach the center of the solutions), as the wormhole throat is approached the infinitesimal volume transported by the congruence goes instead as $V(u) \sim 1/u^{1/3}$ [65], which diverges as $u \to 0$. Moreover, this happens irrespective of the existence of curvature divergences at the throat, in agreement with our discussion above on the lack of correlation between the behaviour of curvature and the extensibility of geodesics. This result on the volume invariant is not formally included within the standard classification of strong/weak singularities above, so it seems that we have come to a dead end this way.

A more physically-intuitive approach is to consider whether the stretching and subsequent contraction undergone by the reference volume in the congruence as the potentially problematic region is approached and left behind has any physical impact upon every observer crossing it. For instance, in the Schwarzschild solution the radial direction undergoes an infinite stretching while the angular directions undergo an infinite contraction (a process mockingly known as spaghettization). One can capture the heart of this problem by determining whether causal contact among the geodesics in the congruence is lost at any moment. Should the latter happen, then one would conclude that the observer has been destroyed by the presence of a strong curvature singularity. The implementation of this approach in specific scenarios is technically daunting, and so far has only been achieved in the quadratic/EiBI gravity case coupled to a Maxwell field. To make a long story short, one finds that the elements in the congruence never lose causal contact among them despite the formally infinite stretching experienced as the wormhole throat is approached, which is then followed by a rapid contraction as the throat is left behind [65]. In this example, there are no observable destructive effects due to this process, no matter the behaviour (divergent or finite) of the curvature scalars, because the stretching and contraction occurs so fast that it ends up been unobservable.

6.4.5 *Completeness of Accelerated Paths*

The principle of general covariance imbued in GR and in most extensions of it dictates the lack of privileges to any set of observers. For our interest here, this means that the completeness of paths of accelerated observers, in addition to the geodesic ones, must be guaranteed for a given space-time to give an accurate description of the physical world. This becomes yet another test for the regularity of a given space-time: to verify that every non-geodesic observer with arbitrary motion and, in particular, with an arbitrary (but bound) acceleration finds infinitely extended paths towards the future and the past. For instance, there are examples (admittedly, toy-like) in which a geodesically complete space-time contains time-like curves with (bound) acceleration but finite total proper length [39], and one would like to make sure that such a thing does not occur in their beloved "regular" space-times.

Under the presence of external forces, the right-hand side of the geodesic equation (6.121) for a massive particle no longer vanishes:

$$\frac{Du^\mu}{d\tau} = \frac{d^2x^\mu}{d\tau^2} + \Gamma^\mu_{\alpha\beta}\frac{dx^\alpha}{d\tau}\frac{dx^\beta}{d\tau} = a^\mu \neq 0 \tag{6.135}$$

where τ is the proper time of a time-like observer. The acceleration undergone by our observer can be computed by considering an orthonormal Frenet-Serret basis $\{u_{(0)}, u_{(1)}, u_{(2)}, u_{(3)}\}$ [38] (parenthesis are introduced here to stress the coordinate-free character of these vectors) to the tangent space satisfying the equation [66]

$$\frac{Du_{(a)}}{ds} = u_{(b)}A^{(b)}{}_{(a)} \;\; ; \;\; A^{(b)}{}_{(a)} = \begin{pmatrix} 0 & k(s) & 0 & 0 \\ k(s) & 0 & \tau_1(s) & 0 \\ 0 & -\tau_1(s) & 0 & \tau_2(s) \\ 0 & 0 & -\tau_2(s) & 0 \end{pmatrix}, \tag{6.136}$$

where the curvature $k(u)$ is interpreted as the linear acceleration experienced by the observer in the direction of $u_{(1)}$, while the first $\tau_1(u)$ and the second $\tau_2(u)$ torsions of the curve correspond to the rotational acceleration $\sqrt{\tau_1^2 + \tau_2^2}$ along the axis given by the vector $\omega = \tau_2\lambda_{(1)} + \tau_1\lambda_{(3)}$. Obviously, when there is neither linear acceleration nor torsion, Eq. (6.135) recovers the usual geodesic equation. In a spherically symmetric space-time written in suitable coordinates as $ds^2 = -A(y)dt^2 + A^{-1}(y)dy^2 + r^2(y)d\Omega^2$, making use of the conservation of energy and angular momentum, and restricting ourselves to linear accelerations only for simplicity, Eq. (6.135) can be written, after a straightforward algebraic exercise, as [66]

$$(u^y)^2 + A(y)\left(1 + \frac{L^2}{r^2(y)}\right) = \left[E + \int_{y_0}^{y} \frac{k(s)dy'}{\sqrt{1 + \frac{L^2}{r^2(y')}}}\right]^2 . \tag{6.137}$$

where $u^y = dy/du$ and $k(y)$ is the total acceleration of the curve in this simplified model. This is the natural extension of the Eq. (6.122) to accelerated observers, covering every space-time considered in this chapter. We can thus check the behaviour of any such set of accelerated observers in any pick of geodesically complete sample found above, and for convenience we shall consider (once again) the quadratic/EiBI gravity by its agreeable properties. By considering a test charged (Maxwell) particle which experiences an acceleration (in modulus) of size $k(y) = \frac{Qq}{mr^2(y)}$, and switching by convenience to the variable x defined by the relation $dy = dx/\Omega_+$, then Eq. (6.137) becomes

$$\left(\frac{u^x}{\Omega_+}\right)^2 + A(x)\left(1 + \frac{L^2}{r^2(x)}\right) = \left(\tilde{E} + I_L^{BI}(r)\right)^2 , \tag{6.138}$$

where $\tilde{E} = E - I_L^{BI}(r_i)$, with r_i the initial location of the particle and the function

$$I_L^{BI}(r) = \frac{Qq}{m}\int^r \frac{dr}{r\Omega_-^{\frac{1}{2}}\sqrt{r^2+L^2}} = -\frac{Qq}{m}\frac{\text{EllipticF}\left[\sin^{-1}\left(\sqrt{\frac{L^2+r_c^2}{L^2+r^2}}\right), \frac{L^2-r_c^2}{L^2+r_c^2}\right]}{\sqrt{r_c^2+L^2}}\,. \tag{6.139}$$

encodes the acceleration term in this scenario. Let us recall the constant character of the radial function at the bounce $x = 0$ ($r = r_c$), which in turn makes the function $I_l^{BI}(r_c)$ to become a constant there and, therefore, the single effect induced by the acceleration of our point-like charge is to shift its total energy, therefore having no effect in the extensibility of geodesics. We thus conclude the completeness of all null and time-like paths (in the latter both geodesic and accelerated) in these geometries.

Let us note that in those cases in which the bounce lies at the future (past) boundary of the space-time (quadratic $f(\mathcal{R})$ case coupled to different fluids), one expects that the introduction of bounded accelerations will not alter the conclusion on every time-like geodesic being bounced back by the infinite potential barrier. And what if, by the sake of curiosity, we consider an unbounded acceleration? In such a case, a similar computation as above reveals that such observers would be able to get to the bounce location and gain access to the other side of the wormhole. Such scenarios are, of course, unlikely to be physically achievable.

6.4.6 *Further Tests?*

Up to now we have considered tests with idealized point-like particles (completeness of geodesic and of accelerated paths), and the effects of tidal forces from unbound curvatures upon extended observers. Is there any other further test one could bring to the table? There is, of course, the important and physically relevant case of the interaction of test waves with potentially pathological space-time regions. In a regular geometry it seems natural to demand that the propagation of any wave through it should be well defined at all times, i.e., that one can establish a well-posed problem of transmission and reflection coefficients. This includes, in particular, the case of gravitational waves, where we find the notable (and perhaps non-intuitive) fact that in Palatini theories of gravity, these waves propagate on the effective gravitational geometry provided by $q_{\mu\nu}$ rather than the space-time one, see [23] for details. For the sake of this chapter, however, we shall simplify the problem down to scalar waves propagating upon the latter geometry, a scenario in which relevant insights on the regularity of these space-times can be obtained.

Let us thus consider a massive scalar field with equation of motion $(\Box - m^2)\phi = 0$. Using the usual mode decomposition, $\phi_{\omega ml} = e^{-i\omega t}Y_{lm}(\theta)f_{\omega l}(x)/r(x)$, one can find the expression of the equation for its radial component in any space-time of interest. Considering again the quadratic/EiBI gravity case, and formulating the problem in terms of the coordinate y given by $dy/dx = 1/(A\Omega_+)$, one finds a Schrödinger-like equation of the form [64]

$$-f_{yy} + V_{eff}(y)f = \omega^2 f \ ; \ V_{eff} = \frac{r_{yy}}{r} + A(r)\left(m^2 + \frac{l(l+1)}{r^2}\right) \quad (6.140)$$

One can thus proceed to define an incoming wave from past null infinity (for horizonless solutions), or from the event/Cauchy horizon (for black holes) and study its behaviour as the bounce location, $r = r_c$, is approached. This problem is better captured by the computation of the transmission and reflection factors, as well as with the cross section of the former. The final result is that no pathologies over these factors, or any other illness of these space-times, are found [64]. This implies that, in addition to regular black holes outside their event horizons, which is the relevant case for gravitational wave astronomy, within these space-times one can also consider the propagation of waves over regular naked wormholes, thus naturally connecting the theoretical issue of space-time singularities with the astrophysically-relevant one of gravitational wave imprints from these objects.

6.5 Closing Thoughts

Well inside the twentieth-first century, space-time singularities are still a cornerstone upon which many attempts to extend GR are built. The canonical viewpoint within this approach is that space-time singularities are an artifice of the classical GR description, and that an extension of it via additional scalar objects, supposedly emerging as low-energy effective (Planck scale-suppressed) contributions from a quantized version of it, should come to save the day. In this approach one faces the new opportunities offered by a larger flexibility in choosing the main geometrical ingredients imbued in the enlarged theory and, in particular, those associated to the affine connection: curvature, torsion, and non-metricity. In this chapter we have granted the mutually desired divorce to the metric and the affine connection, and considered a theory whose building blocks are made up of these two independent entities. The field equations of the theories constructed this way are thus found by independent variation of the action with respect to each of them, which requires the introduction of workable methods to solve the connection equations, a problem which is missing in the metric formulation. We succeeded at this regard by introducing an auxiliary, connection-compatible metric, in such a way that the field equations can be written as a set of Einstein-like ones for it. This procedure can be systematized for a large class of models built upon scalars out of contractions of the metric with the (symmetric part of the) Ricci tensor, and consequently dubbed them as Ricci-based theories. Revolving around these methods we can solve such field equations either by direct attack, or by using the Einstein frame to construct an equivalent GR action with non-linearly coupled fields, the latter procedure allowing to find new solutions of any RBG theory under consideration using a seed solution from the GR side, via purely algebraic transformations. The latter is dubbed as the mapping method, which should be a basic element in the toolkit of the metric-affine practitioner.

The spectrum of (spherically symmetric) solutions found via any of the two procedures above, is ample both in the gravity and matter sectors. For the former, $f(\mathscr{R})$ gravity, quadratic gravity, and EiBI gravity, are theories extensively studied in the literature (including their higher-dimensional and three-dimensional versions), while in the latter we considered electromagnetic fields (both Maxwell and non-linear), different classes of anisotropic fluids, and scalar fields. We actually found different classes of configurations, from black holes with two and a single (degenerated or not) horizons to naked (horizonless) objects of several types including de Sitter cores. Besides the modified horizon structure, the new gravitational corrections have an impact in the innermost structure, where depending on the gravity/matter combination one may find a bounce in the radial function, which is the most interesting case for the hopes of achieving singularity-removal. Indeed, we identified two main mechanisms for the removal of singularities, namely, either pushing the bouncing region to the future (or past) boundary of the space-time beyond the reach of any geodesics, or by showing that when the bounce is reachable by some sets of geodesics, these can be naturally extended across of it to reach arbitrarily large values of their affine parameter. In the latter case, one can wonder what is then the effect of curvature divergences in those cases where they appear at the bounce, since the extensibility of geodesics is utterly numb to the presence of them. In addition, the de Sitter cores familiar of the GR case, which are free of curvature divergences, come out naturally as a nice surprise within this formalism. Subsequently, we brought to the table four additional criteria as seekers of potential pathologies: the presence of unbound tidal forces upon extended observers, the loss of causal contact upon any two components of a congruence of geodesics, the extension of the geodesic completeness criterion to observers with (bound) acceleration, and tests with (scalar) waves. We extensively discussed these criteria for several of the solutions above, identifying some in which all of them are satisfied, thus representing non-singular geometries from this point of view. Nonetheless, while these are necessary conditions for a space-time to be physically reasonable on its full extent, this does not guarantee that other types of pathological behaviours may arise beyond the limited imagination of the authors. In absence of robust theoretical results telling us when a space-time is *regular* rather than when it is *singular*, this is how far we have been able to go.

To conclude, different opinions can be found in the community regarding the interpretation and the severity of the difficulties (or lack thereof) with the existence of singularities within GR, but undoubtedly they are a great reservoir of ideas to challenge (even if timorously) Einstein's interpretation of the geometrical ingredients and physical principles underlying the gravitational interaction. In the end, only experiment will tell if regular black holes contain any clue to it.

Acknowledgements DRG is funded by the *Atracción de Talento Investigador* programme of the Comunidad de Madrid (Spain) No. 2018-T1/TIC-10431. This work is supported by the Spanish Grants PID2019-108485GB-I00, and PID2020-116567GB-C21, funded by MCIN/AEI/10.13039/501100011033 ("ERDF A way of making Europe" and "PGC Generación de Conocimiento"), the project PROMETEO/2020/079 (Generalitat Valenciana), the project H2020-MSCA-RISE-2017 Grant FunFiCO- 777740, the FCT projects No. PTDC/FIS-PAR/31938/2017 and PTDC/FIS-OUT/29048/2017, and the Edital 006/2018 PRONEX (FAPESQ-PB/CNPQ, Brazil, Grant 0015/2019). This article is based upon work from COST Action CA18108, supported by COST (European Cooperation in Science and Technology).

References

1. B.P. Abbott et al., LIGO Scientific and Virgo Collaborations. Astrophys. J. Lett. **848**, L13 (2017)
2. V.I. Afonso, C. Bejarano, J. Beltran Jimenez, G.J. Olmo, E. Orazi, Class. Quant. Grav. **34**, 235003 (2017)
3. V.I. Afonso, G.J. Olmo, D. Rubiera-Garcia, JCAP **08**, 031 (2017)
4. V.I. Afonso, G.J. Olmo, E. Orazi, D. Rubiera-Garcia, Eur. Phys. J. C **78**, 866 (2018)
5. V.I. Afonso, G.J. Olmo, D. Rubiera-Garcia, Phys. Rev. D **97**, 021503 (2018)
6. V.I. Afonso, G.J. Olmo, E. Orazi, D. Rubiera-Garcia, Phys. Rev. D **99**, 044040 (2019)
7. V.I. Afonso, G.J. Olmo, E. Orazi, D. Rubiera-Garcia, JCAP **12**, 044 (2019)
8. R. Aldrovandi, J.G. Pereira, Teleparallel Gravity: An Introduction (Springer, 2013)
9. S. Ansoldi, arXiv:0802.0330 [gr-qc]
10. S. Bahamonde, J. Gigante Valcarcel, Eur. Phys. J. C **81**, 495 (2021)
11. C. Bambi, A. Cardenas-Avendano, G.J. Olmo, D. Rubiera-Garcia, Phys. Rev. D **93**, 064016 (2016)
12. C. Bambi, D. Rubiera-Garcia, Y. Wang, Phys. Rev. D **94**, 064002 (2016)
13. M. Banados, P.G. Ferreira, Phys. Rev. Lett. **105**, 011101 (2010). [erratum: Phys. Rev. Lett. **113**, 119901 (2014)]
14. M. Banados, C. Teitelboim, J. Zanelli, Phys. Rev. Lett. **69**, 1851 (1992)
15. D. Bazeia, L. Losano, G.J. Olmo, D. Rubiera-Garcia, Phys. Rev. D **90**, 044011 (2014)
16. D. Bazeia, L. Losano, G.J. Olmo, D. Rubiera-Garcia, A. Sanchez-Puente, Phys. Rev. D **92**, 044018 (2015)
17. C. Bejarano, A. Delhom, A. Jiménez-Cano, G.J. Olmo, D. Rubiera-Garcia, Phys. Lett. B **802**, 135275 (2020)
18. C. Bejarano, G.J. Olmo, D. Rubiera-Garcia, Phys. Rev. D **95**, 064043 (2017)
19. J. Beltrán Jiménez, D. De Andrés, A. Delhom, Class. Quant. Grav. **37**, 225013 (2020)
20. J. Beltrán Jiménez, A. Delhom, Eur. Phys. J. C **79**, 656 (2019)
21. J. Beltrán Jiménez, L. Heisenberg, T.S. Koivisto, Universe **5**, 173 (2019)
22. J. Beltran Jimenez, L. Heisenberg, G.J. Olmo, JCAP **11**, 004 (2014)
23. J. Beltran Jimenez, L. Heisenberg, G.J. Olmo, D. Rubiera-Garcia, JCAP **10**, 029 (2017)
24. J. Beltran Jimenez, L. Heisenberg, G.J. Olmo, D. Rubiera-Garcia, Phys. Rept. **727**, 1 (2018)
25. N.D. Birrell, P.C.W. Davies, *Quantum Fields in Curved Space* (Cambridge University Press, Cambridge, England, 1982)
26. M. Born, L. Infeld, Proc. R. Soc. London. A **144**, 425 (1934)
27. C. Burrage, J. Sakstein, Living Rev. Rel. **21**, 1 (2018)
28. R. Carballo-Rubio, F. Di Filippo, S. Liberati, M. Visser, Class. Quant. Grav. **37**, 14 (2020)
29. R. Carballo-Rubio, F. Di Filippo, S. Liberati, M. Visser, Phys. Rev. D **101**, 084047 (2020)
30. C.J.S. Clarke, A. Królak, J. Geom. Phys. **2**, 127 (1985)
31. E. Curiel, Stanford Encyclopedia of Philosophy
32. A. De Felice, S. Tsujikawa, Living Rev. Rel. **13**, 3 (2010)
33. J. Diaz-Alonso, D. Rubiera-Garcia, Phys. Rev. D **81**, 064021 (2010)
34. I. Dymnikova, Int. J. Mod. Phys. D **12**, 1015 (2003)
35. I. Dymnikova, Class. Quant. Grav. **21**, 4417 (2004)
36. G.F.R. Ellis, B.G. Schmidt, Gen. Rel. Grav. **8**, 951 (1977)

37. E.S. Fradkin, A.A. Tseytlin, Phys. Lett. B **163**, 123 (1985)
38. Y. Friedman, T. Scarr, Gen. Rel. Grav. **47**, 121 (2015)
39. R.P. Geroch, Ann. Phys. **48**, 526 (1968)
40. M. Guerrero, D. Rubiera-Garcia, Phys. Rev. D **102**, 024005 (2020)
41. M. Guerrero, G. Mora-Pérez, G.J. Olmo, E. Orazi, D. Rubiera-Garcia, JCAP **11**, 025 (2021)
42. J.B. Hartle, S.W. Hawking, Commun. Math. Phys. **26**, 87 (1972)
43. S.I. Kruglov, Mod. Phys. Lett. A **32**, 1750092 (2017)
44. S. Jana, G.K. Chakravarty, S. Mohanty, Phys. Rev. D **97**, 084011 (2018)
45. J.P.S. Lemos, V.T. Zanchin, Phys. Rev. D **77**, 064003 (2008)
46. J.P.S. Lemos, V.T. Zanchin, Phys. Rev. D **83**, 124005 (2011)
47. F.S.N. Lobo, G.J. Olmo, D. Rubiera-Garcia, JCAP **07**, 011 (2013)
48. S.D. Majumdar, Phys. Rev. D. **72**, 930 (1947)
49. J.W. Maluf, Annalen Phys. **525**, 339 (2013)
50. P. Martin-Moruno, M. Visser, Fundam. Theor. Phys. **189**, 193 (2017)
51. C. Menchon, G.J. Olmo, D. Rubiera-Garcia, Phys. Rev. D **96**, 104028 (2017)
52. J.M. Nester, H.J. Yo, Chin. J. Phys. **37**, 113 (1999)
53. B.C. Nolan, Phys. Rev. D **60**, 024014 (1999)
54. Y.N. Obukhov, J.G. Pereira, Phys. Rev. D **67**, 044016 (2003)
55. G.J. Olmo, Phys. Rev. Lett. **95**, 261102 (2005)
56. G.J. Olmo, Phys. Rev. D **72**, 083505 (2005)
57. G.J. Olmo, Int. J. Mod. Phys. D **20**, 413 (2011)
58. G.J. Olmo, E. Orazi, D. Rubiera-Garcia, Eur. Phys. J. C **80**, 1018 (2020)
59. G.J. Olmo, D. Rubiera-Garcia, Phys. Rev. D **84**, 124059 (2011)
60. G.J. Olmo, D. Rubiera-Garcia, Eur. Phys. J. C **72**, 2098 (2012)
61. G.J. Olmo, D. Rubiera-Garcia, Phys. Rev. D **86**, 044014 (2012)
62. G.J. Olmo, D. Rubiera-Garcia, Universe **1**, 173 (2015)
63. G.J. Olmo, D. Rubiera-Garcia, H. Sanchis-Alepuz, Eur. Phys. J. C **74**, 2804 (2014)
64. G.J. Olmo, D. Rubiera-Garcia, A. Sanchez-Puente, Eur. Phys. J. C **76**, 143 (2016)
65. G.J. Olmo, D. Rubiera-Garcia, A. Sanchez-Puente, Class. Quant. Grav. **33**, 115007 (2016)
66. G.J. Olmo, D. Rubiera-Garcia, A. Sanchez-Puente, Class. Quant. Grav. **35**, 055010 (2018)
67. E. Orazi, Int. J. Mod. Phys. D **29**, 2041010 (2020)
68. A. Ori, Phys. Rev. D **61**, 064016 (2000)
69. A. Papapetrou, Proc. R. Irish Acad. **A51**, 191 (1947)
70. L. Parker, D.J. Toms, *Quantum Field Theory in Curved Spacetime: Quantized Fields and Gravity* (Cambridge University Press, Cambridge, England, 2009)
71. R. Penrose, Phys. Rev. Lett. **14**, 57 (1965)
72. T.A. Roman, Phys. Rev. D **33**, 3526 (1986)
73. J.M.M. Senovilla, D. Garfinkle, Class. Quant. Grav. **32**, 124008 (2015)
74. F.J. Tipler, Phys. Rev. D **15**, 942 (1977)
75. M. Visser, *Lorentzian Wormholes* (Springer, NY, 1996)
76. D.N. Vollick, Phys. Rev. D **69**, 064030 (2004)
77. C.M. Will, Living Rev. Rel. **17**, 4 (2014)
78. M. Wyman, Phys. Rev. D **24**, 839 (1981)

Chapter 7
Regular Black Holes from Loop Quantum Gravity

Abhay Ashtekar, Javier Olmedo, and Parampreet Singh

7.1 Introduction

There is general agreement in the gravity community that black hole singularities of classical general relativity (GR) offer excellent opportunities to probe physics beyond Einstein. However, as of now, there is no consensus on the fate of black hole singularities in full quantum gravity. Indeed, there is an ongoing debate even on a central question in the subject: Will singularities of classical GR be naturally resolved in full quantum gravity, or will they persist? As the very name of this Volume suggests, in many circles an affirmative answer is taken to be a necessary condition for the viability of a proposed quantum gravity theory. But this is not an universally accepted viewpoint. For example, it has been argued that taming of black hole singularities in asymptotically anti-deSitter space-times would violate a "No Transmission Principle" motivated by the AdS/CFT correspondence [1]. More generally, discussions of the black hole evaporation process are often based on the assumption that there is a singularity also in quantum gravity. These expectations are based on the Penrose diagram of an evaporating black hole that Hawking drew over 40 years ago [2], where the singularity persists as part of the future boundary of space-time even after the black hole has completely disappeared (see Fig. 7.4). However, this feature of the diagram was not arrived at from a calculation, and

A. Ashtekar (✉)
Physics Department, and, Institute for Gravitation & the Cosmos, Penn State, University Park, PA 16802, USA
e-mail: ava1@psu.edu

J. Olmedo
Departamento de Física Teórica y del Cosmos, Universidad de Granada, Granada 18071, Spain
e-mail: javolmedo@ugr.es

P. Singh
Department of Physics and Astronomy, Louisiana State University, Baton Rouge, LA 70803, USA
e-mail: psingh@lsu.edu

C. Bambi (ed.), *Regular Black Holes*, Springer Series in Astrophysics and Cosmology,
https://doi.org/10.1007/978-981-99-1596-5_7

indeed such a calculation is not available even today. Furthermore, some forty years later Hawking himself changed his mind: A new Penrose diagram was proposed to represent an evaporating black hole in which there is no singularity (see Fig. 2 of [3]). Nonetheless, interestingly, Hawking's first paradigm continues to feature prominently in discussions on the issue of information loss: see, e.g., Ref. [4] where the persistence of this singularity leads to a non-unitary evolution from $\mathscr{I}^-$ to $\mathscr{I}^+$, and Refs. [5–7] where proposals are made on how unitarity could be rescued in spite of this singularity, thereby preventing information loss.

Loop quantum gravity (LQG) provides a systematic avenue to investigate the fate of singularities of classical GR because it is based on *quantum* Riemannian geometry. Consequently, new physics arises in the Planck regime where the continuum space-time of classical GR becomes inadequate (see, e.g., [8]). Implications of this new physics have been analyzed *in detail* in the commonly used cosmological models. Non-perturbative quantum corrections to Einstein's equations imply that, once a curvature invariant approaches the Planck scale, quantum geometry modifications of Einstein dynamics introduce strong 'repulsive corrections' that dilute that invariant, preventing a blow-up (see, e.g., [9–11]). Thus, the big-bang/big-crunch singularity is replaced by a quantum bounce in loop quantum cosmology (LQC). Once the curvature drops to about $\sim 10^{-4}$ Planck scale, quantum corrections can be neglected and classical GR becomes a good approximation.

A natural question then is whether the same phenomenon occurs at the black hole singularities. Results to date provide considerable evidence that it does. However, technically, the situation is more complicated than that in cosmological models for two reasons. First, even in the Schwarzschild solution, although space-time is homogeneous in the vicinity of the singularity, it is *not* isotropic. Second, the nature of the blow up of curvature is different from that in the commonly used cosmological models: As Penrose has emphasized, while the Weyl curvature vanishes identically at the big-bang in homogeneous isotropic cosmologies, it diverges at the Schwarzschild singularity. As a result, although the singularity is resolved in all LQG investigations, as of now, results in the black hole sector are not as strong as they are in LQC. Nonetheless, a large number of investigations, carried out since 2004, have provided conceptual insights as well as detailed technical results on the nature of the resolution of the Schwarzschild singularity. Our goal is to convey an overall picture at a technical level that is accessible to beginning researchers, emphasizing conceptual issues, novel elements, and problems that remain. We also provide references where details can be found. Also for convenience of non-experts, throughout the Chapter, we pause to summarize the main points after each technical discussion and also at the end of subsections.

In Sects. 7.2 and 7.3 we focus on the quantum extension of the Kruskal space-time. Because the static Killing field is space-like in the 'interior' region—bounded by the singularity in the future and the horizon in the past—the space-time metric is spatially homogeneous (but not isotropic). As is well-known, this portion of Kruskal space-time is isometric with the vacuum Kantowski-Sachs cosmological model. Therefore techniques from LQC have been used to analyze the fate of the Schwarzschild singularity in a number of investigations within LQG (See, e.g., [12–38]). While some of these analyses present us with the equations that dictate the evolution of the *quantum state* of the system, the detailed results are based on the so-called 'effective equations' whose goal is to incorporate the leading order quantum corrections to the classical geometry in sharply peaked quantum states.[1] At a conceptual level, all these investigations follow the same strategy. However, the technical implementation of this procedure differs, leading to different effective geometries in the interior region. Nonetheless, in all these cases, the singularity is resolved due to quantum corrections. We will discuss the strategy and compare and contrast various results in Sect. 7.2. Singularity resolution in the Kruskal space-time provides several sharp results on the causal structure of its quantum extension. In particular, the singularity is replaced by a 'transition surface' to the immediate past of which we have a trapped region and to the immediate future, an anti-trapped region. This geometry is sometimes referred to as depicting 'a black hole to white hole transition'. We will avoid this terminology because it has other connotations that are not realized. In particular, the terms 'black hole' and 'white hole' normally go hand in hand with singularities and event horizons. In LQG, singularities are absent and, in dynamical situations, there are also no event horizons either.

In Sect. 7.3 we consider the Schwarzschild exterior, i.e. the region bounded by the horizon and $\mathscr{I}^{\pm}$. Space-time is again foliated by homogeneous 3-dimensional surfaces but they are now time-like rather than space-like. We discuss a possible extension of the 'interior' geometry to this exterior region, following [28, 29, 39]. This extension has several attractive properties [30], but it also has some puzzling features: while the quantum corrected metric is again asymptotically flat in a precise sense (that suffices to define the ADM mass, for example), the approach to the flat metric is weaker than the one generally used in the physics literature. There *are* alternate proposals to arrive at effective metrics with the standard asymptotic behavior (see, e.g., [33, 38]) but a definitive picture is yet to emerge.

Now, the Kruskal space-time itself is an idealization since it represents an 'eternal black hole'; black holes encountered in nature are formed *dynamically*, e.g., via a gravitational collapse, or compact binary mergers. Nonetheless, one would expect the qualitative features of the causal structure that arises from taming of the singularity due to quantum effects would be robust. In Sect. 7.4 we discuss models of dynamical situations that have been analyzed within LQG and summarize the current status, focusing on the Lemaître-Tolman-Bondi type models of collapse and critical phenomena discovered by Choptuik. In Sect. 7.5 we turn to the issue of black hole evaporation and 'information loss'. The LQG discussion of these issues is characterized by two key features [40]. First, as discussed above, in contrast to the Penrose diagram in Hawking's seminal paper [2], there is no singularity in the space-time interior which can serve as a 'sink of information'. Second, as the LQG Penrose

[1] For the conceptual framework underlying effective equations see, e.g., Section V of [9]. Note that the term 'effective equations' has a very different connotation here than in standard quantum field theory. This has caused occasional confusion in the literature. In LQG one *does not integrate out* 'high energy modes'; Planck scale effects are retained. In LQC, for example, there are states that remain sharply peaked even in the Planck regime and the effective equations capture the evolution of the peak of the quantum wave function in these states, ignoring the fluctuations.

diagram of Fig. 7.6 shows, there is no event horizon: what forms and evaporates is a dynamical horizon [41–43]. Much of the discussion in the literature assumes that there is an event horizon which serves as a boundary of an 'interior' region from which no causal signal can ever be sent to the asymptotic region. One is then led one to either conclude that information is lost, or, to introduce 'exotic' ideas such as quantum Xerox machines, firewalls and fast scramblers to restore unitarity. As we discuss, there is a more direct pathway to unitarity once it is realized that there is no event horizon. However, as in every other approach, important issues remain: the precise nature quantum radiation at the final stages of the evaporation process require full LQG and this analysis has only begun. We summarize the current status in Sect. 7.5. In Sect. 7.6 we collect the key features of regular black holes in LQG compare and contrast the regular LQG black holes with this in other approaches.

Our conventions are the following. Space-time metric g_{ab} has signature $-, +, +, +$ and the curvature tensors are defined by $R_{abc}{}^{d}k_d = 2\nabla_{[a}\nabla_{b]}k_c$; $R_{ac} = R_{abc}{}^{b}$; and $R = g^{ab}R_{ab}$. By macroscopic black holes we mean those for which $GM =: m \gg \ell_{\rm Pl}$.

7.2 The Schwarzschild Interior

Denote by (M, g_{ab}) the Kruskal extension of the Schwarzschild metric (see Fig. 7.1) and by $(M_{\rm II}, g_{ab})$ the quadrant of this space-time that represents the (open) 'interior region' II, bounded by the black hole singularity and future horizons. This region is foliated by the $r_{\rm sch} = \text{const}$ space-like manifolds, with topology $\mathbb{S}^2 \times \mathbb{R}^2$. Each leaf admits 3 rotational Killing fields tangential to its 2-dimensional spherical cross sections that are mapped to one another by the translational Killing field.

Consequently, $(M_{\rm II}, g_{ab})$ is spatially homogeneous, but not isotropic; it is isometric to the (vacuum) Kantowski-Sachs cosmological model. Therefore LQG approaches use the procedure from homogeneous cosmologies. Now, while the big-bang and big-crunch singularities persist in the Wheeler-DeWitt (WDW) theory based on metric variables, they are naturally resolved in LQC because of the

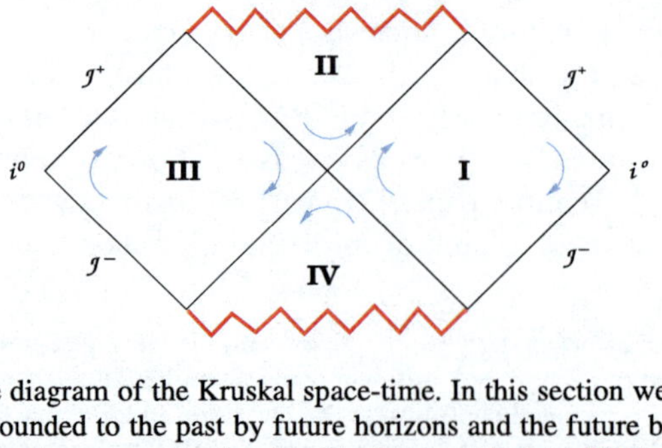

Fig. 7.1 The Penrose diagram of the Kruskal space-time. In this section we discuss the quantum extension of part II, bounded to the past by future horizons and the future by the singularity. The quantum corrected effective geometry of region I is discussed in Sect. 7.2.2

quantum geometry resulting from the use of connection variables (see, e.g., [9]). For the Schwarzschild interior, then, LQG investigations also begin with a $3+1$ decomposition of Einstein's equations using connection variables. In the classical theory, components of the curvature tensor that features in these equations can be obtained by first evaluating holonomies of the gravitational connections around suitable closed loops (called plaquettes) and then taking the limit as the area enclosed by these plackets tends to zero. In LQG, the corresponding quantum operator is obtained by shrinking these plaquettes till the area they enclose reaches the smallest non-zero eigenvalue of the area operator. This eigenvalue is called *the area gap* and denoted by Δ. As a consequence, information about quantum geometry gets encoded in the dynamical equations. Observables such as curvature scalars can acquire *finite* upper bounds on entire dynamical trajectories, whence the singularity is resolved. Δ appears in the denominator of the expressions of these upper bounds; classical singularities emerge as $\Delta \to 0$.

For black holes, while operator equations have been written down [12–15, 34, 37], detailed investigations of the singularity resolution and ensuing quantum corrected geometry have been obtained using 'effective equations' discussed in Sect. 7.1. Solutions to effective equations show that the central singularity is resolved due to quantum corrections. However, different investigations within LQG have made different choices to arrive at the quantum corrected curvature operators. Intuitively these choices represent quantization ambiguities that then affect detailed predictions. For brevity, in Sects. 7.2.1 and 7.2.2 we will present the general framework and results following a recent approach that is free of limitations of the earlier investigations and in Sect. 7.2.3 we will briefly compare and contrast other approaches. Due to space limitation, by and large we will only include motivations behind various constructions and summarize the final results. For detailed derivations and other details, see in particular [12, 13, 19, 20, 29].

7.2.1 The Framework

In connection-dynamics, the initial data for space-time geometry consists of an SU(2)-valued connection A_a^i and its conjugate 'electric field' E_i^a as in Yang-Mills theory. In the final solutions to Einstein's equations, A_a^i has the interpretation of the gravitational connection that parallel transports SU(2) spinors, and E_i^a, represent the ortho-normal spatial triads (with density weight 1). Because of spatial homogeneity of the model, various spatial integrals in the Hamiltonian framework have a trivial divergence. Therefore, one introduces an 'infrared cut-off'. Thus one truncates the homogeneous slices to be finite (rather than infinite) cylinders, with coordinates (θ, ϕ, x) with $x \in (0,\ L_\circ)$ (rather than $x \in (0,\ \infty)$). One has to make sure, of course, that none of the final results depend on $L_\circ$. One can solve the 'kinematical' constraint equations and use gauge-fixing to cast the basic variables in the form

$$A_a^i\, \tau_i\, \mathrm{d}x^a = c/L_\circ\, \tau_3\, \mathrm{d}x + b\, (\tau_2 \mathrm{d}\theta - \tau_1 \sin\theta\, \mathrm{d}\phi) + \tau_3 \cos\theta\, \mathrm{d}\phi,$$
$$E_i^a\, \tau^i \partial_a = p_c\, \tau_3\, \sin\theta\, \partial_x + (p_b/L_\circ)\, \tau_2\, \sin\theta\, \partial_\theta - (p_b/L_\circ)\, \tau_1\, \partial_\phi, \qquad (7.1)$$

where τ_i are SU(2) generators related to Pauli spin matrices σ_i via $\tau_i = -i\sigma_i/2$. Real valued connection components b, c and the triad components p_b, p_c are functions only of time and serve as conjugate coordinates on the 4-dimensional phase space. It is convenient to choose an orientation of the triads so that $b,\ c,\ p_c$ are positive and p_b is negative. It follows from (7.1) that physical quantities can only depend on $b,\ (p_b/L_\circ),\ (c/L_\circ),\ p_c$. Given a time coordinate τ that labels the spatially homogeneous surfaces and the corresponding lapse N_τ, in region II the space-time metric has the form

$$g_{ab}\mathrm{d}x^a\mathrm{d}x^b \equiv \mathrm{d}s^2 = -N_\tau^2 \mathrm{d}\tau^2 + \frac{p_b^2}{p_c L_\circ^2}\mathrm{d}x^2 + p_c(\mathrm{d}\theta^2 + \sin^2\theta \mathrm{d}\phi^2). \qquad (7.2)$$

At the horizon, $b,\ p_b$ vanish and the translation Killing field $X = \partial/\partial x$ becomes null. When p_c vanishes, the radius of the metric 2-spheres shrinks to zero, making the curvature scalars diverge there. This is Schwarzschild singularity.

It turns out that Einstein's equations that govern the dynamics of the basic variables simplify significantly if one uses the lapse $N_{\rm cl} = (\gamma\, \sqrt{p_c})/b$ (which is different from the standard lapse in the Schwarzschild coordinates.) The γ in this expression is the dimensionless Barbero-Immirzi parameter of LQG. It is analogous to the θ-parameter of QCD in that it represents a quantization ambiguity: classical physics is insensitive to the precise value of γ; we only need $\gamma > 0$. In terms of the corresponding time-coordinate $T_{\rm cl}$, the dynamical trajectories are given by:

$$b(T_{\rm cl}) = \gamma\, \left(e^{-T_{\rm cl}} - 1\right)^{1/2} \quad \text{and} \quad p_b(T_{\rm cl}) = p_b^{(\circ)}\, e^{T_{\rm cl}} \left(e^{-T_{\rm cl}} - 1\right)^{1/2}, \qquad (7.3)$$

and

$$c(T_{\rm cl}) = c_{(\circ)}\, e^{-2T_{\rm cl}} \quad \text{and} \quad p_c(T_{\rm cl}) = p_c^{(\circ)}\, e^{2T_{\rm cl}}\, . \qquad (7.4)$$

Here $c_{(\circ)},\ p_b^{(\circ)},\ p_c^{(\circ)}$ are integration constants. Comparison with the standard form of the Schwarzschild solution yields $p_c^{(\circ)} = 4m^2$, $p_b^{(\circ)}/L_\circ = -2m$, and $c_{(\circ)}/L_\circ = \gamma/4m$, where m is related to the mass of the Schwarzschild solution via $m = GM$. At the horizon $T_{\rm cl} = 0$ and at the singularity $T_{\rm cl} = -\infty$.

The dynamical variables are subject to the Hamiltonian constraint

$$H_{\rm cl}[N_{\rm cl}] \equiv -\frac{1}{2G\gamma}\Big((b + \frac{\gamma^2}{b})p_b\, +\, 2c\, p_c\Big) = 0. \qquad (7.5)$$

It is easy to verify that the terms in the b and c sectors on the right side of (7.5) are separately conserved in time, and equal $-m$ and m respectively on solutions. Therefore, if the constraint (7.5) is satisfied at one instant $T_{\rm cl}$, then it holds for all $T \in (-\infty, 0)$.

As explained above, in the passage to quantum theory the spatial curvature is expressed using the holonomoly of the gravitational connection A_a^i around appropriately chosen plaquettes that enclose the minimum non-zero area, $\Delta = 4\sqrt{3}\,\pi\gamma\ell_{\rm Pl}^2$. (Thus, while classical physics is insensitive of the value of the Barbero-Immirzi parameter γ, quantum physics is not. Its value is generally taken to be $\gamma = 0.2375$ via black hole entropy calculation.) As a consequence, the effective equations that capture the leading quantum corrections inherit new *'quantum parameters'*, denoted by δ_b and δ_c, that refer to edge lengths of these plackets, and go to zero in the classical limit, $\ell_{\rm Pl} \to 0$ (or, $\Delta \to 0$, keeping γ fixed). Different choices of these quantum parameters represent quantization ambiguities mentioned above. In this section we will use a strategy [28–30] that is free of the physically undesirable features encountered in other approaches (discussed in Sect. 7.2.3).

A key idea behind this strategy is to use δ_b and δ_c that are 'Dirac observables' i.e. phase space functions that are *constant along dynamical trajectories*.[2] Let us restrict ourselves to such δ_b, δ_c from now on. Then, again, the evolution equations simplify if we include the appropriate quantum corrections in the choice of the lapse, defining it as $N := (\gamma\,\sqrt{p_c})\,\delta_b/\sin(\delta_b b)$. (Note that as the area gap Δ goes to zero, so does δ_b and N reduces to $N_{\rm cl}$.) Denote by T the corresponding time parameter and by 'dot' the derivative with respect to T. Then, as in the classical theory, the effective evolution equations b and the c sectors separate:

$$\dot{b} = -\frac{1}{2}\left(\frac{\sin(\delta_b b)}{\delta_b} + \frac{\gamma^2\delta_b}{\sin(\delta_b b)}\right), \qquad \dot{p}_b = \frac{p_b}{2}\cos(\delta_b b)\left(1 - \frac{\gamma^2\delta_b^2}{\sin^2(\delta_b b)}\right), \tag{7.6}$$

and

$$\dot{c} = -2\,\frac{\sin(\delta_c c)}{\delta_c}, \qquad \dot{p}_c = 2\,p_c\,\cos(\delta_c c). \tag{7.7}$$

But, again as in the classical theory, the two sectors are linked by the (now, effective) Hamiltonian constraint:

$$H_{\rm eff}[N] \equiv -\frac{1}{2G\gamma}\left[\left(\frac{\sin(\delta_b b)}{\delta_b} + \frac{\gamma^2\delta_b}{\sin(\delta_b b)}\right)p_b + 2\frac{\sin(\delta_c c)}{\delta_c}\,p_c\right] = 0. \tag{7.8}$$

A direct calculation shows that the constraint (7.8) is preserved in time.

To summarize, conditions $\dot{\delta}_b = 0$, $\dot{\delta}_c = 0$, the evolution equations (7.6), (7.7) and the constraint equation (7.8) constitute a set of consistent equations that generalize the classical constraint and evolution equations. A notable difference from the classical theory arises because in LQG there is a well-defined operator in the quantum theory

[2] Because the spatial curvature features on the right side of Einstein's evolution equations, the quantum corrected version of the classical dynamical trajectories (7.3) and (7.4) along which δ_b and δ_c are to remain constant themselves feature δ_b and δ_c (see (7.9), (7.10) and (7.11)). Therefore the issue of finding δ_b and δ_c that are Dirac observables is rather subtle conceptually and quite intricate technically. These subtleties has led to some concerns [31]. This issue is analyzed in detail [35–37, 44]. Consistency of the final results directly follows from the effective equations (7.6)–(7.8).

corresponding only to the holonomy defined by the gravitational connection A_a^i, rather than A_a^i itself. As a consequence, only trigonometric functions of $\delta_b\, b$ and $\delta_c\, c$ appear. Hence the domain of these variables is compactified (just as in LQC [45]): they take values in the open interval $(0,\ \pi)$. The momenta $p_b,\ p_c$, by contrast, continue to assume values $p_b < 0$ and $p_c > 0$ as in the classical theory.

To solve the evolution equations, it is convenient to first obtain solutions $c(T)$, $p_c(T)$ and $b(T)$. Now, in the c sector, equations of motion (7.7) immediately imply that $m_c = (\sin(\delta_c c)\, p_c)/(\gamma L_\circ \delta_c)$ is a constant of motion. This fact simplifies the form of the solutions. One obtains

$$\tan\left(\frac{\delta_c\, c(T)}{2}\right) = \frac{\gamma L_o \delta_c}{8m_c} e^{-2T}, \qquad p_c(T) = 4m_c^2\Big(e^{2T} + \frac{\gamma^2 L_o^2 \delta_c^2}{64 m_c^2} e^{-2T}\Big), \tag{7.9}$$

$$\cos\big(\delta_b\, b(T)\big) = b_o \tanh\left(\frac{1}{2}\Big(b_o T + 2\tanh^{-1}\big(\frac{1}{b_o}\big)\Big)\right), \tag{7.10}$$

where there constant b_o is given by $b_o = (1+\gamma^2 {\delta_b}^2)^{1/2}$. One then uses the Hamiltonian constraint to determine $p_b(T)$:

$$p_b(T) = -2\frac{\sin(\delta_c\, c(T))}{\delta_c}\frac{\sin(\delta_b\, b(T))}{\delta_b}\frac{p_c(T)}{\frac{\sin^2(\delta_b\, b(T))}{\delta_b^2} + \gamma^2}. \tag{7.11}$$

Equations (7.9)–(7.10) provide the dynamical trajectories of the effective theory. It is easy to verify that in the limit $\delta_b \to 0,\ \delta_c \to 0$, one recovers the classical trajectories. To summarize, the quantum corrected, effective trajectories are given by (7.9)–(7.11) for any choice of constants of motion $\delta_b,\ \delta_c$. Since these equations only involve the combinations $b,\ (p_b/L_\circ),\ \delta_b;\ (c/L_\circ),\ p_c$, and $L_\circ \delta_c$, the metric (7.2) and all physical results are insensitive to the choice of the infrared cut-off $L_\circ$.

So far $\delta_b,\ \delta_c$ could be any quantum parameters satisfying $\dot{\delta_b} = 0$ and $\dot{\delta_c} = 0$. The following considerations provide a natural avenue to determine them. Recall that on classical solutions, the c part of $H_{\rm cl}[N_{\rm cl}]$ equals m, and the b part equals $-m$. Therefore in the effective theory, one is led to set

$$\frac{1}{2\gamma}\left(\frac{\sin(\delta_b b)}{\delta_b} + \frac{\gamma^2 \delta_b}{\sin(\delta_b b)}\right)\frac{p_b}{L_\circ} = -m_b \quad \text{and} \quad \frac{\sin(\delta_c c)}{\gamma L_\circ \delta_c} = m_c\,. \tag{7.12}$$

Equations of motion (7.6) and (7.7) imply that both m_b and m_c are constants of motion and the effective Hamiltonian constraint reads $m_b = m_c$. On solutions, we will drop the suffix and set $m_b = m_c = m$. The fact that m_b and m_c are constants of motion suggests a natural strategy to restrict the form of $\delta_b,\ \delta_c$: Require that δ_b be a function only of m_b, and δ_c be a function only of m_c. To constrain the functional form requires additional input, summarized in Sect. 7.2.2. Here we only note that the final answer has a rather simple form for large black holes (i.e. for solutions for which $m \gg \ell_{\rm Pl}$): δ_b and δ_c are extremely well-approximated by

$$\delta_b = \Big(\frac{\sqrt{\Delta}}{\sqrt{2\pi}\gamma^2 m_b}\Big)^{1/3}, \quad \text{and} \quad L_o\delta_c = \frac{1}{2}\Big(\frac{\gamma\Delta^2}{4\pi^2 m_c}\Big)^{1/3}. \tag{7.13}$$

(Recall that physical results can only depend on the combination $L_o\delta_c$.)

To summarize, the effective metric in the interior region is given by (7.14), where c, p_c are given by (7.9), b, p_b by (7.10), (7.11), and δ_b, δ_c by (7.13). By inspection we see that as the area gap Δ goes to zero, δ_b and δ_c both go to zero and the effective theory reduces to the classical GR.

7.2.2 *Singularity Resolution, Causal Structure and Curvature Bounds*

Let us explore properties of the space-time metric

$$g_{ab}\mathrm{d}x^a\mathrm{d}x^b \equiv \mathrm{d}s^2 = -N^2\mathrm{d}T^2 + \frac{p_b^2}{p_c L_\circ^2}\mathrm{d}x^2 + p_c(\mathrm{d}\theta^2 + \sin^2\theta\mathrm{d}\phi^2), \tag{7.14}$$

of the effective theory. The past boundary of the open region under consideration is again given by $b = 0,\ p_b = 0$ which occurs at $T = 0$ on every dynamical trajectory. The translational Killing vector X^a becomes null at these points; thus as in the classical theory, this boundary represents the horizon. In the classical theory, the singularity is characterized by the vanishing of the radius of the metric 2-spheres, i.e., of p_c. In the effective theory, however, p_c has a *non-zero* minimum, $p_c^{\min} = \frac{1}{2}\gamma(L_\circ\delta_c)m$ which occurs at $T = \frac{1}{2}\log(\gamma L_\circ\delta_c)/8m$. Note that this minimum radius is of Planck scale but depends on the mass of the initial black hole: $r_{\min} \sim (m\ell_{\rm Pl}^2)^{1/3}$. This is the surface that replaces the classical singularity and the space-time metric (7.14) can be smoothly extended across this 3-manifold.

One can explore the causal structure around this surface by calculating the expansions $\Theta_\pm$ of the two null normals to the metric 2-spheres. To the past of this surface one finds that both expansions are negative. Thus this is a *trapped region* just as the entire region II is in the classical theory. Interestingly, *both* null-expansions vanish on this surface. This is a novel situation that is not encountered in classical GR. Since the metric is smooth across this surface, space-time is well-defined across it and one can analyze the two expansions to the future of this surface. They are both positive, so the region to the future is *anti-trapped*. Thus in the quantum-extended effective space-time, the surface neatly separates a trapped region and an anti-trapped region. Therefore it is called a *transition surface*, denoted by $\mathscr{T}$. It is analogous to the 'bounce surface' in LQC (that replaces the big-bang), to the past of which the expansion of the universe is negative and to the future of which it is positive. However, now the term 'expansion' refers to changes in the areas of metric 2-spheres along its two null normals. How far into the future is the space-time extended by this procedure? The metric is well defined in the open region bounded by the surface

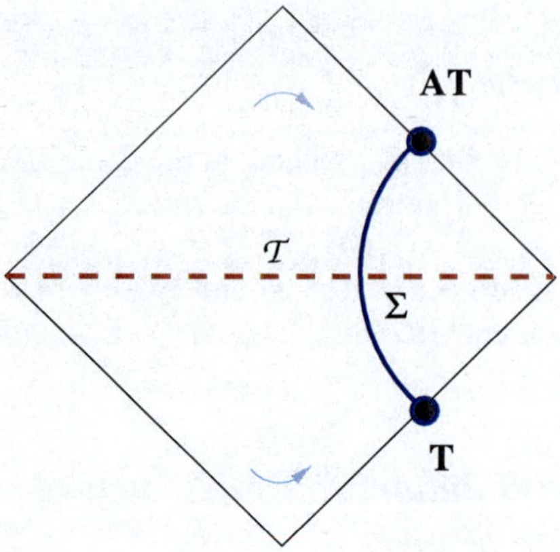

Fig. 7.2 Quantum extension of region II of Fig. 7.1 in the effective theory. The singularity is replaced by the transition surface $\mathscr{T}$. It separates the trapped and anti-trapped regions. The past boundary **T** is a null (black hole type) trapping horizon and the future boundary **AT** is a null (white-hole type) anti-trapping horizon. The time-like 3-manifold Σ joining 2-spheres lying on the two horizons is used in Eq. (7.17)

$T = -(4/b_o)\,\tanh^{-1}(1/b_o)$ where $(\delta_b\, b) = \pi$ and $p_b = 0$. The Killing field X^a is again null on the boundary so it again represents a horizon that bounds the anti-trapped region to the future. In summary, effective dynamics extends the open region II (of Fig. 7.1) to the diamond shaped open region (shown in Fig. 7.2) bounded by Killing horizons. The region is separated by a transition surface $\mathscr{T}$, to the past of which one has a trapped region and to the future of which, an anti-trapped region. This extension is often referred to as the black hole to white hole transition.

In LQC, space-time curvature attains the maximum value on the bounce surface and, furthermore, this upper bound is universal. Does the quantum corrected geometry exhibit the same feature at transition surface $\mathscr{T}$? The answer is in the affirmative. One has:

$$R^2\,|_{\mathscr{T}} \approx \frac{256\pi^2}{\gamma^4\Delta^2} + \cdots \qquad R_{ab}R^{ab}\,|_{\mathscr{T}} = \frac{256\pi^2}{\gamma^4\Delta^2} + \cdots \tag{7.15}$$

$$C_{abcd}C^{abcd}\,|_{\mathscr{T}} = \frac{1024\pi^2}{3\gamma^4\Delta^2} + \cdots \qquad R_{abcd}R^{abcd} = \frac{768\pi^2}{\gamma^4\Delta^2} + \cdots \tag{7.16}$$

where all the correction terms ... have the same form $\mathscr{O}\big((\Delta/m^2))^{1/3}\,\log(m^2/\Delta)\big)$. Recall, first, that the classical limit corresponds to $\Delta \to 0$ (keeping $\gamma > 0$.) Hence in this limit all invariants diverge and $\mathscr{T}$ is replaced by the singularity. Secondly, since leading terms are *mass independent*, the upper bounds are universal. (The numerical coefficients vary simply because the invariants refer to distinct parts of the total curvature.) Third, as one moves away from $\mathscr{T}$, these curvature scalars rapidly approach their classical values even for very small black holes. Thus quantum corrections to space-time geometry are very small away from the transition surface.

For instance, while the horizon radius of the effective solution is always larger than that of its classical counterpart, even for $m = 10^4 \ell_{\rm Pl}$, the relative difference is $\sim 10^{-15}$ and for a solar mass black hole, it is $\sim 10^{-115}$! Finally one can ask for the relation between the radius $r_{\rm T}$ of the trapping horizon that constitutes the past boundary of the diamond, and the radius $r_{\rm AT}$ of the anti-trapping horizon that constitutes the future boundary. Are they approximately the same? The answer is in the affirmative for macroscopic black holes, even though the 'bounce' is not exactly symmetric. For a stellar mass black hole for example, $r_{\rm T} = 3$km and $r_{\rm AT} = (3 + O(10^{-25}))$ km. As we will see in Sect. 7.2.3, these consequences of effective dynamics are non-trivial: it is surprisingly difficult to achieve the singularity resolution without, at the same time, triggering unintended large effects away from the singularity.

Next, note that while the Ricci tensor vanishes identically in classical solutions, it is non-zero in the effective solutions. One can simply set $8\pi G_{\rm N}\, T^{\rm eff}_{ab} := R_{ab} - \frac{1}{2} R g_{ab}$ and interpret $T^{\rm eff}_{ab}$ as the effective stress-energy tensor of the quantum corrected space-time. As one would expect from the above discussion, for macroscopic black holes these quantum corrections are negligible away from $\mathscr{T}$. However, they become large and dominant in the immediate vicinity of $\mathscr{T}$. As one could have anticipated, although it is finite everywhere, the energy density defined by $T^{\rm eff}_{ab}$ becomes large and negative in this region thereby violating the energy conditions, as it must for the singularity resolution to occur.

Interestingly, this fact creates an apparent tension with considerations involving the Komar mass $M_{\rm K}$. Recall that, in the classical theory, $M_{\rm K}$ defined by the translational Killing field X^a is given by (half the) horizon radius. As we saw, for macroscopic black holes the radii $r_{\rm T}$ and $r_{\rm AT}$ are essentially the same. But the difference between the Komar mass evaluated at the anti-trapping horizon and the trapping horizon is given by the integral over a 3-manifold Σ joining a cross-section of the trapping horizon with a cross-section of the anti-trapping horizon (see Fig. 7.2),

$$M_{\rm K}^{\rm AT} - M_{\rm K}^{\rm T} = 2 \int_{\Sigma} \big(T^{\rm eff}_{ab} - \frac{1}{2} T^{\rm eff}\, g^{\rm eff}_{ab}\big)\, X^a {\rm d}\Sigma^b\,, \tag{7.17}$$

and for macroscopic black holes the integrand of the right is *large and negative* near $\mathscr{T}$ (because it represents the effective energy density). How can the two Komar masses be the same, then? It turns out that the integrand of (7.17) is indeed large and negative for macroscopic black holes, but its numerical value is very close to $-2M_K^{\rm T}$. Therefore the Komar mass associated with the anti-trapping horizon is given by $M_{\rm K}^{\rm AT} \approx M_K^T - 2M_K^{\rm T} = -M_{\rm K}^{\rm T}$, and the minus sign is just right because while the translational Killing field is future directed on the trapping horizon **T**, it is *past* directed on the anti-trapping horizon **AT**! (See the (blue) arrows in Fig. 7.3.) This resolution is another example of the conceptually subtle balance achieved with the choice of quantum parameters (7.13).

To summarize, the Schwarzschild singularity is naturally resolved in the effective theory discussed in Sect. 7.2.1 and region II of Fig. 7.1 bounded by the singularity to the future is extended to the singularity free diamond-shaped region shown in Fig. 7.2, bounded in the past by the trapping horizon and to the future by the anti-

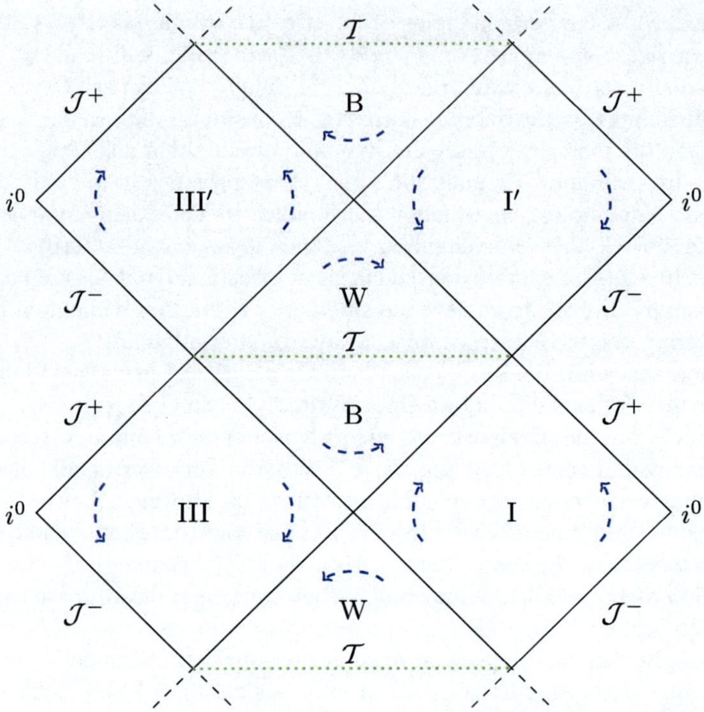

Fig. 7.3 The Penrose diagram of the quantum extended Kruskal space-time in the x, T plane. Arrows show the orientation of the static Killing fields. Since the effective metric is at least C^2 across Killing horizons, space-time continues indefinitely into the future and the past. Successive Killing horizons are trapping and anti-trapping. Each diamond shaped region they bound is divided by a space-like transition surface $\mathscr{T}$ that separates a trapping region (that lies to the past of $\mathscr{T}$) and an anti-trapping region (that lies to its future). Thus, the region B immediately to the past of $\mathscr{T}$ resembles a black hole interior, and the region W immediately to its future resembles a white hole interior. The area-radii of successive horizons are very nearly equal for macroscopic black holes. Only a part of this extension is relevant to black holes formed dynamically through collapse

trapping horizon. The singularity is replaced by a space-like surface $\mathscr{T}$ that marks the transition between trapped and anti-trapped regions. Curvature scalars achieve their maximum values on $\mathscr{T}$ which are universal to the leading order. Although quantum corrections encoded in the area gap Δ dominate near $\mathscr{T}$, they decrease rapidly as one moves away and are completely negligible near horizons for macroscopic black holes. In particular, the radii of the trapping and anti-trapping horizons are indistinguishable for macroscopic black holes.

7.2.3 Summary of LQG Investigations

As we already noted, the LQG investigations of the Schwarzschild singularity follow the same general steps but differ in the selection of the quantum parameters $\delta_b,\ \delta_c$. Since the Schwarzschild interior is isometric to the Kantowski-Sachs cosmological model, discussions have often focused on issues motivated by cosmological considerations such as the behavior of 'scalar factors' and shears, rather than on considerations that are more directly relevant to black holes, in particular properties of the effective geometry that lead to trapping and anti-trapping. We focused on an approach that does [28, 29]. We will now summarize various strategies that have been used to fix $\delta_b,\ \delta_c$ and results they led to. Since our goal is only to present a cohesive picture of the overall status through comparison of results, the discussion will be rather brief; details can be found in the original papers listed in the bibliography.

By and large, these strategies fall into three categories:
(i) The parameters are chosen to be constants. These approaches are often referred to as the μ_o-type schemes because they mimic the strategy of using constant values for the quantum parameter μ used in LQC [46]. Here, the curvature operator is defined using holonomies of the gravitational connection around plaquettes and shrinking them till the *coordinate* area they enclose equals the area gap Δ;
(ii) The parameters are chosen to be phase space functions, using physical considerations. These approaches are often referred to as the $\bar{\mu}$-type schemes, named after the strategy of selecting the quantum parameter μ in LQC [45] in which the curvature operator is defined by shrinking the plaquettes till the *physical* area they enclose equals Δ; and,
(iii) The parameters are chosen to be phase space functions that are constants of motion on the effective dynamical trajectories. The strategy used in the last two sub-sections falls in this class.

The earliest investigations [12–14] used strategy (i); technically it is the simplest to implement. Here the quantum parameters were set to $\delta_b = \delta_c = 2\sqrt{3}$ using 'square' plaquettes in coordinates adapted the symmetries. Predictions of the resulting effective theory were analyzed in detail in [19]. The singularity is again resolved and replaced by a 3-surface at which the symmetry 2-spheres attain the minimum area. However, physical quantities such as the minimum value of the radius and the radius of the anti-trapping horizon now depend on the infrared cutoff $L_\circ$. Another limitation is that quantum effects can become significant even in the low curvature region near the horizons.

In the approaches [17, 18] based on strategy (ii), the quantum parameters were fixed by mimicking the successful $\bar{\mu}$ strategy from LQC. One again adapts the plaquettes to the symmetries of the problem, but shrinks them till the *physical* area they enclose is Δ. Therefore the plaquettes themselves now depend on the phase space point under considerations and change under time evolution. As a consequence, quantum parameters are specific phase space functions that are not constant along dynamical trajectories: $\delta_b = \Delta/p_c$ and $L_\circ^2\,\delta_c^2 = (L_\circ^2\,p_c\Delta)/p_b^2$. In these definitions, the dependence of $L_\circ$ is exactly the one that is needed to assure that physical results are independent of the fiducial choice of $L_\circ$. This is a significant improvement over results from strategy (i). However, a technical complication arises because δ_b depends on p_c and δ_c on p_c: the equations in the b and c sectors no longer decouple. Consequently, it has not been possible to write down analytic solutions and all explorations to date have been performed numerically. These calculations show that the framework has two types of limitations. First, as in (i), there are large deviations from the classical theory even when the curvature is low. Second, when one evolves beyond the transition surface, the dynamical trajectory enters a region of the phase space where the metric 2-spheres have area that is less than the area gap Δ, making the scheme internally inconsistent. Perhaps not surprisingly, then, some of the properties of the extended space-time are difficult to understand physically.

Strategy (iii) was first adopted to improve on this situation by making δ_b, δ_c phase space functions that remain constant along dynamical trajectories [19, 20]. Then the considerations of the first part of Sect. 7.2.1 are applicable, the b and the c sectors separate, dynamical trajectories can be written down analytically, and m_b and m_c are constants of motion. In the first investigation, δ_b and δ_c were chosen by dimensional considerations and by taking into account the fact that it is only the combination $(L_\circ\delta_c)$ that is invariant under the change of the infrared cutoff $L_\circ$. The simplest expressions satisfying these requirements were then selected, $(\delta_b)^2 := \Delta/4m^2$ and $L_\circ(\delta_c)^2 = \Delta$, without the considerations of plaquettes and holonomies of the gravitational connection around them [19]. The physical results are now invariant under rescalings of $L_\circ$ as desired. There is again a transition surface $\mathscr{T}$ that separates the trapped and anti-trapped regions, and the quantum corrected space-time is a diamond bounded by a trapping horizon in the past and an anti-trapping horizon in the future. Furthermore, unlike the μ_o and $\bar{\mu}$-type schemes, quantum corrections are small in regions near the horizons where the curvature is low. However, detailed examination revealed two limitations. First, at the transition surface the Kretchmann scalar of (initially) macroscopic black holes now goes as $1/m$; whence it *decreases* as the mass of m increases. Therefore for astrophysical black holes, large quantum corrections at the heart of the 'bounce' at $\mathscr{T}$ occur at low curvature. A second counter-intuitive result is involves 'mass inflation' across $\mathscr{T}$. The radius $r_{\rm AT}$ of the horizon in the future of $\mathscr{T}$ now goes as $r_{\rm AT} = (r_{\rm T}) \times (r_{\rm T}/\ell_{\rm Pl})^3$. Therefore, if the initial black hole has solar mass with $r_T = 3\,{\rm km}$, one has $r_{\rm AT} \approx 10^{93}{\rm Gpc}$! The physical mechanism responsible for this huge magnification has remained unclear. Therefore, subsequently, more general choices of the quantum parameters were explored by introducing new dimensionless constants α and β, setting $(\delta_b)^2 := (\alpha^2\,\Delta)/4m^2$ and $L_\circ(\delta_c)^2 = \beta^2\Delta$ and varying α and β to ensure $r_{\rm AT} \approx (r_{\rm T})$ for large black holes. Two choices satisfying this condition were found numerically and one analytically. The analytic expression implies that the leading term in Kretchmann scalar at $\mathscr{T}$ is not universal but grows rapidly with m as m^4/Δ^4.

The expressions (7.13) used in Sect. 7.2.2 to discuss results also fall under strategy (iii). However, now the quantum parameters δ_b and δ_c are obtained using certain plaquettes, holonomies around which are used to define the curvature. These plaquettes are tailored to the symmetries of the problem, and enclose physical area Δ as in the strategy (ii). The key difference is that these loops are restricted to lie on the transition surface $\mathscr{T}$ [28, 29]. Since each dynamical trajectory intersects $\mathscr{T}$ once and only one, the prescription is unambiguous and, by construction, makes δ_b and δ_c constants of motion. This choice automatically leads to the result $r_{\text{AT}} \approx (r_{\text{T}})$, without recourse to any additional free parameters (such as α, β discussed above). Furthermore, now the transition surface necessarily lies in the region where curvature is Planckian and the leading terms in the expressions of all curvature invariant are universal.

As this discussion shows, the task of choosing appropriate quantum parameters δ_b, δ_c is a very subtle. While is it not difficult to make a 'reasonable' choice that resolves the singularity, the resulting quantum corrected geometry has to satisfy several non-trivial constraints to be physically admissible. Over the years, several choices have been proposed but the subsequent careful scrutiny by the LQG community showed that they lead to results that are physically unsatisfactory in one way or the other. The choice discussed in the last two subsections passes all the checks known to date. While this is satisfying, the analysis is still incomplete in one respect. In LQC the effective equations could be derived systematically starting from the operator equations of the quantum theory, showing that there are states that remain sharply peaked even in the deep quantum regime, and using expectation values of observables in these states [47, 48] (and Section V of [9]). Thus the LQC effective equations encode the dynamics of the peaks of these wave functions. For black holes, the successful LQC techniques have been used in conjunction with an extended phase space framework (introduced in [29]) to arrive at the desired operator equations and to select physical states in [37]. A systematic derivation of effective equations from this quantum theory remains an interesting open issue in LQG.

7.3 The Schwarzschild Exterior

For the discussion of singularity resolution, it suffices to consider just the region II of Fig. 7.1. Therefore, initially the focus on LQG investigations was on this region. However, for a complete understanding of the quantum corrected space-time, one also has to connect the effective space-time geometry of region II to that of region I. In Sect. 7.3.1 we present an approach to carry out this task. Section 7.3.2 summarizes the properties of the near-horizon quantum corrected geometry it provides, and 7.3.3 discusses the asymptotic structure of the effective space-times. As expected, for macroscopic black holes the near horizon geometry exhibits physically expected features because quantum corrections are small there. In the asymptotic region, on the other hand, this effective geometry has an unforeseen feature: while the quantum corrected metric is asymptotically flat in a precise sense, the approach to flatness is weaker than what one might have a priori expected. We will discuss this issue and summarize its current status in Sect. 7.6.

7.3.1 The Underlying Framework

Recall that the analysis of the Schwarzschild interior was greatly facilitated by the fact that this region is foliated by *homogeneous, space-like* slices. The exterior region on the other hand does not admit such a foliation. However, the four Killing fields do provide a natural foliation of this region by homogeneous, *time-like* slices. Indeed the textbook derivation of the classical Schwarzschild metric can be interpreted as solving the 'evolution equation' in the r direction together with the 'Hamiltonian' constraint on the $r = \text{const}$ homogeneous slices, mirroring the procedure used in the Schwarzschild interior (or, Kantowski-Sachs space-times). The main difference is that the signature of the intrinsic 3-metric on the homogeneous slices is now $-, +, +$ rather than $+, +, +$. Therefore in the connection framework one has to change the internal group that acts on the orthonormal triads from SU(2) to SU(1,1).[3] The generators τ_i that provide a basis for the Lie algebra of SU(2) are now replaced by $\tilde{\tau}_i$ that constitute a basis for the Lie algebra of SU(1,1). The relation between the two is

$$\tilde{\tau}_1 = i\tau_1, \quad \tilde{\tau}_2 = i\tau_2, \quad \text{and} \quad \tilde{\tau}_3 = \tau_3. \tag{7.18}$$

Hence, for exterior region we can choose our basic variables to be

$$\begin{aligned} A^i_a\, \tilde{\tau}_i\, \mathrm{d}x^a &= \frac{\tilde{c}}{L_o}\, \tau_3\, \mathrm{d}x + i\tilde{b}\, \tau_2 \mathrm{d}\theta - i\tilde{b}\, \tau_1 \sin\theta\, \mathrm{d}\phi + \tau_3 \cos\theta\, \mathrm{d}\phi, \\ E^a_i\, \tilde{\tau}^i\, \partial_a &= \tilde{p}_c\, \tau_3\, \sin\theta\, \partial_x + \frac{i\tilde{p}_b}{L_o}\, \tau_2\, \sin\theta\, \partial_\theta - \frac{i\tilde{p}_b}{L_o}\, \tau_1\, \partial_\phi. \end{aligned} \tag{7.19}$$

Comparison with (7.1) reveals that one can arrive at solutions to the 'constraint' and 'evolution' equations in the exterior region simply by using the substitutions

$$b \to i\tilde{b}, \quad p_b \to i\tilde{p}_b \qquad \text{and} \qquad c \to \tilde{c}, \quad p_c \to \tilde{p}_c$$

in the solutions of the interior region. Indeed, one can explicitly check that if one makes these substitutions in the classical solutions (7.3) and (7.4), one obtains the Schwarzschild metric in the exterior region. Therefore we can use these substitutions in the solutions (7.9), (7.10), (7.11) to the effective equations in the interior to obtain the desired dynamical trajectories in the exterior region, $T > 0$. They yield

[3] This strategy of using time-like 3-manifolds to specify fields and then 'evolving' them in space-like directions was proposed and pursued in [39] for the Hamiltonian framework of full LQG. As discussed there, in the full theory one encounters certain non-trivial technical difficulties associated with the fact that SU(1,1) is non-compact. These issues do not arise in the homogeneous context discussed here.

$$\tan\left(\frac{\delta_c\,\tilde{c}(T)}{2}\right) = \frac{\gamma L_o \delta_c}{8\,m} e^{-2T}, \qquad \tilde{p}_c(T) = 4\,m^2\left(e^{2T} + \frac{\gamma^2 L_o^2 \delta_c^2}{64\,m^2} e^{-2T}\right), \tag{7.20}$$

$$\cosh\left(\delta_b\,\tilde{b}(T)\right) = \tilde{b}_o \tanh\left(\frac{1}{2}\left(\tilde{b}_o T + 2\tanh^{-1}\left(\frac{1}{\tilde{b}_o}\right)\right)\right), \tag{7.21}$$

where $\tilde{b}_o = (1+\gamma^2\delta_b^2)^{1/2}$ δ_b, δ_c are given by (7.13) as in Sect. 7.2, and,

$$\tilde{p}_b(T) = -2\,\frac{\sin(\delta_c\,\tilde{c}(T))}{\delta_c}\,\frac{\sinh(\delta_b\,\tilde{b}(T))}{\delta_b}\,\frac{|\tilde{p}_c(T)|}{\gamma^2 - \frac{\sinh^2(\delta_b\,\tilde{b}(T))}{\delta_b^2}}\,. \tag{7.22}$$

Thus, the explicit solutions in the c-sector have the same form as their counterparts (7.9) in the interior region ($T < 0$) while in the b-sector the trigonometric functions of $(b\delta_b)$ are replaced by their hyperbolic analogs. Details of derivations and a discussion of the comparison between the classical and effective descriptions of the exterior region can be found in [29].

Let us conclude by specifying space-time geometry in the exterior region. The translational Killing field—which is time-like in the exterior region—is still given by $\partial/\partial x$ and T is a radial coordinate that vanishes on the horizon and is positive in the exterior region. For $T > 0$, the effective metric is given by

$$\tilde{g}_{ab}\mathrm{d}x^a\mathrm{d}x^b = -\frac{\tilde{p}_b^2}{\tilde{p}_c L_o^2}\mathrm{d}x^2 + \frac{\gamma^2\,\tilde{p}_c\,\delta_b^2}{\sinh^2(\delta_b\tilde{b})}\mathrm{d}T^2 + \tilde{p}_c(\mathrm{d}\theta^2 + \sin^2\theta\,\mathrm{d}\phi^2). \tag{7.23}$$

The metric is well-defined in this region and has signature -,+,+,+. It fails to be well-defined at $T = 0$ because b and p_b vanish there. However, as we show below, this is just a reflection of the breakdown of the coordinate system. In the limit $\ell_{\rm Pl} \to 0$ (or, $\Delta \to 0$, keeping γ positive), the quantum parameters δ_b and δ_c vanish and the metric (7.23) reduces to the Schwarzschild metric in the exterior region. Properties of the geometry induced by this effective metric are discussed in the next two subsections.

7.3.2 *Quantum Corrected, Near Horizon Geometry*

In this subsection we will briefly discuss two features of the near horizon geometry: Matching of the effective metric across the horizon and corrections to the Hawking temperature, computed using Euclidean (or rather, Riemannian) geometry. Further details can be found in [30].

Matching across horizon $T = 0$. Recall that in the classical theory, although the metric appears to be ill-defined across the horizon, one can introduce Eddington-Finkelstein type coordinates to make its regularity explicit. The same strategy can be adopted at the horizon $T = 0$ of the effective metric. As in the classical case, one can ignore the angular part of the metric. Then the relevant 2-metrics in interior and the exterior can be respectively written in the form

$$\mathrm{d}S_2^2 = f_1(T)\mathrm{d}x^2 - f_2(T)\mathrm{d}T^2; \quad \text{and} \quad \mathrm{d}\tilde{S}_2^2 = -\tilde{f}_1(T)\mathrm{d}x^2 + \tilde{f}_2(T)\mathrm{d}T^2, \tag{7.24}$$

where

$$f_1(T) = \frac{p_b^2}{p_c L_o^2}, \quad f_2(T) = \frac{\gamma^2 p_c\, \delta_b^2}{\sin^2(\delta_b b)} \quad \text{and} \quad \tilde{f}_1(T) = \frac{\tilde{p}_b^2}{\tilde{p}_c\, L_o^2}, \quad \tilde{f}_2(T) = \frac{\gamma^2 \tilde{p}_c\, \delta_{\tilde{b}}^2}{\sinh^2(\delta_{\tilde{b}}\tilde{b})}. \tag{7.25}$$

As in the Eddington-Finkelstein extension in the classical case, one can approach the horizon from the exterior region. Remembering that the coordinates (T, x) used for the effective metric are the analogs, respectively, of the Schwarzschild coordinates (r, t), one defines an advanced null coordinate $v = x + \tilde{T}_\star$ where

$$\mathrm{d}\tilde{T}_\star = \left(\tilde{f}_2/\tilde{f}_1\right)^{\frac{1}{2}} \mathrm{d}T. \tag{7.26}$$

Then the metric in the exterior region becomes

$$\mathrm{d}\tilde{S}_2^2 = \tilde{f}_1\, \mathrm{d}v^2 - 2\,(\tilde{f}_1\, \tilde{f}_2)^{\frac{1}{2}}\, \mathrm{d}v\, \mathrm{d}T. \tag{7.27}$$

Since $\tilde{f}_1$ vanishes at $T = 0$, the space-time metric is well-defined at the horizon with signature -,+,+,+ if and only if $\tilde{f}_1$ is smooth, and $\tilde{f}_1\tilde{f}_2$ is smooth and positive in a neighborhood of $T = 0$. This is indeed the case. In particular, $\lim_{T\to 0} \tilde{f}_1\tilde{f}_2 = 4m^2$. In the standard Schwarzschild coordinates (r, t) used in the classical theory, the product is 1, and since $r = 2m\, e^{T_{\text{class}}}$, it is again $4m^2$ in the (T_{class}, x) coordinates. In this sense the product is the 'same' for the classical and the effective metric. The first derivative of $\tilde{f}_1$ differs from its classical values by terms of the order $0(\varepsilon_m)$ where $\varepsilon_m = (\gamma^2 L_0^2 \delta_c^2)/64m^2$ and the second derivative by terms of the order $0(\varepsilon_m, \delta_{\tilde{b}}^2)$. For the metric coefficient $(\tilde{f}_1\, \tilde{f}_2)^{\frac{1}{2}}$, they are given by $2m$ and $m\big(2 + \gamma^2\delta_{\tilde{b}}^2\big)$, respectively. If one approaches the horizon from the interior, one finds that the limits of f_1 and $(f_1 f_2)^{\frac{1}{2}}$ and their first two derivatives exist and match with those coming from the exterior. Thus, the effective metric is (at least) C^2 across the horizon $T = 0$. Furthermore, the corrections to the metric coefficients are negligible for macroscopic black holes.

In summary, although the effective 4-metric is constructed in the interior region $T < 0$ using spatial homogeneity of a space-like foliation and in the exterior region $T > 0$ using temporal homogeneity of a time-like foliation, and the x coordinates becomes ill-defined at the horizon, as in the classical theory, there is a well-defined Eddington-Finkelstein type chart (v, T) in which $\mathrm{d}\tilde{S}_2^2$ is well-defined also at $T = 0$. Therefore the effective metric can be extended across both the future and past horizons as in the classical Kruskal case shown in Fig. 7.1. Furthermore, since the singularity is resolved, one can extend the metric also across the new, anti-trapping horizons shown in Fig. 7.2. One can continue these extensions to arrive at the Penrose diagram of 7.3 which extends indefinitely to the future and to the past.

Quantum corrections to the Hawking temperature. In the classical theory one can arrive at the Hawking temperature by passing to the Riemannian section via wick rotation of the metric in the exterior region. In these considerations, it suffices to restrict oneself to the $r,\ t$ plane where the Riemannian metric $\tilde{\mathfrak{g}}_{ab}$ has the form

$$\tilde{\mathfrak{g}}_{ab}\mathrm{d}x^a\mathrm{d}x^b = \tilde{f}_1(r)\,\mathrm{d}t_{\mathrm{E}}^2 + \tilde{f}_2(r)\mathrm{d}r^2\,. \tag{7.28}$$

Since the norm of the translation Killing vector $\tilde{f}_1(r)$ vanishes at the horizon in the Lorentzian section and since the only vector that has vanishing norm is the zero vector in Riemannian signature, the horizon shrinks to a point where the Killing vector vanishes. In a neighborhood of this point, the static Killing field resembles a rotation, whence t_{E} becomes periodic with period $\mathscr{P}$. This 'rotational' character of t_{E}^a becomes manifest if we set $R = (\tilde{f}_1(r))^{1/2}$ so that the metric on the $r - t_{\mathrm{E}}$ plane becomes

$$\tilde{\mathfrak{g}}_{ab}\mathrm{d}x^a\mathrm{d}x^b = R^2\,\mathrm{d}t_{\mathrm{E}}^2 + 4\,\frac{\tilde{f}_1\tilde{f}_2}{(\tilde{f}_1')^2}\,\mathrm{d}R^2. \tag{7.29}$$

The requirement that the metric be free of a conical singularity at the point $R = 0$ (where the Killing Field vanishes) constrains the period $\mathscr{P}$ of t_E to be

$$\mathscr{P} = \lim_{R\to 0}\frac{4\pi(\tilde{f}_1\tilde{f}_2)^{\frac{1}{2}}}{\tilde{f}_1'} = \lim_{R\to 0}\frac{4\pi(\tilde{f}_1)^{\frac{1}{2}}}{||D\tilde{f}_1||} \tag{7.30}$$

where the last step brings out the invariant nature of $\mathscr{P}$ since it involves only the norm $\tilde{f}_1$ of the Killing field and the norm of its covariant derivative. This periodicity implies that Green's functions satisfying standard boundary conditions in the Riemannian sector have the same periodicity, which is used to endow the temperature $T_{\mathrm{H}} = \hbar/(K\mathscr{P})$ to the black hole through the relation between Lorentzian field theories and their Wick rotated versions [49, 50]. For the classical Schwarzschild solution, we have $\mathscr{P} = 8\pi m$, which yields $T_{\mathrm{H}} = \hbar/(8\pi\ K\ m)$

This strategy can be directly applied to the effective metric (7.23) in the exterior region. The Wick rotated, positive-definite metric in the $(r,\ t)$ plane—i.e., now in the $(T,\ x)$ plane—becomes:

$$\tilde{\mathfrak{g}}_{ab}\mathrm{d}x^a\mathrm{d}x^b = \tilde{f}_1(T)\mathrm{d}x^2 + \tilde{f}_2(T)\mathrm{d}T^2 \quad\text{with}\quad \tilde{f}_1 = \frac{\tilde{p}_b^2}{\tilde{p}_c\,L_o^2} \quad\text{and}\quad \tilde{f}_2 = \frac{\gamma^2\,\tilde{p}_c\,\delta_{\tilde{b}}^2}{\sinh^2(\delta_{\tilde{b}}\tilde{b})}\,.$$

The horizon is at $T = 0$, where $\tilde{p}_b$ and $\tilde{b}$ vanish in the effective solution. Regularity of the metric follows from the properties of $\tilde{f}_1$ and $\tilde{f}_1\tilde{f}_2$ discussed above. The period $\mathscr{P}$ of (7.30) is now given by $\mathscr{P} = 8\pi m(1 + \varepsilon_m)$ where, as before, $\varepsilon_m = (\gamma^2 L_0^2\delta_c^2)/64m^2$. Therefore, the Hawking temperature of the quantum corrected black hole horizon is

$$T_{\mathrm{H}} = \frac{\hbar}{8\pi K m}\,\frac{1}{(1+\varepsilon_m)} \tag{7.31}$$

The mass dependent correction $1/(1+\varepsilon_m)$ due to quantum geometry effects is very small for macroscopic black holes. For a solar mass black hole it is of the order of $\sim 4 \times 10^{-106}$. Indeed, even for a black hole of $\sim 10^6 M_{\rm Pl}$, the correction is of the order 10^{-21}. (Because there are inherent approximations in arriving at the effective theory, further extrapolation to even smaller black holes would not be appropriate.)

As discussed in Sect. 7.2, the quantum corrections to various curvature invariants are very small near the horizon of macroscopic black holes. The correction ε_m to the Hawking temperature provides another facet of that general phenomenon.

7.3.3 Asymptotic Properties of the Effective Geometry

As we saw, the quantum gravity corrections are very small near horizons of macroscopic black holes. Exact calculations have been done using MATHEMATICA in a (large) neighborhood of the horizon as one recedes outwards and they show that quantum corrections to the geometry become even smaller, as one would expect. However, as one recedes further to asymptotic regions $r \gg 2m$, the trend does not continue. The main issue is tied with certain subtleties related to asymptotic flatness and the associated Arnowitt, Deser, Misner (ADM) energy that are not widely appreciated and can lead to confusion (for details, see [30]).

Let us therefore begin by recalling the elementary notion of asymptotic flatness. A given metric g_{ab} is said to be asymptotically flat at spatial infinity if there exists a flat metric $\mathring{\eta}_{ab}$ such that in a Cartesian chart defined by $\mathring{\eta}_{ab}$, components of g_{ab} approach the components of $\mathring{\eta}_{ab}$ at least as fast as $1/r$ as $r \to \infty$, keeping t, θ, φ constant (where (t, r, θ, φ) refer to $\mathring{\eta}_{ab}$). However, $\mathring{\eta}_{ab}$ may not be the 'obvious' flat metric suggested by the coordinates in which g_{ab} is presented. An obvious example is the 2-dimensional metric $\bar{g}_{ab}$ with the line element $\mathrm{d}\bar{s}^2 = -r^2\mathrm{d}t^2 + \mathrm{d}r^2$. The fact that $\partial/\partial t$ is the Killing vector of the metric suggests that the coordinates t, r are 'natural', whence one may be led to consider the flat metric $\bar{\eta}_{ab}$ with the line element $\bar{\eta}_{ab}\mathrm{d}x^a\mathrm{d}x^b = -\mathrm{d}t^2 + \mathrm{d}r^2$. One would then conclude that the given metric $\bar{g}_{ab}$ is *not* asymptotically flat because it does not approach $\bar{\eta}_{ab}$. Indeed, this conclusion may be further re-enforced by the fact that the norm of the static Killing field diverges as $r \to \infty$. But not only is $\bar{g}_{ab}$ asymptotically flat, it is in fact flat because $\bar{g}_{ab}$ is just the Minkowski metric in the Rindler wedge. This example brings out the fact that even a flat metric is generically not asymptotically flat w.r.t. other flat metrics even in the elementary sense! Note, however, that for a given metric g_{ab} to be asymptotically flat, it suffices to find *one* flat metric, say $\mathring{\eta}_{ab}$, to which it approaches; it need not approach a pre-selected flat metric, like $\bar{\eta}_{ab}$ in the above example. A more subtle example is provided by the Levi-Civita solution to Einstein's equation (known as the 'c-metric') [51] that, it turned out, represents the gravitational field of two accelerating black holes [52]. In this solution, the norm of the Killing field $\partial/\partial t$ also diverges at spatial infinity, and it too seems not to be asymptotically flat in the coordinates it is normally presented in. (This feature led to considerable confusion on whether this space-time admits gravitational radiation.) But the c-metric is in fact asymptotically flat in the standard sense [53] (and does admit radiation); the form of the flat metric $\mathring{\eta}_{ab}$ it approaches at infinity is not obvious in the coordinates the c-metric is presented in.

With these preliminaries out of the way, let us return to the effective metric $\tilde{g}_{ab}$ of Eq. (7.23) in the asymptotic region and ask if it asymptotically flat, keeping in mind the subtleties discussed above. Now, b, c, p_b, p_c that enter the expression of $\tilde{g}_{ab}$ are complicated functions of T. To make the asymptotic structure transparent, let us first set

$$r_S := 2m, \qquad r := r_S\, e^T, \qquad \text{and} \qquad \varepsilon := 1 - b_0 \equiv 1 - (1 + \gamma^2 \delta_b^2)^{\frac{1}{2}} \tag{7.32}$$

and replace x by t so the translational Killing field is now $\partial/\partial t$ (rather than $\partial/\partial x$). For macroscopic black holes the dimensionless parameter ε is very small; for example $\varepsilon = 10^{-26}$ for a star mass black hole. Let us therefore assume that $\varepsilon \ll 1$. Then in the asymptotic region, where $r_S/r \ll 1$ and $(\gamma^2\, r_S\, \Delta)^{1/3}/2r \ll 1$, the exact expression (7.23) of the quantum corrected metric simplifies significantly: $\tilde{g}_{ab} \approx \tilde{g}^{\circ}_{ab} \mathrm{d}x^a \mathrm{d}x^b = \tilde{g}^{\circ}_{tt} \mathrm{d}t^2 + \tilde{g}^{\circ}_{rr} \mathrm{d}r^2 + r^2\, \mathrm{d}\omega^2$, where,

$$\tilde{g}^{\circ}_{tt} = -\left(\frac{r}{r_S}\right)^{2\varepsilon} \left(1 - \left(\frac{r_S}{r}\right)^{1+\varepsilon}\right) \quad \text{and} \quad \tilde{g}^{\circ}_{rr} = \left(1 - \left(\frac{r_S}{r}\right)^{1+\varepsilon}\right)^{-1}. \tag{7.33}$$

Now, since $\tilde{g}^{\circ}_{tt}$—and hence $\tilde{g}_{tt}$—diverges as $r \to \infty$, it is clear that the 'obvious' metric does not approach the flat metric $\tilde{\eta}_{ab} \mathrm{d}x^a \mathrm{d}x^b = -\mathrm{d}t^2 + \mathrm{d}r^2 + r^2 \mathrm{d}\omega^2$. Therefore, one may be tempted to conclude that $\tilde{g}^{\circ}_{ab}$—and hence $\tilde{g}^{\circ}_{ab}$—is not asymptotically flat [54]. However, as the examples of the Rindler and the c-metric show, the conclusion does not follow. Rather, the question is whether there *exists* a flat metric $\tilde{\eta}^{\circ}_{ab}$ to which $\tilde{g}_{ab}$ approaches as $r \to \infty$; this $\tilde{\eta}^{\circ}_{ab}$ need not be the 'obvious' flat metric $\tilde{\eta}_{ab}$. The answer turns out to be in the affirmative [30]. To display its form, one has to replace t with $\tau = t(r/r_S)^{\varepsilon}$ (note that τ agrees with t for $\varepsilon = 0$). Then, setting $\tilde{\eta}^{\circ}_{ab} \mathrm{d}x^a \mathrm{d}x^b = -\mathrm{d}\tau^2 + \mathrm{d}r^2 + r^2 \mathrm{d}\omega^2$ one finds that components of $\tilde{g}_{ab}$ approach those of $\tilde{\eta}^{\circ}_{ab}$ as $1/r$, ensuring asymptotic flatness of $\tilde{g}_{ab}$. As one would expect from this property, all curvature invariants of g_{ab} vanish as $r \to \infty$. Furthermore, this fall-off is sufficient to ensure that the ADM energy is well-defined. It can be computed using the spatial Ricci tensor $\tilde{\mathscr{R}}_{ab}$ using an expression [55] that is often used in the recent geometric analysis literature on the subject (see, e.g., [56]). One finds

$$E_{\text{Ricci}} := \lim_{r \to \infty} \frac{1}{8\pi G} \oint_r \mathrm{d}^2 V\; r\, N\, \tilde{\mathscr{R}}_{ab} \hat{r}^a \hat{r}^b \;\equiv M(1 + \varepsilon), \tag{7.34}$$

where $\mathrm{d}^2 V$ is the area element of the $r = \text{const}$ 2-sphere of integration, $\hat{r}^a$ a unit radial vector, and M is the Schwarzschild mass of the classical solution. Thus there *is* a quantum correction to the Schwarzschild mass, but it is minuscule for macroscopic black holes.

However, the fact that $\tilde{g}_{ab}$ does not approach the 'obvious' flat metic $\tilde{\eta}_{ab}$ reflects a limitation of its asymptotic behavior: the approach to flatness is not as strong as assumed in the standard treatments of asymptotics (see, e.g. [55]) because, while the metric components approach their flat space values as $1/r$, not all components of the connection $\tilde{\nabla}$ defined by $\tilde{g}_{ab}$ fall-off as $1/r^2$. As a consequence several components of the space-time curvature have weaker fall-offs than in the standard context. In particular, the curvature invariants fall off only as $1/r^4$ rather than $1/r^6$. These deviations from standard asymptotic behavior have some subtle consequences.

Let us illustrate these subtleties with examples. As we just saw, the expression E_{Ricci} of the ADM energy continues to be well-defined, and yields $E_{\text{Ricci}} = M(1+\varepsilon)$. One can also carry out the calculation using the more familiar expression involving the 3-metric, paying attention to the lapse defined by the Killing field [57]. One then finds $E_{3\text{-metric}} = M$, without any corrections. Similarly one can also evaluate the mass at the horizon using its area A_{hor}, $M_{\text{hor}} = (A_{\text{hor}}/16\pi)^{1/2}$ to find $M_{\text{hor}} = M(1+\varepsilon_m)$ where ε_m is the mass dependent term that enters the expression (7.31) of the corrected Hawking temperature we found in Sect. 7.3.2. For a solar mass black hole $\varepsilon_m \approx 10^{-106}$, much smaller than the correction $\varepsilon \approx 10^{-26}$ that enters (7.34). All these quantities agree for the classical Schwarzschild solution because the asymptotic fall-off is the standard one [55]. Now, it often happens that notions that agree in a limiting theory (e.g., Newtonian gravity) become ambiguous in a more complete theory (e.g., GR) and are thus replaced by several different notions. It remains to be seen whether these findings associated with the notion of energy are conceptually similar for the transition from GR to quantum gravity, or if they are blemishes that point to a genuine limitation of the effective metric $\tilde{g}_{ab}$ in the exterior region, that will be cured by a better candidate. As we will discuss in Sect. 7.6, this issue is under active investigation in LQG.

7.4 Quantum Geometric Effects in Gravitational Collapse: Illustrations

In Sect. 7.2, we saw that the isometry between the Kantowski-Sachs space-time and the Schwarzschild interior allows one to apply tools from LQC to the Schwarzschild spacetime and permits one to study detailed physical implications. However, these studies have an inherent limitation: they can not capture the dynamics of a gravitational collapse, resulting in a black hole. Models of gravitational collapse are significantly richer: in contrast to eternal black holes, one now has a field theory, in which the time evolution of geometry and matter is coupled and governed by non-linear equations [33, 58–68]. In this class of models, several investigations have been carried out to understand the resolution of singularities associated with the dynamical collapse of homogeneous dust in Oppenheimer-Snyder scenarios, in which the interior is modeled by a Friedmann, Lemaître, Robertson, Walker (FLRW) cosmology [68–75]. This allows the application of LQC techniques for the study of the fate of the classical singularity and yields similar results on non-viability of certain quantization schemes. In particular, it turns out that the 'μ_o scheme' on which early LQC was based—but subsequently ruled out on cosmological viability criteria [45, 76]—has novel limitations in the black hole sector: it does not permit formation of trapped surfaces unless one chooses rather unnatural features of quantum geometry [75]. This is an illustration of the fact that these models can provide valuable insights, despite the limitations associated with their simplicity.

Another category of investigations considers dynamics of shells where the interior regions is usually a patch of Minkowski spacetime, while the exterior is a Schwarzschild geometry. They allow for the study of black hole formation, modeling the interior of the star as a simple, empty, flat spacetime. At the quantum level, there is considerable literature on this topic (see for eg. [77] for a review). To understand quantum geometry effects in this setting, a reduced phase space quantization of thin shells has been performed [78–80]. One of these works shows that the classical singularity is eliminated, where the shell either emerges through a white hole type geometry or tunnels into a baby universe inside the black hole [78]. Another work proposes an effective semiclassical description motivated by LQC quantization techniques for the study of a Lemaître-Tolman-Bondi (LTB) spacetime, focusing on the dynamics of the outermost shell of matter [80]. Here, the singularity inside the black hole is resolved. Moreover, after black hole formation, matter bounces, eventually 'evaporating' the black hole and dispersing towards infinity. There are also studies that focus their attention to the search of an effective constraint algebra that is free of anomalies, and include the so-called 'inverse triad corrections' [81, 82], and 'holonomy corrections' [83, 84]. Finally, there have been studies to understand quantum geometric effects on critical phenomena in the scalar field collapse discovered by Choptuik [85] in classical GR [86–92].

Given the richness and complexities of the underlying physics, at the present stage these attempts aim at providing insights on specific aspects of the problem, rather than a complete picture. To illustrate the overall status we will discuss two concrete examples in some detail: the dust collapse scenario, and the critical collapse of a scalar field. The first category of results focus on singularity resolution and therefore use horizon penetrating coordinates. On the other hand, in the second category the focus is primarily on the exterior region, whence it suffices to use coordinates that cover only that part of the space-time. These examples are complementary in the following sense. In the first category, geometry is treated quantum mechanically to start with, and induces quantum effects on matter via field equations. In the second category, to begin with only matter is treated quantum mechanically, and subsequently quantum features descend on geometry from matter, again through field equations.

7.4.1 Dust Field Collapse Models

In this subsection, we consider a few recent investigations [61, 62, 65, 67] that illustrate the quantum modifications of classical dynamics. They use a reduced phase space quantization with certain gauge fixing conditions in spherically symmetric space-times, minimally coupled to an inhomogeneous dust field. The focus is on the family of spherically symmetric Lemaître–Tolman-Bondi (LTB) spacetimes, and its sub-family of Oppenheimer-Snyder (OS) models where the dust field is homogeneous. The approach is inspired by the 'improved dynamics' strategy of LQC. In these models the matter sector—dust—is not quantized but its dynamics is deeply influenced by the quantum nature of underlying geometry, once it enters the high curvature regime.

The metric of LTB space-times is given by [65, 67]

$$\mathrm{d}s_{\rm cl}^2 = -N^2 \mathrm{d}t^2 + \frac{E^{\varphi}(t,x))^2}{E^x}\left(\mathrm{d}x + N_{\rm cl}^x(t,x)\,\mathrm{d}t\right)^2 + x^2 \mathrm{d}\Omega^2, \tag{7.35}$$

where $x \in [0, \infty)$ is the radial coordinate and φ is the azimuthal coordinate in spatial slices. Let us restrict ourselves to the 'marginally bound case' where the spatial slices are flat. In the Hamiltonian framework, one can gauge fix the momentum (or, diffeomorphism) constraint by setting $E^x = x^2$. Preservation of this gauge-fixing condition in time determines the shift $N_{\rm cl}^x$ in terms of the canonical variables: $N_{\rm cl}^x(t,x) = -N(K_\varphi(t,x)/\gamma)$ where K_φ is the momentum conjugate to E^φ. (Because the spatial slices are flat, K_φ equals the connection component A_φ.) One can fix the lapse function N without loss of generality; let us set $N = 1$ so that t represents proper time. The Hamiltonian constraint relates these geometric variables to the matter density and determines evolution equations for E^φ, K_φ through Poisson brackets [67].

To pass to the effective theory, one sets $\beta(t,x) = (\sqrt{\Delta}/x)\,K_\varphi(t,x)$ and, motivated by known results in LQC, one makes the ansatz:

$$N^x(t,x) = -\frac{x}{\gamma\sqrt{\Delta}} \sin\left(\beta(t,x)\right)\; \cos\left(\beta(t,x)\right) \tag{7.36}$$

(so that, in the limit area gap $\Delta \to 0$ (keeping $\gamma > 0$), we recover the classical shift $N_{\rm cl}^x$). In the Painlevé-Gullstrand like coordinates (for unit lapse), E^φ is time independent, given by $E^\varphi(x,t) = x$. The Hamiltonian constraint and the evolution equation for K_φ—which is now encoded in β—are non-trivial:

$$\rho = \frac{1}{8\pi G\gamma^2\Delta x^2}\,\partial_x\left(x^3 \sin^2\beta\right) \quad \text{and} \quad \partial_t\beta = -4\pi G\gamma\sqrt{\Delta}\,\rho. \tag{7.37}$$

As mentioned earlier, ρ is a classical field throughout this analysis; nonetheless it now acquires an upper bound because of its coupling to quantum geometry.

Within this family of LTB spacetimes, it is interesting to analyze the subfamily of OS solutions, those in which the energy density is homogeneous. The star is bounded by the surface $x = L(t)$, outside of which $\rho(t)$ vanishes and inside of which $\rho(t)$ is a positive constant for each t. Thus, there is a *finite* discontinuity in ρ all along the boundary $x = L(t)$. Equation (7.37) implies that β is continuous across the boundary but its time derivative has a finite discontinuity there. One can now solve for the function $\rho(t)$ to obtain

$$\rho(t) = \frac{3GM}{4\pi\left(L(t)\right)^3} \quad \text{for } x < L(t), \qquad \text{and} \quad \rho(t) = 0 \quad \text{for } x > L(t). \tag{7.38}$$

The form of $\rho(t)$ inside the star immediately implies an interesting relation that is reminiscent of the quantum corrected Friedmann equation of LQC [45]:

$$\left(\frac{\dot{L}}{L}\right)^2 = \frac{8\pi G}{3}\rho\left(1 - \frac{\rho}{\rho_c}\right), \tag{7.39}$$

for $x(t) < L(t)$, where $\rho_c = 3/8\pi G\gamma^2\Delta$, is again a *universal constant.* (A similar equation of motion for the homogenous dust collapse was obtained in Refs. [68, 71, 75, 80].) At the bounce, one has $L_{\text{bounce}} = (2GM\gamma^2\Delta)^{1/3}$; the value of the radius at which the bounce occurs grows linearly with the mass of the star. In particular, while the density at the bounce is of Planck scale irrespective of the mass of the star, for macroscopic black holes, the radius at the bounce is not. For a solar mass black hole, for example, $L_{\text{bounce}} \approx 10^{13}\ell_{\text{Pl}}$. This distinction is a robust feature of LQG.

Since the bounce of the effective theory replaces the classical singularity, one might expect the subsequent dynamics to display richer structure. This is indeed the case. Soon after the bounce, $\beta(x,t)$ develops a discontinuity at the boundary. Therefore, it follows from (7.37) that $\rho(x,t)$ acquires a new term that is proportional to the delta distribution $\delta(L(t) - x)$. Consequently, after the bounce the evolution equations have to be solved in the distributional sense; one has weak solutions that solve integral equations obtained by integrating the evolution equation w.r.t. x. When the shock wave meets the dynamical horizon [41–43], it ceases to be a trapping horizon. Taking this instant of the time as the end of the black hole, one can calculate its life time as the proper time interval, *measured by a distant observer*, between the instant of formation of the dynamical horizon and its disappearance. One finds:

$$T_{\text{lifetime}} \sim \frac{8\pi G^2 M^2}{3\gamma\sqrt{\Delta}}. \tag{7.40}$$

Although in the above discussion we used the OS solutions to obtain this result, the scaling $T_{\text{lifetime}} \propto M^2$ is more general in LQG. For example, it holds also for shell collapse and the collapse of inhomogeneous dust (up to corrections linear in M) [67]. This life-time contrasts with the suggestions of $T_{\text{lifetime}} \propto M$ that have appeared in the literature [93–96], motivated by general quantum gravity considerations but based on less detailed arguments. This possibility is ruled out by the LIGO/Virgo discoveries of black hole mergers. However, even with the M^2 scaling, one is led to the some surprising conclusions. Recall first that the life time of the black hole due to Hawking radiation goes as M^3. Therefore, if $T_{\text{lifetime}} \propto M^2$ were to be a firm prediction of a fully developed quantum gravity theory, one would have to conclude that the Hawking evaporation process is physically unimportant since the black hole would disappeared before there is significant Hawking radiation. Secondly, from an astrophysical standpoint, one knows that black holes were formed quite early in the history of the universe. If there were any that formed with, say, lunar mass, they would have disappeared and left us a signature of the shock wave accompanying the bounce. It is more likely that the M^2 scaling will be modified by more complete analyses in the future. For example, the shift is chosen using an educated prescription and not arrived at using some fundamental principles. In fact, recent investigations indicate that this prescription differs from the one that arises from considerations of dynamical stability of the effective gauge fixing conditions under the effective dynamics generated by the 'polymerized' canonical Hamiltonian [97]. The usefulness of the current LQG investigations lies precisely in the fact they provide strong and concrete motivation to make the models more and more realistic.

7.4.2 *Quantum Geometric Effects in the Critical Phenomena*

In the classical theory, there are two possible fates for the gravitational collapse of a spherically-symmetric, minimally coupled, massless scalar field depending on the initial data. One possible end state is that the field collapses to form a black hole, and the other is that the field disperses to infinity. One can label each family of initial data of the field by suitable parameters $\underline{p}$, such that for $\underline{p} > \underline{p}^*$ the collapse leads to a black hole, and for $\underline{p} < \underline{p}^*$ no black hole forms, i.e., the collapsing scalar field eventually disperses towards infinity. For $\underline{p} \simeq \underline{p}^*$, it is possible to form black holes through a second order phase transition with masses as close to zero as desired [85].

More precisely, Choptuik demonstrated that the mass of the black hole depends on the difference $(\underline{p} - \underline{p}^*)$ via a universal power law $m_{\mathrm{BH}} \propto \left|\underline{p} - \underline{p}^*\right|^{\beta}$, and there exists a discrete self-similar behavior for $\underline{p} = \underline{p}^*$. It turns out that $\beta \approx 0.37$ is a universal exponent which is independent of the initial data. Further investigations have brought out a finer structure over and above this power law relation [98]. Due to the discrete self-similarity one can numerically observe echoes with a period whose ratio with β determines the periodicity in the fine structure. Due to the scale invariance of the underlying equations there is no mass gap for the formation of black holes in the classical theory; black holes can form with arbitrarily small mass.

It is natural to ask: How does this universal phenomenon change when modifications due to quantum geometric effects are included? In LQG investigations of such models, the quantum modifications to the gravitational sector have different origins. The first possibility is to replace the inverse powers of triads using a classical identity to write them as Poisson brackets between holonomies of the gravitational connection and the triads, and then passing to the quantum theory by replacing the Poisson brackets with commutators [99]. These quantum corrections are often referred to as 'inverse triad modifications'. The second possibility, explained in Sect. 7.2, is to express the field strength of the connection using holonomies around closed loops. These modifications are the ones responsible for the bounce of the background effective geometry. In addition one can also treat the matter sector using a polymer quantization [100]. While a complete treatment to study the critical behavior of the scalar field including all these effects is yet to be performed, explorations have been carried out to understand the modifications of the critical behavior by including only the inverse triad modifications in Refs. [86–89], and by considering LQG quantization of the scalar field in Refs. [90–92]. In all these models one assumes the validity of the effective spacetime description resulting in dynamical equations encoding quantum geometry modifications. Due to inverse triad effects, the behavior of matter-energy modifies the geometry in such a way that there is no divergence and, as a result, the singularity is tamed [101]. Since inclusion of these modifications inevitably introduces a length scale, the scale-invariance is broken. With these modifications, critical phenomena is recovered albeit with a mass gap, below which a black hole can not form. The value of this gap is determined by the discreteness scale in quantum geometry [87]. The existence of mass gap on inclusion of inverse triad modifications can also be seen in a more general collapse of the scalar field [101].

In contrast, if one considers a quantum scalar field á la LQG, one obtains a set of *scale-invariant* effective equations of motion [90–92]. Then the mass gap disappears, allowing one to study of the effects of 'polymer quantization' of the scalar field during the formation of black holes of very small masses. Since this treatment closely mirrors the classical theory and, at the same time, captures 'polymerization'effects in the matter sector, we discuss it in some detail. The spacetime line element studied in [91] is given by

$$\mathrm{d}s^2 = -N^2(t,x)\mathrm{d}t^2 + \frac{\left(E^\varphi(t,x)\right)^2}{E^x(t,x)}\mathrm{d}x^2 + E^x(t,x)\,\mathrm{d}\Omega^2, \tag{7.41}$$

where one gauge fixes $E^x = x^2$ to parallel the classical treatment by Choptuik. Its conjugate variable $K_x(t,x)$ is fixed by the diffeomorphism constraint. The shift vector is determined by demanding preservation of the gauge fixing condition in time. One also uses the gauge freedom to set $K_\varphi(t,x) = 0$ to maintain the diagonal form of the metric. The dynamical variables are the triad $E^\varphi(t,x)$ and the lapse function N. With the matter content as a scalar field $(\phi(t,x),\ P_\phi(t,x))$, the effective equations of motion are obtained by 'polymerizing' the scalar field via $\phi \to \frac{\sin(k\phi)}{k}$:

$$\frac{N'}{N} - \frac{(E^\varphi)'}{E^\varphi} + \frac{2}{x} - \frac{(E^\varphi)^2}{x^3} = 0, \tag{7.42}$$

$$\frac{(E^\varphi)'}{E^\varphi} - \frac{3}{2x} + \frac{(E^\varphi)^2}{2x^3} - 2\pi x\left(\frac{\left(P_\phi\right)^2}{x^4} + \left(\phi'\right)^2\cos^2(k\phi)\right) = 0, \tag{7.43}$$

$$\dot{\phi} = \frac{4\pi N}{E^\varphi x}P_\phi, \tag{7.44}$$

$$\dot{P}_\phi = \frac{4\pi x^2}{E^\varphi}\left[\left(\frac{3NE^\varphi - xN(E^\varphi)' + N'E^\varphi x}{E^\varphi}\right)\phi'\cos^2(k\phi) + xN\phi''\cos^2(k\phi) - xNk\left(\phi'\right)^2\cos(k\phi)\sin(k\phi)\right], \quad (7.45)$$

The lapse function can be determined from Eq. (7.42) (which is obtained by imposing preservation in time of the gauge fixing condition $K_\varphi(t,x) = 0$.) Finally, the Hamiltonian constraint Eq. (7.43) determines the triad $E^\varphi(t,x)$. (Note that for $k \to 0$, (and expressing $E^\varphi(t,x) = xa(t,x)$), one obtains the classical equations of motion of [85].) One can see that these effective equations remain invariant under the transformation $x \to cx$ and $t \to ct$ for constant c. Hence, there will be no mass gap, as in the classical theory. The coordinate system used here cannot penetrate the horizon. Instead, the collapse of the lapse function, namely $N(t,x) \to 0$, is used to signal the formation of a black hole horizon. Numerical simulations with these equations reveal existence of "wiggles" and "echoes" as in the classical description [90, 91]. One finds that this effective theory shares the universality of the scaling of the mass observed in the classical theory, up to small departures for large values of the 'polymer parameter' k. The period of the discrete self-similarity seems to be independent of the 'polymerization parameter' which indicates that the polymer effective theory has a critical solution with the same periodicity as in the classical theory.

Let us conclude this section with a few remarks. In the investigations of the Kruskal space-time reported in Sects. 7.2 and 7.3, detailed analysis of quantum corrections to the geometry and their physical implications was made possible, thanks to the presence of a 4-dimensional symmetry group. Dynamical problems discussed in this section have only spherical symmetry and therefore are much more difficult. Thus, various questions remain unexplored. For instance, the quantization scheme (called 'K-quantization' [102, 103]), used in [61, 62, 65, 67, 68] to arrive at effective equations governing the dust collapse, is only valid for marginally bound cases. Secondly, there are indications [97] that one may have to revisit the assumptions made while 'polymerizing' the Hamiltonian constraint, choices made in 'polymerization' of lapse and shift, and the issue of consistency of gauge fixing conditions. Further, the choice of shift vector made in [61, 62, 65, 67] and also in [33] seems to be problematic from the covariance of the effective geometries [33]. Finally, there are also studies where another ('non-polymeric') quantization of these classical models has been studied [104–111]. A detailed comparison of both quantization schemes could add clarity on the physical viability and mathematical consistency of these two complementary approaches. Similarly, in the investigations of the critical collapse of scalar field, the role of quantum geometry in the gravitational sector is yet to be included [90, 91]. If one were to introduce 'polymerization'of the gravitational connection as in the models for dust collapse, one will very likely introduce a length scale, breaking the scale invariance and a mass gap would appear as in other works incorporating inverse triad modifications [87, 101]. In explorations of the critical collapse, a more complete picture, including quantum geometric effects in the gravitational sector, is not yet available. This is an important gap as it is these quantum geometry effects that lead to singularity resolution. Despite such limitations, it is encouraging that these models have already provided new perspectives on how quantum effects can manifest themselves in the *dynamical* process of black hole formation and evolution, in the resolution of the classical singularity, and in critical phenomena.

7.5 Black Hole Evaporation

Investigations reported in Sect. 7.4 provide interesting insights into the nature of quantum effects in dynamical situations leading to gravitational collapse. However, because of their underlying assumptions, they cannot address the issue of black hole evaporation. In this section we turn to the LQG investigations of the Hawking process and the associated issue of 'information loss'.

In his original discussion [2] Hawking considered a test, scalar quantum field on a classical space-time depicting gravitational collapse of a spherical star. Heuristic considerations of the inclusion of the back reaction on space-time geometry led to the Penrose diagram of Fig. 7.4 that is still widely used. In this diagram $\mathscr{I}^+$ fails to be the complete future boundary since the singularity is also a part of this boundary. One is then led to the startling conclusion that quantum gravity considerations would force us to generalize quantum physics by abandoning unitarity [112]. However, this line of reasoning has important limitations. The first comes from an elementary observation. For a self-consistent discussion of unitarity, one needs a *closed system.* Thus, the incoming collapsing matter in the distant past has to be represented by *quantum fields*, and the outgoing quantum state in the distant future should refer to the *same* fields. This rather basic point is overlooked in space-time diagram of Fig. 7.4 because the asymptotic Hilbert spaces do not include the quantum state of

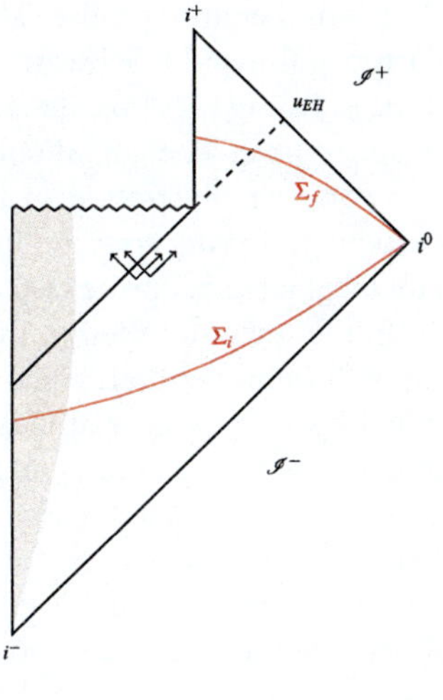

Fig. 7.4 Commonly used Penrose diagram to depict black hole evaporation, including back reaction. Modes are created in pairs, one escaping to $\mathscr{I}^+$ and its partner falling into the black hole. The dashed line is the continuation of the Event Horizon that meets $\mathscr{I}^+$ at retarded time u_{EH}. If this were an accurate depiction, the evolution from $\mathscr{I}^-$ to $\mathscr{I}^+$ would fail to be unitary because the future singularity would act as a 'sink of information' [115]

matter in the star. The second issue is more subtle. In much of the discussion on the subject, challenges and paradoxes arise because one assumes that the quantum corrected space-time has an event horizon that encloses a trapped region which is causally disconnected from the asymptotic region. This seems natural from the perspective of the traditional Penrose diagram of Fig. 7.4. However, event horizons are teleological and, as Hajicek pointed out already in 1987 [113], they can be shifted arbitrarily, and even completely removed, by changing the space-time geometry in a Planck scale neighborhood of the singularity. Now, there is general consensus that classical GR cannot be trusted in such neighborhoods. Therefore the assumption that the event horizon will persist in quantum gravity has no obvious support. Indeed, LQG considerations suggest that it will not.

In Sect. 7.5.1 we explain how these two issues are addressed in the LQG literature. In Sect. 7.5.2 we summarize the current status of LQG investigations in semi-classical gravity and expectations in full quantum gravity. In broad terms these investigations provide closely related avenues to realize the paradigm introduced in [40] based on singularity resolution. *Thus, from LQG perspective, non-singular black holes play a central role in the discussion of the information loss issue.* To anchor the discussion we will use the approach developed in [114, 115]. A complementary discussion can be found in [139].

7.5.1 Setting the Stage

A precise formulation of the issue of 'information loss' is provided by the question of whether the S-matrix from $\mathscr{I}^-$ to $\mathscr{I}^+$ is unitary which, as we discussed, is relevant only for closed systems. The simplest such system is a massless Klein-Gordon field coupled to gravity. Consider, then, gravitational collapse of a spherically symmetric, massless scalar field ϕ from $\mathscr{I}^-$. In the classical theory, if the infalling pulse of ϕ is narrow, the collapse is prompt and analysis is not overly contaminated by the details of the pulse profile. The solution has *Minkowski metric* η_{ab} to the past of this narrow pulse and a Schwarzschild black hole to its future. It is clear from the lower portion of Fig. 7.5 that the *event horizon first forms and grows in the flat portion of space-time*. The actual collapse could occur billions of years to the future! This is a concrete illustration of the teleological nature of the event horizon (EH). In particular, it brings out the fact that the growth of the area of the EH is not tied to any local physical process.

In the quantum theory, the pulse is replaced by a coherent state of the field $\hat{\phi}$ on $\mathscr{I}^+$. In the semi-classical regime—which is expected to be valid in the region in which space-time curvature is much smaller than the Planck scale—one can continue to describe the quantum corrected geometry using a smooth metric. This portion of space-time is depicted in Fig. 7.5, the region with Planck scale curvature in the future being excised. Let us first focus on this region. The Hawking quanta of the quantum field $\hat{\phi}$ are emitted in pairs; one escapes to $\mathscr{I}^+$ and its partner falls into the black hole. The quantum state on a Cauchy surface Σ of the semi-classical portion of space-time

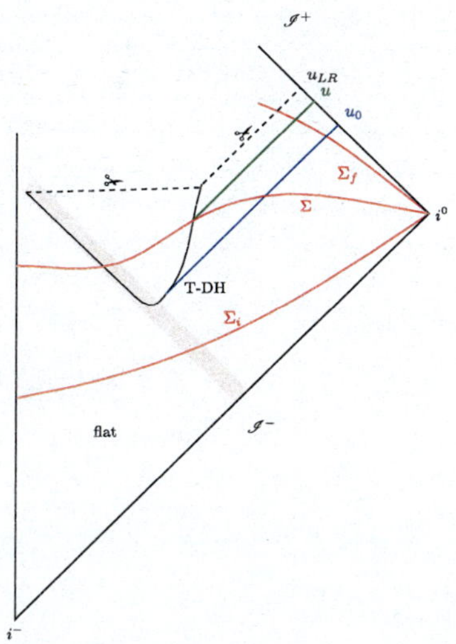

Fig. 7.5 Semiclassical space-time: Black hole is formed by gravitational collapse of a pulse of scalar field, depicted by the (gray) shaded region, incident from $\mathscr{I}^-$. A trapping Dynamical horizon *T-DH* is formed. During the collapse, it is space-like and its area increases (in the outward direction). It becomes time-like during evaporation and its area decreases (in the future direction). Hawking radiation starts in earnest at $u = u_0$. The dashed line with scissors that includes the last ray $u = u_{LR}$ represents the future boundary of the semi-classical region [115]

continues to be pure but there is entanglement between the infalling and outgoing quanta. As for geometry, the space-time metric to the past of the infalling pulse continues to be η_{ab}. But to the future, it is no longer given by the static Schwarzschild solution. The metric is dynamical not only within the pulse but also to its future. Because of its dynamical nature, new structures emerge that are directly relevant to the evaporation process: *dynamical horizons*. These are the dynamical analogs of the trapping and anti-trapping horizons of the quantum corrected Kruskal space-time discussed in Sect. 7.2. They turn out to be more relevant than EHs in discussions of black hole formation and mergers in numerical simulations in classical GR and for the evaporation process in the quantum theory [41–43].

Let us therefore briefly recall this notion. A *dynamical horizon* (*DH*) is a 3-dimensional space-like or time-like submanifold that is foliated by 2-dimensional, surfaces S with 2-sphere topology, such that the expansion of one of the null normals to each leaf S is zero and that of the other null normal is either positive or negative everywhere. Thus, each S is a marginally trapped surface (MTS). In an asymptotically flat space-time, we can distinguish between the two null normals to S. Let us denote by l^a the outgoing null normal and by n^a the ingoing null normal. On a black hole type *DH*, the expansion $\Theta_{(\ell)}$ of the outgoing null normal vanishes (it is positive immediately outside and negative immediately inside the MTS), while the expansion $\Theta_{(n)}$ of the ingoing null normal is negative (both outside and inside). Thus, immediately inside a black hole type *DH*, both expansions are negative and we

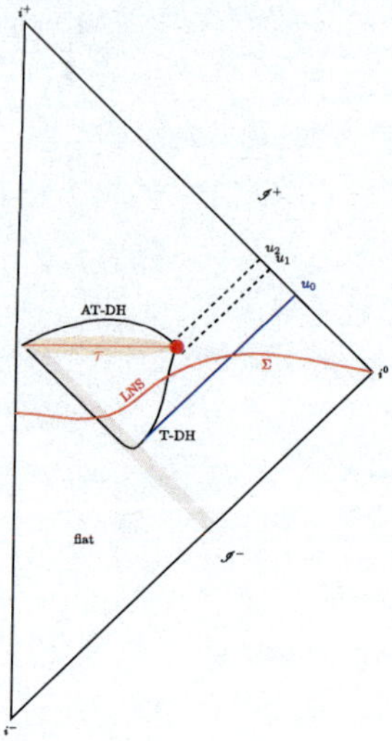

Fig. 7.6 Quantum extension of the space-time in LQG. The classical singularity is replaced by a *Transition surface* τ, to the past of which we have a trapped region, bounded in the past by a trapping dynamical horizon *T-DH*, and to the future of which we have an anti-trapped region bounded by an anti-trapping dynamical horizon *AT-DH*. Cauchy surfaces Σ develop astronomically long necks already in the semi-classical region. The dark (red) blob at the right end of τ is a genuinely quantum region [115]

have a trapped region. Therefore, these *DHs* are called *trapping dynamical horizons, T-DHs*. A *T-DH* is space-like when the area of the MTS increases along the projection of l^a on *DH*, i.e. in the outward direction. In fact, there is an explicit, precise relation between the growth of the area of a *DH* and the flux of energy (carried by matter and/or gravitational waves) flowing *into* it [41]. Thus, not only does the second law of black hole mechanics hold on *T-DHs* but the growth of the horizon area is *directly* related to local physical processes. This is in striking contrast with the situation for *EHs*, where we only have a qualitative statement of growth in classical GR: Area of EHs cannot decrease. Indeed, it is not possible to directly trace the growth back to the infall of energy locally because, as we just saw, *EHs* can form and *grow* in flat space-time where there is nothing at all falling across it.

During the evaporation process, by contrast, the MTSs on the *T-DH* shrink, now in response to the local *negative* energy flux across it, and the *T-DH* is time-like. Recall that in the quantum corrected Kruskal space-time, we also have (white hole type) anti-trapping horizons. But they emerge only when the space-time is extended across the transition surface $\mathscr{T}$ on which curvature is of Plank scale. Therefore one would expect that in dynamical situations, *anti-trapping dynamical horizons AT-DH* would also emerge only when one extends space-time across a transition surface that replaces the classical singularity. This expectation is correct. There is no *AT-DH* in the semi-classical space-time Fig. 7.5 where the region with Planck scale curvature was excised by hand. However, it is present in the quantum extended space-time depicted in Fig. 7.6.

On an *AT-DH* it is the expansion $\Theta_{(n)}$ of the *ingoing* null normal that vanishes and the expansion $\Theta_{(\ell)}$ of the outgoing null normal is positive. Thus, immediately inside these horizons, both expansions are positive: we have an anti-trapped region. Since it is $\Theta_{(n)}$ that vanishes on any *AT-DH*, it is natural to investigate what happens to the area of the MTSs as one moves along the projection of n^a on the *AT-DH*. If the *AT-DH* is space-like, its area *decreases* (now in the inward direction) and if it is time-like its area increases (now in the future direction).

Thus, the key differences between *EHs* and *DHs* can be summarized as follows. First, *EHs* are teleological and can be located only after one has evolved the metric to infinite future. *DHs* by contrast can be located quasi-locally and their properties have direct relation to physical processes at their location. Second, *EHs* are null while *DHs* can be space-like or time-like, and become null only when they become 'isolated' i.e. there is no flux of energy across them. Third, nothing can *ever* escape to the 'exterior' region from the trapped region enclosed by a black hole type *EH* and nothing can *ever* enter the anti-trapped region bounded by a white hole type *EH*. While there are trapped surfaces immediately inside a *T-DH*, one can send causal signals across a *T-DH* from inside to outside (see Fig. 7.5). Similarly, there are future directed causal curves that traverse an *AT-DH* from outside to inside (see Fig. 7.6). Finally, while there is no natural notion of mass and angular momentum for cross-sections of *EHs*, there is one for the canonically defined marginally trapped surfaces on *DHs* which, furthermore lead to the first and second laws of black hole mechanics [41]. Discussions of quantum dynamics in LQG focus on *DHs* . Much of the confusion about the evaporation process and 'purification' of the quantum state melts away once EHs are deemphasized.

7.5.2 Black Hole Evaporation in LQG

LQG investigations of the semi-classical part of the quantum corrected space-time are based on Fig. 7.5 and, although some of the detailed calculations are still in progress, the overall understanding of structures in this space-time is quite satisfactory at a conceptual level. To understand the structure of the future of this region one needs full quantum gravity and, as in every other approach, several questions remain. But there is a general consensus on a majority of issues. In this subsection we summarize this status.

Semi-classical Regime: Consider a coherent state Ψ of a quantum scalar field $\hat{\phi}$, peaked around an infalling classical pulse on $\mathscr{I}^-$ and undergoing a prompt collapse. Let us suppose that the ADM energy in the incoming state is of a solar mass, $M_\odot$. When the radius of the pulse has become sufficiently small, a trapping dynamical horizon *T-DH* forms. In classical GR, this *T-DH* would only have a space-like component that grows from zero radius till it has radius of 3km and then joins on to the null event horizon of the same radius. Once the Hawking radiation starts and the back reaction is included, the black hole shrinks. Initially the process is very slow because the ingoing negative energy flux is *extremely* small. It takes some 10^{64} years for the black hole to shrink to lunar mass $M_{\rm moon}$. However, even at the end of this long, adiabatic process, the black hole is *macroscopic*. Therefore, from our discussion in Sect. 7.2 one would expect the quantum gravity corrections to be sufficiently small for semi-classical considerations to suffice. Let us focus on this phase of evaporation. In this phase, dynamics should be well-described by equations

$$G^{sc}_{ab} = 8\pi G_{N} \langle \hat{T}_{ab} \rangle_{ren} \quad \text{and} \quad \Box\hat{\phi} = 0\,, \tag{7.46}$$

where G^{sc}_{ab} is the Einstein tensor of the semi-classical metric g^{sc}_{ab} and the expectation value of the renormalized stress-energy tensor is computed using the Heisenberg state Ψ. The metric g^{sc}_{ab} does include quantum corrections but they are induced by quantum matter (rather than being dictated by the area gap considerations of quantum geometry).

These corrections to geometry are adiabatic and small. But the infalling negative energy flux introduces a qualitative difference in the horizon structure: Now the expanding, space-like branch *T-DH* of the dynamical horizon joins on, not to a null event horizon as in the classical case, but to the outer, *time-like* branch whose area *decreases* to the future due to the negative energy flux carried by the Hawking 'infalling partner modes'. These two branches of the *T-DH* serve as the past boundary of a trapped region of the semi-classical space-time (M_{sc}, g^{sc}_{ab}) of Fig. 7.5. During this long adiabatic process of $\sim 10^{64}$ years, pairs of Hawking quanta are continually created, one going to $\mathscr{I}^+$ and its partner falling into the trapped region. These modes will be entangled whence, if one uses the usual observable algebra based just at $\mathscr{I}^+$, the state would seem mixed, close to a thermal state.

The issue of unitarity leads one to ask: When will the correlations be restored? For this to happen, the partner modes would have to emerge from the trapped region and propagate outward, restoring correlations at $\mathscr{I}^+$ and 'purifying' the state there. Since the outer part of the boundary *T-DH* of the trapped region is time-like, there is no causal obstruction for these modes to continuously exit the trapped region across *T-DH* throughout the long evaporation process; the standard causal obstructions associated with EHs do not apply. This fact has been used in the literature to argue that there is no information loss issue at all [116, 117]; purification could have occurred all along the evaporation process. But this seemingly easy explanation is flawed. Examination of the renormalized energy flux shows that throughout this process, there is only infall across *T-DH* in semi-classical gravity. Thus, the lack of a causal obstruction for the partner modes to exit the trapped region is not sufficient for the purification to occur in the semi-classical space-time.

In fact there is an apparent puzzle associated with the issue of information loss that is already relevant in the semi-classical regime. Since $M_{moon} \sim 10^{-7}\, M_{\odot}$, at the end of this long evaporation process most of the initial ADM mass is carried away to $\mathscr{I}^+$ by the Hawking quanta. A back of the envelope calculation shows that a very large number $\mathscr{N}$ ($\sim 10^{75}$) of quanta escape to $\mathscr{I}^+$ and all of them are correlated with the ones that fell across *T-DH* . Therefore, at the end of the semi-classical process under consideration, one would have to have a huge number $\mathscr{N}$ of quanta both at $\mathscr{I}^+$ and in the trapped region, but the mass associated with the trapped region is only 10^{-7} times that carried away by the N quanta going out to $\mathscr{I}^+$. Furthermore, the radius of the outer part of *T-DH* has shrunk to only 0.1 mm—the Schwarzschild radius of a lunar mass black hole. How can a *T-DH* with just a 0.1mm radius accommodate all these $\mathscr{N}$ quote? Even if we allowed each mode to have the (apparently maximum) wavelength of 0.1 mm, heuristically one would need the horizon to have a huge mass—some 10^{22} times the lunar mass! While these considerations are quite heuristic, one needs to face the conceptual tension: At the end of the process under consideration, the trapped region has simply too many quanta to accommodate, with a tiny energy budget. Such considerations have led to suggestions that somehow 'purification' must begin already by Page time [118] when the *T-DH* has lost only half its original mass of $M_\odot$, and essentially completed by the time the *T-DH* has shrunk down to the lunar mass. But this would imply that semi-classical considerations must fail in apparently tame regimes due to unforeseen quantum gravity effects that are relevant outside the horizons of astrophysical black holes! As we discussed in previous sections, in LQG quantum gravity corrections are completely negligible near horizons of macroscopic black holes.

The way out of the apparent paradox is that semi-classical theory itself predicts that the geometry of the trapped region has some rather extraordinary features that had not been noticed until relatively recently and not fully appreciated by the wider community even now. Calculations of the stress-energy tensor on the Schwarzschild space-times confirm the idea that, in semi-classical gravity there is a negative energy flux across the time-like portion of *T-DH* such that $M_{T\text{-}DH}$ would decrease according to the standard Hawking formula: $\mathrm{d}M_{T\text{-}DH}/\mathrm{d}v = -\hbar/(GM_{T\text{-}DH})^2$. (Indeed, this has been the basis of the standard estimate that the evaporation time goes as $\sim M_{\mathrm{ADM}}^3$.) One can then argue that, in the phase of evaporation from the solar mass to the lunar mass, the form of the space-time metric in the trapped region of Fig. 7.5 is well approximated by the Vaidya metric:

$$\mathrm{d}s^2 = -\big(1 - \frac{2m(v)}{r}\big)\mathrm{d}v^2 + 2\mathrm{d}v\mathrm{d}r + r^2\big(\mathrm{d}\theta^2 + \sin^2\theta\,\mathrm{d}\varphi^2\big), \tag{7.47}$$

with $m(v) = GM_{T\text{-}DH}(v)$ decreasing *very* slowly. As we saw in Sect. 7.2, corrections due to quantum geometry are completely negligible in Schwarzschild interior until one reaches Planck curvature and, as Fig. 7.5 shows, that region is excluded in the semi-classical space-time under consideration. Thus, in the metric (7.47) quantum corrections are all induced by quantum matter and encoded in $m(v)$. To probe the geometry of the trapped region, it is convenient to foliate it and two natural foliations have been used by the LQG community: One defined by constancy of the Kretchmann scalar and the other by constancy of the radius of metric 2-spheres [30, 114]. Each space-like slice is topologically $\mathbb{S}^2 \times \mathbb{R}$ and is itself foliated by round 2-spheres which can be labelled by values of the advanced time coordinate v. Let us set $v = 0$ when $M_{T\text{-}DH} = 1M_\odot$ and $v = v_0$ when $M_{T\text{-}DH} = M_{\mathrm{moon}}$. During this long process, the radius of the MTSs on the outer, time-like part of *AT-DH* decreases from $r|_{v=0} = 3$km to $r|_{v=v_0} = 0.1$mm. The surprising fact is that as v increases the leaves develop longer and longer necks of length ℓ_N along the $\mathbb{R}$ directions [30, 119, 120]. The 'final leaf' for the process under consideration starts at the right end with $v = v_0$. The length ℓ_N of this final leaf is *astonishingly large*: $\ell_N \approx 10^{64}$ *light years* for the first foliation and $\ell_N \approx 10^{62}$ *light years* for the second! These astronomically large lengths can result because the time the process takes is huge; 10^{64}years corresponds to $\sim 10^{53}$ times our cosmic history!

This enormous stretching is analogous to expansion in (an anisotropic) cosmology. Recall that during the cosmic expansion—e.g. during inflation—the wavelengths of modes get stretched enormously. This suggests that partner modes that fall into the trapped region will also get enormously stretched during evolution from $v = 0$ to $v = v_0$, as in quantum field theory on an expanding cosmological space-time, and become infrared. Can this phenomenon resolve the quandary of 'so many quanta with so little energy'? The answer is in the affirmative. With such infrared wavelengths, it is easy to accommodate them in the trapped region with the energy budget only of $M_{\rm moon}$. Thus, even though the outgoing modes carry away almost all of the initial mass $M_\odot$ to $\mathscr{I}^+$, there is no obstruction to housing all their partners in the trapped region on a slice Σ of Fig. 7.5 with the small energy budget of just $10^{-7} M_\odot$. This argument removes the necessity of starting purification by Page time. In the LQG perspective, purification can be postponed to a much later stage.

To summarize, in the semi-classical regime, there are apparent paradoxes associated with the process of 'purification' that is necessary for dynamics to be described by a unitary process. These disappear when one shifts the focus from event horizons to trapping dynamical horizons *and* takes into account the time evolving geometry of the trapped region. By and large the LQG community has adopted this view.

Beyond the semi-classical regime: When do quantum geometry effects become significant making the semi-classical approximation inadequate? The viewpoint in LQG is that this happens when physically observable quantities such as curvature scalars and matter density enter the Planck regime. This expectation was borne out in the investigation of the Schwarzschild interior in Sect. 7.2. Therefore, one would expect semi-classical considerations to be valid well beyond the time when *T-DH* has shrunk to $M_{T\text{-}DH} = M_{\rm moon}$ we considered in the above discussion, all the way till the curvature is, say, 10^{-6} times the Planck curvature which corresponds to $M_{T\text{-}DH} \approx 10^3 M_{\rm Pl}$. LQG explorations of the evaporation process beyond this stage are being carried out by different groups. The main ingredients are: results on causal structure of the Schwarzschild interior summarized in Sect. 7.2, intuition derived from simpler models such as CGHS [121], conclusions drawn from a long series of works (see, e.g., [122–131]) that posit a space-time structure for the entire process and work out its consequences, strong consistency requirements on the ensuing space-time geometry (see, e.g., [96]), and calculations based on the Vaidya metric for the structure of space-time in the *distant* future [127, 132]. While there is broad consensus on the overall picture, many open issues remain. We will now summarize the current status.

To the future of the semi-classical region, curvature can exceed $10^{-6}\ell_{\rm Pl}^{-2}$, whence we need full quantum gravity. This region with Planck scale curvature is depicted by the shaded (pink) region in Fig. 7.6. (To the past and future of this region, semi-classical gravity should yield a reasonable approximation.) In the shaded (pink) region geometry is described by a quantum state $\Psi_{\rm geo}$ and the difficult task is to evolve the quantum field $\hat{\phi}$ on this quantum geometry. Fortunately, prior experience with other systems—such as the propagation cosmological perturbations on the quantum FLRW geometry—suggests a strategy that is applicable during the adiabatic phase of the Planck regime. The evaporation process is adiabatic so long as mass-loss does not occur too rapidly, i.e., until the radius of *T-DH* is $\approx 10^3 \ell_{\rm Pl}$. At the end of this process, one enters a neighborhood of the future endpoint of the *T-DH* depicted by a (red) blob, where the curvature is Planckian *and* the process speeds up very rapidly, violating the adiabatic approximation. Let us first discuss the adiabatic phase and then return to the (red) blob.

Prior results in LQC strongly suggest that the problem of propagating a quantum field $\hat{\phi}$ on a quantum geometry represented by $\Psi_{\rm geo}$ can be greatly simplified during the adiabatic phase: One can construct a smooth $\tilde{g}_{ab}$ that carries all the information in $\Psi_{\rm geo}$ that the dynamics of quantum fields $\hat{\phi}$ is sensitive to (see, e.g., [133]). $\tilde{g}_{ab}$ is called the *dressed metric*. Thus, the difficult task of evolving quantum fields on quantum geometry is reduced to that of evolving them on the space-time of the dressed metric $\tilde{g}_{ab}$. Next, the expectation from results of Sect. 7.2 is that the shaded (pink) region will contain a transition surface $\mathscr{T}$ (w.r.t. $\tilde{g}_{ab}$) that replaces the classical singularity and separates the trapped region that lies to its past and the untrapped region that lies to its future. The metric $\tilde{g}_{ab}$ will capture two distinct effects: those that originate from quantum geometry and feature the area gap Δ (as in Sect. 7.2), and those that are induced on $\tilde{g}_{ab}$ by the falling quantum matter, dominated by the incident pulse of the scalar field at the left end of the (pink) shaded region, and by the infalling Hawking quanta carrying negative energy as one moves towards the right end.

Discussion of Sect. 7.2 strongly suggests that the first set of effects will decay rapidly as we move away from Planck curvature into the semi-classical region. Therefore in the semi-classical region, $\tilde{g}_{ab}$ will be well approximated by $g^{\rm sc}_{ab}$ used there. As we move to the future of the (pink) shaded region, one would encounter an anti-trapping dynamical horizon *AT-DH* (see Fig. 7.6). The region enclosed by the transition surface $\mathscr{T}$ to the past and *AT-DH* to the future would be anti-trapped as in Fig. 7.2. But now *AT-DH* would be space-like rather than null and its area would not be constant, but decrease as one moves left. Qualitatively this change in the structure of *AT-DH* from the one of Fig. 7.2 is parallel to the change in the structure of the trapping horizon *T-DH* that we already discussed in some detail.

Finally, the region to the future of *AT-DH* would also be well approximated by a Vaidya metric, but now the outgoing one, expressed in terms of the retarded time coordinate u in place of the advanced time coordinate v of Eq. (7.47). It will describe the propagation of the infrared modes that will emerge from the *AT-DH* and arrive at $\mathscr{I}^+$ at very late times. (The metric in this region will be nearly flat because the total energy in the scalar field is small and dispersed over very large spatial regions.) Recall that these are the partner modes that fell into the horizon and were therefore entangled with the outgoing modes that carried away most of the initial ADM mass. In the LQG scenario, then, correlations are finally restored at $\mathscr{I}^+$ where, in the end, the infalling modes also arrive. The total energy carried by the two sets of modes is very different. But this is not an obstruction for restoring correlations, i.e., for the 'purification' to occur. The timescale of this purification process is very long, $O(M^4)$ [127, 130, 132]. Purification can occur much later than the Page time because of the LQG singularity resolution.

The final picture is rather similar to the process of burning a piece of coal that is often invoked in the discussion of black hole evaporation. Initially the piece of coal is in a pure state. When lit, it emits photons and the energy they carry is well described by a thermal state at high temperature. But the total state is pure because there are correlations between the quantum state of outgoing photons and the left-over coal. As the fire extinguishes, there is very little energy left and the ashes emits photons with lower and lower frequencies for a very long time. At the end of the process the cold ashes are in a pure state and all photons have escaped. The late time, long wavelength photons are able to restore the correlations that were apparently lost in the middle of the process (when the photon spectrum seemed approximately thermal) even though the total energy they carry is small compared to the energy carried by high frequency photons that were emitted earlier.

Note, however, that this scenario is incomplete because one still has to deal with the very last part of the evaporation process, depicted by the red blob and the associated null rays $u = u_1$ and $u = u_2$ of Fig. 7.6. In this region, not only is the curvature of Planck scale, but it is varying extremely rapidly because it lies at the end point of the evaporation process. It is this combination that makes the problem difficult; if we had only one of these features, we could have used known approximation methods. Independent considerations suggest that something very non-trivial must happen in this region. We will conclude with an example. During the semi-classical phase, as the *T-DH* shrinks, the temperature associated with the radiation at $\mathscr{I}^+$ grows. Consequently, the modes become increasingly ultraviolet as one approaches the point $u = u_1$ on $\mathscr{I}^+$. On the other hand, the radiation that emerges from the anti-trapped region is infrared and received at $\mathscr{I}^+$ to the future of $u = u_2$. This is a dramatic transition, strongly suggesting that the physics of the region which is highly dynamical *and* has Planck scale curvature will be very subtle and interesting. For example, it has been suggested that Planck scale 'seeds' may be left behind, scattered in this region [134]. Understanding the nature of this quantum geometry remains an attractive challenge in LQG.

Let us summarize the current status of LQG investigations of black-evaporation. They are distinguished by their emphasis on two features that are generally ignored in other approaches: (i) A shift away from the teleological event horizons *EH* to quasi-locally defined trapping and anti-trapping DHs *T-DH* and *AT-DH* ; and, (ii) replacement of the classical singularity by the transition surface $\mathscr{T}$. As a consequence, the traditionally used Penrose diagram of Fig. 7.4 is replaced by the Penrose diagram of Fig. 7.6.

7.6 Discussion

As the bibliography indicates, LQG literature on regular black holes is very rich. Indeed, even this long list is far from being exhaustive! To make the material accessible to non-experts, we focused on four lines along which advances have occurred, and in each case built the discussion around a few of the mainstream developments. Much of this discussion is based on effective equations, motivated by the fact that high performance computations have shown that effective space-time metrics provide an excellent approximations to the quantum geometry in LQC, also in Bianchi models where the Weyl curvature is non-zero and diverges at the singularity [135, 136].

The first area, discussed in Sect. 7.2, focuses on the 'Schwarzschild interior' that contains the singularity. Since the resolution of this singularity is central to the theme of 'regular black holes' of this Volume, we included an account of several different effective descriptions. These investigations bring out two features: (i) singularity resolution due to the underlying quantum geometry effects of LQG is robust, and does not depend on details of the quantization methods; but, (ii) the precise manner in which quantization is carried out can unleash unintended and physically undesirable effects that are not apparent until a detailed examination is carried out. We summarized a scheme that is free of these drawbacks. The resulting quantum corrected geometry exhibits interesting causal structures: the singularity is replaced by a transition surface, $\mathscr{T}$, to the past of which there is a trapped region, and to the future, an anti-trapped region. Each is bounded by null horizons and, for macroscopic black holes, the area of the future horizon is approximately equal to that of the past (see Fig. 7.2). In each of these effective geometries, curvature scalars attain their maxima at the transition surface which, furthermore, have *universal* values, independent of the mass of the black hole. This universality seems to be a general feature of the singularity resolution due to quantum geometry effects of LQG.

Section 7.3 extended the quantum corrected geometry of Sect. 7.2 to the exterior, asymptotic region of the Schwarzschild space-time by exploring the homogeneity of time-like surfaces $r_{\rm sch} = \text{const}$. For macroscopic black holes (i.e., those with $m \gg \ell_{\rm Pl}$), the near horizon geometry of this exterior has the expected and physically desired features: the quantum corrected, effective metric is smooth across the horizon and corrections to the Hawking temperature, computed using methods from Euclidean quantum field theory, are tiny. More generally, for macroscopic black holes there is excellent agreement between the effective geometry and that of the classical Schwarzschild metric in a vast neighborhood of the horizon in the exterior region. Unfortunately, there is some confusion in the literature on this point arising from the simplified form (7.33) of the metric that holds in the far-asymptotic region, i.e., *only on ignoring terms* $O(r_s/r)$. If one overlooks this key approximation and uses (7.33) in the entire exterior region—as was done in [137]—one obtains "unsettling features", such as non-trivial corrections to the innermost circular orbit. These are consequences not of the actual effective geometry, but of the incorrect use of its simplified form. (Nonetheless, unfortunately, incorrect conclusions of [137] have been repeated in some of the subsequent literature, e.g. [138]). More generally, the near horizon quantum corrections to astrophysical black holes will be very small to have observable relevance in the foreseeable future (at least in the non-rotating case on which most LQG investigations have focused so far).[4]

The full metric in the exterior region is also asymptotically flat with curvature decay that is sufficient for the ADM mass to be well defined (e.g., if one uses the expression (7.34) in terms of the spatial Ricci tensor). For macroscopic black holes, quantum corrections to the classical value are very small. However, the decay is slower than that in the standard notion of asymptotic flatness. Consequently, different expressions of the ADM mass, that must agree with one another exactly if the standard asymptotic conditions hold [55], now differ by quantum corrections. Much more surprising is the feature that the norm of the time-translation Killing field of the effective metric diverges at spatial infinity! One's first reaction would be that such deviations from standard asymptotic flatness must lead to a plethora of physically inadmissible consequences. One test is provided by quasi-normal modes. Do they exhibit a pathological behavior? A detailed investigation [142] has shown that the potential which enters the quasi-normal mode analysis continues to be well-defined everywhere. One can then compute quasi-normal frequencies using an approximation tailored to improving accuracy. The corrections to the classical result are found to be negligibly small. An independent investigation [143] provided expressions for axial and polar perturbations, computed their quasi-normal frequencies and found departures with respect to the classical theory; in particular, isospectrality is broken. However, all these relative deviations from the classical predictions are only a small-percent effect even for black holes as small as $r_S \sim 10^3 \ell_{\rm Pl}$, and they decrease with the mass of the black hole, becoming completely negligible for macroscopic black holes. These investigations also show that the metric passes the stability criterion for tensor and massless scalar field perturbations. However, the infrared behavior of the potential is different from that in the classical Schwarzschild case, leading to a qualitative difference in the power-law tails. These tails play an important role in the mathematical literature but are not astrophysical significant because they occur after the waves are exponentially damped in the quasi-normal ringing phase. In summary, at present it is not clear whether counter-intuitive features associated with the asymptotic behavior of the effective metric of [29] are indications that it may be inadmissible in the asymptotic region $r_S/r \ll 1$, or if they are physically harmless. In view of this uncertainty, several investigations are exploring alternate ways of

---

[4] In this review, we did not touch on the issue of black hole entropy that arises in LQG by counting microstates of the area operator that are compatible with parameters characterizing a given macroscopic black hole (see. e.g., [139]. The possibility of testing discreteness of area using gravitational waves has drawn considerable attention in the literature. It has been argued that the simplest area spectrum with area eigenvalues given by $kn\, \ell_{\rm Pl}^2$ (where n is an integer and k a constant), considered by Bekenstein and Mukhanov [140], could be ruled out using data from a sufficiently large number of compact binary mergers. But in LQG the area spectrum is *not* equidistant, it crowds exponentially, making the continuum an excellent approximation very quickly. However, for small black holes the area eigenvalues are grouped, exhibiting a band structure, and the separation between bands is $O(\ell_{\rm Pl}^2)$. If this structure were to persist for large rotating black holes, each band would serve as a proxy of the Bekenstein-Mukhanov eigenvalues and gravitational observations would then lead to non-trivial constraints [141]. However, currently there is no evidence that points to the persistence of bands for macroscopic areas.

arriving at an effective metric that has the standard asymptotic behavior (see, e.g., [33, 38]).

Another conceptual issue concerns covariance. There *is* a 4-metric in the full quantum extended Kruskal space-time and results on singularity resolution, for example, refer to curvature invariants; these considerations are all 4-dimensionally covariant. But to arrive at the effective Einstein's equations with quantum geometry modifications, one uses symmetry reduction. The question is whether there is a covariant action for the full theory *without symmetry reduction* whose equations of motion reduce to those in Sects. 7.2 and 7.3 in its static, spherically symmetric sector [144]. This is a technically difficult issue that is still open. Indeed, it took some time to show that the much simpler effective equations of the homogeneous, isotropic sector of LQC [29] can be obtained in this way, but finally the answer turned out to be in the affirmative [145]. A similar situation arose also in string theory where it seemed for quite some time that the exact 1+1 dimensional stringy black hole [146] did not arise from the symmetry reduction of a covariant action [147]. In the end, it was shown that there is such an action but it requires inclusion of additional fields [148]. There are some concrete indications that the situation is likely to be similar in the LQG black hole sector we discussed; see e.g., [38] that introduces a covariant action in the 'mimetic gravity' setting that adds a scalar field with a specific potential and uses the same time-like homogeneous slices as in Sect. 7.3 in the symmetry reduced sector, and [84] that uses a reasoning based on the constraint algebra to argue for covariance.

Discussion in Sects. 7.2 and 7.3 was confined to the LQG treatment of the eternal black hole and arrived at the Penrose diagram of Fig. 7.3 for the quantum extension of the Kruskal space-time. This entire space-time is non-singular. In classical relativity as well as in the discussion of black hole evaporation, Kruskal space-time of Fig. 7.1 provides useful mathematical tools as well as physical intuition. The same is true of its quantum extension. However, realistic and more interesting situations involve formation of black holes by gravitational collapse (for which only a part of the full Kruskal space-time is relevant). In Sect. 7.4 we focused on two complementary issues that have been investigated in dynamical situations featuring gravitational collapse. The first involves the resolution of singularity for collapsing dust models. Here, the emphasis is on quantum geometry effects because the matter is characterized by dust rather than a fundamental quantum field. Thus, non-classical features associated with matter—such as boundedness of the dust density—are induced on matter by the quantum nature of geometry. These investigations show that, as in cosmology, the singularity is replaced by a bounce; this is a robust result. The second class of investigations focuses on critical phenomena. Now the strategy is the opposite in that it is matter that is represented by a quantum scalar field of LQG [100] while geometry is classical to begin with; corrections to classical effects on geometry are induced by quantum matter through field equations. The overall finding is that the quantum corrections to the classical results are small for macroscopic black holes, just as one would hope. However, while there is no 'mass-gap' in the classical theory—i.e. a black hole can be formed with arbitrarily small mass—a mass gap can develop if one uses a quantization scheme that leads to effective equations violating scale invariance.

While dynamics is at the heart of these investigations, they do not encompass the Hawking process because, in the first set of analyses matter is classical, and the second focuses on critical behavior in gravitational collapse, rather than on the scalar field quanta going out to $\mathscr{I}^+$, or the issue of entanglement. LQG investigations of the evaporation process—including the issue of back reaction on geometry—were discussed in Sect. 7.5. They reflect a broad consensus that the arguments that lead to the traditional Penrose diagram of Fig. 7.4 are flawed in two important respects.

First, they assume that a part of classical singularity persists in the quantum theory while it is resolved in LQG. Second, the event horizon plays a key role in Fig. 7.4 even though it is teleological and can be made to disappear by changing space-time geometry in a Planck scale neighborhood of the singularity [113]. The traditional Penrose diagram is replaced by a new LQG Penrose diagram shown in Fig. 7.6. There is consensus that there is no information loss: The S-matrix from $\mathscr{I}^-$ to $\mathscr{I}^+$ is unitary provided, of course, we consider a closed system in which the black hole forms by the gravitational collapse of a quantum field from $\mathscr{I}^-$ and we use the quantum state of the same field at $\mathscr{I}^+$. That the singularity would be resolved by quantum geometry effects is motivated by two considerations: (i) Quantum geometry effects discussed in Sect. 7.2 that provide universal upper bounds to curvature scalars because of a non-zero value of the area gap; and, (ii) detailed numerical simulations in the CGHS case that show that even in the semi-classical theory, the singularity is significantly weakened when back reaction effects are included, which already suffice to make the metric continuous there [149]. There is no *EH* in the final picture; *what forms classically in the gravitational collapse and evaporates through quantum processes is a DH.*

The evaporation process of LQG can be described as follows. The Hawking quanta are created in pairs, the outgoing quanta go out to $\mathscr{I}^+$ as in Hawking's original paradigm, and their partner quanta fall across the trapping dynamical horizon *T-DH*. In the semi-classical regime depicted in Fig. 7.5, the outgoing quantum state is well approximated by a thermal state at $\mathscr{I}^+$ (at sufficiently late times), and the partner modes carry a negative energy flux into the trapped region that is bounded by *T-DH* in this figure. When the back reaction is included, the geometry in the trapped region changes adiabatically, and space-like surfaces Σ of Fig. 7.6 get stretched and become long necked surfaces (*LNS*). Suppose that at its formation, the black hole has solar mass $M_\odot$. Although the process of elongation of necks is very slow, it continues for a very long time since the semi-classical phase lasts some 10^{64} years. At the end of this phase, the necks become astronomically long, stretched to some 10^{62} light years! Therefore the modes that have fallen in the trapped region also get enormously stretched (as they do during inflation) and become infrared. They can continue to be entangled with their partner modes that went out to $\mathscr{I}^+$ during this long semi-classical phase—even though the total energy carried by the outgoing modes is almost $M_\odot$ and that carried by the trapped modes is tiny—precisely because the trapped modes are infrared. Thus, the quantum state on a surface such as Σ continues to be pure. Since the singularity is resolved and replaced by a transition surface $\mathscr{T}$ that lies in the shaded (pink) region, these modes can evolve across the transition surface $\mathscr{T}$ and emerge on the other side. Then they propagate to the approximately flat region that lies to the future of the anti-trapped dynamical horizon *AT-DH*, and arrive at $\mathscr{I}^+$ restoring the correlations with the partner modes that reached $\mathscr{I}^+$ much earlier, during the semi-classical phase. (As explained towards the end of Sect. 7.5.2, this situation is qualitatively similar to that of burning a piece of coal where correlations are restored at late times when the large wavelength modes emerge from ashes as they cool down, restoring correlations with short wavelength modes emitted earlier.)

What remains largely unexplored so far is the (red) 'blob' at the right end to the shaded (pink) region in Fig. 7.6 and how it affects the physics at $\mathscr{I}^+$. As discussed at the end of Sect. 7.5, the problem is hard because one simultaneously encounters two difficulties: Planck scale curvature and rapid changes that make adiabatic approximation inadequate. But feasible calculations may suffice to reveal whether most, if not all, of the correlations are restored when the infrared modes traverse the anti-trapped region and emerge at $\mathscr{I}^+$. If they are restored, then the fully quantum 'blob' would not be that relevant for the issue of information loss and the S-matrix would be unitary. Although the consensus in LQG favors this possibility, this issue is open. There *are* arguments involving a fully quantum evolution from past of the blob to its future, but so far they are inconclusive because of the underlying assumptions. At this stage, one cannot, for example, rule out the possibility that the 'blob' joins on to a baby universe whose states are inaccessible from $\mathscr{I}^+$ of Fig. 7.6. If this were ti happen, from the perspective of $\mathscr{I}^\pm$ of this figure, information may be lost, although the 'total' S-matrix would be unitary. Showing that this does not happen remains a fascinating challenge in the LQG community.

Let us summarize. The LQG community has explored different aspects of the many fascinating properties of black holes. The distinguishing feature of these investigations is their emphasis on quantum geometry that is directly responsible for replacement of the singularity by a transition surface with interesting causal properties, and boundedness of physical observables such as curvature scalars and matter density. As discussed in Sect. 7.1, these features are not shared by other approaches: Using the AdS/CFT correspondence as motivation, it is sometimes argued that singularities should persist also in quantum gravity, and indeed, much of the literature uses the Penrose diagram of Fig. 7.4 in which a singularity features as part of the future boundary of space-time. On the other hand, because quantum geometry effects become important only in the Planck regime, LQG corrections to the classical results are very small near the horizons of astrophysical black holes; examples we discussed include corrections to the Hawking temperature using the near horizon geometry and the machinery of Euclidean quantum field theory, corrections to quasi-normal frequencies of astrophysical black holes, and to results associated with critical collapse. This is also in striking contrast to some other approaches, e.g., the 'firewall scenario' that emerged from string theory considerations. More generally, LQG does not lead to violations of semi-classical expectations of physics near horizons of astrophysical black holes that had been advocated before the LIGO/Virgo discoveries showed that predictions of classical GR, without such major corrections, are realized in compact binary mergers. As mentioned in Sect. 7.5.2, there is also a large body of investigations that posit a space-time structure for the entire process and work out its consequences. By and large one solves *classical* Einstein's equations (with suitable stress-energy tensors) in various patches, and joins them consistently. Some of these space-time diagrams resemble Fig. 7.6. While these investigations do pay close attention to consistency conditions, and often also to energy considerations, the issue of *quantum correlations and unitarity* has received little attention in these works. The LQG line of reasoning of Sect. 7.5 fills this conceptually important gap that is key to the issue of 'information loss'.

There is also considerable discussion on the issue of young versus old black holes, and long lived remnants. In LQG, there is indeed an important difference between a young and an old black hole. As a concrete example, let us consider two lunar mass black holes—a young one that is freshly formed from gravitational collapse, and an old one what started out as a solar mass black hole and then evaporated down to the lunar mass, as discussed in Sect. 7.5.2. While their dynamical horizons will have the same radius, 0.1mm, and mass $M_{T\text{-}DH} = M_{\text{Moon}}$, their external environment as well as internal structure will be *very* different. In the second case, the evaporation process would have gone on for some 10^{64} years. Therefore, there will be a very large number of outgoing Hawking quanta in the exterior region, and an equal number of ingoing quanta in the trapped region, the two being entangled. Therefore, the small area of *T-DH* will not be a measure of the entropy of what is in the interior (or exterior). However, in LQG, in both cases the area is a measure of the surface degrees of freedom of the horizon, i.e., degrees of freedom that can communicate both the outside and inside regions. But it is sometimes argued that there is a potential problem with this scenario: because old black holes can have small energy but an enormous number of modes, it should be easy to produce them in particle accelerators. But these arguments seem to take into account only the conservation laws normally used in computing scattering amplitudes; since old black holes have astronomically long necks, it is hard to imagine how such changes in space-time structure can occur on time scales of accelerator physics [150].

Finally, let us discuss some of the limitations of the current LQG investigations. The analyses we summarized make a strong use of symmetry reduced models and effective equations that capture the leading order quantum corrections. There *have been* a number of interesting investigations that aim at arriving at these effective equations starting from full LQG (see, e.g., [151, 152]). But they are still in a rather preliminary stage, and further and more detailed investigations are needed. Another key limitation is that so far the LQG investigations have focused primarily on non-rotating black holes, where the classical singularity is space-like. But for rotating black holes the inner horizons would be unstable and therefore the singularity would be null. So far quantum geometry considerations have not been applied to null singularities. This is an outstanding open problem. Indeed, inclusion of rotating black holes in the discussion of 'information loss' during evaporation remains a fascinating problem in *all* approaches to quantum gravity.

Acknowledgements This work was supported in part by the NSF grant PHY-1806356, PHY-1912274 and PHY-2110207, Penn State research funds associated with the Eberly Chair and Atherton professorship, and by Projects PID2020-118159GB-C43, PID2019-105943GB-I00 (with FEDER contribution), by the Spanish Government, and also by the "Operative Program FEDER2014-2020 Junta de Andalucía-Consejería de Economía y Conocimiento" under project E-FQM-262-UGR18 by Universidad de Granada. We would like to thank Eugenio Bianchi, Kristina Giesel, Muxin Han, Bao-Fei Li, Guillermo Mena, Sahil Saini and Ed Wilson-Ewing for discussions, and Tommaso De Lorenzo for Figures 4–6.

References

1. N. Englehardt, G.T. Horowitz, Int. J. Mod. Phys. D **25**, 1643002 (2016)
2. S.W. Hawking, [erratum: Commun. Math. Phys. **46**, 206 (1976)]. https://doi.org/10.1007/BF02345020
3. S.W. Hawking, M.J. Perry, A. Strominger **116**, 231301 (2016)
4. W.G. Unruh, R.M. Wald, Information loss. Rep. Prog. Phys. **80**, 092002 (2017)
5. A. Almheiri, D. Marolf, J. Polchinski, D. Stanford, J. Sully, JHEP **09**, 018 (2013)
6. S. Giddings, Phys. Lett. B **738**, 92 (2014)
7. D. Marolf, Rep. Prog. Phys. **80**, 092001 (2017)
8. A. Ashtekar, J. Pullin (eds.), *Loop Quantum Gravity: The First 30 Years* (World Scientific, Singapore, 2017)
9. A. Ashtekar, P. Singh, Class. Quantum Gravity **28**, 213001 (2011)
10. P. Singh, Class. Quantum Gravity **26**, 125005 (2009)
11. I. Agullo, P. Singh, in *Loop Quantum Gravity: The First 30 Years*, ed. by A. Ashtekar, J. Pullin (World Scientific, Singapore, 2017). arXiv: 1612.01236
12. A. Ashtekar, M. Bojowald, Class. Quantum Gravity **23**, 391 (2006)
13. L. Modesto, Class. Quantum Gravity **23**, 5587 (2006)
14. M. Campiglia, R. Gambini, J. Pullin, AIP Conf. Proc. **977**, 52–63 (2008)
15. R. Gambini, J. Pullin, Phys. Rev. Lett. **110**, 211301 (2013)
16. D. Cartin, G. Khanna, Phys. Rev. D **73**, 104009 (2006)
17. C.G. Boehmer, K. Vandersloot, Phys. Rev. D **76**, 1004030 (2007)
18. D.W. Chiou, Phys. Rev. D **78**, 064040 (2008)
19. A. Corichi, P. Singh, Class. Quantum Gravity **33**, 055006 (2016)
20. J. Olmedo, S. Saini, P. Singh, Class. Quantum Gravity **34**, 225011 (2017)
21. J. Cortez, W. Cuervo, H.A. Morales-Técotl, J.C. Ruelas, Phys. Rev. D **95**, 064041 (2017)
22. A. Yonika, G. Khanna, P. Singh, Class. Quantum Gravity **35**, 045007 (2018)
23. A. Joe, P. Singh, Class. Quantum Gravity **32**, 015009 (2015)
24. D.-W. Chiou, Phys. Rev. D **78**, 044019 (2008)
25. J. Brannlund, S. Kloster, A. DeBenedictis, Phys. Rev. D **79**, 084023 (2009)
26. R. Gambini, J. Olmedo, J. Pullin, Class. Quantum Gravity **31**, 095009 (2014)
27. N. Dadhich, A. Joe, P. Singh, Class. Quantum Gravity **32**, 185006 (2015)
28. A. Ashtekar, J. Olmedo, P. Singh, Phys. Rev. Lett. **121**, 241301 (2018)
29. A. Ashtekar, J. Olmedo, P. Singh, Phys. Rev. D **98**, 126003 (2018)
30. A. Ashtekar, J. Olmedo, Int. J. Mod. Phys. **29**, 2050076 (2020)
31. N. Bodendorfer, F.M. Mele, J. Münch, Class. Quantum Gravity **36**, 187001 (2019)
32. C. Zhang, Y. Ma, S. Song, X. Zhang, Phys. Rev. Phys. Rev. D **105**, 024069 (2020)
33. R. Gambini, J. Olmedo, J. Pullin, Front. Astron. Space Sci. **8**, 74 (2021)
34. C. Zhang, Y. Ma, S. Song, X. Zhang, Phys. Rev. D **105**, 024069 (2022)
35. B.E. Navascués, A. García-Quismondo, G.A. Mena Marugán, Phys. Rev. D **106**, 063516 (2022)
36. A. García-Quismondo, G. A. Mena Marugán, Phys. Rev. D **106**, 023532 (2022)

37. B. E. Navascués, A. García-Quismondo, G. A. Mena Marugán, Phys. Rev. D**106**, 043531 (2022)
38. M. Han, H. Liu. arXiv:2212.04605
39. H. Liu, K. Noui, Class. Quantum Gravity **34**, 135008 (2017)
40. A. Ashtekar, M. Bojowald, Class. Quantum Gravity **22**, 3349 (2005)
41. A. Ashtekar, B. Krishnan, Phys. Rev. D **68**, 104030 (2003)
42. A. Ashtekar, B. Krishnan, Liv. Rev. (Relativ.) **7**(10) (2004)
43. I. Booth, Can. J. Phys. **83**, 1073 (2005)
44. G. Ongole, H. Zhang, T. Zhu, A. Wang, B. Wang, Universe **8**, 543 (2022)
45. A. Ashtekar, T. Pawlowski, P. Singh, Phys. Rev. D **74**, 084003 (2006)
46. A. Ashtekar, M. Bojowald, J. Lewandowski, Adv. Theo. Math. Phys. **7**, 233–268 (2003)
47. J. Willis, On the low energy ramifications and a mathematical extension of loop quantum gravity; Ph.D. Dissertation, The Pennsylvaina State University (2004)
48. V. Taveras, Phys. Rev. D **78**, 064072 (2008)
49. G.W. Gibbons, M.J. Perry, Proc. Roy. Soc. Lond. A **358**, 467 (1978)
50. S.A. Fulling, S.N.M. Ruijsenaars, Phys. Rep. **152**, 135 (1987)
51. T. Levi-Civita, Il sottocaso B2: soluzioni oblique, Atti. Accad. Naz. Lincei Cl. Sci. Fis. Mat. Nat. Rend. **27**, 343 (1918)
52. W. Kinnersley, M. Walker, Phys. Rev. D **2**, 1359 (1970)
53. A. Ashtekar, T. Dray, Commun. Math. Phys. **79**, 581 (1981)
54. M. Bouhmadi-López, S. Brahma, C.-Y. Chen, P. Chen, D.-H. Chen, Phys. Dark Univ. **30**, 100701 (2020)
55. A. Ashtekar, A. Magnon, J. Math. Phys. **25**, 2682 (1984)
56. R. Schoen, Localizing solutions of the Einstein equations, Lectures at the Institut Henri Poincaré, General Relativity: A Celebration of the 100th Anniversary (2015). https://philippelefloch.files.wordpress.com/2015/11/2015-ihp-richardschoen.pdf
57. T. Thiemann, Class. Quantum Gravity **12**, 181 (2003)
58. M. Bojowald, R. Swiderski, Class. Quantum Gravity **23**, 2129 (2006)
59. D.-W. Chiou, W.-T. Ni, A. Tang. arXiv:1212.1265
60. J. Ben Achour, F. Lamy, H. Liu, K. Noui, Euro. Phys. Lett. **123**, 20006 (2018)
61. J.G. Kelly, R. Santacruz, E. Wilson-Ewing, Phys. Rev. D **102**, 106024 (2020)
62. J.G. Kelly, R. Santacruz, E. Wilson-Ewing, Class. Quant. Grav. **38**, 04LT01 (2021)
63. R. Gambini, J. Olmedo, J. Pullin, Class. Quantum Gravity **37**, 205012 (2020)
64. M. Han, H. Liu, Class. Quantum Gravity **39**, 035011 (2022)
65. V. Husain, J.G. Kelly, R. Santacruz, E. Wilson-Ewing, Phys. Rev. Lett. **128**, 121301 (2021)
66. C. Zhang, Phys. Rev. D **104**, 126003 (2021)
67. V. Husain, J.G. Kelly, R. Santacruz, E. Wilson-Ewing, Phys. Rev. D **106**, 024014 (2022)
68. K. Giesel, M. Han, B. Li, H. Liu, P. Singh, Phys. Rev. D **107**, 044047 (2023)
69. L. Modesto, Int. J. Theor. Phys. **47**, 357 (2008)
70. S. Hossenfelder, L. Modesto, I. Premont-Schwarz, Phys. Rev. D **81**, 044036 (2010)
71. Y. Tavakoli, J. Marto, A. Dapor, Int. J. Mod. Phys. D **23**, 1450061 (2014)
72. C. Bambi, D. Malafarina, L. Modesto, Phys. Rev. D **88**, 044009 (2013)
73. Y. Liu, D. Malafarina, L. Modesto, C. Bambi, Phys. Rev. D **90**, 044040 (2014)
74. J. Ben Achour, S. Brahma, J.-P. Uzan, JCAP **2003**, 041 (2020)
75. B.F. Li, P. Singh, Universe **7**(11), 406 (2021)
76. A. Corichi, P. Singh, Phys. Rev. D **78**, 024034 (2008)
77. P. Hájíček, Quantum Gravity: Quantum Theory of Gravitational Collapse. Lecture Notes on Quantum Conchology, Springer Berlin Heidelberg, pp. 255–299 (2003)
78. M. Campiglia, R. Gambini, J. Olmedo, J. Pullin, Class. Quantum Gravity**33**, LT01 (2016)
79. J. Ziprick, J. Gegenberg, G. Kunstatter, Phys. Rev. D **94**, 104076 (2016)
80. K. Giesel, B.-F. Li, P. Singh, Phys. Rev. D **104**, 106017 (2021)
81. M. Bojowald, T. Harada, R. Tibrewala, Phys. Rev. D **78**, 064057 (2008)
82. M. Bojowald, J.D. Reyes, R. Tibrewala, Phys. Rev. D **80**, 084002 (2009)
83. A. Alonso-Bardaji, D. Brizuela, Phys. Rev. D **104**, 084064 (2021)

84. A. Alonso-Bardaji, D. Brizuela, R. Vera, Phys. Lett. B **829**, 137075 (2022)
85. M.W. Choptuik, Phys. Rev. Lett. **70**, 9 (1993)
86. V. Husain, O. Winkler, Class. Quantum Gravity **22**, L127 (2005)
87. V. Husain. arXiv:0808.0949
88. J. Ziprick, G. Kunstatter, Phys. Rev. D **80**, 024032 (2009)
89. A. Kreienbuehl, V. Husain, S.S. Seahra, Class. Quantum Gravity **29**, 095008 (2012)
90. F. Benitez, R. Gambini, L. Lehner, S. Liebling, J. Pullin, Phys. Rev. Lett. **124**, 071301 (2020)
91. F. Benitez, R. Gambini, S. Liebling, J. Pullin, Phys. Rev. D **104**, 024008 (2021)
92. F. Benitez, R. Gambini, J. Pullin, Universe **8**, 10 (2022)
93. C. Barceló, R. Carballo-Rubio, L.J. Garay, Int. J. Mod. Phys. D **23**, 1442022 (2014)
94. C. Barcelo, R. Carballo-Rubio, L.J. Garay, G. Jannes, Class. Quantum Gravity **32**, 035012 (2015)
95. C. Barceló, R. Carballo-Rubio, L.J. Garay, JHEP **01**, 157 (2016)
96. R. Carballo-Rubio, F. Di Filippo, S. Liberati, M. Visser, Class. Quantum Gravity **37**, 14 (2020)
97. K. Giesel, B.F. Li, P. Singh, S.A. Weigl, Phys. Rev. D **105**, 066023 (2022)
98. S. Hod, T. Piran, Phys. Rev. D **55**, 440 (1997)
99. T. Thiemann, Class. Quantum Gravity **15**, 839 (1998)
100. A. Ashtekar, J. Lewandowski, H. Sahlmann, Class. Quantum Gravity **20**, L11 (2003)
101. M. Bojowald, R. Goswami, R. Maartens, P. Singh, Phys. Rev. Lett. **95**, 091302 (2005)
102. K. Vandersloot, Phys. Rev. D **75**, 023523 (2007)
103. P. Singh, E. Wilson-Ewing, Class. Quantum Gravity **31**, 035010 (2014)
104. T. Schmitz, Phys. Rev. D **101**, 026016 (2020)
105. R.J. LeVeque, *Finite Volume Methods for Hyperbolic Problems* (Cambridge University Press, Cambridge, 2002)
106. C. Vaz, L. Witten, Gen. Rel. Gravity **43**, 3429 (2011)
107. C. Kiefer, T. Schmitz, Phys. Rev. D **99**, 126010 (2019)
108. W. Piechocki, T. Schmitz, Phys. Rev. D **102**, 046004 (2020)
109. T. Schmitz, Phys. Rev. D **103**, 064074 (2021)
110. J. Münch, Class. Quantum Gravity **38**, 175015 (2021)
111. J. Ben Achour, S. Brahma, S. Mukohyama, J.P. Uzan, JCAP **09**, 020 (2020)
112. S.W. Hawking, Phys. Rev. D **14**, 2460 (1976)
113. P. Hajicek, Phys. Rev. D **36**, 1066 (1987)
114. A. Ashtekar, The issue of information loss: the current status, ILQG seminar of February 23rd (2015). http://relativity.phys.lsu.edu/ilqgs/ashtekar022316.pdf
115. A. Ashtekar, Universe **6**, 21 (2020)
116. S. Hayward, arXiv:gr-qc/0504038
117. S. Hayward, arXiv:gr-qc/0504037
118. D. Page, Phys. Rev. Lett. **71**, 3743–3746 (1993)
119. M. Christodoulou, T. De Lorenzo, Phys. Rev. D **94**, 104002 (2016)
120. M. Christodoulou, C. Rovelli, Phys. Rev. D **91**, 064046 (2015)
121. A. Ashtekar, V. Taveras, M. Varadarajan, Phys. Rev. Lett. **100**, 211302 (2008)
122. S.A. Hayward, Phys. Rev. Lett. **96**, 031103 (2006)
123. V.P. Frolov, JHEP **1405**, 049 (2014)
124. J.M. Bardeen, arXiv:1406.4098
125. C. Rovelli, F. Vidotto, Int. J. Mod. Phys. D **23**, 1442026 (2014)
126. E. Bianchi, M. Smerlak, Gen. Relativ. Gravit. **46**, 1809 (2014)
127. E. Bianchi, T. De Lorenzo, M. Smerlak, JHEP **06**, 180 (2015)
128. H.M. Haggard, C. Rovelli, Phys. Rev. D **92**, 104020 (2015)
129. M. Christodoulou, Geometry transition in covariant LQG: black to white. http://relativity.phys.lsu.edu/ilqgs/christodoulou102417.pdf
130. E. Bianchi, M. Christodoulou, F. D'Ambrosio, H. M Haggard, C. Rovelli, Class. Quantum Grav. **35**, 225003 (2018)
131. P. Martin-Dessuad, C. Rovelli, Class. Quantum Gravity **36**, 245002 (2019)
132. A. Ori, Personal communication

133. I. Agullo, A. Ashtekar, W. Nelson, Class. Quantum Gravity **30**, 085014 (2013)
134. L. Amadei, A. Perez, Phys. Rev. D **106**, 063528 (2022)
135. P. Diener, B. Gupt, P. Singh, Class. Quantum Gravity **31**, 105015 (2014)
136. P. Diener, A. Joe, M. Megevand, P. Singh, Class. Quantum Gravity **34**, 094004 (2017)
137. V. Faraoni, A. Giusti, Symmetry **12**, 1264 (2020)
138. M. Bojowald, Universe **6**, 125 (2020)
139. A. Perez, Rept. Prog. Phys. **80**, 126901 (2017)
140. J.D. Bekenstein, V.F. Mukhanov, Phys. Lett. B **360**, 7–12 (1995)
141. I. Agullo, V. Cardoso, A. del Rio, M. Maggiore, J. Pullin Phys. Rev. Lett. **126**, 041302 (2021)
142. R.G. Daghigh, M.D. Green, G. Kunstatter, Phys. Rev. D **103**, 084031 (2021)
143. D. del-Corral, J. Olmedo, Phys. Rev. D **105**, 064053 (2022)
144. M. Bojowald. arXiv: 1906.04650
145. G. Olmo, P. Singh, JCAP **0901**, 030 (2009)
146. R. Dijkgraaf, H. Verlinde, E. Verlinde, Null. Phys. B **371**, 269–314 (2002)
147. D. Grumiller, JHEP **0211**, 018 (2002)
148. D. Grumiller, JHEP **0505**, 028 (2005)
149. A. Ashtekar, F. Pretorius, F. Ramazanoglu, Phys. Rev. D **83**, 044040 (2011)
150. A. Ori, Gen. Relativ. Gravit. **48**, 9 (2016)
151. W.C. Gan, G. Ongole, E. Alesci, Y. An, F.W. Shu, A. Wang, Phys. Rev. D. **106**, 126013 (2022)
152. M. Assanioussi, A. Dapor, K. Liegener, Phys. Rev. D **101**, 026002 (2020)

Chapter 8
Gravitational Vacuum Condensate Stars

Emil Mottola

8.1 The Issue of the Final State of Gravitational Collapse

The possibility that a gravitationally bound system could be so compact that not even light can escape its surface was first imagined in the eighteenth century by Rev. Michell [138], and Laplace [105, 118]. Each reasoned (apparently independently) that since the escape velocity v from a spherical mass M of radius r is given by $\mathrm{v}^2 = 2GM/r$ in Newtonian gravitation, light traveling at speed c would not be able to escape the surface if emitted from a radius less than $r_M = 2GM/c^2$. Since this radius is much smaller than the actual size of astronomical bodies known at the time, such as the sun or the planets, and with no astronomical evidence of the existence of the highly compact objects for which r_M would be relevant, the possibility of such 'dark stars' from which light itself could not escape was not taken very seriously, nor developed further for well over a century.

In 1915 A. Einstein introduced the theory of general relativity (GR). Soon thereafter, K. Schwarzschild found an exact solution of the sourcefree Einstein's equations in the case of spherical symmetry. The general static, spherically symmetric metric line element in GR can be expressed in the form

$$ds^2 = -f(r)\, c^2 dt^2 + \frac{dr^2}{h(r)} + r^2 \left(d\theta^2 + \sin^2\theta\, d\phi^2\right) \tag{8.1}$$

in terms of two functions of radius, $f(r)$ and $h(r)$. For the sourcefree asymptotically flat Schwarzschild solution, $f(r)$ and $h(r)$ are equal and given by

$$f(r) = h(r) = 1 - \frac{2GM}{c^2 r} = 1 - \frac{r_M}{r} \tag{8.2}$$

E. Mottola (✉)
Department of Physics and Astronom, University of New Mexico, Albuquerque, NM, USA
e-mail: emottola@unm.edu

C. Bambi (ed.), *Regular Black Holes*, Springer Series in Astrophysics and Cosmology,
https://doi.org/10.1007/978-981-99-1596-5_8

determining the static geometry of empty sourcefree space exterior to a spherically symmetric body of mass M. In these (t, r) coordinates the metric has two apparent singularities, one at the origin $r = 0$ of spherical coordinates, and the other of quite a different character on a sphere of finite radius $r = r_{_M}$, the same radius discussed by Michell and Laplace in Newtonian gravity.

The singularity at $r = 0$ is similar to that of a point charge in Coulomb's law where the electric field diverges. Likewise for (8.1)–(8.2) the Riemannian curvature and Kretschmann scalar $R_{\alpha\beta\gamma\lambda}R^{\alpha\beta\gamma\lambda} = 12r_{_M}^2/r^6$ diverge as $r \to 0$. In contrast the curvature remains finite at $r = r_{_M}$, and of order $1/r_{_M}^2$, which is relatively small for large M. However a light wave emitted from any $r \geq r_{_M}$ with local frequency $\omega_{\rm loc}(r)$ is gravitationally redshifted according to the relation

$$\omega_\infty = \omega_{\rm loc}(r)\, f^{\frac{1}{2}} = \omega_{\rm loc}(r)\sqrt{1 - \frac{r_{_M}}{r}} = const. \tag{8.3}$$

Thus light emitted at $r = r_{_M}$ with any finite local frequency $\omega_{\rm loc}(r_{_M})$ becomes redshifted to zero frequency and cannot propagate outwardly at all. The radius $r = r_{_M}$ therefore defines the critical surface or *event horizon* from beyond which no event can ever be observed by an outside observer. Thus the wave description of light in GR recovers exactly the 'dark star' surface that Michell and Laplace had imagined within a Newtonian framework. Conversely and equivalently, (8.3) also implies that an inwardly directed light wave with the finite frequency ω_∞ far from the horizon is blueshifted to an *infinite* local frequency upon reaching $r = r_{_M}$.

One would expect to be able to remove at least the singularity at $r = 0$, and the infinite curvatures it produces by allowing for a stress tensor source to Einstein's equations of finite extent, rather than one concentrated all at the origin. Indeed Schwarzschild also found such a regular *interior* solution [172], shortly after finding the well-known exterior solution (8.1)–(8.2) most often associated with his name.

To simplify matters, Schwarzschild assumed a model of the interior matter composed of an incompressible fluid with isotropic pressure and constant mass density

$$\rho = \bar{\rho} \equiv \frac{3M}{4\pi r_{_S}^3} \equiv \frac{3H^2}{8\pi G} \quad \text{so that} \quad H^2 = \frac{r_{_M}}{r_{_S}^3} \tag{8.4}$$

where $r_{_S}$ is the radius of the star's surface.[1] At this surface the pressure $p(r_{_S}) = 0$, and the interior solution for $r \leq r_{_S}$ is matched to the exterior one (8.1)–(8.2) for $r \geq r_{_S}$ by requiring $f(r_{_S}) = h(r_{_S}) = 1 - r_{_M}/r_{_S} > 0$ to be continuous there (although their derivatives need not be). With these boundary conditions the metric for Schwarzschild's interior solution can be expressed in the form (8.1), but with

[1] Here and often in the following, units are employed where the speed of light $c = 1$, unless otherwise needed for emphasis.

$$h(r) = 1 - H^2 r^2 \tag{8.5a}$$

$$f(r) = \frac{D^2}{4} \geq 0 \tag{8.5b}$$

where

$$D \equiv 3\sqrt{1 - H^2 r_s^2} - \sqrt{1 - H^2 r^2} \tag{8.5c}$$

is assumed to be non-vanishing, so that the pressure

$$p(r) = \frac{\bar{\rho}}{D}\left[\sqrt{1 - H^2 r^2} - \sqrt{1 - H^2 r_s^2}\right] \tag{8.5d}$$

is also finite (and positive) on the interval $r \in [0, r_s]$. This assumption holds for

$$r_s > \frac{9}{8} r_M \tag{8.6}$$

in which case the solution (8.5) is everywhere non-singular and finite, and the first derivative $f'(r_s) = r_M / r_s^2$ also turns out to be continuous at the surface $r = r_s$.

Conversely, if the inequality (8.6) is *not* satisfied, then D vanishes at

$$r_0 = 3r_s \sqrt{1 - \frac{8}{9}\frac{r_s}{r_M}} \in [0, r_s] \quad \text{for} \quad r_M \leq r_s \leq \frac{9}{8} r_M \tag{8.7}$$

where $f(r_0) = 0$, and the pressure (8.5d) $p(r_0) \to \infty$ diverges. In this case the solution (8.5) becomes singular and may no longer be deemed acceptable. Since $M = 4\pi \bar{\rho} r_s^3 / 3$, the lower bound (8.6) on r_s translates to an *upper* bound on the mass

$$\overline{M}_{\rm crit} = \frac{4c^3}{9G}\left(\frac{1}{3\pi G \bar{\rho}}\right)^{\frac{1}{2}}, \tag{8.8}$$

that is, $M < \overline{M}_{\rm crit}$ is the condition for a non-singular Schwarzschild star of mass M to exist with a given fixed constant density $\rho = \bar{\rho}$ interior.

An incompressible fluid would seem to be matter that is the most resistant to further gravitational contraction, and it is already extreme at that for implying an infinite speed of sound ($v_s > c$), violating relativistic causality. Thus $\overline{M}_{\rm crit}$ would seem to be an upper bound for the mass of any self-gravitating body to contain a non-singular interior. If $\bar{\rho}$ is taken to be 5 times the density 2.3×10^{14} gm/cm^3 of normal nuclear matter, about the value expected in the cores of neutron stars, then $\overline{M}_{\rm crit} \simeq 3.4\, M_\odot$. Any higher values of $\bar{\rho}$ lead only to smaller values of $\overline{M}_{\rm crit}$ from (8.8). Thus any self-gravitating system composed of ultra dense nuclear matter with a mass larger than a few solar masses would appear to be unstable and liable to complete gravitational collapse to the singular state of the original solution (8.1)–(8.2), in which all the matter is crushed to infinite density at the central singularity.

J. A. Wheeler and his collaborators carried out calculations utilizing models of the equations of state of nuclear matter, and their stability, obtaining upper bounds of order of a solar mass $M_\odot$ [96]. Current theoretical values for this Tolman–Oppenheimer–Volkoff (TOV) limit have been estimated to lie between 2.2 $M_\odot$ and 2.9 $M_\odot$ [113], the range due to the still considerable uncertainty in the high density nuclear equation of state. The LIGO/LSC observation of GW170817, the first gravitational wave (GW) event due to merging neutron stars, placed the limit in the range of 2.01 $M_\odot$ to 2.17 $M_\odot$ [69, 163] suggesting that higher central nuclear densities and the lower end of the theoretically allowed mass range is preferred by the GW data.

The key point Wheeler made clear, and emphasized, is the reason that such an upper TOV bound exists at all in GR is due to the fact that unlike Newtonian gravity, positive internal pressure $p > 0$ only increases gravitational attraction in Einstein's theory [191]. Formally taking the limit $c \to \infty$ in (8.8) shows that the upper bound on M is removed to infinity in the non-relativistic Newtonian limit where this pressure effect disappears. Thus in full GR even very 'hard' equations of state of the internal matter such as Schwarzschild's incompressible fluid only serve to promote rather than resist gravitational collapse, making singularities of the kind exhibited in (8.1)–(8.2) for the final state seem to be inevitable. Wheeler hoped that consistent inclusion of quantum effects might provide an escape from this difficulty, since as he said: *"proper physical variables do not and cannot go to infinity"* [191].

Wheeler thus brought to many physicists' attention the seriousness of the issue of the final state of complete gravitational collapse in classical GR. Despite strong misgivings, but unable to see any apparent way out of the difficulty, Wheeler came to believe in the inevitability of singularities in gravitational collapse, and then popularized the term *black hole* to both physicists and non-physicists alike.

The nature of the Schwarzschild singularity at $r = r_M$ as a causal boundary only became clear from the work of Finkelstein [87], who showed that the coordinate singularity in the line element (8.1)–(8.2) at $r = r_M$ could be removed by a change of coordinates associated with the trajectories of freely falling light rays (a set of coordinates actually found earlier by Eddington [83]). Since the worldlines of timelike trajectories starting at finite $r > r_M$ also arrive at the horizon at a finite proper time, it seems natural to extend the coordinates through the horizon to the interior, $r < r_M$, so that the proper time of such a timelike observer could continue uninterrupted. This leads to the maximal analytic extension of the Schwarzschild geometry in Kruskal–Szekeres coordinates [139], illustrated in the Carter–Penrose conformal diagram of Fig. 8.1. The causal boundary at which light hovers indefinitely, unable to escape to infinity, which in the Schwarzschild solution (8.1)–(8.2) is the spherical surface of radius $r = r_M$, at $T = X$ in Fig. 8.1 is the (future) event horizon, following terminology introduced by Rindler [166].

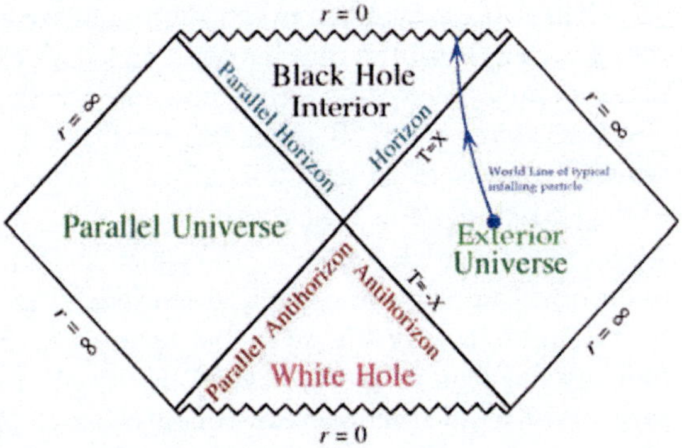

Fig. 8.1 The Carter–Penrose conformal diagram of the maximal Kruskal analytic extension of the Schwarzschild geometry. Radial light rays are represented in this diagram as 45° lines. The angular coordinates θ, ϕ are suppressed

In order to remove the coordinate singularity at $r = r_M$, the transformation from (8.1) to Eddington-Finkelstein or Kruskal–Szekeres (T, X) coordinates must itself be singular at the horizon. Although not often stated explicitly, this means that the geometries before and after the 'singular coordinate transformation' are not necessarily physically equivalent at $r = r_M$, or if extended beyond it. This is because the mathematical procedure of analytic continuation through a null hypersurface involves a *physical assumption*, namely that the stress-energy tensor $T_\mu{}^\nu$ is exactly vanishing there. Even in classical GR, the hyperbolic character of Einstein's equations allows generically for sources and discontinuities on the horizon which would violate this assumption, invalidating analytic continuation and potentially altering also the geometry of the interior from what the analytic continuation of Fig. 8.1 would predict.

It is relevant here to recall that Einstein's original conception of the Equivalence Principle between gravitational and inertial mass, and its subsequent mathematical formulation in terms of Riemannian geometry, requires physics to be independent of *real* and *regular* local coordinate transformations [85]. Analytic continuation through BH horizons appends to this the much stronger mathematical hypothesis of continuation in the space of *complex* metrics, admitting *singular* coordinate transformations that lead to globally extended spacetimes as in Fig. 8.1, which may or may not be realized in Nature. Occasional statements that free fall through the event horizon implied by continuation of worldlines to $r < r_M$ is required by the Equivalence Principle are incorrect, just because of the possibility of sources and discontinuities on the horizon, which would interrupt that free fall. The Equivalence Principle does not prevent Einstein's elevator from coming to an abrupt end of its free fall at the surface of the earth, or a surface at $r = r_M$, if a surface exists there.

When R. Kerr found an exact solution of the sourcefree Einstein equations for a rotating BH [114], the interior was found to contain not only singularities where the curvature diverges, located now on a ring of finite radius, but also *closed timelike curves* [105]. It is significant that this paradoxical and acausal feature, widely believed to be unphysical, occurs at *macroscopic* distance scales of order of the horizon scale $r_M \simeq 3\,(M/M_\odot)$ km, far larger than the microscopic Planck scale $L_{Pl} = \sqrt{\hbar G/c^3} \simeq 1.6 \times 10^{-33}$ cm, at which quantum gravity effects might be expected to become important or come to the rescue. The maximal analytic extension of the Kerr solution also results in global features such as an infinite number of asymptotically flat regions [105], beyond the two in Fig. 8.1. Thus endowing collapsing matter with angular momentum not only does not avoid the appearance of spacetime singularities, as Einstein apparently had hoped [86], but rather leads to additional unphysical features of analytically extended BH solutions.

An open question at the time was whether singularities might be a special property of highly symmetric exact solutions of Einstein's equations To address this question, in 1965 R. Penrose introduced the notion of a *closed trapped surface* [154], within which outgoing light waves are actually bent backwards by gravity and forced to become ingoing, such as occurs in (8.1)–(8.2) for $r < r_M$. He then showed that *if* such a trapped surface exists, and if matter within it satisfies the weak energy condition $\rho + p_i \geq 0$, then Einstein's equations imply that a singularity would necessarily result, independently of any symmetries. Here $p_i, i = 1, 2, 3$ are any of the three principal pressure components of the matter stress tensor. The important assumption made in [154] is that some continuation of the geometry beyond the horizon into the interior is necessary to have a trapped surface at all. Related singularity theorems were subsequently proven by S. W. Hawking, and Hawking and Penrose, with slightly different assumptions, but also relying either on the existence of a closed trapped surface or a stronger condition $\rho + \sum_{i=1}^{3} p_i \geq 0$ on the matter stress tensor (the strong energy condition) [105]. Since these singularity theorems are independent of any symmetries and the existence of trapped surfaces in gravitational collapse was widely assumed, they reinforced the conviction that BH singularities will arise inevitably in the final state of gravitational collapse in classical GR.

Also in 1965, a quite different possibility for the final state of gravitational collapse was proposed by E. Gliner, based on his speculation that superdense matter could arrive at an equation of state with all three principal pressures equal and *negative,* i.e. $p_i = -\rho < 0$ [93]. This corresponds to a cosmological Λ term in Einstein's equations, but localized within the high density matter, rather than everywhere constant. With this equation of state and negative pressure, the strong energy condition is *violated,* i.e. $\rho + 3p < 0$. This has the consequence that timelike geodesics become *defocused* rather than focused down to a singularity, or in more colloquial terms, that gravitational attraction in GR becomes effectively *repulsion* instead. The key point about such an equation of state is that it reverses the sign of the pressure effect that led previously to the critical mass bound (8.8). The $p = -\rho < 0$ equation of state is that which A. D. Sakharov had speculated also could arise to avoid a singularity at the initial conditions of the universe [171], implying instead a non-singular expanding de Sitter phase (well before the term 'inflation' was coined). Today this equation of state is believed to be that of the cosmological dark energy driving the Hubble expansion to accelerate, rather than decelerate, again intuitively a kind of repulsion.

With an interior $p = -\rho > 0$ equation of state, BH singularities and the conclusions of the Penrose singularity theorem can be avoided, but if Einstein's equations hold, only if no true closed trapped surface is ever formed. This shows that the crucial assumption of the Penrose theorem involves the physics of the near-horizon region, since once any trapped surface at any $r < r_M$ is assumed, it is 'game over,' and a singularity must occur in GR. The logical loopholes in all the singularity theorems and resolution of Wheeler's difficulty for the final state of gravitational collapse can be realized in Einstein's theory, or its semi-classical extension, only if the stress tensor of quantum matter can produce a large enough effective positive $\Lambda_{\rm eff}$ term with negative pressure, *and* prevent the formation of a true trapped surface in the collapse.

8.2 What's the (Quantum) Matter with Black Holes?

Additional difficulties beyond the singularities and acausal closed timelike curves, arising in strictly classical GR, appear when quantum effects in BH spacetimes are considered. The genesis of these is Hawking's argument that a non-rotating, uncharged BH would emit thermal radiation, at a temperature [101, 102]

$$T_H = \frac{\hbar\kappa}{2\pi c k_B} = \frac{\hbar c^3}{8\pi G k_B M} \tag{8.9}$$

where $\kappa = c^4/4GM$ is the surface gravity. With the temperature (8.9), it was argued that thermodynamics could be applied to BHs, and a 'First Law' [26, 98]

$$dE = dMc^2 = \frac{\kappa\, c^2}{8\pi G}\, dA_H = T_H\, dS_{BH} \tag{8.10}$$

was deduced, with the BH assigned the Bekenstein-Hawking BH entropy,

$$S_{BH} = k_B \frac{A_H}{4\, L_{Pl}^2}, \tag{8.11}$$

equal to 1/4 of its area $A_H = 4\pi r_M^2$ in Planck units. This identification of entropy with area supported J. Bekenstein's speculation that BHs be assigned an entropy in order to avoid violations of the second law of thermodynamics [28], since ordinary matter with non-zero entropy apparently can fall into BHs and so disappear without a trace. The hypothesis that BH entropy is proportional to the area of its event horizon was guided by Christodoulou's classical result that horizon area can never decrease in any process of interaction with matter, and will positively increase if the process irreversibly sends matter with positive mass-energy into the BH [71].

The curious feature of (8.10) is that $\hbar$ cancels out between T_H and dS_{BH}. This is because the first form of (8.10) is nothing but the *classical* Smarr formula [176]

$$dM = \frac{\kappa}{8\pi G}\, dA_H \tag{8.12}$$

relating the change of a BH mass M to the change in area A_H of its horizon, a simple geometric relation in which $\hbar$ does not appear at all. As in Christodoulou's area law [71], quantum mechanics plays no role in the classical Smarr relation (8.12). Multiplying and dividing by $\hbar$ should not be expected to turn a strictly classical relation into a valid quantum one.

Related to this, if the identification of the classical area rescaled by $k_B/4L_{Pl}^2$ with entropy and the thermodynamic interpretation of (8.10) is to be valid in the quantum theory, then the classical limit $\hbar \to 0$ (with M fixed) which yields an arbitrarily low Hawking temperature (8.9), assigns to the BH an arbitrarily large entropy, completely unlike the zero temperature limit of any other cold quantum system.

Another paradoxical consequence of temperature in (8.9) inversely proportional to $M = E/c^2$ is the implication that the heat capacity of a Schwarzschild BH

$$\frac{dE}{dT_H} = -\frac{8\pi G k_B M^2}{\hbar c} = -\frac{Mc^2}{T_H} < 0 \tag{8.13}$$

is *negative*, as was pointed out by Hawking himself [103], and incidentally also very large in magnitude. In statistical equilibrium the heat capacity of a system is related to the energy fluctuations about its mean value $\langle E\rangle$ by

$$c_V = \left(\frac{d\langle E\rangle}{dT}\right)_V = \frac{1}{k_B T^2}\left\langle \left(E - \langle E\rangle\right)^2\right\rangle > 0 \tag{8.14}$$

at constant volume V. If pressure or some other thermodynamic variable is held fixed there is an analogous relation. The positivity of the statistical average in (8.14) requires only the existence of a well defined ground state upon which the thermal equilibrium canonical ensemble is defined, but is otherwise quite independent of the details of the system or its interactions. Hence on general grounds of quantum statistical mechanics, the heat capacity of any system in stable equilibrium at a fixed temperature must be positive. Although a negative heat capacity can be observed in self-gravitating systems in the microcanonical ensemble, that situation is temporary and not a true equilibrium state, but one subject to large fluctuations and indicative of a phase transition involving a large latent heat [123, 148].

The physical basis of the instability of a BH in equilibrium with its own Hawking radiation at $T = T_H$ is easy to see. If by a small thermal fluctuation the BH should absorb slightly more radiation in a short time interval than it emits, its mass would increase, $\Delta M > 0$, and hence from (8.9) its temperature would decrease, $\Delta T < 0$, so that it would now be cooler than its surroundings and be favored to absorb more energy from the heat bath than it emits in the next time step, decreasing its temperature further and driving it further from equilibrium. In this way a runaway process of the BH growing to absorb all of the surrounding radiation in the heat bath would ensue. Likewise, if the initial fluctuation has $\Delta M < 0$, the BH temperature would increase, $\Delta T > 0$, so that it would now be hotter than its surroundings and favored to emit more energy than it absorbs from the heat bath in the next time step, increasing its temperature further. Then a runaway process toward hotter and hotter evaporation of all its mass to its surroundings would ensue. In either case, the initial equilibrium is clearly unstable, and hence cannot be a candidate for the quantum equilibrium ground state for the system. The instability has also been verified from the negative eigenvalue of the fluctuation spectrum of a BH in a box of (large enough) finite volume [7, 94, 193].

The time scale for this unstable runaway process to grow exponentially is the time scale for fluctuations away from the mean value of the Hawking flux, *not* the much longer time scale associated with the lifetime of the BH under continuous emission of that flux. The time scale for thermal fluctuations is easily estimated to be the typical time between emissions of a single quantum with typical energy (at infinity) of $k_B T$, of a source whose energy emission per unit area per unit time is of order $(k_B T_H)^4/\hbar^3 c^2$. Multiplying by the area and dividing by the typical energy $k_B T$, the average number of quanta emitted per unit time is found. The inverse of this, namely

$$\Delta t \sim \frac{1}{A_H} \frac{(\hbar c)^3}{(k_B T_H)^3} \sim \frac{r_M}{c} \sim 10^{-5} \left(\frac{M}{M_\odot}\right) \text{s} \tag{8.15}$$

is the typical time interval (as measured by a distant observer) between successive emissions of individual Hawking quanta. This time scale is quite short. Any tendency for the system to become unstable or undergo a phase transition to a stable phase would be expected to show up on this short a time scale, governing the fluctuations in the mean flux, which is of order of the collapse time itself and before a steady state flux could even be established. The interpretation of the classical relation of energy to area (8.12) as an equilibrium thermodynamic relation (8.10) of any kind is therefore placed in serious doubt.

It is also instructive to evaluate S_{BH} for typical astrophysical BHs. For a solar mass $M_\odot \simeq 2 \times 10^{33}$ gm

$$S_{BH} \simeq 1.05 \times 10^{77}\, k_B \left(\frac{M}{M_\odot}\right)^2 \tag{8.16}$$

which is truly an enormous entropy. For comparison, the entropy of the sun as it is, a hydrogen burning main sequence star, is given to good accuracy by the entropy of a non-relativistic perfect fluid. This is of the order Nk_B where N is the number of nucleons in the sun $N \sim M_\odot/m_N \sim 10^{57}$, times a logarithmic function of the density and temperature profile which may be estimated to be of the order of 20 for the sun. Hence the entropy of the sun is approximately $S_\odot \sim 2 \times 10^{58}\, k_B$ or nearly 19 orders of magnitude smaller than (8.16).

A simple scaling argument shows that the entropy of a gravitationally bound relativistic fluid should scale like $M^{3/2}$ and be of order of [141, 194]

$$S \sim k_B \left(\frac{M}{M_{Pl}}\right)^{\frac{3}{2}} \sim 10^{57}\, k_B \left(\frac{M}{M_\odot}\right)^{\frac{3}{2}} \tag{8.17}$$

which is less than $S_\odot$ because the relativistic radiation pressure in the sun is small compared to the non-relativistic fluid pressure. However, the entropy from the relativistic radiation pressure (8.17) grows with the $3/2$ power of the mass, whereas the non-relativistic fluid entropy $S_\odot$ grows only linearly with M. For stars with masses greater than about $50\, M_\odot$ which are hot enough for their pressure to be dominated by the photon T^4 radiation pressure, (8.17) gives the correct order of magnitude estimate of such a star's entropy at a few times $10^{59}\, k_B$ [194].

On the other hand, the BH entropy (8.16) is proportional to M^2, so the discrepant factor with (8.17) is $(M/M_{Pl})^{1/2} \sim 10^{19}$ for $M = M_\odot$. Since (8.16) makes no reference to how the BH was formed, and a BH may always be theoretically idealized as forming from an adiabatic process, which keeps the entropy constant, (8.16) states that this entropy must suddenly jump by a factor of order 10^{19} for a solar mass BH at the instant the horizon forms at $r = r_M$. When Boltzmann's formula

$$S = k_B \ln W(E) \tag{8.18}$$

is recalled, relating the entropy to the total number of microstates in the system $W(E)$ at the fixed energy E, the number of such microstates of a BH satisfying (8.16) must jump by $\exp(10^{19})$ at that instant that the event horizon is reached, a truly staggering proposition. How and from where this enormous number of microstates can appear, when the horizon is supposed to be nothing but a mathematical surface only, with no independent degrees of freedom or stress tensor of its own, has remained perplexing and the subject of numerous investigations spanning five decades [8–10, 92, 126, 130, 145, 161], often with the suggestion that new physics or modifying the laws of physics themselves might be required.

Hawking radiation emerging from the BH at temperature (8.9) in a mixed thermal state also seems to imply a breakdown of quantum unitary evolution, that is difficult if not impossible to recover at the late or final stages of the BH evaporation process, resulting in a severe 'information paradox' [104, 149, 161, 183].

As a consequence of the gravitational redshift (8.3), it also follows that although the Hawking temperature T_H of the radiation far from the BH is very low, the local temperature of the radiation

$$T_{\rm loc}(r) = \frac{T_H}{\sqrt{1 - \frac{r_M}{r}}} \tag{8.19}$$

is arbitrarily high when extrapolated back to the vicinity of the horizon $r \rightarrow r_M$, even becoming transplanckian in this limit. Unlike classical test particles, when $\hbar \neq 0$ such extremely blue shifted photons are necessarily present in the vacuum as virtual quanta. Their effects upon the geometry depend upon the quantum state of the vacuum, defined by boundary conditions on the wave equation in a non-local way over all of space, and $\langle T_\mu^{\ \nu} \rangle$ may be large at $r = r_M$, notwithstanding the smallness of the local classical curvature there [35, 70]. Since the limits $r \rightarrow r_M$ and $\hbar \rightarrow 0$ do not commute, non-analytic behavior near the event horizon, quite different from that in the strictly classical ($\hbar \equiv 0$) situation is possible in the quantum theory. Since gravity couples to all energies, thermal fluctuations at large transplanckian temperatures and energies would be expected to have significant backreaction effects on the classical spacetime geometry near the horizon, giving rise to just the sources and discontinuities that would violate the analytic continuation to the interior and global geometry in Fig. 8.1 it leads to. Thus it is by no means clear why assuming a rigidly fixed classical BH background, neglecting quantum fluctuations in the stress tensor down to $r = r_M$, as in Hawking's original semi-classical treatment, should be valid [112].

These myriad difficulties and BH paradoxes suggest that some important element is missing, and needed to describe quantum effects or degrees of freedom on the horizon scale in BH spacetimes, which are not present in classical GR. Since the various problems and paradoxes arise at the macroscopic scale of the BH horizon, these additional degrees of freedom require an addition or modification of Einstein's theory at that scale, well before the microscopic Planck scale $L_{\rm Pl}$ or Planck scale curvatures are reached. The problem of the magnitude of cosmological Λ vacuum energy on the very largest macroscopic Hubble scale of the universe carries with it a similar implication. Together with the observation that the $p = -\rho$ equation of state can provide a resolution of both the BH and Big Bang singularities, this indicates that the two problems at the interface of classical GR and quantum theory, involving the nature of the quantum vacuum and vacuum energy, at macroscopic scales are related [143].

8.3 The Proposed Solution: Gravitational Condensate Stars

8.3.1 *Background and Motivations for the First Gravastar Model*

One model that attempted to take backreaction of Hawking radiation on the BH into account, imagined it to be immersed in a Hawking radiation atmosphere, with an effective equation of state, $p = \kappa\rho$ [178]. It was found that due to the blue shift effect (8.19), the backreaction of such an atmosphere on the metric near $r = r_M$ is enormous, with the interior region quite different from the vacuum Schwarzschild solution. A large entropy of order of S_{BH} is obtained from the hot fluid alone in such a model, with $S = 4\,\frac{\kappa+1}{7\kappa+1}\,S_{BH}$, becoming equal to the BH entropy (8.11) for $\kappa = 1$. This suggested that the maximally stiff equation of state consistent with the causal limit $p = +\rho$ may play a role in the quantum theory of fully collapsed objects.

Despite the interesting result of [178], indicating the importance of backreaction on the geometry, the model of [178] cannot be viewed as a satisfactory solution to the final state of the collapse problem, since it involves huge Planckian energy densities near $r = r_M$, and a negative mass singularity at $r = 0$, indicating the breakdown of the semi-classical approximation in both regions. The negative mass singularity arises because a repulsive core is necessary to counteract the self-attractive gravity of the dense relativistic fluid with positive energy. This suggests again that an effectively repulsive equation of state with a negative pressure $p < 0$ (rather than a negative mass-energy singularity) should play a role in the BH interior.

A different proposal of a quantum phase transition at or near $r = r_M$ was made in [63], based on a suggestive condensed matter analogy with the liquid-vapor critical point of a non-relativistic Bose fluid, where the wave equation for acoustic excitations mimics the wave equation for light in the vicinity of a BH horizon. In this condensed matter analog system the position dependent speed of sound $v_s(r)$ takes the place of a position dependent speed of light $c_{\rm eff}(r)$, both of which formally vanish at the critical surface, where $c^2_{\rm eff}(r) = c^2 f(r)$ is the g_{tt} metric coefficient of dt^2 in (8.1). The authors of [61, 63] also argued for a negative pressure on the interior side of the critical surface, and an effective $p = -\rho$ equation of state, as Gliner had 35 years earlier, but did not tackle the delicate issue of joining a de Sitter interior to a Schwarschild exterior, in a fully consistent treatment of the problem in GR.

Matching the Schwarzschild exterior solution to a non-singular de Sitter (dS) interior had a long previous history. Continuous transitions between the two were studied e.g. in [81], while it was recognized that joining the exact Schwarzschild and dS geometries directly at their mutual horizons H^{-1} and $2GM$, requires some discontinuity or interposition of 'non-inflationary material' [157]. In addition to uncertainties of the physics involved, the earlier GR formalism [109, 110, 117, 147] for dealing with singular hypersurfaces when the normal to hypersurface becomes null, as it does at a BH horizon, were recognized to be inadequate [27]. The necessity of some anisotropic matter at the joining of the interior to exterior geometries was made explicit in [56].

Motivated by these considerations, the first fully relativistic gravitational vacuum condensate star model was proposed in [133, 137], with the critical surface of Ref. [63] replaced by a thin shell of ultra-relativistic fluid with the maximally stiff equation of state $p = \rho$ and vacuum condensate interior with negative pressure, $p = -\rho$. Partly for the reason of avoiding the technical difficulties associated with singular null hypersurfaces, the proposal in the original paper [133, 137] made use of two timelike hypersurfaces at r_1 and r_2 with an interposed fluid boundary layer of 'non-inflationary material' obeying the equation of state $p = \rho$. In addition to the model of [178], the choice of this equation of state at the causal limit where the speed of sound coincides with the speed of light, was motivated by physical considerations of a quantum phase transition produced by the infrared effects of dimensional reduction from $D = 4$ to $D = 2$ dimensions. Nevertheless the choice of $p = \rho$ in [137] is certainly an ansatz, illustrating a proof of principle, but without a rigorous basis in fundamental physics. It therefore would be subject to modification as that fundamental physics came more clearly into view by subsequent developments [141, 143, 144], and it also became possible to treat joining at null horizon hypersurfaces directly [30, 136].

8.3.2 The First Gravastar Model

The general form of the stress-energy tensor in the static, spherically symmetric geometry of (8.1) for a non-rotating body is

$$T_\mu{}^\nu = \begin{pmatrix} -\rho & 0 & 0 & 0 \\ 0 & p & 0 & 0 \\ 0 & 0 & p_\perp & 0 \\ 0 & 0 & 0 & p_\perp \end{pmatrix} \tag{8.20}$$

so that the Einstein equations in the static spherical coordinates of (8.1) are

$$-G^t{}_t = \frac{1}{r^2}\frac{d}{dr}\big[r\,(1-h)\big] = -8\pi G T^t_{\ t} = 8\pi G\,\rho, \tag{8.21a}$$

$$G^r{}_r = \frac{h}{rf}\frac{df}{dr} + \frac{1}{r^2}\big(h-1\big) = 8\pi G T^r_{\ r} = 8\pi G\,p \tag{8.21b}$$

together with the conservation equation

$$\nabla_\lambda\, T^\lambda_{\ r} = \frac{dp}{dr} + \frac{\rho+p}{2f}\frac{df}{dr} + \frac{2}{r}\,(p - p_\perp) = 0 \tag{8.22}$$

which ensures that the other components of Einstein's equations are satisfied. In (8.22) the transverse pressure $p_\perp \equiv T^\theta_{\ \theta} = T^\phi_{\ \phi}$ is allowed to be different from the radial pressure $p \equiv T^r_{\ r}$. For a perfect fluid $p_\perp = p$ and the last term of (8.22) vanishes.

In that case (8.21)–(8.22) are three first order equations for the four functions, f, h, ρ, and p, which become closed when an equation of state for the fluid relating p and ρ is specified.

Because of the considerations of the previous sections, the first gravastar model was assumed to contain three different regions with the three different equations of state

$$\begin{array}{lll} \text{I. de Sitter Interior:} & 0 \le r < r_1, & \rho = -p, \\ \text{II. Thin Shell:} & r_1 < r < r_2, & \rho = +p, \\ \text{III. Schwarzschild Exterior:} & r_2 < r, & \rho = p = 0. \end{array} \tag{8.23}$$

At the interfaces $r = r_1$ and $r = r_2$, the metric functions f and h are required to be continuous, although the first derivatives of f, h and p are generally discontinuous from the first order equations (8.21) and (8.22).

In the interior region $\rho = -p$ is a constant from (8.22). Labelling this constant $\rho_v = 3H^2/8\pi G$, and requiring that at the origin the solution is free of any mass singularity determines the interior to be a region of dS spacetime in static coordinates, i.e.

$$\text{I.} \quad f(r) = C\, h(r) = C\left(1 - H^2 r^2\right), \quad 0 \le r \le r_1 \tag{8.24}$$

where C and H are constants, which at this point are arbitrary. Note that the static dS horizon where both f and h vanish is at $r = r_H = H^{-1}$.

The unique solution in the exterior vacuum region where $T_\mu{}^\nu = 0$ that approaches flat Minkowski space as $r \to \infty$ is the Schwarzschild solution (8.2)

$$\text{III.} \quad f(r) = h(r) = 1 - \frac{2GM}{r} = 1 - \frac{r_M}{r}, \quad r_2 \le r \tag{8.25}$$

where the mass M can take on any (positive) value.

The only non-vacuum region is the thin shell interface region II. In this region it proves useful to define the dimensionless variable w by

$$w \equiv 8\pi G r^2 p \tag{8.26}$$

so that Eqs. (8.21)–(8.22) with $\rho = p$ may be recast in the form

$$\frac{dr}{r} = \frac{dh}{1 - w - h}, \tag{8.27a}$$

$$\frac{dh}{h} = -\frac{1 - w - h}{1 + w - 3h}\,\frac{dw}{w}. \tag{8.27b}$$

together with $pf \propto wf/r^2$ a constant. The first Eq. (8.27a) is equivalent to the definition of the (rescaled) Misner-Sharp mass function $\mu(r) = 2Gm(r)$, with $h = 1 - \mu/r$ and $d\mu(r) = 8\pi G\, \rho r^2\, dr = w\, dr$ within the shell. The second Eq. (8.27b) can be solved only numerically in general. However, it is possible to obtain an analytic solution in the thin shell limit $0 < h \ll 1$, since in this limit h may be set to zero on the right side of (8.27b) to leading order. Assuming w remains finite, (8.27b) can then be integrated immediately to obtain

$$h \equiv 1 - \frac{\mu}{r} \simeq \varepsilon \, \frac{(1+w)^2}{w} \ll 1 \tag{8.28}$$

in region II, where ε is an integration constant. Because of the condition $h \ll 1$, we require $\varepsilon \ll 1$, if w is of order unity. Making use of Eqs. (8.27) and (8.28) then gives

$$\frac{dr}{r} \simeq -\varepsilon \, dw \, \frac{(1+w)}{w^2} \tag{8.29}$$

so that because $\varepsilon \ll 1$ the radius r hardly changes within region II, and dr is of order $\varepsilon \, dw$. The final unknown function f is given by (8.22) to be $f = (r/r_1)^2(w_1/w) f(r_1)$ so that

$$f \simeq \frac{w_1}{w} \, f(r_1) \tag{8.30}$$

to leading order in ε for $\varepsilon \ll 1$.

Continuity ($\mathcal{C}^0$) of the metric functions f and h at r_1 and r_2 gives the conditions

$$f(r_1) = C\, h(r_1) = C\,(1 - H^2 r_1^2) \simeq C\, \varepsilon \, \frac{(1+w_1)^2}{w_1} \tag{8.31a}$$

$$f(r_2) = h(r_2) = 1 - \frac{2GM}{r_2} \simeq \varepsilon \, \frac{(1+w_2)^2}{w_2} \tag{8.31b}$$

which together with (8.30) evaluated at $r = r_2, w = w_2$ provides

$$C\,(1+w_1)^2 = (1+w_2)^2 \tag{8.32}$$

and hence three independent relations among the eight integration constants $(r_1, r_2, w_1, w_2, H_0, M, C, \varepsilon)$. Assuming that $(r_1, r_2, w_1, w_2, H_0, M, C)$ all remain finite as $\varepsilon \to 0$, i.e. they are all of order ε^0, then $r_1 \to r_{_H} = H^{-1}$ and $r_2 \to r_{_M}$ with $r_2 - r_1 = \Delta r$ of order ε, so that $r_{_H} \simeq r_{_M}$ to leading order in ε. Thus the boundary layer II straddles the location of the classical Schwarzschild and dS horizons, and $r_1 \to r_2$ coincide at $r_{_H} = r_{_M}$, becoming no longer independent in the limit $\varepsilon \to 0$. Since the mass M is a free parameter there remain three undetermined integration constants C, w_1, w_2 which satisfy the one relation (8.32) in addition to $\varepsilon \ll 1$ itself.

The important feature of this solution is that for any $\varepsilon > 0$ both f and h are of order ε but nowhere vanishing. Hence there is no event horizon or trapped surface, and t is a global Killing time. A photon experiences a very large, $\mathcal{O}(\varepsilon^{-\frac{1}{2}})$ but finite blue shift in falling into the shell from infinity.

The physical proper thickness of the shell in the metric (8.1) is

$$\ell = \int_{r_1}^{r_2} \frac{dr}{\sqrt{h}} = r_M\sqrt{\varepsilon}\int_{w_2}^{w_1} dw\, w^{-\frac{3}{2}} = 2r_M\sqrt{\varepsilon}\left(w_2^{-\frac{1}{2}} - w_1^{-\frac{1}{2}}\right) \tag{8.33}$$

to leading order in ε, and hence ℓ is $\mathcal{O}(\varepsilon^{\frac{1}{2}})$ and small compared to r_M. The magnitude of ε and hence of ℓ can be fixed only by consideration of the quantum effects that give rise to the phase transition boundary layer, which will be discussed in Sect. 8.6.

The entropy of the thin shell is obtained from its equation of state, which is that of a relativistic fluid in $1 + 1$ dimensions, and can be written in the form $p = \rho = (a^2/8\pi G)(k_B T/\hbar)^2$, where G is introduced for dimensional reasons, so that a^2 is a dimensionless constant. By the standard Gibbs relation, $Ts = p + \rho$ for a relativistic fluid with zero chemical potential, the local specific entropy density is

$$s(r) = \frac{a^2 k_B^2 T(r)}{4\pi\hbar^2 G} = \frac{a k_B}{\hbar}\left(\frac{p}{2\pi G}\right)^{\frac{1}{2}} = \frac{a k_B}{4\pi\hbar G r}\, w^{\frac{1}{2}} \tag{8.34}$$

for local temperature $T(r)$. The entropy of the fluid within the shell is

$$S = 4\pi\int_{r_1}^{r_2} \frac{s\, r^2\, dr}{\sqrt{h}} = \frac{a k_B r_M^2}{\hbar G}\sqrt{\varepsilon}\,\ln\left(\frac{w_1}{w_2}\right) \sim a\, k_B \frac{M\ell}{\hbar} \tag{8.35}$$

and of order $k_B M\ell/\hbar$ to leading order in ε, assuming a, w_1, w_2 are $\mathcal{O}(1)$. Since the interior region I has $\rho_V = -p_V$, $(Ts)_V = p_V + \rho_V = 0$ there. This is in accord with a gravitational Bose–Einstein condensate (GBEC) being a single macroscopic quantum state with zero entropy. Thus the entropy of the entire compact quasi-black hole (QBH) is given by the entropy of the shell alone. By (8.35) this is of order $k_B(r_M/L_{\rm Pl})^{\frac{3}{2}}$ for $\ell \sim \sqrt{L_{\rm Pl} r_M}$, or $S \sim \sqrt{\varepsilon} S_{BH} \ll S_{BH}$, far smaller than the Bekenstein-Hawking entropy (8.11). The $M^{3/2}$ scaling of (8.35) furthermore makes it comparable to the entropy of typical stellar progenitors of mass M, in the range of $10^{57} k_B$ to $10^{59} k_B$ for a solar mass and $M_\odot/m_N \sim 10^{57}$ nucleons. Thus there is no information paradox arising from an enormous entropy unaccountably associated with a BH horizon, if the horizon is replaced by a thin boundary layer of this kind. Since w is of order unity in the shell, the *local* temperature of the fluid within the shell is of order $T_H \sim \hbar/k_B r_M$ and quite cold, so that the typical quanta are *soft* with wavelengths of order r_M, and there is no transplanckian problem.

Because of the global timelike Killing field $K_{(t)} = \partial_t$ and absence of either an event horizon or an interior singularity, a gravitational condensate star shows no loss of unitarity, information paradox, or any conflict with either quantum theory or general principles of statistical mechanics. As a static solution, neither the interior nor the thin shell boundary layer emit Hawking radiation. A gravitational condensate star is both cold and dark, and hence in its appearance to distant observers and in its external geometry, in most respects indistinguishable from a BH.

8.3.3 The Lanczos–Israel Conditions at the r_1 and r_2 Boundaries

In the original gravastar model of [133, 135, 137] the Lanczos–Israel junction conditions [109, 110, 116, 147]

$$[K_a^{\ b}] \equiv K_{a+}^{\ b} - K_{a-}^{\ b} = -4\pi G\left(2S_a^{\ b} - \delta_a^{\ b} S_c^{\ c}\right) \tag{8.36}$$

were used to relate the discontinuity in the extrinsic curvature

$$K_{ab} = -n_\mu \upsilon_{(b)}^{\nu} \nabla_\nu \upsilon_{(a)}^{\mu} = -n_\mu \upsilon_{(b)}^{\nu}\left(\partial_\nu \upsilon_{(a)}^{\mu} + \Gamma^{\mu}_{\ \nu\lambda}\upsilon_{(a)}^{\lambda}\right) \tag{8.37}$$

to the surface stress tensor $S_a^{\ b}$, where n^μ is the spacelike normal to the surface at fixed r, normalized to

$$n^\mu n_\mu = 1 \tag{8.38}$$

in the full four-dimensional spacetime. The indices a, b in (8.36) are intrinsic to the surface and thus range over t, θ, ϕ only, while $\upsilon_{(a)}$ are a set of three mutually orthogonal basis vectors, orthogonal also to n_μ, i.e. satisfying $\upsilon_{(a)}^{\mu} n_\mu = 0$, and normalized so as to project the four-dimensional metric onto the hypersurface of fixed r. Thus

$$\gamma_{ab} = \upsilon_{(a)}^{\mu}\upsilon_{(b)}^{\nu} g_{\mu\nu} \tag{8.39}$$

is the induced metric on the three-dimensional hypersurface, the inverse of which must be used to raise the indices a, b, c. Since the normal to the surface at fixed r and basis vectors so defined have components

$$n^\mu = \sqrt{h(r)}\, \delta^\mu_{\ r}, \qquad \upsilon_\mu^{(a)} = \begin{cases} \delta_\mu^{\ a}, & \mu = t, \theta, \phi \\ 0, & \mu = r \end{cases} \tag{8.40}$$

in the coordinates (8.1), the non-vanishing components of $K_a^{\ b} = K_{ac}\gamma^{cb}$ in these coordinates are

$$K_t^{\ t} = \frac{\sqrt{h}}{f}\,\Gamma_{rtt} = \frac{\sqrt{h}}{2f}\frac{df}{dr} \tag{8.41a}$$

$$K_\theta^{\ \theta} = K_\phi^{\ \phi} = -g^{\theta\theta}\sqrt{h}\,\Gamma_{r\theta\theta} = \frac{\sqrt{h}}{r} \tag{8.41b}$$

with the result that the non-vanishing components of the surface stress tensor from (8.36) are

$$S_t^{\ t} = \frac{1}{4\pi G}\left[\frac{\sqrt{h}}{r}\right] \tag{8.42a}$$

$$S_\theta^{\ \theta} = S_\phi^{\ \phi} = \frac{1}{8\pi G}\left(\left[\frac{\sqrt{h}}{2f}\frac{df}{dr}\right] + \left[\frac{\sqrt{h}}{r}\right]\right) \tag{8.42b}$$

on the timelike surface interfaces at r_1 and r_2. Since the metric function h is continuously matched at the interfaces, $[K_\theta^{\ \theta}]$ and $S_t^{\ t}$ vanish, while $S_\theta^{\ \theta}$ is of order $\varepsilon^{-\frac{1}{2}}$. Making use of the Eqs. (8.28)–(8.31)

$$\frac{\sqrt{h}}{2f}\frac{df}{dr} \simeq \frac{1}{2r_{_M}}\sqrt{\frac{w}{\varepsilon}} \qquad \text{in region II} \tag{8.43}$$

to leading order in ε, which enables evaluation of the discontinuities in (8.42). The non-zero angular components are then [137] [2]

$$S_\theta^{\ \theta}\big|_{r=r_1} = S_\phi^{\ \phi}\big|_{r=r_1} \equiv -\sigma_1 = \frac{1}{32\pi G^2 M}\frac{(3+w_1)}{(1+w_1)}\left(\frac{w_1}{\varepsilon}\right)^{\frac{1}{2}} \tag{8.44a}$$

$$S_\theta^{\ \theta}\big|_{r=r_2} = S_\phi^{\ \phi}\big|_{r=r_2} \equiv -\sigma_2 = -\frac{1}{32\pi G^2 M}\frac{w_2}{(1+w_2)}\left(\frac{w_2}{\varepsilon}\right)^{\frac{1}{2}} \tag{8.44b}$$

respectively, to leading order in ε, at r_1 and r_2.

The signs of these surface stresses correspond to the inner surface at r_1 exerting an outward force and the outer surface at r_2 exerting an inward force, i.e. both surfaces exert a confining force on the thin shell layer in region II. Clearly these large transverse surface stresses violate the perfect fluid ansatz at the interfacial boundaries. Nevertheless, as it will turn out from semi-classical estimates in Sect. 8.6, $\varepsilon^{-\frac{1}{2}} \sim (M/M_{\rm pl})^{\frac{1}{2}}$, so that the surface tensions (8.44) are of order $M^{-\frac{1}{2}}$ and far from Planckian. Thus the matching of the metric at the phase interfaces r_1 and r_2, analogous

[2] The sign conventions in [133, 135] are such that $\sigma_{1,2}$ there are the *negative* of the surface stress tensors $S_\theta^{\ \theta} = S_\phi^{\ \phi}$ properly defined here. Equations (C5) and (C7) of [136] also have an overall sign change from the Lanczos–Israel equation (8.36) for $S_a^{\ b}$, such that η, σ of (C7) have the same values as η, σ in [133, 135].

to that across stationary shocks in hydrodynamics, should be reliable in the mean field semi-classical approximation. The time component of the surface stress tensor at r_1 and r_2 vanishes and makes no contribution to the Misner-Sharp mass function $\mu(r) = 2Gm(r)$ at either of the two interfaces.

The Misner-Sharp mass-energy within the shell

$$E_{\rm II} = 4\pi \int_{r_1}^{r_2} \rho\, r^2 dr = \varepsilon M \int_{w_2}^{w_1} \frac{dw}{w}\,(1+w) = \varepsilon M \Big[\ln\left(\frac{w_1}{w_2}\right) + w_1 - w_2\Big] \tag{8.45}$$

to leading order in ε, is of order $M_{\rm Pl}$ and also extremely small. In this accounting essentially all of the mass of the object comes from the energy density of the vacuum condensate in the interior, even though the shell is responsible for all of its entropy.

8.3.4 Thermodynamic Stability

In [135, 137] a thermodynamic argument for the stability of the first gravastar model was provided. This was based on analysis of the entropy functional which can be expressed in the form

$$S = \frac{ak_B}{\hbar G} \int_{r_1}^{r_2} r\, dr \left(\frac{d\mu}{dr}\right)^{\frac{1}{2}} \left(1 - \frac{\mu(r)}{r}\right)^{-\frac{1}{2}} \tag{8.46}$$

in the thin shell region II where the equation of state $p = \rho$, (8.34), and the relation $d\mu(r) = 8\pi G\rho r^2 dr = w dr$ has been used. The first variation of (8.46) with respect to $\mu(r)$ with the endpoints r_1 and r_2 fixed vanishes, i.e. $\delta S = 0$ by the Einstein equations (8.21) for a static, spherically symmetric star. Thus any solution of Eqs. (8.21)–(8.22) is guaranteed to be an extremum of S [72]. This is consistent with regarding Einstein's equations as an effective field theory (EFT) of low energy hydrodynamics, strictly valid only at long wavelengths.

The second variation of (8.46) is

$$\delta^2 S = \frac{ak_B}{4\hbar G} \int_{r_1}^{r_2} r\, dr \left(\frac{d\mu}{dr}\right)^{-\frac{3}{2}} h^{-\frac{1}{2}} \left\{ - \left[\frac{d(\delta\mu)}{dr}\right]^2 + \frac{(\delta\mu)^2}{r^2 h^2}\frac{d\mu}{dr}\left(1 + \frac{d\mu}{dr}\right)\right\} \tag{8.47}$$

when evaluated on the solution. Associated with this quadratic form in $\delta\mu$ is a second order linear differential operator $\mathcal{L}$ of the Sturm–Liouville type, viz.

$$\mathcal{L}\chi = \frac{d}{dr}\left\{ r \left(\frac{d\mu}{dr}\right)^{-\frac{3}{2}} h^{-\frac{1}{2}} \frac{d\chi}{dr}\right\} + \frac{h^{-\frac{5}{2}}}{r}\left(\frac{d\mu}{dr}\right)^{\frac{1}{2}}\left(1 + \frac{d\mu}{dr}\right)\chi. \tag{8.48}$$

This operator possesses two solutions satisfying $\mathcal{L}\chi_{_0} = 0$, obtained by variation of the classical solution, $\mu(r; r_1, r_2)$ with respect to the parameters (r_1, r_2). Indeed by changing variables from r to w and using the explicit solution (8.28)–(8.29) it is readily verified that one solution to $\mathcal{L}\chi_{_0} = 0$ is $\chi_{_0} = 1 - w$, from which the second linearly independent solution $(1 - w) \ln w + 4$ may be obtained. Since these correspond to varying the positions of the r_1, r_2 interfaces, neither $\chi_{_0}$ vanishes at (r_1, r_2) and neither is a true zero mode. However, by setting $\delta\mu = \chi_{_0}\, \psi$, where ψ does vanish at the endpoints and inserting this into the second variation (8.47) one obtains

$$\delta^2 S = -\frac{ak_B}{4\hbar G}\int_{r_1}^{r_2} r\, dr\, \left(\frac{d\mu}{dr}\right)^{-\frac{3}{2}} h^{-\frac{1}{2}} \chi_{_0}^2 \left(\frac{d\psi}{dr}\right)^2 < 0 \tag{8.49}$$

after integration by parts and using $\delta\mu = 0$ at the endpoints and $\mathcal{L}\chi_{_0} = 0$.

Since (8.49) is negative definite, the entropy of the solution is maximized with respect to radial variations that vanish at the endpoints, with fixed total energy. Since deformations with non-zero angular momentum decrease the entropy even further, stability under radial variations is sufficient to demonstrate that the solution is stable to all small perturbations with the given boundary conditions. On the other hand, the boundary conditions that the radial variations $\delta\mu$ vanish at the interface boundaries at r_1 and r_2 may themselves be questioned. This apparent shortcoming of the thermodynamic stability proof provided in [135, 137] can only be addressed within a a fully dynamical theory where the boundary conditions follow from the dynamical variational principle itself, rather than being imposed, as they were in the original gravastar model of [135, 137]. The EFT introduced in [143] and discussed in Sect. 8.7 is expected to settle the question of stability with a full dynamical treatment based on an effective Lagrangian whose Euler–Lagrange equations possess a gravitational condensate star solution, removing any additional assumptions of boundary conditions beyond those of a non-singular interior and an asymptotically flat exterior.

Even from the preliminary analysis of [135, 137], it should be clear that the main qualitative results do not depend crucially on the specific equation of state used in the thin shell transition region II. The important features are the interior dS and exterior Schwarzschild regions and the location of the thin shell at or straddling the mutual Schwarzschild and dS horizons, where they are joined. The assumption of the specific equation of state $p = \rho$ of the thin shell in the first gravastar model of [133, 135] should therefore be regarded as an ansatz, illustrating a proof of principle of joining the interior dS geometry of a GBEC condensate with $\Lambda_{\rm eff} > 0$ to the exterior Schwarzschild solution, with the possibility left open to modification of the thin shell transition region when a more complete or fundamental theoretical framework became available [137]. A proposal for that more fundamental EFT framework will be described in Sects. 8.6 and 8.7.

8.4 The Schwarzschild Constant $\bar{\rho}$ Interior Solution Revisited: Evading the Buchdahl Bound and Determination of C

A major step and refinement of the original gravastar proposal took place in 2015, with the re-analysis of the Schwarzschild constant density interior solution (8.5) in the limit of $r_s \to r_M$, i.e. for $\varepsilon = 0$, leading also to an improved understanding and formulation of the Lanczos–Israel conditions for null hypersurfaces.

In 1959 H. A. Buchdahl proved that in order for a static, spherically symmetric solution of Einstein's equations of mass M and radius r_s to remain everywhere finite, its compactness $GM/c^2 r_s$ must not be greater than 4/9, or equivalently

$$r_s \geq \frac{9}{8} r_M \tag{8.50}$$

the same as from the limiting case (8.6) of the constant density interior Schwarzschild solution (8.5). The Buchdahl bound (8.50) depends upon three necessary conditions:

(i) The pressure is everywhere isotropic: $p_\perp(r) = p(r)$,
(ii) The density profile is monotonically non-increasing outward: $\rho' = \frac{d\rho}{dr} \leq 0$,
(iii) The metric functions $f(r)$, $h(r)$ and the derivative $\frac{df}{dr}$ are continuous at $r = r_s$,

in addition to Einstein's equations holding everywhere. The first condition (i) is satisfied by all ordinary fluid equations of state. The second condition (ii) is physically reasonable since an outer layer of higher density material would generally result in an instability to the denser material falling towards the center. The third condition (iii) precludes a discontinuity at the surface of the star.

When the inequality (ii) is saturated, i.e. $\rho = \bar{\rho}$ is a constant, one recovers the Schwarzschild interior solution (8.5), which actually disallows the equality in (8.50), since in that marginal case the pressure (8.5d) diverges at $r_0 = 0$. This is consistent with the bound on the central pressure [127]

$$p(0) \geq \bar{\rho} \left[\frac{1 - \sqrt{1 - r_M/r_s}}{3\sqrt{1 - r_M/r_s} - 1} \right] \tag{8.51}$$

when condition (ii) holds, since this lower bound diverges when $r_s = \frac{9}{8} r_M$.

Because the bound for the general spherically symmetric static solution obeying conditions (i)–(iii) is saturated by the limiting case of $\rho' = 0$, the behavior of the constant density solution (8.5) itself is of fundamental interest as the limit (8.50) is reached and r_s is then reduced further.

When one does consider the constant density solution (8.5) for $r_s < \frac{9}{8} r_M$, some of its rather remarkable features quickly become apparent. First, the pressure divergence which first appears at the origin moves out to a spherical surface of finite radius r_0 given by (8.7), and a *new regular solution* for $0 \leq r < r_0$ opens up behind it, with $D < 0$, *negative* pressure, and $\bar{\rho} + 3p < 0$, violating the strong energy condition.

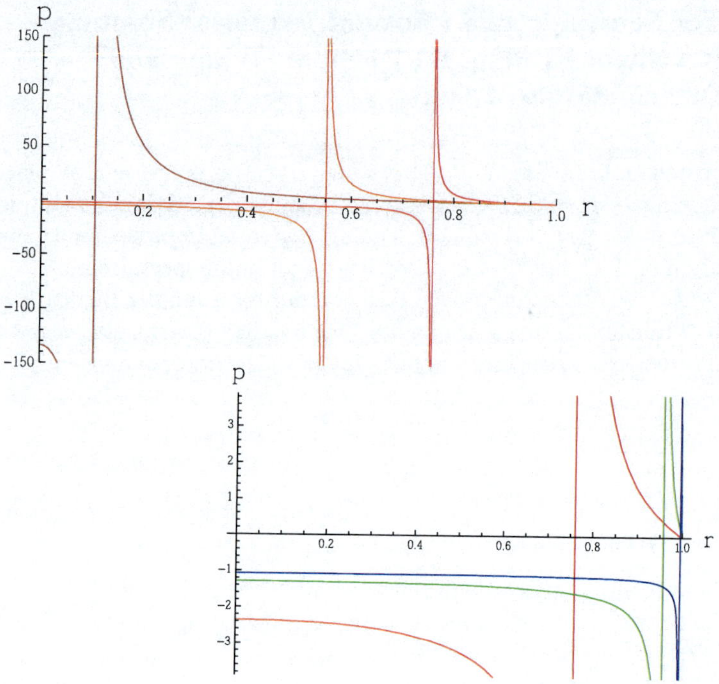

Fig. 8.2 Pressure (in units of $\bar{\rho}$) as a function of r (in units of r_s) of the interior Schwarzschild solution (8.5) for various values of the parameter $r_s/r_M < 9/8 = 1.125$. The upper plot shows $p(r)$ for the values $r_s/r_M = 1.124, 1.087, 1.053$ (brown, orange, red curves), where the divergence in the pressure occurs at $r_0/r_s = 0.106, 0.552, 0.761$ respectively. The lower plot shows $p(r)$ for the values $r_s/r_M = 1.053, 1.010, 1.001$ (red, green, blue curves), where the divergence in the pressure occurs at $r_0/r_s = 0.761, 0.959, 0.996$ respectively. For $r < r_0$ the pressure and $\bar{\rho} + 3p$ are negative. Note the change of vertical scale in the plots (the red curves are the same in each) and the pointwise approach of the negative interior pressure $p \to -\bar{\rho}$ as r_s approaches the Schwarzschild radius r_M from above and r_0 approaches r_M from below

The metric functions f, h remain non-negative in the interior, with $f(r_0) = 0$ at $r = r_0$ only, cf. Figs. 8.2, 8.3 and 8.4.

As the star is compressed beyond the Buchdahl bound and its radius approaches the Schwarzschild radius $r_s \to r_M^+$ from outside, (8.7) shows that the radius of the sphere where the pressure diverges and $f(r_0) = 0$ moves from the origin to the outer edge of the star, i.e. $r_0 \to r_M^-$, and in that limit the interior solution with negative pressure comes to encompass the entire interior region $0 \leq r < r_s$, excluding only the outer boundary at $r_s = r_M$. Finally, and most remarkably of all, since in this limit $H^2 r_s^2 = r_M/r_s \to 1$, inspection of (8.5d) shows that the entire interior solution then has *constant* negative pressure

$$p = -\bar{\rho}, \qquad \text{for} \qquad r < r_s = r_0 = r_M = 2GM \tag{8.52}$$

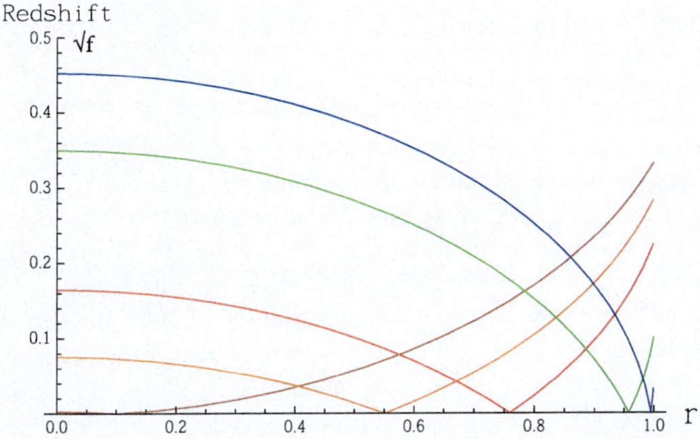

Fig. 8.3 The redshift factor $\sqrt{f}$ as a function of r (in units of r_s) of the interior Schwarzschild solution (8.5) for the same values of the parameter $r_s/r_M < 9/8 = 1.125$ as in Fig. 8.2. The brown, orange, red, green and blue curves are for the values $r_s/r_M = 1.124, 1.087, 1.053, 1.010, 1.001$ respectively. Note the approach of the zero of $\sqrt{f}$ at r_0 towards r_s from below as r_s approaches the Schwarzschild radius r_M from above. In this limit $\sqrt{f(0)} \to 1/2$

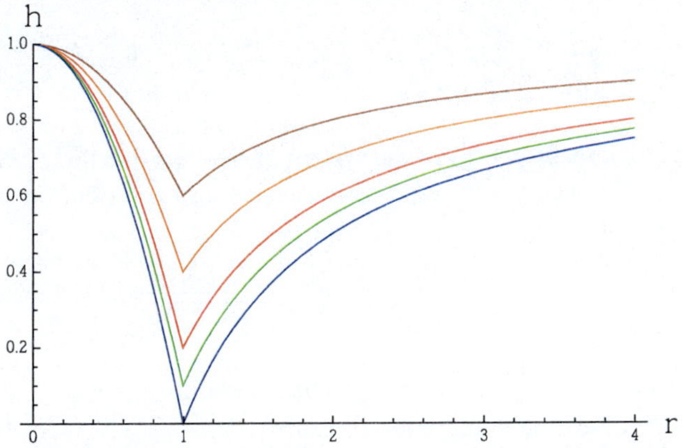

Fig. 8.4 The metric function h as a function of r (in units of r_s) of the interior and exterior Schwarzschild solution for the values of the parameter $r_s/r_M = 2.500, 1.667, 1.250, 1.111, 1.000$ (brown, orange, red, green and blue curves) respectively. The minimum of h approaches zero at $r = r_s$, as r_s approaches the Schwarzschild radius r_M from above. Note the cusp-like discontinuity of the derivatives dh/dr at r_s and df/dr at r_0 in this and the previous figure, which coincide at $r_0 = r_s$, when $r_s = r_M$

with the metric functions

$$f(r) = \frac{1}{4}\,(1 - H^2 r^2) = \frac{1}{4}\,h(r) = \frac{1}{4}\left(1 - \frac{r^2}{r_M^2}\right), \qquad H = \frac{1}{r_M} \tag{8.53}$$

corresponding to a regular static patch of pure dS space, although one in which g_{tt} is $\frac{1}{4}$ its usual value, so that the passage of time in the interior is modified from what would be expected in the usual static coordinates of dS space. In other words the constant density interior solution which Schwarzschild found in 1916 becomes essentially the gravitational condensate star solution of 2001 and Sect. 8.3, in which the thin shell boundary layer is of infinitesimal thickness ($\varepsilon = 0$), residing exactly at the null hypersurface $r = r_M$, fixing the value of $C = \frac{1}{4}$ of (8.24) in the interior dS patch unambiguously.

Typical profiles of the pressure $p(r)$ and $f^{\frac{1}{2}}$ for values of the radius r_s in the range $r_M < r_s < \frac{9}{8} r_s$ and the approach of $r_s \to r_M$ are shown in Figs. 8.2 and 8.3. Insight into the pressure divergence at $r = r_0$ and vanishing of $f(r_0)$ is obtained by a careful examination of the covariant Komar mass-energy for a time independent geometry possessing the static Killing field $K_{(t)} = \partial_t$, with components $K^\mu_{(t)} = \delta_t{}^\mu$ in coordinates (8.1). This Komar mass is defined by

$$M = \int_V \left(-\,2\,T_\mu{}^\nu + T_\lambda{}^\lambda\,\delta_\mu{}^\nu\right) K^\mu_{(t)}\, d^3\Sigma_\nu \;+\; \frac{1}{4\pi G}\int_{\partial V_-} \kappa\, dA \tag{8.54}$$

expressing the total mass-energy of the system M in terms of a three-volume integral of the matter stress-energy, plus a possible surface flux contribution from the inner two-surface, where

$$\kappa = -\nabla^\mu K^\nu_{(t)}\, e^0{}_{[\mu} e^1{}_{\nu]} = \frac{1}{2}\sqrt{\frac{h}{f}}\,\frac{df}{dr} \tag{8.55}$$

is the surface gravity, defined in terms of the vierbeins $e^0{}_t = \sqrt{-g_{tt}} = f^{\frac{1}{2}}$, $e^1{}_r = \sqrt{g_{rr}} = h^{-\frac{1}{2}}$ with the areal integration measure

$$dA = r^2 \sin\theta\; d\theta\; d\phi \tag{8.56}$$

on the spherical two-surface of constant t and r.

Since the volume integration measure in (8.54) is

$$d^3\Sigma_\nu = \delta^t{}_\nu \sqrt{\frac{f}{h}}\; r^2 \sin\theta\; dr\; d\theta\; d\phi \tag{8.57}$$

it contains a factor of $\sqrt{f}$ which vanishes at exactly the same $r = r_0$ as where $p(r_0)$ diverges, so that the pressure singularity at $r = r_0$ and in fact the total volume integrand in (8.54) is [136]

$$\frac{1}{G}\frac{d}{dr}\big[r^2\kappa(r)\big] = 4\pi\sqrt{\frac{f}{h}}\,r^2\left(\rho + p + 2p_\perp\right) = 4\pi r^2\bar{\rho}\,\mathrm{sgn}(D) + \frac{8\pi}{3}\,{r_0}^3\,\bar{\rho}\,\delta(r - r_0) \tag{8.58}$$

with a well-defined δ-function contribution having support at $r = r_0$, which when integrated over dr gives a finite result. This δ-function contribution to the volume integral in (8.54) is a result of the discontinuity in the surface gravities

$$\kappa_\pm \equiv \lim_{r\to r_0^\pm} \kappa(r) = \pm\frac{4\pi G}{3}\,\bar{\rho}\,r_0 = \pm\frac{r_M\,r_0}{2r_s^3} \tag{8.59}$$

on either side of the $r = r_0$ surface, which results in a transverse pressure and contribution to the integrand (8.58) of

$$8\pi\sqrt{\frac{f}{h}}\,r^2\left(p_\perp - p\right) = \frac{8\pi}{3}\,\bar{\rho}{r_0}^3\,\delta(r - r_0) = 2M\left(\frac{r_0}{r_s}\right)^3\delta(r - r_0) \tag{8.60}$$

localized on the spherical surface at $r = r_0$, and hence a breakdown of the isotropic pressure assumption on that surface. This gives the finite contribution

$$E_s = 2M\left(\frac{r_0}{r_s}\right)^3 \to 2M \tag{8.61}$$

to the Komar energy (8.54) of the surface, which becomes $2M$ in the limit $r_0 \to r_s$.

The Schwarzschild constant density interior solution (8.5) in the limit $r_s \to r_M$ therefore has the physical interpretation of a gravastar with a well defined surface tension energy (8.61) localized at $r = r_M = r_H$, precisely where the BH and dS horizons would be, and the two geometries are joined instead.

The dynamical stability of this sequence of constant density Schwarzschild stars beyond the Buchdahl limit has been studied in [166].

8.4.1 Redshift Modified Boundary Conditions on a Null Hypersurface

It is clear that the presence of the $\sqrt{f} \to 0$ factor in the volume measure (8.57) is critical in rendering the Komar energy on the null surface at $r = r_0$ finite. This points to the modification of the Lanczos–Israel junction conditions necessary to give a finite, well-defined and physical surface stress tensor on a null surface where the normal n^μ itself becomes null and cannot be normalized as in (8.38). To handle this case one should define the redshifted extrinsic curvature tensor

$$\mathbf{K}_a^{\ b} = -\gamma^{bc}\mathbf{n}_\mu \upsilon_{(c)}^{\ \nu} \nabla_\nu \upsilon_{(a)}^{\ \mu} = -\gamma^{bc}\mathbf{n}_\mu \upsilon_{(c)}^{\ \nu} \Gamma^\mu_{\ \nu\lambda} \upsilon_{(a)}^{\ \lambda} \tag{8.62}$$

where the normal $\mathbf{n}$ to the surface at constant r approaching r_0 satisfies

$$\mathbf{n} \cdot \mathbf{n} = (\mathbf{n}^r)^2 g_{rr} = f(r), \qquad \text{so that} \qquad \mathbf{n}^r = \sqrt{fh} \tag{8.63}$$

which goes to zero on the null surface, instead of (8.38) and the first member of (8.40). This gives a redshift modified extrinsic curvature tensor (8.62) with the non-vanishing components

$$\mathbf{K}_t^{\ t} = \sqrt{\frac{h}{f}}\, \Gamma_{rtt} = \frac{1}{2}\sqrt{\frac{h}{f}}\frac{df}{dr} = \kappa \tag{8.64a}$$

$$\mathbf{K}_\theta^{\ \theta} = \mathbf{K}_\phi^{\ \phi} = -g^{\theta\theta}\sqrt{fh}\, \Gamma_{r\theta\theta} = \frac{\sqrt{fh}}{r} \to 0 \tag{8.64b}$$

instead of (8.41). This redshifted extrinsic curvature has the finite discontinuities

$$\left[\mathbf{K}_t^{\ t}\right] = [\kappa] \tag{8.65a}$$

$$\left[\mathbf{K}_\theta^{\ \theta}\right] = \left[\mathbf{K}_\phi^{\ \phi}\right] = 0 \tag{8.65b}$$

on the null surface, so that applying the standard Lanczos–Isreal conditions, but to this redshift modified extrinsic curvature $\mathbf{K}_a^{\ b}$, gives the surface stress tensor density

$${}^{(\Sigma)}T_a^{\ b}\sqrt{\frac{f}{h}} = \mathcal{S}_a^{\ b}\,\delta(r - r_0), \qquad 8\pi G\, \mathcal{S}_a^{\ b} = -\left[\mathbf{K}_a^{\ b}\right] + \delta_a^{\ b}\left[\mathbf{K}_c^{\ c}\right] \tag{8.66}$$

instead of (8.42), and a finite surface tension

$$\mathcal{S}_\theta^{\ \theta} = \mathcal{S}_\phi^{\ \phi} = \tau_s = \frac{\Delta\kappa}{8\pi G} = \frac{M\, r_0}{8\pi\, r_s^3} \tag{8.67}$$

of the null surface at $r = r_0$, instead of (8.44) which diverges as $\varepsilon \to 0$.

The factor of $\sqrt{f/h}$ in (8.67) and modified extrinsic curvature (8.62) is clearly necessitated by the covariant integration measure (8.57) in the Komar energy integral (8.54). Once $\mathbf{K}_a^{\ b}$ is defined as in (8.62), with the normal $\mathbf{n}$ normalized as in (8.63) so as to become a null vector in the horizon limit, the surface stress tensor (8.67) on a null hypersurface can be determined unambiguously by application of the junction conditions (8.66), which are of the same form as the standard Lanczos–Israel conditions, but for $\mathbf{K}_a^{\ b}$ rather than $K_a^{\ b}$ of (8.37), and crucially for $\mathbf{K}_a^{\ b}$ with one covariant and one contravariant index. Raising or lowering these indices would introduce factors of f or $1/f$ that would again render the formalism empty or ambiguous on a null hypersurface where $f(r_0)$ vanishes. There is no need for introducing a 'transverse' normal and 'transverse' curvature tensor as proposed in [27].

This procedure can also be justified and verified explicitly from the form of the Einstein tensor in the spacetime metric (8.1). Since the metric functions f, h must be equal on the null hypersurface approached from either side, but their derivatives may not be, the first derivatives f', h' can contain Heaviside step functions. Thus Dirac δ-functions signaling integrable surface contributions can only appear from second derivative terms f'', h''. The angular components $G_\theta^{\ \theta} = G_\phi^{\ \phi}$ in the metric (8.1) indeed contain a second derivative of f with respect to r, but they contain singular $f'/f, h'/h$ and h/f terms as well, which become infinite or ill-defined on the null hypersurface where $f(r_0) = 0$. However, the tensor *densities*

$$\sqrt{\frac{f}{h}}\, G_\theta^{\ \theta} = \sqrt{\frac{f}{h}}\, G_\phi^{\ \phi} = \frac{1}{2}\frac{d}{dr}\left(\sqrt{\frac{h}{f}}\frac{df}{dr}\right) + \frac{1}{2r}\left(\sqrt{\frac{f}{h}}\frac{dh}{dr} + \sqrt{\frac{h}{f}}\frac{df}{dr}\right) \tag{8.68}$$

$$= \left[\frac{1}{2}\sqrt{\frac{h}{f}}\frac{df}{dr}\right]\delta(r - r_0) + \cdots \tag{8.69}$$

combine these terms into a total r derivative. It is clear that the remaining terms of (8.68), although discontinuous, do not contain any δ-function contributions to the surface stress tensor at $r = r_0$, while the δ-function term coming from the total derivative first term is precisely the well-defined discontinuity in the surface gravity κ of (8.55). The δ-function comes automatically with the correct measure factor of the Komar mass integral, so that the surface stress tensor in the angular components on the null hypersurface at $r = r_0$ is defined by (8.66), directly from Einstein's equations

Thus once the technical issue of the correct modification of the Lanczos–Israel junction conditions on a null hypersurface is determined, with the finitely integrable distributional surface tensor at $r = r_0$ given by (8.66), the Schwarzschild constant density interior solution becomes not a pathological case arguing for the necessary collapse to a BH singularity for masses violating the Buchdahl bound (8.50), but on the contrary a well-defined and physically sensible solution to Einstein's equations for $r_{_S} = r_{_M}$, which shows that evading the Buchdahl bound requires discarding the first condition (i) of isotropy assumed in its proof, and the appearance of a physical surface with surface tension (8.67), as well as allowing a negative pressure interior. Furthermore and remarkably, in the limit $r_{_S} \to r_{_M}^+$ from above, the spherical null surface becomes coincident with the Schwarzschild radius itself $r_0 \to r_{_M}^-$ from below, and the negative pressure of the interior also becomes a constant $p = -\bar{\rho}$, which is just the $\varepsilon \to 0$ classical limit of the gravitational condensate star of [135, 137]. Thus this viable non-singular alternative for the final state of gravitational collapse might have been found just a few months after Einstein introduced GR.

In the gravastar limit $r_{_S} \to r_{_M}$ the surface stress tensor and surface tension becomes

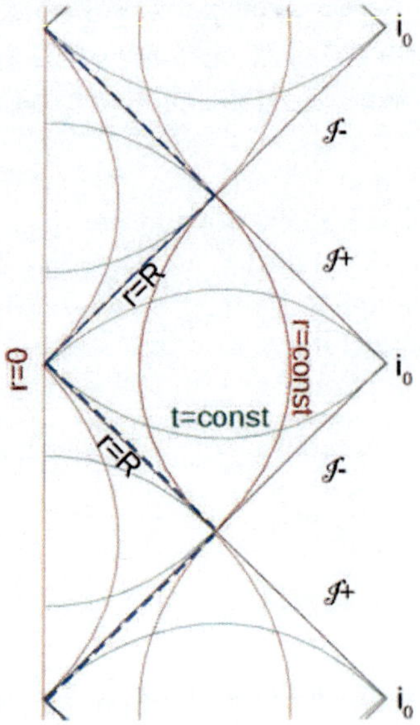

Fig. 8.5 Conformal diagram for the gravastar with angular variables (θ, ϕ) suppressed, in the idealized case ε=0. Schwarzschild exterior (diamonds) and regular de Sitter static patch interior (triangles), are joined at their mutual null horizons at $r= R(=r_M=r_H)$. A few typical constant r timelike worldlines (red curves) and constant t spacelike hypersurfaces (green curves) are also shown

$$\mathcal{S}_\theta^{\ \theta} = \mathcal{S}_\phi^{\ \phi} = \tau_s = c^2 \frac{\Delta\kappa}{8\pi G} = \frac{Mc^2}{8\pi\, r_{_M}^2} = \frac{c^6}{32\pi\, G^2 M} \tag{8.70}$$

after restoring the factors of c. The discontinuity of the derivatives f', h' at $r = r_{_M}$ and non-zero stress tensor there in this limit clearly does not satisfy the analytic continuation assumption, or 'uneventful free fall' hypothesis used to obtain Fig. 8.1. The conformal diagram of the idealized classical gravastar in which $\varepsilon = 0$ identically is shown in Fig. 8.5 instead.

It is important that in this limit $f(r_{_M}) = h(r_{_M}) = 0$, vanishing at one radius only, but never becoming negative. Hence light rays can hover there indefinitely, but there is no trapped surface on which outgoing light rays are bent inward, so this critical assumption of the 1965 Penrose theorem is not satisfied. There is no singularity in the interior, only a regular static patch of dS space. Since $p = -\bar{\rho}$ also violates the strong energy condition, the singularity theorems relying on this condition instead of a trapped surface are also evaded.

More realistically one may expect that f, h become small, $\mathcal{O}(\varepsilon)$, but never exactly vanishing when quantum effects are included, as in the first gravastar model. Then light rays can also escape the interior in principle, and the multiple copies of the conformal diagram of Fig. 8.5 would no longer appear. This shows again that global properties are very sensitive to small changes in the near-horizon region. The conformal diagram of a gravastar with any finite $\varepsilon > 0$ (however small) is one triangular wedge only of Fig. 8.5, resembling that of any regular body such as a star, albeit with a highly redshifted timelike tube, which is the thin transitional boundary layer between the Schwarzschild exterior and dS interior.

To relate (8.70) to (8.44) of the earlier first gravastar model of Sect. 8.3, one observes that multiplying (8.43) by the redshift factor of $\sqrt{f(r)}$ needed in the definitions of **K** and κ for null hypersurfaces gives

$$\kappa_{\rm II} \simeq \frac{1}{2r_{_M}}\sqrt{\frac{wf}{\varepsilon}} \simeq \frac{1}{2r_{_M}}\sqrt{\frac{w_1 f(r_1)}{\varepsilon}} \simeq \frac{1+w_1}{4r_{_M}}$$

in the entire region II, which is both finite and *constant* as $\varepsilon \to 0$. Thus adding the contributions to the surface stress tensor from the discontinuities of $\mathbf{K}_a^{\ b}$ at the two surfaces at r_1 and r_2, give equal and opposite contributions from the intermediate region II, which cancel. This leaves only the total discontinuity from the interior dS region I to exterior Schwarzschild region III, i.e.

$$\begin{aligned} \sqrt{f(r_1)}\, S_\theta^{\ \theta}\big|_{r=r_1} + \sqrt{f(r_2)}\, S_\theta^{\ \theta}\big|_{r=r_2} &= \frac{3+w_1}{64\pi G^2 M} - \frac{w_2}{64\pi G^2 M}\frac{1+w_1}{1+w_2} \\ &= \frac{1}{16\pi G^2 M} = \frac{\Delta\kappa}{8\pi G} \end{aligned} \tag{8.71}$$

where (8.32) with $C = 1/4$, implying $w_1 = 1 + 2w_2$ has been used. Hence the total surface tension is the very thin boundary layer on the null horizon is obtained, to leading order in $\varepsilon \ll 1$ (even if non-zero), in which no reference to the intermediate region II, its equation of state or integration constants w_1, w_2 at all appear in the final result. Thus the multiplication of the extrinsic curvature and surface stress tensor defined by (8.36)–(8.42) by an additional factor of $\sqrt{f}$, as required by the volume measure in the the Komar energy integral gives a finite result when the normal in the definition (8.37) declines into tangency with the hypersurface and **n** itself becomes a null vector [27, 30]. This result is finite and universal, in the sense of being entirely independent of any assumptions of the equation of state or any other characteristics of the thin shell boundary layer interposed between the exterior Schwarzschild and interior dS geometries, joined at their mutual null horizons, in the classical limit $\varepsilon \to 0$.

8.4.2 The 'First Law' for Spherically Symmetric Gravastars

For the sourcefree exterior BH solution of (8.1), the volume term in (8.54) vanishes if integrated from $r = r_{_M}$ to $r = \infty$, but since

$$\kappa_{ext}(r) = \frac{GM}{r^2} \to \kappa_+ = \frac{1}{4GM} = \frac{1}{2r_{_M}} \qquad as \qquad r \to r_{_M} \tag{8.72}$$

the surface integral at the horizon boundary gives the entire mass M of the BH. Already at the level of the Smarr relation (8.12), the coefficient of $dA_{_H}$ appears to be some sort of surface tension, but this interpretation is unclear since the BH horizon is assumed to be a causal boundary only, with vanishing surface stress tensor.

On the other hand, for the full interior + exterior Schwarzschild solution in the gravastar limit with $r_s = r_M$, there is a surface stress tensor and physical surface tension (8.70) at $r = r_M$. Since the solution is regular at the origin, one may integrate (8.54) from $r = 0$ to $r = \infty$. There is no contribution from the last explicit surface area integral in (8.54), but the integrand (8.58) of the volume integral in (8.54) has a Dirac δ-function contribution at $r = r_M$ from the discontinuity (8.59) in the surface gravities there, which becomes

$$\Delta\kappa \equiv \kappa_+ - \kappa_- = \frac{1}{2GM} = \frac{1}{r_M} \tag{8.73}$$

for $r_s = r_M$. This surface tension itself gives rise to a surface energy (8.61), which becomes just $2M$ in the gravastar limit. The first term in (8.58) gives the volume contribution $-M$ to the Komar mass integral from the negative pressure in the interior. The sum of these and result from (8.54) is

$$E = E_v + E_s = -M + 2M = M \tag{8.74}$$

giving again just the total mass $+M$. The differential relation

$$dE = dE_v + \tau_S \, dA_H = -dMc^2 + \tau_S \, dA_H \tag{8.75}$$

for the volume and surface terms in the Komar mass may be derived from this [136], showing that identification of τ_s, the coefficient of dA_H with the surface tension of the physical surface located at $r = r_M = r_H$ for the full (interior + exterior) Schwarzschild solution is fully justified. Since τ_s depends on the discontinuity $\Delta\kappa = \kappa_+ - \kappa_- = 2\kappa_+$ of equal and opposite surface gravities at the surface it contributes twice the value from the Smarr BH formula (8.12), with the volume term of the interior, absent in (8.12) making up the difference to the total M. Despite the different local attributions of energy to the volume and surface in the Komar vis-a-vis the Misner-Sharp definition of energy, the total mass of the gravastar is M in either definition, as it must be.

The constant density interior solution is the critical case that saturates and exceeds the Buchdahl bound. Its behavior including negative pressure and the dark energy vacuum equation of state $p_v = -\rho_v$ are inherent in and can be described quite satisfactorily in strictly classical GR, providing an explicit counterexample to the singularity theorems. This maximally compact classical solution has no trapped surface, assumed in the Penrose singularity theorem. The discontinuity (8.73) of the surface gravities at the would-be horizon is instead the location of the thin shell of a gravitational condensate star, with a non-zero surface tension of a physical surface. This possibility for a non-singular endpoint of gravitational collapse is and was always allowed in classical GR, awaiting a more complete analysis of possible discontinuities at null surfaces, and physical realization of a quantum phase transition at the horizon capable of generating a change in the local vacuum to $p_v = -\rho_v < 0$.

A summary of the main results of the reanalysis of the Schwarzschild constant density interior solution presented in this section is as follows.

- The Buchdahl bound (8.50) on the compactness of a spherically symmetric self-gravitating mass M of any kind subject to Einstein's equations is evaded and $r_s \to r_M < 9r_M/8$, if the assumption of isotropy of the pressure everywhere is removed, and instead $p_\perp - p \neq 0$ exhibiting a δ-function corresponding to a physical surface tension at $r_0 = r_M$ when $r_s = r_M$ coincide. The constant negative pressure interior with $p = -\bar{\rho}$ is a non-singular patch of static dS space, which is responsible for effective repulsion in general relativity rather than the fatal attraction that leads to the critical mass (8.8), thereby resolving Wheeler's 'difficulty' of the final state of complete gravitational collapse, which need not be a BH.
- The $\varepsilon \to 0$ classical GR limit may be considered the *universal* gravastar limit, in the sense that it is independent of any ansatz or additional assumptions of an equation of state of an interposed boundary layer of 'non-inflationary material' with a well-defined surface tension τ_S of (8.70). This is a clear specific example of the generalization of the Lanczos–Israel matching conditions to null hypersurfaces, also dispensing with the need to restrict to matching on timelike boundaries, as in the original gravastar proposal, and subsequent papers.
- The determination of $C = \frac{1}{4}$ is just that required to make the surface gravities on the Schwarzschild exterior and dS interior sides of the surface at $r = r_M$ equal and opposite, i.e. $\kappa_- = -\kappa_+$, which is the physical condition of equality of the forces pressing inward and outward on the surface in static equilibrium. This is seen also in the Euclidean formulation as the equality of the Euclidean time periodicities $\beta_+ = 2\pi\kappa_+ = 8\pi GM$ and $\beta_- = 2\pi\kappa_- = 4\pi/H$, so that the inferred Rindler–Hawking 'temperatures' $1/\beta_+ = 1/8\pi GM$ and $1/\beta_- = H/4\pi$ (rather than $H/2\pi$) are *equal* at $r_H = H^{-1} = r_M = 2GM$, just as would be expected in equilibrium.
- Because $\tau_S > 0$, the 'First Law' of non-rotating gravastars shows that increasing the area by A_H by allowing for surface mode perturbations with non-zero angular momenta costs energy, so that the static equilibrium of the surface is stable to such perturbations, independently of any equation of state assumptions. Similar to the Bohr atom, these angular perturbations are expected to have a *discrete spectrum* which will distinguish gravastars with a surface from BH horizons in gravitational wave (GW) signatures.
- The differential relation (8.75) between gravastar solutions parametrized by M is purely a *classical mechanical* relation, entirely within the domain of classical GR, rather than a quantum or thermodynamic one. The area A_H is simply the geometrical area of the condensate star surface with no

implication of entropy. As (8.76) shows there is no entropy at all associated with a macroscopic condensate at zero temperature. The Planck length $L_{Pl} = \sqrt{\hbar G/c^3}$, Planck mass $M_{Pl} = \sqrt{\hbar c/G}$, or the Boltzmann constant k_B do not enter these classical relations at all at $\varepsilon = 0$.

- Further, by the Gibbs relation

$$p + \rho = s\,T + \mu\, n \tag{8.76}$$

and $\mu = 0$ here, since no chemical potential corresponding to a conserved quantum number has entered our classical considerations, the interior Schwarzschild-de Sitter solution with $p + \rho = 0$ is a *zero entropy* density $s = 0$ and *zero temperature* single macroscopic state, justifying its designation as a gravitational Bose–Einstein condensate (GBEC).

- The matching of the metric interior to the exterior solution for $r = r_M$ has the cusp-like behavior shown in Figs. 8.3 and 8.4, with discontinuous first derivative with respect to the original Schwarzschild radial coordinate r. This is symptomatic of non-analytic behavior, invalidating the assumption of metric analyticity needed for deriving periodicity in complexified imaginary time $t \to -i\tau$ [98]. Unlike in the analytically extended vacuum Schwarzschild solution, where $f(r)$ becomes negative and the Killing vector $K = \partial_t$ becomes spacelike in the interior $r < r_M$ of a BH, in the negative pressure gravastar solution $f(r)$ is everywhere non-negative, and there is no requirement of any fixed periodicity in imaginary time of either the geometry or Green's functions of quantum fields in this geometry. The surface gravity κ therefore carries no implication of temperature or thermal radiation. The zero temperature vacuum state in the exterior Schwarzschild geometry is expected to be the Boulware vacuum, which is the usual Minkowski vacuum in the asymptotically flat region $r \to \infty$, whose Green's functions also have non-analytic cusp-like behavior at $r = r_M$ [35], and which has no Hawking radiation [70].
- Non-analyticity at $r = r_M$ is exactly the property suggested by the analogy of BH horizons to phase boundaries and quantum critical surfaces in condensed matter physics [61, 63], although the equality of $|\kappa_\pm|$ was not obtained in the condensed matter analogies, requiring as it does a proper general relativistic treatment of matching on null hypersurfaces, which is given also for horizons with non-zero angular momentum in [30].
- Since $f(r) = 0$ corresponds to the 'freezing' of local proper time at $r = r_s$, it also suggests critical slowing down characteristic of a phase transition. The vanishing of the effective speed of light $c_{eff}^2 = c^2 f(r)$ is analogous to the behavior of the sound speed determined by the low energy excitations at a critical surface or phase boundary. This and experience with Bose–Einstein condensates in other contexts suggest that gravitation and spacetime itself

are 'emergent' phenomena of a more fundamental microscopic many-body theory [64, 119, 131].

- The interpretation of $c_{eff}^2 \geq 0$ as the effective speed of light squared, which must remain always non-negative, calls to mind Einstein's original papers on the local relativity principle for static gravitational fields, which led him to the general theory from the Minkowski metric $ds^2 = -c^2\, dt^2 + dx^2 + dy^2 + dz^2$, by allowing first the time component $-g_{tt} = c^2$ and eventually all other components of the metric to be functions of space (and then also time) [85]. Thus it could be argued that the non-negativity of $c_{eff}^2 = c^2 f(r)$ in a static geometry is more faithful to Einstein's original conception of the Equivalence Principle, realized by real continuous coordinate transformations, rather than complex analytic extension around a square root branch point that would allow $c_{eff}^2 < 0$.
- At the minimum, the matching of the $p = -\bar{\rho}$ dS interior to Schwarzschild exterior provides a consistent alternative to analytic extension, entirely within the framework of classical GR, provided only that surface boundary layers on null boundary surfaces are admitted. A phase boundary and non-singular 'BH' interior require no violation of the Equivalence Principle, at least in its weak form.
- Finally, since $K_{(t)}$ remains timelike for a gravastar, t is a global time and unlike in the analytic continuation hypothesis, the spacetime is truly globally static. The $t =$ const. hypersurface is a Cauchy surface and is everywhere spacelike. This is exactly the property of a static spacetime necessary to apply standard quantum theory, for the quantum vacuum to be defined as the lowest energy state of a Hamiltonian bounded from below, and for the Schrödinger equation to describe unitary time evolution, thus avoiding any conflict with unitarity or an 'information paradox.'

8.5 Slowly Rotating Gravastars: Junction Conditions and Moment of Inertia

A serendipitous consequence of the reanalysis of the 1916 Schwarzschild interior solution via the Komar mass integral (8.54) is that it provides an explicit example of 'gluing' of two different geometries at their mutual null horizons, in which the surface stresses can be unambiguously determined, providing a clear interpretation of the surface tension of the null surface at $r_{_M} = r_{_H}$. This example serves as a general template to provide the general matching conditions for null surface with non-zero angular momentum, necessary to describing rotating gravastars as well [30].

Consider the general axisymmetric and stationary metric line element [25, 57, 58, 99, 121]

$$ds^2 = -e^{2\nu}dt^2 + e^{2\psi}\left(d\phi - \omega\, dt\right)^2 + e^{2\alpha}dr^2 + e^{2\beta}d\theta^2 \tag{8.77}$$

in coordinates adapted to the two Killing symmetries of time translation $K_{(t)} = \partial_t$ and axisymmetry $K_{(\phi)} = \partial_\phi$ around the axis of rotation. The five functions $\nu, \psi, \omega, \alpha, \beta$ are functions of the remaining (r, θ) coordinates only. The choice of those remaining two coordinates is still subject to some coordinate freedom, which if fully exploited reduces the five functions appearing in (8.77) to just four independent functions. However it is convenient to leave the remaining coordinate freedom unfixed, in order to encompass various choices that may be made.

Corresponding to the two Killing symmetries and Killing vectors $K_{(t)}$ and $K_{(\phi)}$ are two conserved quantities which may be constructed by Komar's general method, viz. the total mass-energy

$$\begin{aligned} M &= \frac{1}{4\pi G}\int_{\partial V_+}(\kappa + \omega \mathcal{J})\, dA \\ &= \int_V \sqrt{-g}\,\left(-T_t^{\ t} + T_r^{\ r} + T_\theta^{\ \theta} + T_\phi^{\ \phi}\right) dr\, d\theta\, d\phi + \frac{1}{4\pi G}\int_{\partial V_-}(\kappa + \omega\mathcal{J})\, dA \end{aligned} \tag{8.78}$$

and angular momentum

$$J = \frac{1}{8\pi}\int_{\partial V_+}\mathcal{J}\, dA = G\int_V \sqrt{-g}\; T_t^{\ \phi}\, dr\, d\theta\, d\phi + \frac{1}{8\pi}\int_{\partial V_-}\mathcal{J}\, dA \tag{8.79}$$

where [30]

$$\kappa = \frac{1}{2}\, e^{-\alpha-\nu}\,\frac{\partial e^{2\nu}}{\partial r} \tag{8.80a}$$

$$\mathcal{J} = -\frac{1}{2}\, e^{2\psi}\, e^{-\alpha-\nu}\,\frac{\partial \omega}{\partial r} \tag{8.80b}$$

are the surface gravity and (8π times) the angular momentum density per unit area respectively. The volume measure factor and area element are given by

$$\sqrt{-g} = \exp\left(\nu + \psi + \alpha + \beta\right) \tag{8.81}$$

$$dA = e^{\beta+\psi}\, d\theta\, d\phi. \tag{8.82}$$

in the coordinates of (8.77). In the case of spherical symmetry ω, $T_t^{\ \phi}$ and $\mathcal{J}$ vanish, (8.78)–(8.80a) reduce to (8.54)–(8.55), and (8.82) becomes $\sqrt{f/h}\, r^2 \sin\theta$.

A null horizon in the geometry (8.77) occurs when $e^{2\nu} \to 0$, which without loss of generality may be taken to be at a fixed $r = r_{_H}$. The induced metric on the horizon

$$ds^2\Big|_{r=r_H} = e^{2\psi_H}\left(d\phi - \omega_{_H} dt\right)^2 + e^{2\beta_H}\, d\theta^2 \tag{8.83}$$

involves the functions ψ, ω, β which must be continuous there, whereas ν and α need not be. It is the inclusion of the $\sqrt{-g}$ factor of the volume integration measure in the Komar integrals that leads to total second derivatives of ν, α with respect to r. These total derivative terms give rise both to the areal surface terms in (8.78)–(8.79), and the distributional Dirac δ-functions on null hypersurfaces from the discontinuities of κ and $\mathcal{J}$ there. Thus, as in (8.68)–(8.69) it is necessary to consider the tensor density $\sqrt{-g}G_\mu^{\ \nu}$ and in particular its potentially singular part $e^{\nu+\alpha}G_\mu^{\ \nu}$, to find these δ-function and surface contributions [30].

This is the fundamental physical basis for defining the redshifted extrinsic curvature for stationary, axisymmetric geometries by

$$\mathbf{K}^a_{\ b} = -\gamma^{ac} n^\mu \Gamma_{\mu bc} = \frac{1}{2} e^{-\alpha+\nu} \gamma^{ac} \left(\frac{\partial g_{bc}}{\partial r} - \frac{\partial g_{rc}}{\partial x^b} - \frac{\partial g_{rb}}{\partial x^c} \right), \quad \{a, b, c\} = t, \theta, \phi \tag{8.84}$$

with the normal vector having the components

$$\mathbf{n}_\mu = \delta^r_{\ \mu}\, e^{\alpha+\nu}, \qquad \mathbf{n}^\mu = \delta^\mu_{\ r}\, e^{-\alpha+\nu} \tag{8.85}$$

normalized to

$$\mathbf{n}\cdot\mathbf{n} = e^{2\nu} \to 0 \tag{8.86}$$

on the null hypersurface, instead of (8.38) for a timelike hypersurface. These relations and the inverse γ^{ac} of the induced metric are to be evaluated for finite $e^{2\nu}$ and the horizon limit $e^{2\nu} \to 0$ taken at the end.

With $\mathbf{n}$ defined by (8.85), $\mathbf{K}^a_{\ b}$ has the components

$$\mathbf{K}^t_{\ t} = \frac{1}{2} e^{-\alpha-\nu} \frac{\partial}{\partial r} e^{2\nu} - \frac{1}{2}\, \omega\, e^{2\psi}\, e^{-a-\nu} \frac{\partial \omega}{\partial r} = \kappa + \omega \mathcal{J} \tag{8.87a}$$

$$\mathbf{K}^t_{\ \phi} = \frac{1}{2} e^{2\psi}\, e^{-a-\nu} \frac{\partial \omega}{\partial r} = -\mathcal{J} \tag{8.87b}$$

$$\begin{aligned} \mathbf{K}^\phi_{\ t} &= \frac{1}{2}\, \omega\, e^{-\alpha-\nu} \frac{\partial}{\partial r} e^{2\nu} - \frac{1}{2}\, \omega^2 e^{2\psi}\, e^{-a-\nu} \frac{\partial \omega}{\partial r} - \frac{1}{2} e^{-2\psi}\, e^{-a+\nu} \frac{\partial}{\partial r}\left(\omega e^{2\psi}\right) \\ &= \omega\kappa + \omega^2 \mathcal{J} + \mathcal{O}(\epsilon) \end{aligned} \tag{8.87c}$$

$$\mathbf{K}^\theta_{\ \theta} = \frac{1}{2} e^{-\alpha+\nu-2\beta} \frac{\partial}{\partial r} e^{2\beta} = \mathcal{O}(\epsilon) \tag{8.87d}$$

$$\mathbf{K}^\phi_{\ \phi} = \frac{1}{2}\, \omega\, e^{2\psi}\, e^{-a-\nu} \frac{\partial \omega}{\partial r} + \frac{1}{2} e^{-2\psi}\, e^{-a+\nu} \frac{\partial}{\partial r} e^{2\psi} = -\omega \mathcal{J} + \mathcal{O}(\epsilon) \tag{8.87e}$$

with the non-listed components vanishing. Thus the discontinuities on the null horizon surface are

$$[\mathbf{K}^t{}_t] = [\kappa] + \omega_H [\mathcal{J}] \tag{8.88a}$$
$$[\mathbf{K}^t{}_\phi] = -[\mathcal{J}] \tag{8.88b}$$
$$\left[\mathbf{K}^\phi{}_t\right] = \omega_H [\kappa] + \omega_H^2 [\mathcal{J}] \tag{8.88c}$$
$$[\mathbf{K}^\theta{}_\theta] = 0 \tag{8.88d}$$
$$\left[\mathbf{K}^\phi{}_\phi\right] = -\,\omega_H [\mathcal{J}] \tag{8.88e}$$

where $\omega = \omega_H$ on the horizon. The null horizon junction conditions are then of the same form as the Lanczos–Israel conditions (with one contravariant and one covariant index), namely

$$[\mathbf{K}_a{}^b] = -4\pi G \left(2\mathcal{S}_a{}^b - \delta_a{}^b \mathcal{S}_c{}^c\right) \tag{8.89a}$$
$$8\pi G\, \mathcal{S}_a{}^b = -[\mathbf{K}_a{}^b] + \delta_a{}^b [\mathbf{K}_c{}^c] \tag{8.89b}$$

but with this physical stress tensor on the null horizon hypersurface related to the original Lanczos–Israel (LI) one by

$$\left(\mathcal{S}_a{}^b\right)_{\text{here}} = \lim_{e^\nu \to 0} \left\{ e^\nu \left(S_a{}^b\right)_{\text{LI}} \right\} \tag{8.90}$$

which is finite in the horizon limit $e^\nu \to 0$ since the discontinuities of $\mathbf{K}^a{}_b$ in (8.88) are finite in that limit. The surface stress tensor density is then found to be

$$^{(\Sigma)}T_a{}^b\, e^{\alpha+\nu} = \mathcal{S}_a{}^b\, \delta(r - r_H) \tag{8.91}$$

localized on the surface $r = r_H$ with the correct volume measure factor (to be multiplied by the continuous $e^{\psi+\alpha}$ factors) for the Komar mass and angular momentum integrals.

This improved understanding and generalization of the junction condition formalism to null hypersurfaces and in particular rotating null horizons appropriate for the Kerr geometry opens the way to finding rotating gravastar solutions. The formalism has been applied in the case of slow rotation following methods of [59, 97, 100] who express the slowly rotating line element in the (8.77) form with

$$e^{2\nu} = f(r)\Big[1 + 2h_0(r) + 2h_2(r)\, P_2(\cos\theta)\Big] \tag{8.92a}$$
$$e^{2\psi} = r \sin\theta \Big[1 + 2k_2(r)\, P_2(\cos\theta)\Big] \tag{8.92b}$$
$$e^{2\alpha} = \frac{1}{h(r)} \left[1 + 2\, \frac{m_0(r) + m_2(r)\, P_2(\cos\theta)}{r - 2m(r)}\right] \tag{8.92c}$$
$$e^{2\beta} = r\Big[1 + 2k_2(r)\, P_2(\cos\theta)\Big]. \tag{8.92d}$$

The frame dragging angular velocity $\omega = \omega(r)$ is independent of θ to lowest (first) order in the slow rotation expansion about the spherically symmetric solution, while h_0, m_0 are the monopole ($\ell = 0$) and h_2, m_2, k_2 the quadupole ($\ell = 2$) perturbations, which are second order in that expansion, and multiplied by the appropriate Legendre polynomials P_0, P_2. The Einstein equations linearized around the previously determined spherically symmetric solution are then solved assuming that the equation of state for the fluid composing a slowly rotating star is unchanged from its form in the non-rotating solution. In the case of a slowly rotating gravastar the solutions are separated again into an exterior sourcefree region, and interior dS region with the same $p = -\rho$ equation of state, and the solutions are matched at their mutual horizons using the discontinuities and surface stress tensor (8.89b).

Requiring that the solution be asymptotically flat as $r \to \infty$, the induced metric on the horizon (8.83) to be continuous ($\mathcal{C}^0$), and the interior solution not to contain δ-function source terms at the origin $r = 0$, severely restricts the solutions for the functions h_0, m_0, h_2, m_2, k_2. With these conditions one finds [29] [3]

$$r > r_H: \qquad \omega = \frac{2J}{r^3} \tag{8.93a}$$

$$m_0 = J^2\left(\frac{1}{r_H^3} - \frac{1}{r^3}\right) \tag{8.93b}$$

$$h_0 = -\frac{m_0}{r - r_H} = -\frac{J^2}{r_H^3 r^3}\left(r^2 + r r_H + r_H^2\right) \tag{8.93c}$$

$$m_2 = \frac{J^2}{r^3}\left(r - r_H\right)\left(\frac{5}{r} - \frac{2}{r_H}\right) \tag{8.93d}$$

$$h_2 = \frac{J^2}{r^3}\left(\frac{1}{r} + \frac{2}{r_H}\right) \tag{8.93e}$$

$$k_2 = -\frac{2J^2}{r^3}\left(\frac{1}{r} + \frac{1}{r_H}\right) \tag{8.93f}$$

for the exterior solution $r > r_H$, and

[3] Here J is the angular momentum in geometric units GJ/c^3, with dimensions of (length)^2.

$$r < r_{_H}: \qquad \omega = \frac{2J}{r_{_H}^3} = \omega_H \tag{8.94a}$$

$$m_0 = 0 \tag{8.94b}$$

$$h_0 = C_{_I} \tag{8.94c}$$

$$m_2 = -\frac{2J^2(r^2 - r_{_H}^2)^2}{r^2 r_{_H}^5} \tag{8.94d}$$

$$h_2 = \frac{2J^2}{r r_{_H}}\left(\frac{1}{r^2} - \frac{1}{r_{_H}^2}\right) \tag{8.94e}$$

$$k_2 = -\frac{2J^2}{r r_{_H}}\left(\frac{1}{r^2} + \frac{1}{r_{_H}^2}\right) \tag{8.94f}$$

for the interior solution, with $r_{_H} = r_{_M}$ unchanged from the non-rotating solution to this order.

With these requirements the exterior solution for a slowly rotating gravastar (8.93) is in fact the same as that of a slowly rotating Kerr BH, up to a coordinate transformation from the (r, θ) Hartle–Thorne coordinates used in slow rotation expansion of the metric in (8.92) to the Boyer–Lindquist $(r_{_{\rm BL}}, \theta_{_{\rm BL}})$ coordinates in which the Kerr solution is more commonly expressed, and then expanded to second order in J. This coordinate transformation is given explicitly by Eq. (4.12) of [29]. In particular the ergosphere boundary is given by $r_{\rm ergo}(\theta) = r_{_H}[1 + 4J^2 \sin^2\theta / r_{_M}^4] \geq r_{_H}$ in the Hartle–Thorne coordinates of (8.92). The reason for the two exterior solutions of a Kerr BH and slowly rotating gravastar to be identical is that although the metric functions α, ν, ω have discontinuous derivatives on the gravastar null horizon surface, the induced surface metric (8.83) is continuous, and all the metric functions have been required to be finite there, which are the same conditions required for the horizon of a Kerr BH, that lead to the 'no hair' theorems [53–55, 167].

The solution (8.93)–(8.94) is such that the discontinuities of two quantities which appear in (8.88) and the surface stress-energy tensor (8.91) are

$$[\kappa] = \frac{1}{r_{_M}}\left(1 - \frac{3J^2}{r_{_M}^4} + \frac{C_{_I}}{2}\right) \tag{8.95a}$$

$$[\mathcal{J}] = \frac{3J}{r_{_M}^2}\,\sin^2\theta \tag{8.95b}$$

with the result that surface energy of the horizon boundary at $r = r_{_H}$ in the Komar mass-energy is

$$E_S = \frac{1}{4\pi G}\int_H \Big([\kappa] + \omega[\mathcal{J}]\Big) dA = \frac{[\kappa]}{4\pi G}\,A_H + \frac{3\omega_H\, J}{2G}\int_0^\pi d\theta\,\sin^3\theta \tag{8.96}$$

$$= \frac{[\kappa] r_{_M}^2}{G} + \frac{2\omega_H\, J}{G} = 2M\left(1 + \frac{C_{_I}}{2} + \frac{J^2}{r_{_M}^4}\right) \tag{8.97}$$

where (8.94a) has been used. The Komar volume energy of the interior solution is

$$E_V = -M\left(1 + C_I\right) \tag{8.98}$$

so that the total energy of the slowly rotating gravastar specified by (8.93)–(8.94) is

$$E \equiv \mathcal{M} = E_V + E_S = M\left(1 + \frac{2J^2}{r_M^4}\right) \tag{8.99}$$

where M is the mass of the non-rotating solution. Since this is the irreducible mass for a non-rotating BH, (8.99) will be recognized as equivalent to the Christodoulu formula

$$\mathcal{M}^2 = M^2 + \frac{J^2}{4G^4M^2} = M^2\left(1 + \frac{4J^2}{r_M^4}\right) \tag{8.100}$$

when $E = \mathcal{M}$ is expanded to first order in J^2/r_M^4. Thus C_I drops out of the total mass-energy $\mathcal{M}$ of the slowly rotating solution, when the volume and surface contributions are added, which then obeys exactly the same relation as a Kerr BH of the same mass and angular momentum to this order in J^2.

For the Komar angular momentum of the gravastar, the integrand of the volume term is proportional to $T_\phi^{\ t}$, which vanishes in the interior under the assumption of the $p = -\rho$ equation of state being unchanged from the non-rotating case in order to apply the formalism of [59, 97, 100]. Using (8.95b), the entire angular momentum is carried by the surface contribution

$$J_{\text{tot}} = J_S = \frac{1}{8\pi}\int_H [\mathcal{J}]\, dA = \frac{3J}{4}\int_0^\pi d\theta\, \sin^3\theta = J \tag{8.101}$$

which is due to the surface stress tensor $\mathcal{S}_\phi^{\ t}$. From (8.36) and (8.95) the components of the surface stress tensor are

$$8\pi\,\mathcal{S}^t{}_t = -\frac{6J^2}{r_H^5}\,\sin^2\theta \tag{8.102a}$$

$$8\pi\,\mathcal{S}^t{}_\phi = \frac{3J}{r_H^2}\,\sin^2\theta \tag{8.102b}$$

$$8\pi\,\mathcal{S}^\phi{}_t = -\frac{2J}{r_H^4} \tag{8.102c}$$

$$8\pi\,\mathcal{S}^\phi{}_\phi = \frac{1}{r_H} + \frac{C_I}{2r_H} - \frac{3J^2}{r_H^5} + \frac{6J^2}{r_H^5}\,\sin^2\theta \tag{8.102d}$$

$$8\pi\,\mathcal{S}^\theta{}_\theta = \frac{1}{r_H} + \frac{C_I}{2r_H} - \frac{3J^2}{r_H^5}. \tag{8.102e}$$

by a straightforward application of the modified Lanczos–Israel junction conditions on the null horizon hypersurface at $r_H = r_M$ [29, 30].

The moment of inertia I is defined as the ratio of its angular momentum J/G to angular velocity ω_H, and given by

$$I = \frac{J/G}{\omega_H} = M r_H^2 = 4G^2 M^3 \tag{8.103}$$

identical to its value for a Kerr BH [31, 165]. This might have been expected from the fact that the external geometry is identical to the Kerr BH geometry, and the interior dS condensate carries no angular momentum, all of J being concentrated on the rotating null horizon surface from (8.101). The result (8.103) may be compared to the moment of inertia of a thin shell of uniform surface density with total mass M and radius r_H in Newtonian mechanics, which is $2/3$ of (8.103). The factor of $3/2$ larger for the moment of inertia of a rotating gravastar is evidently a relativistic effect due to the surface density contributing to the mass-energy localized on the surface being $2M$ rather than M and the non-uniform distribution of the surface stress tensor components $-\mathcal{S}^t{}_t + \mathcal{S}^\theta{}_\theta + \mathcal{S}^\phi{}_\phi$ (8.102) contributing to the Komar surface energy (8.61). in the non-rotating inertial frame. Nonetheless, with the moment of inertia (8.103) the contribution of the rotational kinetic energy to the total mass $\mathcal{M}$ in (8.99) is

$$E_{\rm rot} = 2M\,\frac{J^2}{r_M^4} = \frac{1}{2}\,I\omega_H^2 \tag{8.104}$$

exactly what would be expected from the Newtonian mechanics of a rotating body.

Unlike a BH where it is supposed that there is no mass-energy at all at the horizon to rotate or to give rise to the large moment of inertia (8.103), the rotating gravastar has a well-defined Komar stress-energy (8.102) and angular momentum on the rotating surface at $r = r_H$ (8.101), where all the angular momentum resides, and this gives rise to (8.103) and (8.104) by straightforward evaluation of the relevant surface integrals (8.61), (8.101).

The results for the external solution (8.93), (8.103) and (8.104) being identical to that of a slowly rotating Kerr BH is consistent with the BH ‘no hair’ theorems being extended to rotating gravastars as well [24, 53, 55, 132], at least in the strictly classical limit of an infinitesimally thin shell ($\varepsilon \to 0$). This again is the result of the boundary conditions on the metric perturbations in (8.92) at the null horizon being finite and identical in the two cases. Thus in this limit slowly rotating gravastars are expected to be indistinguishable from Kerr BH’s in their external geometry, and have the same tidal deformability and Love number as a Kerr BH, namely none at all

[67]. Any gravastar 'hair' would be limited to the very thin quantum phase transition boundary layer of thickness ℓ (8.33), and be very short.

As welcome as these first results for rotating gravastars are, the interior solution (8.94) contains a singularity at $r = 0$, which leads to divergences in the Weyl tensor at the origin proportional to J^2/r^5 in both the monopole and quadrupole terms. For that reason (8.94) cannot be considered fully satisfactory. Since all the perturbations to order J^2 are fully determined by the boundary conditions of asymptotic flatness, matching of the induced metric (8.83) on the null horizon surface, and absence of δ-function singularities in the Komar M or J at the origin, the singular result at the origin cannot be avoided in the Hartle–Thorne method, which requires that the equation of state of the rotating matter, in this case $p = -\rho$, be unchanged from the non-rotating solution. This assumption may be questioned, and indeed considerations of rotating superfluid condensates suggests that a non-vanishing vortex density should arise in a rotating gravitational condensate as well. This would require a different interior solution for a rotating gravastar, with a non-vanishing torsion, in which the $p = -\rho$ condition of [59, 97, 100] is relaxed.

To summarize the results of this section, the conditions of:

(i) Asymptotic flatness,
(ii) Matching conditions on the slowly rotating null horizon,
(iii) Unchanged $p = -\rho$ equation of state from the non-rotating gravastar, and
(iv) Absence of δ-function singularities at the origin in the Komar mass and angular momentum,

determine the slowly rotating gravastar solution to order J^2, up to one undetermined constant C_I. The exterior solution is identical to that of a Kerr BH to this order, including the moment of inertia (8.103) which is consistent with all of the angular momentum being carried by the rotating null hypersurface at this lowest order in J^2. However the interior solution (8.94) under these four requirements is singular at the origin, and therefore not fully satisfactory. It appears that relaxing at least condition (iii) is necessary to remove the $1/r^5$ power law singularity at $r = 0$. It is possible that then allowing a distribution of angular momentum within the interior would also modify the exterior solution for a rapidly rotating gravastar from that of a Kerr BH.

8.6 Macroscopic Effects of the Quantum Conformal Anomaly

The considerations of the previous sections are entirely classical, and show that an endpoint of complete gravitational collapse different from a BH is possible even in

Einstein's classical GR, with no violation of the Equivalence Principle. However, it does not provide a mechanism by which a phase transition from one value of $\Lambda_{\rm eff}$ outside a gravastar to another value inside can occur, elucidate the special role of the event horizon in triggering this transition, nor specify the physics determining the value of $\varepsilon > 0$. Addressing these questions requires consideration of the effects of the quantum theory of the matter/radiation stress tensor on the near-horizon geometry, and in particular, the macroscopic effects of the conformal anomaly of the quantum stress tensor source for Einstein's equations in the near-horizon region.

At the smallest microscopic scales probed, the principles of quantum field theory (QFT) hold, and matter under extreme pressures and densities in the standard model (SM) of particle physics is clearly quantum in nature. Yet Einstein's classical GR remains unreconciled to quantum theory, and the tension between quantum matter and classical gravity comes to the fore both in the puzzles and paradoxes of BHs reviewed in Sect. 8.2, and in the nature and value of cosmological dark energy driving the accelerated expansion of the universe [2, 156, 165].

The problems of reconciling classical GR with QFT first appear with the stress-energy tensor $T^{\mu\nu}$, which is treated as a completely classical source in Einstein's equations, whereas $\hat{T}^{\mu\nu}$ is a UV divergent operator in QFT. Thus the minimal accomodation necessary to couple quantum matter to Einstein's theory is to replace the divergent $\hat{T}^{\mu\nu}$ operator by its renormalized expectation value $\langle\hat{T}^{\mu\nu}\rangle$, in a semi-classical approximation. Defining this expectation value by regularizing and renormalizing the contributions of matter loops leads inevitably to consideration of the effects of quantum matter on gravity itself, or in other words, to the effective field theory (EFT) treatment of gravity, keeping track of quantum corrections to GR.

The quantum corrections are of two quite distinct kinds. The first are the strictly short distance/high energy corrections contained in higher order local curvature invariants which are needed in any case to renormalize $\langle\hat{T}^{\mu\nu}\rangle$ [33, 77]. The second are the effects of higher point stress tensor correlators, which can contain light cone singularities, that extend over macroscopic distance scales and provide quantum corrections to GR even at low energy scales, and in particular on null horizons.

Usually only quantum corrections of the first kind are considered [40, 78, 79], and the additions to the classical action of GR are assumed to involve only an expansion in the local curvature invariants, such as $R_{\alpha\beta\mu\nu}R^{\alpha\beta\mu\nu}$, $R_{\alpha\beta}R^{\alpha\beta}$, R^2. Since these are fourth order in derivatives of the metric, they become significant only at the extreme UV Planck scale, but are negligibly small at macroscopic distance scales or weak curvatures. The expansion in local invariants is based on the assumption of decoupling of UV degrees of freedom from the low energy EFT and strict separation of scales, familiar in other EFT approaches [11, 19, 120].

On the other hand it has also been known for some time, even in flat space QFT, that anomalies are not captured by such an expansion in higher order local invariants, nor are they suppressed by any UV scale. Anomalies are associated instead with the fluctuations of massless fields which do not decouple, and which lead to $1/k^2$ poles in momentum space correlation functions, that grow large on the light cone $k^2 \to 0$ rather than the extreme UV regime $k^2 \sim M_{\rm Pl}^2$. Such massless light cone poles are found in explicit calculations of the triangle anomaly diagrams of

$\langle \hat{J}_5^\lambda \hat{J}^\alpha \hat{J}^\beta \rangle$, $\langle \hat{T}^{\mu\nu} \hat{J}^\alpha \hat{J}^\beta \rangle$ in massless QED$_4$ [20, 90], and in the stress tensor three-point correlator $\langle \hat{T}^{\alpha\beta} \hat{T}^{\gamma\lambda} \hat{T}^{\mu\nu} \rangle$ of a general conformal field theory (CFT), by solution of the conformal Ward Identities in momentum space [41, 73].

Quite contrary to the decoupling hypothesis, quantum anomalies lead instead to the principle of *anomaly matching* from UV to low energy EFT [177]. In the strong interactions, the chiral anomaly of the UV theory, QCD, survives to low energies, requiring a specific Wess–Zumino (WZ) addition to the low energy meson EFT [120, 189], which is not suppressed by any high energy scale, and without which the low energy $\pi^0 \to 2\gamma$ decay rate, which helped establish QCD as the UV theory of the strong interactions, does not come out correctly [32, 89, 179].

In gravitation theory the relevant anomaly requiring attention is the conformal anomaly of $T_\mu^{\ \nu}$. The classical stress-energy tensor of conformal matter or radiation is traceless $T_\mu^{cl\,\mu} = 0$. The conformal (or trace) anomaly arises because it is impossible to maintain this traceless condition at the quantum level, in an arbitrary metric background, if the covariant conservation equation $\nabla_\nu T_\mu^{\ \nu} = 0$ is also to be maintained. Since covariant conservation of $T_\mu^{\ \nu}$ is a necessary requirement for the consistent coupling of matter to Einstein's equations, it is the trace condition on $T_\mu^{\ \mu}$ that is given up and which is 'anomalous.' This clash of symmetries, forcing one to choose between them at the quantum level, is what is meant by an 'anomaly.' Although perhaps a surprise when they were first discovered, it eventually became clear that 'anomalies' are a natural and necessary feature of QFT, and lead to interesting and essential connections between the low energy (infrared:IR) and high energy (ultraviolet:UV) limits of the theory, both in the SM and for GR and the EFT of low energy gravity.

Intuitively, a conformal symmetry of the classical theory cannot be maintained at the quantum level because QFT in all interesting cases involves regularization and renormalization of UV divergent Feynman diagrams, which introduce a scale, if only through logarithms, that necessarily breaks the conformal and scale invariance of the classical theory. This is in fact the reason that coupling 'constants' run with energy scale [60, 90], a well-verified experimental fact [5].

Since event horizons are null hypersurfaces with geometric significance as the locus of points at which the timelike Killing field $K_{(t)}$ (or $K = K_{(t)} + \omega K_{(\phi)}$ in the rotating case [30]) become null, signaling the critical condition for which a trapped surface can form if K^2 changes sign, the light cone singularities of anomalous quantum correlators such $\langle \hat{T}^{\alpha\beta} \hat{T}^{\gamma\lambda} \hat{T}^{\mu\nu} \rangle$ can have large macroscopic effects there. Such light cone singularities imply the existence of at least one additional light scalar (*a priori massless*) degree of freedom in the *low-energy* EFT of macroscopic gravity, that is clearly not accounted for in the classical action of GR, nor by the addition of higher order local curvature invariants to the effective action.

It is $f(r) = -K_{(t)}.K_{(t)} \to 0$ (or $e^{2\nu} = -K^2 \to 0$ in the rotating case of Sect. 8.5 which leads to the infinite blueshifting of frequencies (8.3) or energies (8.19), as soon as $\hbar \neq 0$ is admitted. The result is that the local quantum 'vacuum' near a BH horizon is infinitely shifted in energy scales with respect to a state asymptotically far from the BH, violating a naive expectation of decoupling of UV from IR scales. The specification of the vacuum state involves *non-local* boundary conditions of the entire Cauchy constant time slice of (8.1), which are not determined by any simple expansion in the local curvature. On the contrary these boundary conditions span a large range of scales between IR and UV, and are sensitive to the conformal transformation properties of the effective action of quantum matter coupled to gravity. The enormous entropy (8.16), the Hawking effect upon which it is predicated, and the various BH paradoxes and conundra rely crucially upon the specification of the quantum vacuum state for arbitrarily high local frequencies and energies in the near-horizon region, precisely where masses become irrelevant compared to local energies typified by (8.19), making all finite mass scales negligible and the conformal anomaly relevant there [17, 141].

These observations lead to the important step in an EFT of gravity by the taking into account the macroscopic effects of the conformal anomaly on horizons in [144], where it was shown that the energy-momentum tensor derived from the effective action of the conformal trace anomaly of massless fields in curved space becomes large (indeed formally infinite) for *generic* quantum states at both the Schwarzschild BH and dS static horizons.

8.6.1 *The Two Dimensional Conformal Anomaly and Stress Tensor*

In order to see the macroscopic effects of the conformal anomaly on horizons in a simple case, it is useful to consider first $D = 2$ spacetime dimensions. In 2D it is straightforward to show by any covariant method preserving coordinate invariance [33], that the renormalized expectation value

$$\langle T_\mu{}^\mu \rangle \Big|_{2\text{D}} = \frac{N\hbar}{24\pi}\, R \tag{8.105}$$

is non-vanishing in a general curved 2D spacetime, for $N = N_s + N_f$ the number of free conformal scalar and Dirac fermion fields, i.e. in a conformal field theory (CFT). By taking a variation of (8.105) with respect to the arbitrary background metric, and then evaluating the result in flat space, one can relate the anomaly to the existence of a massless $1/k^2$ pole in the one-loop vacuum polarization diagram of $\Pi_{\mu\nu\alpha\beta} = i\langle T_{\mu\nu} T_{\alpha\beta}\rangle$, cf. Fig. 8.6.

This tensor has the form in momentum space

$$\Pi_{\mu\nu\alpha\beta}(k)\Big|_{2D} = (\eta_{\mu\nu}k^2 - k_\mu k_\nu)(\eta_{\alpha\beta}k^2 - k_\alpha k_\beta)\,\frac{N\hbar}{12\pi k^2} \tag{8.106}$$

$$\Pi^{\ \mu}_{\mu\ \alpha\beta}(k)\Big|_{2D} = \frac{N\hbar}{12\pi}\,(\eta_{\alpha\beta}k^2 - k_\alpha k_\beta) \tag{8.107}$$

showing that the non-zero trace and coefficient on the right side of (8.105) is directly related to the existence and residue of the $1/k^2$ pole in $\Pi_{\mu\nu\alpha\beta}$. In fact, once the tensor index structure indicated in (8.106) is fixed, as required by symmetries and the covariant conservation law Ward identities $k^\mu \Pi_{\mu\nu\alpha\beta}(k) = 0$ on any index, the one-loop diagram of Fig. 8.6 is *UV finite* and completely determined, with (8.107) the result [31]. This shows that the conformal anomaly and pole is independent of the regularization scheme and detailed UV behavior of the quantum theory, provided that the identities following from the covariant conservation law are maintained.

The essential point now is that the massless pole in (8.106) is a lightlike singularity, indicating significant effects on the light cone, which extends to macroscopic distance scales, and is particularly relevant on null horizons. The $1/k^2$ pole can be expressed as the propagator of an effective scalar degree of freedom arising from the fluctuations and correlations of massless (or sufficiently light) quantum fields in the vicinity of BH horizons. Note that the classical theory of 2D gravity has no propagating degrees of freedom at all, so this $1/k^2$ propagator arises entirely from the quantum effect of the anomaly [141].

This effective scalar degree of freedom and its consequences can be derived from the effective action corresponding to (8.105), viz. [159, 160]

$$S^{2D}_{\rm anom}[g] = -\frac{N\hbar}{96\pi}\int d^2x\sqrt{-g(x)}\int d^2y\sqrt{-g(y)}\;R(x)\left(\Box^{-1}\right)_{xy} R(y) \tag{8.108}$$

where $\left(\Box^{-1}\right)_{xy}$ denotes the Green's function inverse of the scalar wave operator, that becomes the $1/k^2$ pole of (8.106) in momentum space. This scalar degree of freedom can be made explicit by expressing the non-local anomaly effective action (8.108) in the local form

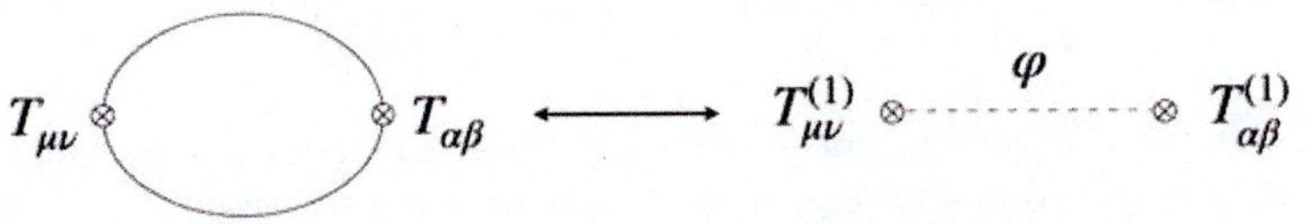

Fig. 8.6 The one-loop stress tensor vacuum polarization of a 2D CFT, which exhibits the massless $1/k^2$ pole of (8.106). In the equivalent tree graph at right the conformalon scalar vertex $T^{(1)}{}_{munu}$ is given by the term linear in phi in (8.111), and the phi propagator is obtained from (8.109), cf. Ref. [147]

$$S_A^{2D}[g;\varphi] = -\frac{N\hbar}{96\pi}\int d^2x\sqrt{-g}\,\left\{g^{\mu\nu}(\nabla_\mu\varphi)(\nabla_\nu\varphi) - 2R\varphi\right\} \tag{8.109}$$

by the introduction of the scalar field φ describing the collective spin-0 degree of freedom, which is linearly coupled to R, and whose massless propagator gives rise to the light cone singularities of the underlying massless CFT [34]. Variation of (8.109) with respect to φ gives its equation of motion

$$-\Box\varphi = R \tag{8.110}$$

which when solved for $\varphi = -\Box^{-1}R$ and substituted back into (8.109) returns the non-local form of the effective action (8.108) up to a surface term. On the other hand variation of (8.109) with respect to the metric $g_{\mu\nu}$ yields the stress tensor

$$T_A^{\mu\nu}\Big|_{2D} = \frac{N\hbar}{48\pi}\left(2\nabla^\mu\nabla^\nu\varphi - 2g^{\mu\nu}\,\Box\varphi + \nabla^\mu\varphi\nabla^\nu\varphi - \frac{1}{2}g^{\mu\nu}\,\nabla_\alpha\varphi\nabla^\alpha\varphi\right) \tag{8.111}$$

which is covariantly conserved in 2D, by use of (8.110) and by virtue of the vanishing of the Einstein tensor in two dimensions. The trace of this energy-momentum tensor gives the 2D trace anomaly (8.105) by use of (8.110).

To see the effect of the anomaly and φ on horizons, consider the 2D line element of the form

$$ds^2 = -f(r)dt^2 + \frac{dr^2}{f(r)} = f(r)\big(-dt^2 + dx^2\big), \qquad dx = \frac{dr}{f(r)} \tag{8.112}$$

where $x = r^*$ is the Regge–Wheeler tortoise coordinate. This metric has the static time Killing vector $K_{(t)}$ with invariant norm $-K_{(t)}.K_{(t)} = f(r)$, analogous to the Schwarzschild BH and dS static patch in coordinates (8.1). The 2D Ricci scalar is

$$R = -f'' = -\frac{d^2f}{dr^2} \tag{8.113}$$

and (8.110) in this case is

$$\Box\varphi = -\frac{1}{f}\frac{\partial^2\varphi}{\partial t^2} + \frac{\partial}{\partial r}\left(f\frac{\partial\varphi}{\partial r}\right) = \frac{1}{f}\left(-\frac{\partial^2}{\partial t^2} + \frac{\partial^2}{\partial x^2}\right)\varphi = f''. \tag{8.114}$$

A particular solution to this linear inhomogeneous equation is $\varphi = \ln f(r)$. The associated homogeneous wave equation has general wave solutions $e^{ik(x\pm t)}$. For stationary states $k = 0$, which implies that only linear functions of t and x are allowed for such states. Thus the general stationary state solution of (8.114) is

$$\varphi = Pt + Qx + \ln f \tag{8.115}$$

where an irrelevant constant is set to zero because (8.109) and (8.111) depend only upon the derivatives of φ. The values of the integration constants P, Q are dependent upon the boundary conditions and quantum state of the underlying QFT that give rise to the anomaly (8.105).

Substituting the solution (8.115) into the stress tensor (8.111) gives

$$T_t^{\ t} = \frac{N\hbar}{24\pi}\left\{-\frac{1}{4f}\left(P^2+Q^2-f'^2\right)-f''\right\} \tag{8.116a}$$

$$T_x^{\ t} = -\frac{N\hbar}{48\pi}\frac{PQ}{f} \tag{8.116b}$$

$$T_x^{\ x} = \frac{N\hbar}{96\pi}\frac{1}{f}\left(P^2+Q^2-f'^2\right) \tag{8.116c}$$

in the (t, x) coordinates. Hence if one takes as in (8.1)

$$f(r) = 1-\frac{r_M}{r}, \qquad f' = -\frac{r_M}{r^2}, \qquad f''' = -\frac{2r_M}{r^3} \tag{8.117}$$

to model the 2D analog of the Schwarzschild geometry, (8.116) becomes

$$T_t^{\ t} = \frac{N\hbar}{24\pi}\left\{-\frac{1}{4f}\left(\frac{p^2+q^2}{r_M^2}-\frac{r_M^2}{r^4}\right)+\frac{2r_M}{r^3}\right\} \tag{8.118a}$$

$$T_x^{\ t} = -\frac{N\hbar}{48\pi r_M^2}\frac{pq}{f} \tag{8.118b}$$

$$T_x^{\ x} = \frac{N\hbar}{96\pi}\frac{1}{f}\left(\frac{p^2+q^2}{r_M^2}-\frac{r_M^2}{r^4}\right) \tag{8.118c}$$

where the constants $P = p/r_M$ and $Q = q/r_M$, with (p, q) dimensionless. This result for the 2D anomaly stress tensor shows divergences generically $\propto f^{-1}$ as $f \to 0$, and $r \to r_M$. The divergences can be arranged to cancel precisely on the future horizon by choosing $p = -q = \pm 1/2$, or on the past horizon by $p = q = \pm 1/2$, corresponding to the future or past Unruh states [182], or on both horizons by $p = 0, q = \pm 1$, corresponding to the Hartle–Hawking thermal state [91, 98, 111] at the price of being non-vanishing as $r \to \infty$, and thermodynamically unstable due to negative heat capacity (8.13).

Any other values for (p, q) result in divergences on the horizon. In particular, if one requires a time independent truly static solution $p = 0$, and asymptotic flatness as $r \to \infty$ so that $q = 0$ also, then

$$\left(T^{\mu}_{\ \nu}\right)_{\mathcal{A}} \to -\frac{N\hbar}{96\pi r_M^2 f}\begin{pmatrix}-1 & 0\\ 0 & 1\end{pmatrix} \to \infty \quad \text{as} \quad r \to r_M \tag{8.119}$$

which diverges on the two-dimensional horizon as $f \to 0$. There is no value of q which yields a time independent regular solution for φ and (8.118) on both the horizon and $r \to \infty$ [147].

Thus the stress tensor of the 2D conformal anomaly has potentially large effects on a 2D model of a BH horizon in generic states which can be quickly and efficiently computed from the local form of the effective action (8.109) in terms of the independent conformalon scalar field φ, by choosing different values of the integration constants (p, q). Although attention has usually been focused solely on the special vacuum states for which the divergences in the stress tensor (8.118) are arranged to cancel precisely as in [75, 182], inspection of (8.118) shows that these special states are of measure zero and large backreaction effects on the horizon are to be expected in any other state, irrespective of the small value of the local curvature (8.113) there. The φ field propagator is the $1/k^2$ light cone pole associated with the conformal anomaly, whose coherent classical solutions, which are bi-linear condensates of the underlying quantum theory [34], describe the macroscopic effects of the anomaly stress tensor on BH null horizons.

8.6.2 *The Conformal Anomaly Effective Action and Stress Tensor in Four Dimensions*

The general property of anomalies described in the 2D case carries over to four (and higher even) dimensions, although the algebraic details becomes somewhat more complicated. The general form of the conformal anomaly in 4D in the presence of background curvature or gauge fields is [33, 42, 77, 125]:

$$\langle T_\mu{}^\mu \rangle \equiv \frac{\mathcal{A}}{\sqrt{-g}} = b\,C^2 + b'\left(E - \tfrac{2}{3}\Box R\right) + \sum_i \beta_i\,\mathcal{L}_i \tag{8.120}$$

again even when the classical theory is conformally invariant, and one might have expected $\langle T^\mu{}_\mu \rangle$ to vanish. In (8.120)

$$E = R_{\alpha\beta\gamma\lambda}R^{\alpha\beta\gamma\lambda} - 4R_{\alpha\beta}R^{\alpha\beta} + R^2, \qquad C^2 = R_{\alpha\beta\gamma\lambda}R^{\alpha\beta\gamma\lambda} - 2R_{\alpha\beta}R^{\alpha\beta} + \tfrac{1}{3}R^2 \tag{8.121}$$

are the Euler–Gauss–Bonnet (EGB) term and the square of the Weyl tensor respectively, and the $\mathcal{L}_i$ are dimension-4 invariants of gauge fields, with coefficients determined by the β-function of the corresponding gauge coupling. The b, b', β_i coefficients are dimensionless numbers multiplied by $\hbar$ [33, 125], so that the conformal anomaly is a quantum effect with no intrinsic length scale, applying at all scales at which any masses of the underlying quantum fields can be neglected, and determined by the light fields of the SM minimally coupled to gravity.

The non-local form of the effective action for the 4D conformal anomaly (8.120) analogous to (8.108) was found first in [164]. The local form of the effective action of the 4D conformal anomaly analogous to (8.109) in 2D requires the introduction of at least one scalar field, and is also given in a number of papers [16, 18, 38, 144, 164, 175], the simplest form for which is [142], i.e.

$$S_{\mathcal{A}}[\varphi] = \frac{b'}{2}\int d^4x\sqrt{-g}\left\{-\left(\Box\varphi\right)^2 + 2\left(R^{\mu\nu} - \tfrac{1}{3}Rg^{\mu\nu}\right)(\partial_\mu\varphi)(\partial_\nu\varphi)\right\} + \frac{1}{2}\int d^4x\,\mathcal{A}\,\varphi. \tag{8.122}$$

in terms of a single scalar (conformalon) field φ. This is quite analogous to the WZ action of the axial anomaly in QCD which must be added to the low energy EFT of mesons [189]. However its connection to $1/k^2$ poles took some time to recognize, beginning with [20, 90, 141] by analogy with the axial anomaly in massless QED, which first appears in $D = 4$ in 3-point correlation function triangle diagrams. This is because in $D = 4$ the anomaly involves quadratic curvature invariants which require *two* variations with respect to the metric to yield a non-zero result in flat space. The $1/k^2$ light cone singularities of the stress tensor three-point correlator $\langle\hat{T}^{\alpha\beta}\hat{T}^{\gamma\lambda}\hat{T}^{\mu\nu}\rangle$ of a general CFT in 4D has been verified explicitly only relatively recently by solution of the anomalous conformal Ward Identities in momentum space [41, 73], which are precisely generated by the anomaly effective action (8.122).

It is important that $S_{\mathcal{A}}$ as in $D = 2$ is not purely a local functional of higher order curvature invariants (although $\mathcal{A}$ itself is), and either has a non-local form in terms of the original metric and curvature variables, or requires a new φ field to be recast in the local form (8.122). As a result, the effects described by φ do *not* decouple and can have macroscopic effects on null horizons, depending on non-local boundary conditions, notwithstanding the smallness of local curvatures there [143, 144].

The Euler–Lagrange equation for φ following from the variation of (8.122) is

$$\Delta_4\varphi = \frac{1}{2}\left(E - \tfrac{2}{3}\Box R\right) + \frac{1}{2b'}\left(bC^2 + \sum_i \beta_i \mathcal{L}_i\right) \tag{8.123}$$

in which the fourth order Panietz-Riegert differential operator is [150, 164]

$$\Delta_4 \equiv \nabla_\mu(\nabla^\mu\nabla^\nu + 2R^{\mu\nu} - \tfrac{2}{3}Rg^{\mu\nu})\nabla_\nu = \Box^2 + 2R^{\mu\nu}\nabla_\mu\nabla_\nu - \tfrac{2}{3}R\Box + \tfrac{1}{3}(\nabla^\mu R)\nabla_\nu \tag{8.124}$$

the unique fourth order scalar differential operator that is conformally covariant and the analog of the scalar wave operator $\Box$ in two dimensions.

The full form of the 4D anomaly stress tensor following by variation of $S_{\mathcal{A}}$ with respect to the spacetime metric has been given in [142, 144]:

$$T_{\mathcal{A}}^{\mu\nu}[g;\varphi] \equiv \frac{2}{\sqrt{-g}}\frac{\delta}{\delta g_{\mu\nu}}\,S_{\mathcal{A}}[g;\varphi] = b'E^{\mu\nu} + b\,C^{\mu\nu} + \sum_i \beta_i T^{(i)\,\mu\nu} \tag{8.125}$$

The first term here is

$$
\begin{aligned}
E_{\mu\nu} = & -2\,(\nabla_{(\mu}\varphi)(\nabla_{\nu)}\Box\varphi) + 2\nabla^{\alpha}\big[(\nabla_{\alpha}\varphi)(\nabla_{\mu}\nabla_{\nu}\varphi)\big] - \tfrac{2}{3}\,\nabla_{\mu}\nabla_{\nu}\big[(\nabla_{\alpha}\varphi)(\nabla^{\alpha}\varphi)\big] \\
& + \tfrac{2}{3}\,R_{\mu\nu}\,(\nabla_{\alpha}\varphi)(\nabla^{\alpha}\varphi) - 4\,R^{\alpha}{}_{(\mu}\,\big[(\nabla_{\nu)}\varphi)(\nabla_{\alpha}\varphi)\big] + \tfrac{2}{3}\,R\,(\nabla_{(\mu}\varphi)(\nabla_{\nu)}\varphi) \\
& + \tfrac{1}{6}\,g_{\mu\nu}\,\big\{-3\,(\Box\varphi)^2 + \Box\big[(\nabla_{\alpha}\varphi)(\nabla^{\alpha}\varphi)\big] + 2\,(3R^{\alpha\beta} - Rg^{\alpha\beta})(\nabla_{\alpha}\varphi)(\nabla_{\beta}\varphi)\big\} \\
& - \tfrac{2}{3}\,\nabla_{\mu}\nabla_{\nu}\Box\varphi - 4\,C_{\mu\ \nu}^{\ \alpha\ \beta}\,\nabla_{\alpha}\nabla_{\beta}\varphi - 4\,R^{\alpha}{}_{(\mu}\nabla_{\nu)}\nabla_{\alpha}\varphi + \tfrac{8}{3}\,R_{\mu\nu}\,\Box\varphi + \tfrac{4}{3}\,R\,\nabla_{\mu}\nabla_{\nu}\varphi \\
& - \tfrac{2}{3}(\nabla_{(\mu}R)\nabla_{\nu)}\varphi + \tfrac{2}{3}g_{\mu\nu}\big[\Box^2\varphi + 3R^{\alpha\beta}\,\nabla_{\alpha}\nabla_{\beta}\varphi - 2R\Box\varphi + \tfrac{1}{2}(\nabla^{\alpha}R)\nabla_{\alpha}\varphi\big] \qquad (8.126)
\end{aligned}
$$

which is the metric variation of all the b' terms in (8.122), both quadratic and linear in φ [141, 142, 144], while

$$
\begin{aligned}
C_{\mu\nu} &\equiv -\frac{1}{\sqrt{-g}}\,\frac{\delta}{\delta g^{\mu\nu}}\left\{\int d^4x\sqrt{-g}\,C^2\,\varphi\right\} \\
&= -4\,\nabla_{\alpha}\nabla_{\beta}\left(C_{(\mu\ \nu)}^{\ \ \alpha\ \ \beta}\,\varphi\right) - 2\,C_{\mu\ \nu}^{\ \alpha\ \beta}\,R_{\alpha\beta}\,\varphi \qquad (8.127\text{a}) \\
T_{\mu\nu}^{(i)} &\equiv -\frac{1}{\sqrt{-g}}\,\frac{\delta}{\delta g^{\mu\nu}}\left\{\int d^4x\sqrt{-g}\,\mathcal{L}_i\,\varphi\right\} \qquad (8.127\text{b})
\end{aligned}
$$

are the metric variations of the last two b and β_i terms in (8.125), both of which are linear in φ. With these explicit forms it is straightforward to compute the stress tensor in 4D geometries with horizons.

8.6.3 *The Stress Tensor of the Conformal Anomaly on the Schwarzschild and de Sitter Horizons*

The general solution to (8.123) for $\varphi = \varphi(r)$ and $\mathcal{L}_i = 0$ that is finite as $r \to \infty$ is readily found by quadratures for the Schwarzschild metric (8.1)–(8.2) to be [141, 144]

$$
\frac{d\varphi_S}{dr} = \frac{c_S\, r_M}{r(r - r_M)} - \frac{2}{3\,r_M}\left(\frac{r}{r_M} + 1 + \frac{r_M}{r}\right)\ln\left(1 - \frac{r_M}{r}\right) - \frac{2}{3\,r_M} - \frac{1}{r} \qquad (8.128\text{a})
$$

$$
\varphi_S(r) = c_S\,\ln\left(1 - \frac{r_M}{r}\right) + \int_{r/r_M}^{\infty} dx\left\{\frac{2}{3x}\,(x^2 + x + 1)\ln\left(1 - \frac{1}{x}\right) + \frac{2}{3} + \frac{1}{x}\right\} \qquad (8.128\text{b})
$$

in terms of the dimensionless integration constant c_S. This solution has the limits

$$
\varphi_S(r) \to \begin{cases} c_S\,\ln\left(1 - \frac{r_M}{r}\right) + c_i - 2\left(1 - \frac{r_M}{r}\right)\left[\ln\left(1 - \frac{r_M}{r}\right) + \frac{2}{3}\right] + \cdots, & r \to r_M \\ -\left(c_S + \frac{11}{9}\right)\frac{r_M}{r} - \left(2c_H + \frac{13}{9}\right)\frac{r_M^2}{4r^4} + \cdots, & r \to \infty \end{cases} \qquad (8.129)
$$

where the constant $c_i = -(7/6 + \pi^2/9) = -2.26329...$ is the finite integral in (8.128b) evaluated at the lower limit $r = r_M$, $x = 1$. The addition of an arbitrary integration constant to (8.128b) does not substantially change the results. Substituting the solution (8.128) into the anomaly stress tensor (8.125)–(8.126), one finds

$$\left(T^{\mu}{}_{\nu}\right)_{\mathcal{A}} \to \frac{c_s^2}{6r_M^4}\,\frac{b'}{f^2}\begin{pmatrix} -3 & 0 & 0 & 0 \\ 0 & 1 & 0 & 0 \\ 0 & 0 & 1 & 0 \\ 0 & 0 & 0 & 1 \end{pmatrix} \to \infty \quad \text{as} \quad r \to r_M \tag{8.130}$$

diverging quadratically ($\propto f^{-2}$) on the Schwarzshild BH horizon for any $c_s \neq 0$. This is analogous to the linear divergence ($\propto f^{-1}$) of (8.119) in the 2D case.

The divergent behavior of the local stress tensor (8.130) shows that the anomaly stress tensor can become important near the horizon of a BH and even dominate the classical terms in the Einstein equations, the smallness of the curvature tensor there notwithstanding. Even with $c_s = 0$ in (8.128), which can be arranged by specific choice of the state of the underlying QFT, to remove the leading f^{-2} divergence in (8.130), there remain subleading divergences proportional to f^{-1}, $(\ln f)^2$ and $\ln f$. In fact, there is no solution of (8.123) in Schwarzschild spacetime with $\varphi = \varphi(r)$ only, corresponding to a fully Killing time t invariant and spherically symmetric quantum state, with a finite stress tensor at both singular points $r = r_M$ and $r = \infty$ of the differential Eq. (8.123). This result, following simply and directly from the conformal anomaly effective action, confirms results of previous studies of the stress tensor expectation value in specific states in Schwarzschild spacetime [70].

The divergences on either the future or past BH horizon (but not both) can be cancelled by allowing linear time dependent solutions of (8.123), which give rise to a Hawking flux $\langle T^{t}{}_{r}\rangle$ [144], such as in the Unruh states [182]; or by relaxing the regularity condition at infinity which gives rise to a non-zero stress tensor there, as in the Hartle–Hawking state [91, 98, 111]. This thermal state is both incompatible with asymptotically flat boundary conditions and unstable due to its negative heat capacity (8.13) [103]. The only other possibility is that divergence on the fixed Schwarzschild BH horizon (8.130) is uncancelled.

Usually the *assumption* of regularity of the semi-classical $\langle T_\mu{}^\nu\rangle$ on the horizon is used to argue for the necessity of Hawking radiation flux $\langle T_r{}^t\rangle > 0$ [88, 168]. However the converse is also true, namely if the final state of gravitational collapse is:

1. asymptotically flat,
2. truly static (implying with $\langle T^t{}_r\rangle = 0$), and
3. thermodynamically stable,

then quantum effects at the horizon imply the breakdown of regularity there.

The stringency of these conditions and the non-local topological obstruction to regular behavior on the horizon they imply is well illustrated by the impossibility of finding a static solution only of r of the conformalon Eq. (8.123) that satisfies the three conditions above and is also regular on the horizon in the Schwarzschild background. It is natural that the effective action of the conformal anomaly contains information about the global quantum state properties in the solutions of (8.123) that relates the behavior of the state on the horizon with that at infinity. The definition of the vacuum state in QFT relies on a clear and Lorentz invariant separation between positive and negative energy states, whereas all finite energy scales on the horizon can be neglected. This means that the mass gap $2mc^2$ between the positive and negative energy excitations effectively vanishes and the local vacuum at the horizon is radically different from that in the asymptotic region, and susceptible to breakdown. The coordinate invariant large quantum effects in the energy-momentum-stress tensor $T_\mathcal{A}^{\mu\nu}$ due to the conformal anomaly in (8.130) is the signal of this vacuum breakdown– and an incipient quantum phase transition. A large stress tensor would necessarily produce large backreaction effects on the near-horizon geometry, and could in principle prevent the classical BH horizon or any trapped surface from ever forming.

The breakdown of regularity on the horizon and potentially large quantum effects on the Schawrzschild BH horizon, were first found by D. Boulware [35], by defining the vacuum state by the positive frequency modes with respect to the Schwarzschild Killing time t. Rather than being special to the Boulware state, or 'pathological,' the anomaly stress tensor (8.125) makes it clear that vacuum stresses growing without bound as $r \to r_M$, as in (8.130), are the *generic* case for the static geometry, while states for which these divergences are cancelled and are regular on the horizon with a Hawking flux are the special ones and a set of measure zero.

The f^{-2} behavior of (8.130), where $\kappa_+ = 1/2r_M$ is the surface gravity, can be understood on dimensional grounds from the $T_{\rm loc}^4$ conformal behavior of the local Tolman temperature (8.19) of virtual massless fields near the horizon, where all but the heaviest fields may be treated as effectively massless. Indeed the near-horizon geometry of the static metric (8.1) at constant t is isomorphic to three dimensional Euclidean anti-dS (Łobachewsky) space $EAdS_3$ [17, 134, 169], where operators of conformal weight w behave as $f^{-\frac{w}{2}}$. The stress energy tensor being of weight $w = D$

in D dimensions has the power law behavior f^{-1} in $D = 2$ as in (8.118), f^{-2} in $D = 4$ as in (8.130). The effective action of the anomaly with the $\ln f$ behavior of the dimensional conformalon field φ carries this conformal behavior to the stress tensor in the horizon region. No other part of the quantum effective action has this conformal property [134]. This is another way of seeing that the anomaly $S_{\mathcal{A}}$ is relevant for BH physics [141], at the macroscopic horizon scale, no matter how small the local curvature is there.

As the importance of the anomaly stress tensor on the exterior BH horizon is demonstrated by (8.130), a similar behavior is observed in dS spacetime, approaching the static dS horizon from the interior of a gravastar. Indeed the f^{-2} behavior in any spherically symmetric static spacetime (8.1) with a horizon at which $-K^{\mu}K_{\mu} = f(r) \to 0$ [12]. With Λ positive, the static patch of dS space is of this form with $f(r) \propto h(r) = 1 - H^2r^2 = 1 - \Lambda r^2/3$. The general spherically symmetric static solution of (8.123) for $\varphi = \varphi(r)$ which is regular at the origin in this case is [141, 144]

$$\varphi_{dS}(r) = \ln\left(1 - H^2r^2\right) + c_0 + \frac{q}{2}\ln\left(\frac{1-Hr}{1+Hr}\right) + \frac{2c_H - 2 - q}{2Hr}\ln\left(\frac{1-Hr}{1+Hr}\right)$$
$$\to \left[c_H + \left(c_H - 1 - \frac{q}{2}\right)(1 - Hr) + \cdots\right]\ln(1 - Hr) + c' + \mathcal{O}(1 - Hr) \qquad (8.131)$$

where the constant $c' = c_0 + (2 - c_H)\ln 2$. Substituting this into the anomaly stress tensor (8.125) gives

$$(T^{\mu}{}_{\nu})_{\mathcal{A}} \to \frac{2}{3}c_H^2\,H^4\frac{b'}{(1-Hr)^2}\begin{pmatrix} -3 & 0 & 0 & 0\\ 0 & 1 & 0 & 0\\ 0 & 0 & 1 & 0\\ 0 & 0 & 0 & 1\end{pmatrix} \to \infty \quad \text{as} \quad r \to r_H \equiv H^{-1} \qquad (8.132)$$

which also diverges as f^{-2} for any $c_H \neq 0$ as the dS static horizon is approached (Fig. 8.7).

As in the Schwarzschild case this divergence can be removed if a thermal state is considered, but then only if the temperature is precisely matched to the Hawking temperature $T_H = H/2\pi$ associated with the horizon, which in the dS case leads to the maximally $O(4,1)$ symmetric state [39, 65]. However, this state is not a vacuum state of QFT and is unstable to particle pair creation, much as a uniform, constant electric field is [13–15, 140, 158], and for essentially the same reason. Due to the non-existence of a global static time by which positive and negative frequency (particle and anti-particle) solutions can be invariantly distinguished, a time independent Hamiltonian bounded from below and stable vacuum state cannot be defined. This is similar to the non-existence of a globally static time in the full Kruskal extension of the Schwarzschild solution in Fig. 8.1.

In both cases the horizon where the Killing vector of time translation ∂_t becomes null and $f(r) = 0$ in either fully analytically extended Schwarzschild or dS space is the sign of this, so that Λ cannot be globally constant and positive everywhere in space in QFT. The conformal anomaly shows this through its sensitivity to lightlike

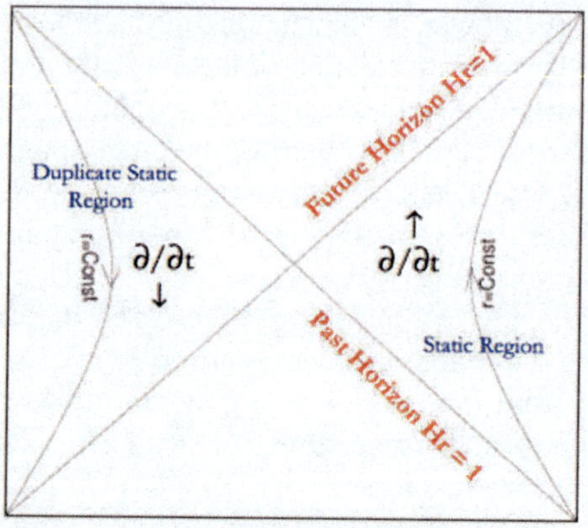

Fig. 8.7 The Carter–Penrose conformal diagram of the maximal extension of de Sitter space, showing the two static patches where the static time Killing field ∂_t runs in opposite directions. As in the maximal analytic extension of the Schwarzschild geometry of Fig. 8.1, this renders it impossible to define a conserved quantum Hamiltonian, invariantly separating positive and negative frequencies that is bounded from below

correlations on the horizon and non-local boundary conditions on the quantum state. In each case one is restricted to one of the static patches only, and their joining together at their mutual horizons as in the realization of the gravastar as a limiting case of the Schwarzschild constant density solution in Sect. 8.4. This resolves the conflict each of these globally extended solutions of Einstein's equations has with QFT.

8.6.4 *Determination of ε and ℓ for a Gravastar*

It is important to recognize that in a full solution the stress tensor will not diverge. Instead the $1/f^2$ behavior of the anomaly stress tensor will grow until (8.130) becomes large enough to affect the classical Schwarzschild geometry (8.1)–(8.2). At that point their backreaction effects on the geometry must be taken into account in a self-consistent solution of the Einstein equations together with the conformalon field $\varphi(r)$. As long as $f(r)$ is finite, the states with large stress tensors on the horizon are continuously connected to the Hilbert space of the usual Minkowski vacuum.

Since the quantum anomaly terms are parametrically suppressed by a factor of $L_{\rm Pl}^2/r_M^2$ compared to classical terms, only when

$$f \sim \epsilon \equiv \frac{M_{\rm Pl}}{M} \simeq 1.1 \times 10^{-38}\, \frac{M_{\odot}}{M} \ll 1 \tag{8.133}$$

or at $\Delta r = r - r_M \sim L_{\rm Pl}$, does the anomaly stress tensor become comparable to the classical terms, for c_s in (8.130) of $\mathcal{O}(1)$. This gives a *physical* thickness of the critical boundary layer of

$$\ell \sim \frac{\Delta r}{\sqrt{h(r)}}\bigg|_{r=r_M+\Delta r} \simeq \sqrt{\epsilon}\, r_M \simeq 2.2 \times 10^{-14} \sqrt{\frac{M}{M_\odot}}\ \text{cm} \tag{8.134}$$

a factor of 10^{19} greater than the Planck scale $L_{\rm Pl}$, making a mean field EFT treatment possible, but well justifying the thin shell approximation of the original gravastar model reviewed in Sect. 8.3, and determining ℓ in (8.33).

The result of these considerations is that effective action $S_{\mathcal{A}}$ of (8.122) amounts to a specific addition to Einstein's GR, consistent with, and in fact required by first principles of QFT, general covariance, and the general form of the conformal anomaly (8.120). It is a relevant addition in both the mathematical and physical sense [134], capturing the macroscopic light cone singularities of anomalous correlation functions, and a necessary part of the low energy EFT of gravity, particularly relevant for spacetimes with BH or cosmological horizons. It therefore should be added to the Einstein-Hilbert action of classical GR, much as the WZ term must be added to the low energy meson theory to account for the chiral anomaly of QCD [32, 90, 134, 141–144, 189]. The effective action (8.122) and stress tensor (8.125) generated by it generically produce the large quantum (but still semi-classical) backreaction effects on BH and dS horizons, necessary to trigger the quantum phase transition hypothesized in the original gravastar paper.

8.7 The EFT of Gravity, and Dynamical Λ Vacuum Energy

In addition to the conformal anomaly, the second essential element needed for the formation of a gravastar is the ability of vacuum energy, or effective cosmological term $\Lambda_{\rm eff}$, to change rapidly at the horizon. The EFT that describes this quantum phase transition boundary layer at the horizon also provides a resolution of the 'naturalness' problem of Λ dark energy in cosmology [6, 95, 143]. That the physics of BHs is closely related to the 'fine tuning' issue of cosmological vacuum energy is inherent in the gravastar proposal, which requires a dynamical vacuum energy $\rho_v = 3H^2/8\pi G$ in the dS interior to adjust to the total mass M of the gravastar to be joined at their mutual horizons $H^{-1} = 2GM$.

The basis of the solution to both problems begins with the observation that the constant Λ term can be reformulated in terms of an abelian gauge theory of a 4-form field strength [21, 22, 80, 107]

$$F = \frac{1}{4!}\, F_{\alpha\beta\gamma\lambda}\, dx^\alpha \wedge dx^\beta \wedge dx^\gamma \wedge dx^\lambda \tag{8.135}$$

which is the curl of a totally anti-symmetric 3-form gauge potential

$$F_{\alpha\beta\gamma\lambda} = 4\,\partial_{[\alpha} A_{\beta\gamma\lambda]} = 4\,\nabla_{[\alpha} A_{\beta\gamma\lambda]} = \nabla_\alpha A_{\beta\gamma\lambda} - \nabla_\beta A_{\alpha\gamma\lambda} + \nabla_\gamma A_{\alpha\beta\lambda} - \nabla_\lambda A_{\alpha\beta\gamma} \tag{8.136}$$

that is, F is an exact 4-form

$$F = dA, \qquad A = \frac{1}{3!} A_{\alpha\beta\gamma}\, dx^\alpha \wedge dx^\beta \wedge dx^\gamma . \tag{8.137}$$

As a natural generalization of ordinary electromagnetism where $F = dA$ is an exact 2-form exterior derivative of the 1-form vector gauge potential $A = A_\mu\, dx^\mu$, let F be provided with the 'Maxwell' action

$$S_F = -\frac{1}{2\varkappa^4}\int F\wedge \star F = -\frac{1}{48\,\varkappa^4}\int d^4x\sqrt{-g}\; F_{\alpha\beta\gamma\lambda}F^{\alpha\beta\gamma\lambda} = \frac{1}{2\varkappa^4}\int d^4x\sqrt{-g}\;\widetilde{F}^2 \tag{8.138}$$

where

$$\widetilde{F} \equiv \star F = \frac{1}{4!}\,\varepsilon_{\alpha\beta\gamma\lambda}\,F^{\alpha\beta\gamma\lambda}, \qquad F_{\alpha\beta\gamma\lambda} = -\varepsilon_{\alpha\beta\gamma\lambda}\,\widetilde{F} \tag{8.139}$$

is the scalar Hodge star $\star$ dual to F, and $\varkappa$ is a free parameter whose significance as a topological susceptibility of the gravitational vacuum is discussed in [143].

When the rank of the D-form is matched to the number of $D=4$ spacetime dimensions in this way, the free 'Maxwell' theory (8.138) has two very special properties:

(i) $\widetilde{F}$ is constrained to be a constant, with no propagating degrees of freedom, and
(ii) Its stress tensor $T_F^{\mu\nu}$ is proportional to the metric $g^{\mu\nu}$, hence equivalent to a cosmological Λ term.

That F and its dual $\widetilde{F}$ are constant follows from the *sourcefree* 'Maxwell' equation obtained by variation of (8.138) with respect to $A_{\alpha\beta\gamma}$, viz.

$$\nabla_\lambda F_{\alpha\beta\gamma}{}^{\lambda} = 0, \qquad \text{for} \qquad J^{\alpha\beta\gamma} = 0 \tag{8.140}$$

and $\partial_\lambda \widetilde{F} = 0$, so that $\widetilde{F} = \widetilde{F}_0$ is a spacetime constant—in the complete absence of any sources $J = 0$. At the same time the stress tensor corresponding to (8.138)

$$T_F^{\mu\nu} = \frac{2}{\sqrt{-g}}\frac{\delta S_F}{\delta g_{\mu\nu}} = -\frac{1}{4!\,\varkappa^4}\left(\frac{1}{2}g^{\mu\nu}F^{\alpha\beta\gamma\lambda}F_{\alpha\beta\gamma\lambda} - 4F^{\mu\alpha\beta\gamma}F^{\nu}{}_{\alpha\beta\gamma}\right) = -\frac{1}{2\varkappa^4}\,g^{\mu\nu}\,\widetilde{F}^2 \tag{8.141}$$

is proportional to the metric tensor, if the convention that $F_{\alpha\beta\gamma\lambda}$ with all lower indices is independent of the metric is adopted.

Hence (8.135)–(8.141) are completely equivalent to a cosmological term in Einstein's equations in $D = 4$ dimensions, with the identification

$$\Lambda_{\rm eff} = \frac{4\pi G}{\varkappa^4}\,\widetilde{F}^2 \geq 0 \tag{8.142}$$

the effective (necessarily non-negative) cosmological constant term for $\varkappa$ and $\widetilde{F}$ real constants. In this way one can freely trade a positive cosmological constant Λ of classical GR for a new fundamental constant $\varkappa$ of the low energy EFT, together with an integration constant $\widetilde{F}_0$ of the constraint $\partial_\lambda \widetilde{F} = 0$.

The integration constant $\widetilde{F}_0$ can be fixed by a classical global boundary condition in flat space, without any reference to quantum zero-point energy, UV divergences, or cutoffs. A vanishing $\Lambda_{\rm eff}$ corresponds instead to the vanishing of the sourcefree 'electric' field strength $\widetilde{F} = F^{0123}$ in infinite 3-dimensional empty flat space, analogous to the vanishing of the electric field in the vacuum of ordinary electromagnetism in the absence of sources. In either case this is simply the classical state of lowest energy, as well as the unique state that is even under the discrete symmetry of space parity inversion.

Moreover the setting of the value of the free constant $\widetilde{F} = \widetilde{F}_0$, which is a priori independent of geometry, to zero in empty flat space, is *required* by the sourcefree Einstein's equations

$$\left[R_{\mu\nu} - \frac{R}{2} g_{\mu\nu}\right]_{\rm flat} = 0 = -\Lambda_{\rm eff}\Big|_{\rm flat}\, \eta_{\mu\nu} \tag{8.143}$$

viewed as a low energy EFT. This shows that flat space QFT estimates of vacuum energy in any way dependent upon UV cutoffs or heavy mass scales are *inconsistent* with Einstein's equations It is well-known that QFT in flat space is sensitive only to *differences* in energy. Hence the absolute value of quantum zero point energy in flat space, and its dependence upon cutoffs or UV regularization schemes is arbitrary and of no physical significance [155]. The value of $\Lambda_{\rm eff}$ is significant only through its gravitational effects, and hence cannot be evaluated in isolation, but only within the context of a gravitational EFT, 'on shell' as in (8.143), and only if each side of (8.143) can be evaluated *independently*. This becomes possible only if $\Lambda_{\rm eff}$ is a *free* constant of integration, as it is in (8.142), and not a fixed parameter of the Lagrangian.

For the longstanding problem of the cosmological term when QFT is coupled to gravity, the classical state of minimum energy is that of vanishing 4-form classical coherent field or condensate: $\widetilde{F} = \widetilde{F}_0 = 0$. By the identification of $\Lambda_{\rm eff}$ in (8.142), the condition (8.143) automatically sets the value of the cosmological term to zero in infinite flat Minkowski space. By simply allowing a consistent flat Minkowski solution, this removes one oft-stated obstacle and 'no-go theorem' to solution of the 'cosmological constant problem' [115, 188].

In isolation this reformulation of the cosmological constant Λ_{eff} in terms of $\widetilde{F} = \widetilde{F}_0$ and $\varkappa$ shifts the consideration of cosmological vacuum energy away from the UV divergences of QFT to a macroscopic (IR) boundary condition solving the classical constraint Eq. (8.140) and minimization of energy in flat space. In addition to removing the fine tuning or naturalness problem of Λ, introducing an independent 4-form field F in place of constant Λ also allows for the introduction of sources in (8.140) that will enable F (and hence Λ_{eff}) to change, departing from its zero value in infinite sourcefree flat space in finite calculable ways, and eventually to become a full-fledged dynamical variable of the low energy EFT of gravity in its own right.

This is exactly what occurs if the 3-form potential A is identified with the torsional part of the 3-form Chern–Simons potential naturally defined by the Euler–Gauss–Bonnet (EGB) term E (8.121) in the trace anomaly. This term is distinguished by its topological character. Its integral is a topological invariant insensitive to local variations, and is also associated with a 4-form gauge field

$$\mathsf{F} \equiv \varepsilon_{abcd}\, \mathsf{R}^{ab} \wedge \mathsf{R}^{cd} = \frac{1}{4}\epsilon_{abcd}\, R^{ab}_{\ \ \alpha\beta}\, R^{cd}_{\ \ \gamma\lambda}\, dx^\alpha \wedge dx^\beta \wedge dx^\gamma \wedge dx^\lambda \tag{8.144}$$

which is also exact, i.e. $\mathsf{F} = d\mathsf{A}$, where A is the (3,1) Lorentz frame dependent Chern–Simons 3-form [192]

$$\mathsf{A} = \epsilon_{abcd}\left(\omega^{ab} \wedge d\omega^{cd} + \frac{2}{3}\,\omega^{ab} \wedge \omega^{ce} \wedge \omega^{fd}\,\eta_{ef}\right) \tag{8.145}$$

defined in terms of the (3,1) spin connection ω^{ab}. The spin connection is a priori independent of the spacetime metric, becoming locked to it only by the condition of zero torsion, in which case ω^{ab} becomes the usual Christoffel connection in holonomic coordinates [84, 124]. In Einstein-Cartan geometries non-zero torsion as well as non-zero curvature are allowed, as general considerations of minimal coupling of fermions to gravity require [52, 106, 173, 184]. Then ω^{ab} and hence $\mathsf{A} = \mathsf{A}_\mathsf{R} + \mathsf{A}_\mathsf{T}$ acquire a non-Riemannian torsion dependent contribution A_T independent of the metric, in addition to the Riemanian part A_R. The identification of $\mathsf{A}_\mathsf{T} = A$ with the 3-form potential of (8.136)–(8.137), which must be varied independently of the metric, then couples the anomaly scalar φ to Λ_{eff}, making it spacetime dependent.

This is established as follows. Since F is related to E by the Hodge star $\star$ dual

$$\begin{aligned}\star\mathsf{F} &= \varepsilon^{\alpha\beta\gamma\lambda}\left(\frac{1}{4}\epsilon_{abcd}\, R^{ab}_{\ \ \alpha\beta}\, R^{cd}_{\ \ \gamma\lambda}\right) = \frac{1}{4}\epsilon_{abcd}\,\epsilon^{mnrs}\, R^{ab}_{\ \ mn} R^{cd}_{\ \ rs} \\ &= -\left(R_{abcd}R^{abcd} - 4R_{ab}R^{ab} + R^2\right) = -E\end{aligned} \tag{8.146}$$

which in component form means that

$$E = -(\star d\mathsf{A}) = \frac{1}{3!}\,\varepsilon^{\alpha\beta\gamma\mu}\partial_\mu \mathsf{A}_{\alpha\beta\gamma} = \frac{1}{3!}\,\nabla_\mu\left(\varepsilon^{\alpha\beta\gamma\mu}\mathsf{A}_{\alpha\beta\gamma}\right) \tag{8.147}$$

is a total derivative, the torsional part of the $E\varphi$ term in the anomaly effective action is first separated off, by replacing E in (8.120) and the anomaly effective action (8.122) with $E - (\star d\mathsf{A}_\mathsf{T}) = E - \widetilde{F}$, where E is now defined to be the strictly Riemannian non-torsional part. The torsion dependent term in (8.122) can then be integrated by parts

$$-\frac{b'}{2}\int d^4x\sqrt{-g}\,\widetilde{F}\,\varphi = -\frac{b'}{2}\frac{1}{3!}\int d^4x\sqrt{-g}\,A_{\alpha\beta\gamma}\,\varepsilon^{\alpha\beta\gamma\mu}\,\partial_\mu\varphi \tag{8.148}$$

up to a surface term which does not affect local variations. Defining the 3-current

$$J^{\alpha\beta\gamma} \equiv -\frac{b'}{2}\,\varepsilon^{\alpha\beta\gamma\mu}\,\partial_\mu\varphi \tag{8.149}$$

the last term in (8.148) can be expressed in the form

$$S_{\rm int}[\varphi, A] = \frac{1}{3!}\int d^4x\sqrt{-g}\,J^{\alpha\beta\gamma}A_{\alpha\beta\gamma} \tag{8.150}$$

analogous to a $J\cdot A$ interaction of ordinary electromagnetism. Since the 'Maxwell' action (8.138) and interaction term (8.150) must be stationary with respect to independent variation of $A_{\alpha\beta\gamma}$, with $g_{\mu\nu}$ and φ fixed, (8.150) provides a source term for the 'Maxwell' equation of the 4-form gauge field

$$\nabla_\lambda F^{\alpha\beta\gamma\lambda} = \varkappa^4 J^{\alpha\beta\gamma} = -\frac{\varkappa^4 b'}{2}\,\varepsilon^{\alpha\beta\gamma\lambda}\,\partial_\lambda\varphi \tag{8.151}$$

with the source current (8.149). Upon taking its dual, with (8.139), this becomes

$$\partial_\lambda\left(\widetilde{F} - \frac{\varkappa^4 b'}{2}\,\varphi\right) = 0 \tag{8.152}$$

which is an equation of constraint that is immediately solved by

$$\widetilde{F} = \frac{\varkappa^4 b'}{2}\,\varphi + \widetilde{F}_0 \tag{8.153}$$

in which $\widetilde{F}_0$ is a spacetime constant. The result is that $\widetilde{F}$ given by (8.153) is no longer a constant, and will change, as will the effective cosmological term $\Lambda_{\rm eff}$ in (8.142), when φ changes according to (8.123). From (8.128)–(8.130) and (8.131)–(8.132), this most rapid change occurs at the BH and dS horizons. The blueshifting of local frequencies $\sim f^{-\frac{1}{2}}$ as in (8.3) leads to the φ field having an increasingly large radial derivative in the vicinity of $r_{_M} \simeq r_{_H}$, so that (8.149) and (8.125) and the torsional effects become significant there. Thus the addition of the 4-form gauge field 'Maxwell' action (8.138) to the effective action of gravity, which exactly reproduces a positive cosmological term (8.142) in the purely classical theory, allows (and requires) the vacuum energy to change when its torsional part is coupled the conformalon scalar and effective action of the quantum conformal anomaly by (8.150).

The scalar $\widetilde{F}$, dual to the 4-form field strength F is a classical coherent field that provides an explicit realization of a gravitational condensate $\Lambda_{\rm eff}$ interior. When coupled to the conformalon scalar through the 3-form abelian current $J^{\alpha\beta\gamma}$, concentrated on a three-dimensional extended world tube of topology $\mathbb{R} \times \mathbb{S}^2$ where $\partial_\mu \varphi$ grows large, φ, $\widetilde{F}$ and the condensate $\Lambda_{\rm eff}$ all change rapidly in the radial direction. This is precisely the appropriate description of a thin shell phase boundary layer of a gravastar with $\mathbb{S}^2$ spatial topology sweeping out a world tube in spacetime.

The resulting low energy EFT of gravity contains two additional degrees of freedom in addition to the metric, which lead to a coupled set of equations relevant for the near-horizon geometry of a gravastar where φ and $\Lambda_{\rm eff}$ change rapidly [143]. This realizes the quantum phsae transition at the horizon hypothesized in the original gravastar proposal [133, 135, 137].

That the extension of GR to Einstein-Cartan spacetime with torsion should occur where the vierbein 'soldering form' $e^0_{\ t}$ vanishes, and the locking together of the (3,1) tangent space gauge group and $GL(4, \mathbb{R})$ group of coordinates transformations is broken, is related to some earlier studies [74]. However the specific relation to the conformal anomaly and its macroscopic effects on horizons as the locus of where these torsional effects should first appear is introduced only recently in [143].

In the resulting EFT of gravity one can now search for static, rotationally invariant solutions of the EFT equations which are externally Schwarzshild with $\Lambda_{\rm eff}$ vanishing in the exterior region, changing rapidly but continuously near the Schwarzschild r_M or dS r_H classical horizons by (8.153) and (8.123) in the phase boundary region, and then remaining nearly constant $3H^2$ in the interior region. The stability and normal modes of oscillations about this solution can be studied in a systematic way within a Lagrangian framework, for the first time, without reliance upon fluid ansätze. The interactions of SM fields with the quantum phase boundary surface layer of thickness given by (8.33) can be studied in this EFT as well, and the effects of this surface layer and regular dS interior on accretion, binary BH mergers, gravitational waves, ringdown and 'echoes' investigated. The effective Lagrangian framework for gravity also will permit studies of gravastar formation by the triggering of a quantum phase transition in the EFT before any trapped surface can form.

Clearly there is considerable work remaining to be done, but the path seems open to the goal of establishing gravitational condensate stars as the stable endpoint of gravitational collapse consistent with, in fact relying upon quantum field theory.

Acknowledgements It is a pleasure to acknowledge the seminal contributions of Pawel O. Mazur to the inception of the proposal of gravitational condensate stars in [133, 135, 137], as well as numerous insights on the conformal anomaly, revisiting the Schwarzschild interior solution and the bringing Ref. [74] to the author's attention. The author also gratefully acknowledges the other colleagues with whom he has collaborated on various aspects of gravitational condensate stars, de Sitter space and the effective action of the conformal anomaly, including P. R. Anderson, I. Antoniadis, P. Beltracchi, D. Blaschke, R. Carballo-Rubio, G. Chapline, C. Corianò, M. Giannotti, P. Gondolo, M. M. Maglio, C. Molina-Parìs, I. L. Shapiro and R. Vaulin. The recent hospitality of the gravity group at SISSA of S. Liberati, and gravity wave research group at SISSA of E. Barausse in the organization of the workshop "Quantum Effective Field Theory and Black Hole Tests of Einstein Gravity" [https://grams-815673.wixsite.com/september12-16] in September, 2022 is also gratefully acknowledged.

Appendix 1: Thin Shell Versus Thick Shell

The distinguishing feature of the original gravitational condensate star proposal of [133, 137] of the main text is the abrupt change in ground state vacuum energy at the horizon, characteristic of a quantum phase transition there. This should be clear from the essential role of the horizon as a infinite red shift surface in both [63, 133, 137], the assumption of $\varepsilon \ll 1$ and the estimate of ε and ℓ in Sect. 8.6 from the conformal anomaly. The proper length ℓ determined by the stress tensor of the conformal anomaly takes the place of the 'healing length' introduced, but left undetermined in the analogy of the horizon in GR to the non-relativistic quantum critical surface of a sound horizon in [63]. Thus the term 'gravastar' should apply only to the gravitational condensate star model of [133, 137] in the text, described also in [135], and further refined in [136], where the lightlike null horizon clearly plays a privileged role as the locus of joining of interior and exterior classical geometries, with equal and opposite surface gravities, and where the conformal anomaly stress tensor also grows large, and a quantum phase transition can occur.

Despite this physically privileged role of the horizon in the original gravastar proposal [133, 137], a number of papers appeared subsequently that discussed what may be called 'generalized gravastars,' or regular solutions with macroscopically large or 'thick' shells, comparable to the gravitational radius r_M itself, with compactness GM/r differing from the maximal value of $1/2$, by order unity, some time varying, or with timelike surfaces displaced from the Schwarzschild or dS horizons by finite amounts [66, 67, 76, 82, 108, 122, 128, 151–153, 162, 170, 174, 180, 181, 185, 186]. Several authors proceeded to discuss both ergoregion instabilities and observational bounds on such hypothetical objects, with various assumptions about boundary conditions and surface matchings [50, 51, 68]. It should be clear that these instabilities or observational bounds do not apply to gravastars, which *by definition* are static configurations with a *infinitesimally thin shell located at the horizon*, for $\varepsilon = 0$, or straddling and replacing the would-be classical Schwarzschild and dS horizons for very small finite ε, with metric functions $f \sim h = \mathcal{O}(\varepsilon)$ there. Any other regular QBH is not a gravastar.

Appendix 2: Observational Bounds, and Claims of 'Proofs' of Horizons

Because the exterior geometry of a gravastar is identical to that of a classical BH down to the scales of its very thin shell surface boundary layer at ℓ given by (8.33) above the would-be classical horizon, it should be clear that a gravastar will be cold, dark and indistinguishable from a BH by almost all traditional astronomical observations. The first images of the 'BHs' at the center of M87 and Sgr A* at the center of own Milky Way galaxy by the Event Horizon Telescope, impressive as they are, lack the angular resolution to resolve the very near $r = r_M$ horizon region, and are principally viewing the electromagnetic radiation from the light ring or photon sphere ($r = 3GM/c^2$ for a non-rotating BH), significantly farther away from the horizon than the very small ℓ of [72]. If light at these mm wavelengths from behind the 'BH' is either too faint, absorbed by the surface or accreting material, or too defocused to be observed, the EHT images will not be able to distinguish a gravastar from a BH.

If there is a surface which is as deeply redshifted as the semi-classical estimates of ε and ℓ would imply, any radiation emitted from the surface can escape to infinity only if emitted from a tiny 'pinhole' solid angle less than of order ε from the perpendicular, or it will fall back onto the surface. Attempts to 'prove' the existence of a BH horizon or absence of a surface from the absence of thermal radiation and/or absence of X-ray bursts which would be expected *if* the surface is composed of conventional matter, and *if* any advected matter deposited onto the surface is re-radiated rather than absorbed, are therefore bound to fail. This point was succinctly made in [4], soon after the gravastar proposal of [133, 137]. The authors of [4] also recognized that any surface of an ultracompact QBH was bound *not* to be composed of conventional matter, such as a neutron star crust, needed for the thermonuclear reactions that give rise to X-ray bursts. Moreover, in order for the gravastar proposal to be a viable alternative for a BH of *any* mass, a gravastar must be able to absorb accreting baryonic matter and convert it to the interior condensate, thereby growing its mass to any larger value. Any substantial efficiency of absorption and conversion of energy to interior condensate would reduce the energy re-radiated and make the object dark in most if not all the observable electromagnetic spectrum.

The authors of [36] argued for quite stringent limits on what they called 'gravastar' models, in the context of a certain specialized assumption of internal energy of the 'matter' composing the QBH, assuming a conventional thermalization of accreting matter in a steady state emission. Aside from not accounting for the relativistic 'pinhole' effect suppressing all emission from a deeply redshifted surface, and ignoring the possibility of near total absorption of accreting matter without any heating of the QBH, which would all but eliminate any thermal re-emission whatsoever, the observational bounds of [36] are attempts to constrain the condensed matter analog model of Chapline et al. [61, 62], which in any case is not the gravastar described in [135, 137], this article, or [136].

Similar arguments based on thermalization and steady state re-emission of radiation, again ignoring the possibility of absorption by the QBH surface, with claims of strong observational bounds were made in [37, 146]. These and similar unjustifiably strong claims of 'proof' of BH horizons and the assumptions upon which they are based have been critically examined by several authors [43, 44, 48, 49], and shown to be flawed. First, the assumption that thermodynamic and dynamic equilibrium can be established between an accretion disk and the QBH on a reasonably short timescale is incorrect for a deeply redshifted surface for $\varepsilon \to 0$, due to the gravitational lensing 'pinhole' effect, already pointed out in [4]. The best limits one can obtain from the observations of M87 or Sgr A* when this classical GR lensing is taken into account is in the range of $\varepsilon < 10^{-15}$ to 10^{-17}, impressive, but still many orders of magnitude short of 10^{-38} expected for a gravastar. Second, the energy emitted was assumed to be electromagnetic in order to be observable, whereas a sizable fraction of any re-emitted energy could be in the form of neutrinos or other unobserved radiation [48, 49]. Third, and most importantly, as already mentioned, a sizable fraction even approaching unity of the accreting matter may be absorbed by the gravastar, with virtually no re-emission whatsoever. As a result, there are no useful bounds from the non-observation of electromagnetic emission from any astrophysical QBH, and the possibility that they may all be gravastars with $\varepsilon \ll 10^{-17}$ remains open.

The converse claim of a *lower* bound of $\varepsilon \gtrsim 10^{-24}$ in [45] is based on a strong assumption of the restrictive form of the Vaidya metric and stress tensor in the vicinity of the QBH surface, setting to zero all of its components except T_{vv} in advanced null coordinates. This bound also disappears if the assumption upon which it is based is relaxed, as it almost certainly should be.

Appendix 3: Gravitational Waves and Echoes

The observation of gravitational waves (GWs) by LIGO/LSC [1] has opened up a new window on the universe that among many other interesting possibilities provides perhaps the best opportunity for observational tests of the gravastar proposal. The GW data is not yet accurate enough to test the prediction of a *discrete* spectrum of ringdown modes from a non-singular gravastar with a surface made in [136]. Indeed it was quickly realized that sensitivity to the nature of a very compact QBH with $\varepsilon \ll 1$ is obtained only some delay time after the initial GW merger signal, in the ringdown phase [46], where the signal/noise ratio is very much lower. Nevertheless a regular QBH such as a gravastar could produce a GW 'echo' at multiples of the characteristic time

$$\Delta t \sim 2GM\,\ln(1/\varepsilon) \tag{8.154}$$

after the compact object merger event [47]. These may be observable with the improved sensitivities of Advanced LIGO and future detectors.

The basis for such echoes is the expectation that GWs produced in the merger could reflect from the internal centrifugal barrier of a gravastar and re-emerge with a logarithmically long time delay for $\varepsilon \ll 1$, thus in principle opening up the possibility of testing GR and the nature of QBHs on scales very close to the would-be horizon, and their interior. A somewhat different scenario was considered in [3], with a claim of tentative evidence for an echo signal in the LIGO data. However, an analysis of the same data by members of the LIGO/LSC collaboration concluded that the echo signal was just 1.5σ above the noise level [190]. The subject of GW echoes from QBHs such as gravastars continues as an area of active research [23, 129, 186, 187], requiring substantially more data from Advanced LIGO and successor detectors to resolve this question or possibly provide the first evidence of deviation in the universe from the mathematical BHs of present textbooks [49].

References

1. B.P. Abbott et al., Phys. Rev. Lett. **116**, 061102 (2016). https://doi.org/10.1103/PhysRevLett.116.061102
2. T.M.C. Abbott et al., Astrophys. J. **872**, L30 (2019). https://doi.org/10.3847/2041-8213/ab04fa
3. J. Abedi, H. Dykaar, N. Afshordi, Phys. Rev. D **96**, 082004 (2017). https://doi.org/10.1103/PhysRevD.96.082004
4. M.A. Abramowicz, W. Kluźniak, J.-P. Lasota, Astron. & Astrophys. **396**, L31–L34 (2002). https://doi.org/10.1051/0004-6361:20021645
5. P. Achard et al., Phys. Lett. B **623**, 26–36 (2005). https://doi.org/10.1016/j.physletb.2005.07.052
6. R.X. Adhikari et al., Report of the topical group on cosmic probes of fundamental physics for snowmass 2021 (2022). https://doi.org/10.48550/arxiv.2209.11726
7. B. Allen, Phys. Rev. D **30**, 1153–1157 (1984). https://doi.org/10.1103/PhysRevD.30.1153
8. A. Almheiri, T. Hartman, J. Maldacena et al., J. High Ener. Phys. **2020**, 013 (2020). https://doi.org/10.1007/jhep05(2020)013
9. A. Almheiri, T. Hartman, J. Maldacena et al., Rev. Mod. Phys. **93**, 035002 (2021). https://doi.org/10.1103/RevModPhys.93.035002
10. A. Almheiri, D. Marolf, J. Polchinski et al., J. High Ener. Phys. **2013**, 62 (2013). https://doi.org/10.1007/JHEP02(2013)062
11. L. Álvarez-Gaumé, M. Á. Vázquez-Mozo, Effective field theories and naturalness, in *An Invitation to Quantum Field Theory* (Springer, 2012), pp. 231–260
12. P.R. Anderson, W.A. Hiscock, D.A. Samuel, Phys. Rev. D **51**, 4337–4358 (1995). https://doi.org/10.1103/PhysRevD.51.4337
13. P.R. Anderson, E. Mottola, Phys. Rev. D **89**, 104038 (2014). https://doi.org/10.1103/PhysRevD.89.104038
14. P.R. Anderson, E. Mottola, Phys. Rev. D **89**, 104039 (2014). https://doi.org/10.1103/PhysRevD.89.104039
15. P.R. Anderson, E. Mottola, D.H. Sanders, Phys. Rev. D **97**, 065016 (2018). https://doi.org/10.1103/PhysRevD.97.065016
16. I. Antoniadis, P.O. Mazur, E. Mottola, Phys. Rev. D **55**, 4770–4784 (1997). https://doi.org/10.1103/PhysRevD.55.4770

17. I. Antoniadis, P.O. Mazur, E. Mottola, J. Cosmo. Astropart. Phys. **2012**, 024 (2012). https://doi.org/10.1088/1475-7516/2012/09/024
18. I. Antoniadis, E. Mottola, Phys. Rev. D **45**, 2013–2025 (1992). https://doi.org/10.1103/PhysRevD.45.2013
19. T. Appelquist, J. Carazzone, Phys. Rev. D **11**, 2856–2861 (1975). https://doi.org/10.1103/PhysRevD.11.2856
20. R. Armillis, C. Corianò, L. Delle Rose, Phys. Lett. B **682**, 322–327 (2009). https://doi.org/10.1016/j.physletb.2009.11.013
21. A. Aurilia, D. Christodoulou, F. Legovini, Phys. Lett. B **73**, 429–432 (1978). https://doi.org/10.1016/0370-2693(78)90757-8
22. A. Aurilia, H. Nicolai, P. Townsend, Nucl. Phys. B **176**, 509–522 (1980). ISSN 0550-3213. https://doi.org/10.1016/0550-3213(80)90466-6
23. C. Barceló, R. Carballo-Rubio, L.J. Garay, J. High Ener Phys. **2017**, 54 (2017). ISSN 1029-8479. https://doi.org/10.1007/JHEP05(2017)054
24. C. Barceló, R. Carballo-Rubio, S. Liberati, Class. Quan. Grav. **36**, 13LT01 (2019). https://doi.org/10.1088/1361-6382/ab23b6
25. J.M. Bardeen, Astrophys. J. **162**, 71 (1970). https://doi.org/10.1086/150635
26. J.M. Bardeen, B. Carter, S.W. Hawking, Comm. Math. Phys. **31**, 161–170 (1973). https://doi.org/10.1007/BF01645742
27. C. Barrabès, W. Israel, Phys. Rev. D **43**, 1129–1142 (1991). https://doi.org/10.1103/PhysRevD.43.1129
28. J.D. Bekenstein, Phys. Rev. D **7**, 2333–2346 (1973). https://doi.org/10.1103/PhysRevD.7.2333
29. P. Beltracchi, P. Gondolo, E. Mottola, Phys. Rev. D **105**, 024002 (2022). https://doi.org/10.1103/PhysRevD.105.024002
30. P. Beltracchi, P. Gondolo, E. Mottola, Phys. Rev. D **105**, 024001 (2022). https://doi.org/10.1103/PhysRevD.105.024001
31. R. Bertlmann, E. Kohlprath, Ann. Phys. **288**, 137–163 (2001). https://doi.org/10.1006/aphy.2000.6110
32. R.A. Bertlmann, *Anomalies in Quantum Field Theory* (Clarendon Press, 2001)
33. N.D. Birrell, P.C.W. Davies, *Quantum Fields in Curved Space* (Cambridge University Press, 1984)
34. D.N. Blaschke, R. Carballo-Rubio, E. Mottola, JHEP **12**, 153 (2014). https://doi.org/10.1007/JHEP12(2014)153
35. D.G. Boulware, Phys. Rev. D **11**, 1404–1423 (1975). https://doi.org/10.1103/PhysRevD.11.1404
36. A.E. Broderick, R. Narayan, Class. Quant. Grav. **24**, 659–666 (2007). https://doi.org/10.1088/0264-9381/24/3/009
37. A.E. Broderick, R. Narayan, J. Kormendy et al., Astrophys. J. **805**, 179 (2015). https://doi.org/10.1088/0004-637X/805/2/179
38. I.L. Buchbinder, S.D. Odintsov, I.L. Shapiro, *Effective Action in Quantum Gravity* (CRC Press, 1992)
39. T.S. Bunch, P.C.W. Davies, Proc. R. Soc. Lond. A Math. Phys. Sci. **360**, 117–134 (1978). https://doi.org/10.1098/rspa.1978.0060
40. C.P. Burgess, Living Rev. Rel. **7**, 1–56 (2004). https://doi.org/10.12942/lrr-2004-5
41. A. Bzowski, P. McFadden, K. Skenderis, J. High Ener. Phys. **2018**, 1–70 (2018). https://doi.org/10.1007/JHEP11(2018)153
42. D.M. Capper, M.J. Duff, N. Cimento A **23**, 173–183 (1974). https://doi.org/10.1007/BF02748300
43. R. Carballo-Rubio, F. Di Filippo, S. Liberati et al., Phys. Rev. D **98**, 124009 (2018). https://doi.org/10.1103/PhysRevD.98.124009
44. R. Carballo-Rubio, F.D. Filippo, S. Liberati et al., J. Cosmo. Astropart. Phys. **2022**, 055 (2022). https://doi.org/10.1088/1475-7516/2022/08/055

45. R. Carballo-Rubio, P. Kumar, W. Lu, Phys. Rev. D **97**, 123012 (2018). https://doi.org/10.1103/PhysRevD.97.123012
46. V. Cardoso, E. Franzin, P. Pani, Phys. Rev. Lett. **116**, 171101 (2016). https://doi.org/10.1103/PhysRevLett.116.171101
47. V. Cardoso, S. Hopper, C.F.B. Macedo et al., Phys. Rev. D **94**, 084031 (2016). https://doi.org/10.1103/PhysRevD.94.084031
48. V. Cardoso, P. Pani, Nat. Astron. **1**, 586–591 (2017). ISSN 2397-3366. https://doi.org/10.1038/s41550-017-0225-y
49. V. Cardoso, P. Pani, Living Rev. Rel. **22**, 4 (2019). https://doi.org/10.1007/s41114-019-0020-4
50. V. Cardoso, P. Pani, M. Cadoni et al., Phys. Rev. D **77**, 124044 (2008). https://doi.org/10.1103/PhysRevD.77.124044
51. V. Cardoso, P. Pani, M. Cadoni et al., Class. Quant. Grav. **25**, 195010 (2008). https://doi.org/10.1088/0264-9381/25/19/195010
52. É. Cartan, Comptes Rendus. Ac. Sc. Paris **174**, 593–595 (1922)
53. B. Carter, Phys. Rev. Lett. **26**, 331–333 (1971). https://doi.org/10.1103/PhysRevLett.26.331
54. B. Carter, Black hole equilibrium states: lectures at the 1972 Les Houches School. Black Holes **23**, 57 (1973)
55. B. Carter, Gen. Rel. Grav. **42**, 653–744 (2010). https://doi.org/10.1007/s10714-009-0920-9
56. C. Cattoen, T. Faber, M. Visser, Class. Quant. Grav. **22**, 4189 (2005). https://doi.org/10.1088/0264-9381/22/20/002
57. S. Chandrasekhar, *The Mathematical Theory of Black Holes* (Oxford University Press, 1983)
58. S. Chandrasekhar, J.L. Friedman, Astrophys. J. **175**, 379 (1972). https://doi.org/10.1086/151566
59. S. Chandrasekhar, J.C. Miller, Mon. Not. Roy. Astron. Soc. **167**, 63–80 (1974). https://doi.org/10.1093/mnras/167.1.63
60. M.S. Chanowitz, J. Ellis, Phys. Rev. D **7**, 2490–2506 (1973). https://doi.org/10.1103/PhysRevD.7.2490
61. G. Chapline, Int. J. Mod. Phys. A **18**, 3587–3590 (2003). https://doi.org/10.1142/S0217751X03016380
62. G. Chapline, Dark energy stars (2005). https://doi.org/10.48550/arxiv.astro-ph/0503200
63. G. Chapline, E. Hohlfeld, R.B. Laughlin et al., Phil. Mag. B **81**, 235–254 (2001). https://doi.org/10.1080/13642810108221981
64. G. Chapline, P.O. Mazur, Superfluidity and stationary space-times (2009). https://doi.org/10.48550/ARXIV.0911.2326
65. N.A. Chernikov, E.A. Tagirov, Ann. Inst. H. Poincaré Phys. Théor. **9**, 109–141 (1968). http://eudml.org/doc/75603
66. C. Chirenti, L. Rezzolla, Phys. Rev. D **94**, 084016 (2016). https://doi.org/10.1103/PhysRevD.94.084016
67. C.B.M.H. Chirenti, L. Rezzolla, Class. Quant. Grav. **24**, 4191–4206 (2007). https://doi.org/10.1088/0264-9381/24/16/013
68. C.B.M.H. Chirenti, L. Rezzolla, Phys. Rev. D **78**, 084011 (2008). https://doi.org/10.1103/PhysRevD.78.084011. (arXiv:0808.4080)
69. A. Cho, Science **359**, 724–725 (2018). https://doi.org/10.1126/science.359.6377.724. arXiv:https://www.science.org/doi/pdf/10.1126/science.359.6377.724
70. S.M. Christensen, S.A. Fulling, Phys. Rev. D **15**, 2088–2104 (1977). https://doi.org/10.1103/PhysRevD.15.2088
71. D. Christodoulou, Phys. Rev. Lett. **25**, 1596–1597 (1970). https://doi.org/10.1103/PhysRevLett.25.1596
72. W.J. Cocke, Ann. Inst. H. Poincaré, Sec. A Phys. Théor. **2**, 283–306 (1965). http://www.numdam.org/item/AIHPA_1965__2_4_283_0/
73. C. Corianó, M.M. Maglio, E. Mottola, Nucl. Phys. B **942**, 303–328 (2019). https://doi.org/10.1016/j.nuclphysb.2019.03.019
74. R. d'Auria, T. Regge, Nucl. Phys. B **195**, 308–324 (1982)

75. P.C.W. Davies, S.A. Fulling, W.G. Unruh, Phys. Rev. D **13**, 2720–2723 (1976). https://doi.org/10.1103/PhysRevD.13.2720
76. A. DeBenedictis, D. Horvat, S. Ilijić et al., Class. Quant. Grav. **23**, 2303 (2006). https://doi.org/10.1088/0264-9381/23/7/007
77. B.S. DeWitt, Phys. Rep. **19**, 295–357 (1975). https://doi.org/10.1016/0370-1573(75)90051-4
78. J.F. Donoghue, Phys. Rev. D **50**, 3874–3888 (1994). https://doi.org/10.1103/PhysRevD.50.3874
79. J.F. Donoghue, in *AIP Conference Proceedings*, vol. 1483 (2012), pp. 73–94. https://doi.org/10.1063/1.4756964. arXiv:1209.3511
80. M. Duff, P. van Nieuwenhuizen, Phys. Lett. B **94**, 179–182 (1980). ISSN 0370-2693. https://doi.org/10.1016/0370-2693(80)90852-7
81. I. Dymnikova, Gen. Rel. Grav. **24**, 235–242 (1992). ISSN 1572-9532. https://doi.org/10.1007/BF00760226
82. I. Dymnikova, E. Galaktionov, Class. Quant. Grav. **22**, 2331 (2005). https://doi.org/10.1088/0264-9381/22/12/003
83. A.S. Eddington, Nature **113**, 192 (1924). https://doi.org/10.1038/113192a0
84. T. Eguchi, P.B. Gilkey, A.J. Hanson, Phys. Rep. **66**, 213–393 (1980). https://doi.org/10.1016/0370-1573(80)90130-1
85. A. Einstein, Ann. Phys. **343**, 355–369 (1912). https://doi.org/10.1002/andp.19123430704. http://neo-classical-physics.info/uploads/3/4/3/6/34363841/einstein_-_speed_of_light_and_grav.pdf
86. A. Einstein, Ann. Math. (1939) 922–936. https://doi.org/10.2307/1968902
87. D. Finkelstein, Phys. Rev. **110**, 965–967 (1958). https://doi.org/10.1103/PhysRev.110.965
88. K. Fredenhagen, R. Haag, Comm. Math. Phys. **127**, 273–284 (1990). https://doi.org/10.1007/BF02096757
89. H. Fritzsch, M. Gell-Mann, H. Leutwyler, Phys. Lett. B **47**, 365–368 (1973). https://doi.org/10.1016/0370-2693(73)90625-4
90. M. Giannotti, E. Mottola, Phys. Rev. D **79**, 045014 (2009). https://doi.org/10.1103/PhysRevD.79.045014
91. G.W. Gibbons, M.J. Perry, Phys. Rev. Lett. **36**, 985–987 (1976). https://doi.org/10.1103/PhysRevLett.36.985
92. S.B. Giddings, Phys. Today **66**(4), 30–35 (2013). https://doi.org/10.1063/PT.3.1946
93. E.B. Gliner, Sov. J. Exptl. Theor. Phys. (JETP) **22**, 378 (1966). http://jetp.ras.ru/cgi-bin/dn/e_022_02_0378.pdf
94. D.J. Gross, M.J. Perry, L.G. Yaffe, Phys. Rev. D **25**, 330–355 (1982). https://doi.org/10.1103/PhysRevD.25.330
95. D. Harlow et al., TF1 Snowmass Report: Quantum gravity, string theory, and black holes (2022). https://doi.org/10.48550/arxiv.2210.01737
96. B.K. Harrison, K.S. Thorne, M. Wakano et al., *Gravitation Theory and Gravitational Collapse* (University of Chicago Press, Chicago, 1965)
97. J.B. Hartle, Astrophys. J. **150**, 1005–1029 (1967). https://doi.org/10.1086/149400
98. J.B. Hartle, S.W. Hawking, Phys. Rev. D **13**, 2188–2203 (1976). https://doi.org/10.1103/PhysRevD.13.2188
99. J.B. Hartle, D.H. Sharp, Astrophys. J. **147**, 317 (1967). https://doi.org/10.1086/149002
100. J.B. Hartle, K.S. Thorne, Astrophys. J. **153**, 807 (1968). https://doi.org/10.1086/149707
101. S.W. Hawking, Nature **248**, 30–31 (1974). https://doi.org/10.1038/248030a0
102. S.W. Hawking, Comm. Math. Phys. **43**, 199–220 (1975). https://doi.org/10.1007/BF02345020
103. S.W. Hawking, Phys. Rev. D **13**, 191–197 (1976). https://doi.org/10.1103/PhysRevD.13.191
104. S.W. Hawking, Phys. Rev. D **14**, 2460–2473 (1976). https://doi.org/10.1103/PhysRevD.14.2460
105. S.W. Hawking, G.F.R. Ellis, *The Large Scale Structure of Space-Time* (Cambridge University Press, 1973)

106. F.W. Hehl, P. von der Heyde, G.D. Kerlick et al., Rev. Mod. Phys. **48**, 393–416 (1976). https://doi.org/10.1103/RevModPhys.48.393
107. M. Henneaux, C. Teitelboim, Found. Phys. **16**, 593–617 (1986). https://doi.org/10.1007/BF01889624
108. D. Horvat, S. Ilijić, Class. Quant. Grav. **24**, 5637 (2007). https://doi.org/10.1088/0264-9381/24/22/022
109. W. Israel, N. Cimento B **44**, 1–14 (1966). https://doi.org/10.1007/BF02710419
110. W. Israel, N. Cimento B **48**, 463 (1967). https://doi.org/10.1007/BF02712210
111. W. Israel, Phys. Lett. A **57**, 107–110 (1976). ISSN 0375-9601. https://doi.org/10.1016/0375-9601(76)90178-X
112. T. Jacobson, Phys. Rev. D **44**, 1731–1739 (1991). https://doi.org/10.1103/PhysRevD.44.1731
113. V. Kalogera, G. Baym, Astrophys. J. **470**, L61 (1996). https://doi.org/10.1086/310296
114. R.P. Kerr, Phys. Rev. Lett. **11**, 237–238 (1963). https://doi.org/10.1103/PhysRevLett.11.237
115. F. Klinkhamer, G.E. Volovik, JETP Lett. **91**, 259–265 (2010). https://doi.org/10.1134/S0021364010060019
116. K. Lanczos, Physik Z. **23**, 537–539 (1922)
117. K. Lanczos, Ann. Physik **379**, 518–540 (1924). https://doi.org/10.1002/andp.19243791403
118. P.-S. Laplace. Exposition du système du monde, (Reprinted in the Cambridge Library Collection; 2009) (1796)
119. R.B. Laughlin, Int. J. Mod. Phys. A **18**, 831–854 (2003). https://doi.org/10.1142/S0217751X03014071. (arXiv:gr-qc/0302028)
120. H. Leutwyler, Ann. Phys. **235**, 165–203 (1994). ISSN 0003-4916. https://doi.org/10.1006/aphy.1994.1094
121. T. Lewis, G.A. Schott, Proc. Roy. Soc. London A: Math. Phys. Sci. **136**, 176–192 (1932). https://doi.org/10.1098/rspa.1932.0073
122. F.S.N. Lobo, Class. Quant. Grav. **23**, 1525 (2006). https://doi.org/10.1088/0264-9381/23/5/006
123. D. Lynden-Bell, R.M. Lynden-Bell, Mon. Not. Roy. Astron. Soc. **181**, 405–419 (1977). https://doi.org/10.1093/mnras/181.3.405
124. M. Göckeler, T. Schücker, *Differential Geometry, Gauge Theories, and Gravity* (Cambridge University Press, 1989)
125. M.J. Duff, Nucl. Phys. B **125**, 334–348 (1977). https://doi.org/10.1016/0550-3213(77)90410-2
126. D. Marolf, Rep. Prog. Phys. **80**, 092001 (2017). https://doi.org/10.1088/1361-6633/aa77cc
127. D. Martin, M. Visser, Class. Quant. Grav. **20**, 3699 (2003). https://doi.org/10.1088/0264-9381/20/16/311
128. P. Martin-Moruno, N.M. Garcia, F.S. Lobo et al., J. Cosmo. Astropart. Phys. **2012**, 034 (2012). https://doi.org/10.1088/1475-7516/2012/03/034
129. A. Maselli, S.H. Völkel, K.D. Kokkotas, Phys. Rev. D **96**, 064045 (2017). https://doi.org/10.1103/PhysRevD.96.064045
130. S.D. Mathur, Class. Quan. Grav. **26**, 224001 (2009). https://doi.org/10.1088/0264-9381/26/22/224001
131. P.O. Mazur, Quantum theory and the nature of gravitation (2007). https://doi.org/10.48550/arxiv.astro-ph/0701270
132. P.O. Mazur, Unique regular interior solution for the slowly rotating kerr black hole (2020). https://indico.cern.ch/event/948828/contributions/4025466/
133. P.O. Mazur, E. Mottola, Gravitational condensate stars: an alternative to black holes (2001). https://doi.org/10.48550/arXiv.gr-qc/0109035
134. P.O. Mazur, E. Mottola, Phys. Rev. D **64**, 104022 (2001). https://doi.org/10.1103/PhysRevD.64.104022
135. P.O. Mazur, E. Mottola, Proc. Nat. Acad. Sci. (U.S.) **101**, 9545–9550 (2004). https://doi.org/10.1073/pnas.0402717101
136. P.O. Mazur, E. Mottola, Class. Quant. Grav. **32**, 215024 (2015). https://doi.org/10.1088/0264-9381/32/21/215024

137. P.O. Mazur, E. Mottola, Universe **9** (2023). https://doi.org/10.3390/universe9020088
138. J. Michell, Phil. Trans. Roy. Soc. **74**, 35–57 (1784). https://doi.org/10.1098/rstl.1784.0008
139. C.W. Misner, K.S. Thorne, J.A. Wheeler, *Gravitation* (Princeton University Press, 2017)
140. E. Mottola, Phys. Rev. D **31**, 754–766 (1985). https://doi.org/10.1103/PhysRevD.31.754
141. E. Mottola, Acta Phys. Polon. B **41**, 2031–2162 (2010). https://www.actaphys.uj.edu.pl/R/41/9/2031/pdf
142. E. Mottola, J. High Ener. Phys. **2017**, 43 (2017). https://doi.org/10.1007/JHEP07(2017)043
143. E. Mottola, J. High Ener. Phys. **2022**, 37 (2022). https://doi.org/10.1007/JHEP11(2022)037
144. E. Mottola, R. Vaulin, Phys. Rev. D **74**, 064004 (2006). https://doi.org/10.1103/PhysRevD.74.064004
145. E. Mottola, R. Vaulin, Phys. Today **66**(11), 9 (2013). https://doi.org/10.1063/PT.3.2161
146. R. Narayan, J.E. McClintock, New Astron. Rev. **51**, 733–751 (2008). ISSN 1387–6473. https://doi.org/10.1016/j.newar.2008.03.002
147. S. O'Brien, J. Synge, Dublin Inst. Adv. Stud. Ser. A **9** (1952). https://dair.dias.ie/id/eprint/23/1/DIAS-STP-Communications-009-OBrienandSynge.pdf
148. T. Padmanabhan, Phys. Rep. **188**, 285–362 (1990). https://doi.org/10.1016/0370-1573(90)90051-3
149. D.N. Page, Phys. Rev. Lett. **71**, 3743–3746 (1993). https://doi.org/10.1103/PhysRevLett.71.3743
150. S.M. Paneitz et al., SIGMA **4**, 036 (2008). https://doi.org/10.3842/SIGMA.2008.036. (arXiv:0803.4331)
151. P. Pani, Phys. Rev. D **92**, 124030 (2015). https://doi.org/10.1103/PhysRevD.92.124030
152. P. Pani, E. Berti, V. Cardoso et al., Phys. Rev. D **80**, 124047 (2009). https://doi.org/10.1103/PhysRevD.80.124047
153. P. Pani, E. Berti, V. Cardoso et al., J. Phys. Conference Series **222**, 012032 (2010). https://doi.org/10.1088/1742-6596/222/1/012032
154. R. Penrose, Phys. Rev. Lett. **14**, 57–59 (1965). https://doi.org/10.1103/PhysRevLett.14.57
155. J.S. Peracaula, Phil. Trans. Roy. Soc. A: Math. Phys. Eng. Sci. **380** (2022). https://doi.org/10.1098/rsta.2021.0182
156. S. Perlmutter et al., Astrophys. J. **517**, 565–586 (1999). https://doi.org/10.1086/307221
157. E. Poisson, W. Israel, Class. Quant. Grav. **5**, L201 (1988). https://doi.org/10.1088/0264-9381/5/12/002
158. A. Polyakov, Nucl. Phys. B **797**, 199–217 (2008). ISSN 0550-3213. https://doi.org/10.1016/j.nuclphysb.2008.01.002
159. A.M. Polyakov, Phys. Lett. B **103**, 207–210 (1981). https://doi.org/10.1016/0370-2693(81)90743-7
160. A.M. Polyakov, Mod. Phys. Lett. A **2**, 893–898 (1987). https://doi.org/10.1142/S0217732387001130
161. J. Preskill, in *Black Holes, Membranes, Wormholes and Superstrings*, ed. by S. Kalara, D.V. Nanopoulos (World Scientific, 1992), pp. 22–39. https://doi.org/10.1142/9789814536752
162. S. Ray, R. Sengupta, H. Nimesh, Int. J. Mod. Phys. D **29**, 2030004 (2020). https://doi.org/10.1142/S0218271820300049
163. L. Rezzolla, E.R. Most, L.R. Weih, Astrophys. J. Lett. **852**, L25 (2018). https://doi.org/10.3847/2041-8213/aaa401
164. R.J. Riegert, Phys. Lett. B **134**, 56–60 (1984). https://doi.org/10.1016/0370-2693(84)90983-3
165. A.G. Riess et al., Astron. J. **116**, 1009–1038 (1998). https://doi.org/10.1086/300499
166. W. Rindler, Mon. Not. Roy. Astron. Soc. **116**, 662–677 (1956). https://doi.org/10.1093/mnras/116.6.662
167. D.C. Robinson, Phys. Rev. Lett. **34**, 905–906 (1975). https://doi.org/10.1103/PhysRevLett.34.905
168. S.P. Robinson, F. Wilczek, Phys. Rev. Lett. **95**, 011303 (2005). https://doi.org/10.1103/PhysRevLett.95.011303

169. I. Sachs, S.N. Solodukhin, Phys. Rev. D **64**, 124023 (2001). https://doi.org/10.1103/PhysRevD.64.124023
170. N. Sakai, H. Saida, T. Tamaki, Phys. Rev. D **90**, 104013 (2014). https://doi.org/10.1103/PhysRevD.90.104013
171. A.D. Sakharov, Sov. Phys. JETP **22**, 241 (1966). http://www.lutecium.fr/mirror/www.jp-petit.org/papers/cosmo/Sakharov/1966-JETP.pdf
172. K. Schwarzschild, Sitzungberichte königlich preußischen Akad. Wissenschaften Berlin (1916), pp. 424–434. arXiv:physics/9912033
173. D.W. Sciama, Rev. Mod. Phys. **36**, 463–469 (1964). https://doi.org/10.1103/RevModPhys.36.463
174. R. Sengupta, S. Ghosh, S. Ray et al., Phys. Rev. D **102**, 024037 (2020). https://doi.org/10.1103/PhysRevD.102.024037
175. I.L. Shapiro, A.G. Jacksenaev, Phys. Lett. B **324**, 286–292 (1994). ISSN 0370-2693. https://doi.org/10.1016/0370-2693(94)90195-3
176. L. Smarr, Phys. Rev. Lett. **30**, 71–73 (1973). https://doi.org/10.1103/PhysRevLett.30.71
177. G. 't Hooft, NATO Sci. Ser. B **59**, 135–157 (1980). https://doi.org/10.1007/978-1-4684-7571-5_9
178. G. 't Hooft, Nucl. Phys. B - Proc. Suppl. **68**, 174–184 (1998). https://doi.org/10.1016/S0920-5632(98)00151-0. Strings '97
179. S. Treiman, R. Jackiw, D.J. Gross, *Lectures on Current Algebra and Its Applications* (Princeton University Press, 2015)
180. N. Uchikata, S. Yoshida, Class. Quant. Grav. **33**, 025005 (2016). https://doi.org/10.1088/0264-9381/33/2/025005
181. N. Uchikata, S. Yoshida, P. Pani, Phys. Rev. D **94**, 064015 (2016). https://doi.org/10.1103/PhysRevD.94.064015
182. W.G. Unruh, Phys. Rev. D **14**, 870–892 (1976). https://doi.org/10.1103/PhysRevD.14.870
183. W.G. Unruh, R.M. Wald, Rep. Prog. Phys. **80**, 092002 (2017). https://doi.org/10.1088/1361-6633/aa778e
184. R. Utiyama, Phys. Rev. **101**, 1597–1607 (1956). https://doi.org/10.1103/PhysRev.101.1597
185. M. Visser, D. Wiltshire, Class. Quant. Grav. **21**, 1135–1152 (2004). https://doi.org/10.1088/0264-9381/21/4/027
186. S.H. Völkel, K.D. Kokkotas, Class. Quant. Grav. **34**, 125006 (2017). https://doi.org/10.1088/1361-6382/aa68cc
187. Q. Wang, N. Oshita, N. Afshordi, Phys. Rev. D **101**, 024031 (2020). https://doi.org/10.1103/PhysRevD.101.024031
188. S. Weinberg, Rev. Mod. Phys. **61**, 1–23 (1989). https://doi.org/10.1103/RevModPhys.61.1
189. J. Wess, B. Zumino, Phys. Lett. B **37**, 95–97 (1971). https://doi.org/10.1016/0370-2693(71)90582-X
190. J. Westerweck, A.B. Nielsen, O. Fischer-Birnholtz et al., Phys. Rev. D **97**, 124037 (2018). https://doi.org/10.1103/PhysRevD.97.124037
191. J.A. Wheeler, in *Relativity, Groups and Topology: Les Houches Summer School of Theoretical Physics 1963*, ed. by C. DeWitt, B. DeWitt (Gordon and Breach, New York, 1964), pp. 317–522
192. A. Yale, T. Padmanabhan, Gen. Rel. Grav. **43**, 1549–1570 (2011). https://doi.org/10.1007/s10714-011-1146-1
193. J.W. York, Phys. Rev. D **33**, 2092–2099 (1986). https://doi.org/10.1103/PhysRevD.33.2092
194. I.B. Zel'dovich, I. Novikov, *Relativistic Astrophysics: Stars and Relativity*, vol. 1 (University of Chicago Press, 1971)

Chapter 9
Singularity-Free Gravitational Collapse: From Regular Black Holes to Horizonless Objects

Raúl Carballo-Rubio, Francesco Di Filippo, Stefano Liberati, and Matt Visser

9.1 Singularity Regularization in Effective Geometries

In this chapter we shall primarily investigate singularity regularization [14, 19, 24, 47, 52, 53, 63–65, 71, 87, 88, 99] at a purely kinematical level; eschewing for now explicit use of the Einstein equations. The reasons for this are two-fold: First, even in standard general relativity, the Einstein equations [54] only have predictive power once you make (rather strong) assumptions on the nature of the stress-energy—be it vacuum, or some nontrivial stress-energy satisfying some form of (semi-classical) energy condition [25, 35, 49, 55–58, 77, 81, 83, 84, 93, 118, 119]. Second, if for some reason one wishes to step beyond standard general relativity, the status

R. Carballo-Rubio (✉)
CP3-Origins, University of Southern Denmark, Campusvej 55, 5230 Odense, Denmark
e-mail: raul@sdu.dk

Florida Space Institute, University of Central Florida, 12354 Research Parkway, Partnership 1, Orlando, FL 32826, USA

F. Di Filippo
Center for Gravitational Physics, Yukawa Institute for Theoretical Physics, Kyoto University, Kyoto 606-8502, Japan
e-mail: francesco.difilippo@yukawa.kyoto-u.ac.jp

S. Liberati
SISSA - International School for Advanced Studies, Via Bonomea 265, 34136 Trieste, Italy
e-mail: liberati@sissa.it

IFPU - Institute for Fundamental Physics of the Universe, Via Beirut 2, 34014 Trieste, Italy

INFN - Sez, Trieste, Italy

M. Visser
School of Mathematics and Statistics, Victoria University of Wellington,
PO Box 600, Wellington 6140, New Zealand
e-mail: matt.visser@sms.vuw.ac.nz

C. Bambi (ed.), *Regular Black Holes*, Springer Series in Astrophysics and Cosmology,
https://doi.org/10.1007/978-981-99-1596-5_9

of the (modified) Einstein equations and (modified) energy conditions is even more fraught. In view of these observations we shall see just how much we can do using purely kinematical observations. Such questions are of considerable importance in view of recent dramatic advances in both observational techniques [1–4, 6–10] and phenomenological understanding [16, 28, 36, 41, 43, 45, 69, 75, 79, 80, 98, 101, 107, 108, 115, 117].

9.1.1 General Relativity: Singularity Theorems and Geodesic Incompleteness

The existence of singularities is one of the most intriguing aspects of the theory of general relativity. From the conceptual subtleties in their definition [68] to the implicit suggestion of new physics that would avoid their formation [67], singularities have been at the core of numerous developments in classical and quantum gravity [47].

From a mathematical perspective, the formation of singularities in general relativity is unavoidable once certain conditions are met. These conditions are captured by the so-called singularity theorems [74, 94] (see also [35, 55, 58, 95], and see [103] for a recent review).

From a physical perspective, these conditions are expected to be realized in two different kinds of astrophysical situations: the early universe (which results in the Big Bang singularity), and gravitational collapse (which results in black holes, or closely related objects). Here, we will be mostly concerned with the latter situation, adequately encapsulated in Penrose's singularity theorem [94].

In a nutshell, Penrose's theorem demonstrates that, once a closed trapped surface $\mathscr{S}^2$ is formed in a spacetime $\mathscr{M}$, there exists an incomplete geodesic in the causal future $J(\mathscr{S}^2)$ of $\mathscr{S}^2$. This result relies on a number of technical assumptions that are spelled out below. However, before diving into these assumptions, we think it can be useful to discuss in some detail the notions introduced above:

- A closed trapped surface $\mathscr{S}^2$ is a closed and spacelike 2-dimensional surface such that the area of all light fronts propagating through any of the points on the surface is decreasing toward the future. This is clearly associated with strong gravitational fields, as the standard behavior in weak gravitational fields (or in the absence of gravitational field) is that the area can either decrease or increase, depending on the initial conditions considered for light rays (the simplest example is that of exploding/imploding spheres of light in flat spacetime).
- The causal future $J(A)$ of a set $A \subset \mathscr{M}$ is the set of points that are connected by past-directed causal (null and timelike) curves to the points in A. In other words, $J(A)$ is the set of points that can be reached following causal trajectories going through points in A.
- An incomplete geodesic is a geodesic that cannot be followed indefinitely. For instance, an observer following a timelike incomplete geodesic will only be able to record its experience for a finite amount of proper time. A simple example of

a geodesically incomplete spacetime, which illustrates how this notion captures the existence of "holes" in spacetime, is obtained by artificially removing a point from flat spacetime. (An incomplete null geodesic is one which terminates in finite affine parameter "time".)

Hence, the condition that must be met, according to Penrose's theorem, is the formation of closed trapped surfaces. The formation of closed trapped surfaces is reasonable from a conceptual perspective [13, 50, 73], and has also been reproduced numerically [102], so there is no plausible physical reason to doubt that this condition can be satisfied, at least within the framework of classical general relativity.

Moreover, Penrose's theorem relies on the following assumptions (ordered in terms of increasing strength):

1. The weak energy condition is satisfied.
2. The Einstein field equations hold.
3. Global hyperbolicity holds.
4. Pseudo-Riemanniann geometry provides an adequate description of spacetime.

Violating any of these assumptions would open up the possibility of getting rid of the singular behavior in Penrose's theorem. Hence, we can use this list of input assumptions to classify different approaches to this problem.

For instance, it is argued by many authors that spacetime must be fundamentally discrete—a discretium rather than a continuum [11, 12, 100, 109, 110, 116]. This would violate all of the above assumptions: being formulated in the mathematical framework of pseudo-Riemanniann geometry, they lose their meaning if the latter ceases to be applicable. It is therefore clear that fundamental discreteness can be one guiding principle leading to singularity-free theories. Nonetheless, we still lack a definitive theory of quantum gravity (albeit we do have some tentative calculations indicating that the resolution of singularities is indeed achieved [15]), so we shall take here a more humble approach.

More specifically, our working framework will consist of hypothesising that both assumptions {3–4} hold, but we shall relax assumptions {1–2}. This is equivalent to assuming that pseudo-Riemanniann geometry provides a good description of the kinematics, while the only condition on the dynamics is that it leads to a well-posed initial value problem (hence the global hyperbolic condition).

Using the tools provided by pseudo-Riemanniann geometries has several practical advantages due to our familiarity with them [96], but we also believe this procedure has heuristic value from the perspective of understanding the main features to be expected in a framework that supersedes general relativity.

Basically, we are entailing scenarios in which, after the formation of a trapped region, a quantum gravitational description is circumscribed to some finite region of spacetime (possibly associated with Planckian densities). The outcome of this evolution is then a globally hyperbolic geometry, regular and classical everywhere.[1]

[1] This can be relaxed so to admit Planck scale regions still requiring a quantum gravitational description, as long as these are not considered as missing points from the manifold, as indeed in this case

9.1.2 *Beyond General Relativity: Deforming Black Holes into Geodesically Complete Spacetimes*

In this section, we discuss the different kinds of geometries describing singularity-free black holes. Our precise definition of these geometries is based on the following features

- Global hyperbolicity.
- Geodesic completeness.
- Asymptotic flatness.
- Existence of a closed trapped surface $\mathscr{S}^2$ (the boundary of the black hole).
- Finiteness of the curvature invariants.

As explained in the previous sections, these geometries must violate at least one of the assumptions {1–2}. In this sense, Penrose's theorem is a convenient starting point.

We can begin by noticing that in the latter the existence of incomplete geodesics is intimately linked to the formation of a focusing point (defined as a point where a congruence of geodesics is characterized by a vanishing cross-sectional area).

Let us make some of these notions more mathematically precise. We are assuming the existence of a spacelike trapped surface $\mathscr{S}^2$, which can be defined using the two null vector fields that are normal to it. The (3+1) dimensionality of $\mathscr{M}$ and 2 dimensionality of $\mathscr{S}^2$, together with the spacelike character of the latter, imply that there are two linearly independent (future-directed) normal null vectors at each point of $\mathscr{S}^2$. We will call these two independent normal null vectors $\boldsymbol{l}$ (outgoing null normal) and $\boldsymbol{k}$ (ingoing null normal). If we define h_{ab} as the 2-metric induced on $\mathscr{S}^2$, the expansions along these vector fields are given by:

$$\theta^{(\boldsymbol{X})} = \frac{1}{\sqrt{h}} \mathscr{L}_{\boldsymbol{X}} \sqrt{h} = h^{ab} \nabla_a X_b, \qquad \boldsymbol{X} \in \{\boldsymbol{l}, \boldsymbol{k}\}, \tag{9.1}$$

where $\mathscr{L}_{\boldsymbol{X}}$ is the Lie derivative along $\boldsymbol{X}$, $h = \det(h_{ab})$ and ∇ is the 4-dimensional covariant derivative.

According to the definition above, the expansion $\theta^{(\boldsymbol{X})}$ measures the local change in the area of $\mathscr{S}^2$ when the latter undergoes a local deformation along the vector field $\boldsymbol{X}$. The expansion $\theta^{(\boldsymbol{k})}$ being negative is the standard behavior expected for ingoing geodesics, while $\theta^{(\boldsymbol{l})}$ is in a flat spacetime (or a spacetime describing weak gravitational fields) always positive. A trapped surface is defined in terms of these expansions as:

$$\theta^{(\boldsymbol{k})} < 0, \qquad \theta^{(\boldsymbol{l})} < 0. \tag{9.2}$$

and a focusing point is characterized by the divergence of the outgoing expansion: $\theta^{(\boldsymbol{l})} \to -\infty$.

---

one can still consider to cover them by analytically extending the regular geometry describing the rest of the spacetime.

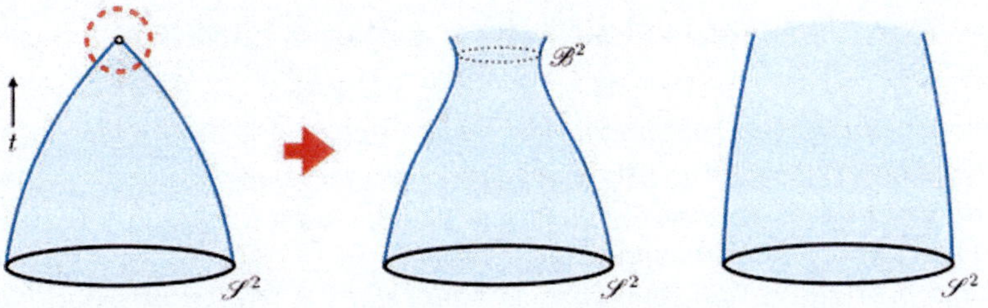

Fig. 9.1 Avoiding that the spacetime be geodesically incomplete entails modifying the geometry in the surroundings of the focusing point in Penrose's theorem, in such a way that either a defocusing point is created—either at a finite affine distance (thus also creating the 2-surface $\mathscr{B}^2$ displayed above) or at infinite affine distance—or the focusing point is displaced to infinite affine distance. The figure on the right is compatible with the two last cases. Ingoing radial null geodesics are not included in this picture as, for each of these cases, these geodesics can display different behaviors that are analyzed in detail in the text

The proof of Penrose's theorem goes schematically as follows: First the Raychaudhuri equation is used to prove that the Einstein field equations and the null energy conditions imply that $\theta^{(l)}$ keeps becoming more negative towards the future, until reaching a focusing point at finite affine parameter $\lambda = \lambda_0$ at which $\theta^{(l)}\big|_{\lambda=\lambda_0} = -\infty$ for all null geodesics. On the other hand, purely geometrical arguments can be used to show that the existence of such focusing point is incompatible with the existence of a non-compact Cauchy hypersurface.[2]

Thus, following Penrose's theorem, in order to make the spacetime geodesically complete we need to modify its geometry in the vicinity of the focusing point, either by creating a defocusing point or by displacing the focusing point to infinite affine distance (see Fig. 9.1). Equivalently, the expansion $\theta^{(l)}(\lambda)$ must remain finite for all the possible values of $\lambda \in [0, \infty)$ (where we are identifying $\lambda = 0$ with $\mathscr{S}^2$ without loss of generality).

These considerations constrain the behavior of outgoing null geodesics: the outgoing expansion can either remain negative but finite (thus having no defocusing points), vanish asymptotically for infinite affine distance without a defocusing point, vanish asymptotically for infinite affine distance with a defocusing point, or vanish at a finite affine distance (thus having a defocusing point at a finite affine distance). Simultaneously, ingoing null geodesics can generally display two qualitatively different behaviors, namely either being negative or being non-negative at the defocusing point. Combining these possibilities shows that the number of possible qualitatively different behaviors around the defocusing point is eight.

However, not all these geometries are regular. If we restrict for simplicity to the spherically symmetric case (no symmetry was required in the discussion up to this point), we can show that there are only four regular classes. The technical details

[2] No assumptions on the topology of the Cauchy hypersurface are actually required if we instead assume the existence of at least one geodesic that does not fall into the black hole [73].

are discussed in Ref. [42], in which the four regular classes of geometries have been identified to be:

1. *Evanescent horizon. The singularity is replaced by an inner horizon at which the* $\theta^{(l)}$ *changes sign while* $\theta^{(k)}$ *stays negative. The inner and the outer horizon merge in a finite time. This purely geometrical classification is blind to the type of the dynamical process or the timescale involved.*
2. *Hidden wormhole. The singularity is replaced by a (global or local) minimum radius hypersurface. Both expansions change signs. Such a structure is reminiscent of a wormhole throat hidden inside a trapping horizon.*
3. *Everlasting horizon. As in the evanescent horizon case, the geometries in this class possess an inner and an outer horizon. In this case, the two horizons never merge. This class can be seen as the limit of the evanescent horizon class when the timescale of merging of the two horizons is pushed out to infinity.*
4. *Asymptotic hidden wormhole. The singularity is replaced by a global minimum radius hypersurface that is reached in an infinite affine time. This class can be obtained from the hidden wormhole class by pushing the wormhole throat out to an infinite affine distance.*

There are two main aspects about these geometries that are essential for our discussion below:

- The spacetimes of interest for us describe the collapse of a regular distribution of matter from a given initial Cauchy surface with topology $\mathbb{R}^3$. Any geometry that satisfies this condition and belong to one of the classes above must be dynamical.
- Classes $\{1, 3\}$ above are simply connected, and differ in their dynamical behavior only. On the other hand, classes $\{2, 4\}$ above are non-simply connected.

Instead of working directly with time-dependent situations, we will start considering static situations and focus on the second property above, namely whether the regularization mechanism results in either simply or non-simply connected spacetimes.

After discussing what static spacetimes with these different topologies look like, we will discuss time-dependent situations. This provides an intuitive and gradual way of understanding the main differences between these classes.

9.2 Geodesically Complete Alternatives to Static Black Holes

As mentioned above, we will start our discussion with static situations, namely with no explicit time dependence in the metric. The reader must keep in mind that the geometries discussed in this section have therefore no relevance for the discussion of gravitational collapse, though they provide a stepping stone towards constructing these geometries, as we will discuss in more detail below.

The most generic static spherically symmetric line element is given in this case by

$$ds^2 = -F(r)\,dt^2 + F^{-1}(r)\,dr^2 + \rho^2(r)\,d\Omega^2\,, \tag{9.3}$$

where $d\Omega^2$ is the usual line element on the 2-sphere. (These are sometimes called Buchdahl coordinates [31, 82].) We can also use the Eddington–Finkelstein form

$$ds^2 = -F(r)\,dv^2 + 2\,dv\,dr + \rho^2(r)\,d\Omega^2\,. \tag{9.4}$$

The line element has a trapping horizon at $r = r_{\rm H}$ whenever $F(r_{\rm H}) = 0$ vanishes. In fact, one can show explicitly that:

$$\theta^{(l)} = 2F(r)\,\frac{\partial_r \rho(r)}{\rho(r)}; \qquad \text{and} \qquad \theta^{(k)} = -2\,\frac{\partial_r \rho(r)}{\rho(r)}\,, \tag{9.5}$$

where we have normalized $g(\boldsymbol{l}, \boldsymbol{k}) = -2$.

This trapping horizon can be [70]:

- *inner* if $F'(r_{\rm H}) < 0$,
- *outer* if $F'(r_{\rm H}) > 0$.

We know that geometries that have an outer trapping horizon and satisfy the assumptions in Penrose's theorem are singular. An example is the Schwarzschild metric

$$F(r) = 1 - \frac{2M}{r}\,, \qquad \rho^2(r) = r^2\,, \tag{9.6}$$

that becomes singular in the limit $r \to 0$.

In order to avoid scalar curvature singularities, either $F(r)$ has an even number of zeros, or $\rho(r)$ has a minimum [42]. It is clear that none of these conditions is satisfied by the Schwarzschild metric above. However, we can consider a specific deformation of the Schwarzschild geometry satisfying both criteria for singularity regularization:

$$F(r) = 1 - \frac{2M\rho^2(r)}{\rho^3(r) + 2M\ell_1^2}\,, \qquad \rho^2(r) = r^2 + \ell_2^2\,. \tag{9.7}$$

This example has the following interesting features:

- It reduces to the so called Simpson–Visser metric for $\ell_1 = 0$ (see [82, 104, 106], see also [59, 86] for its extension to rotating configurations). There is then no singularity due to the existence of a wormhole throat, with a radius proportional to ℓ_2. Whether or not there is an outer horizon depends on the value of ℓ_2 with respect to M.
- It reduces to the Hayward metric for $\ell_2 = 0$ [71]. There is then no singularity due to the existence of an even number of horizons, with an inner horizon radius pro-

portional to ℓ_1. Whether or not there is an outer horizon (and thus an accompanying inner horizon) depends on the value of ℓ_1 with respect to M.
- It reduces to the (singular) Schwarzschild metric for $\ell_1 = \ell_2 = 0$.

In this simple example, we see something that is generic: as anticipated by our local analysis of the defocusing point, there are two qualitatively distinct regularization mechanisms, characterized by the topology of the resulting spacetimes:

- A regularization mechanism that is based in the introduction of at least one inner horizon ($\ell_1 \neq 0$), resulting in simply connected spacetimes.
- A regularization mechanisms based on the introduction of a wormhole throat ($\ell_2 \neq 0$), which produces non-simply connected spacetimes.

This observation continues to hold true for more general static geometries, as well as for time-dependent situations.

For the sake of simplicity, in the following we will continue working with time-independent situations, and we will moreover focus on the example introduced in Eq. (9.7). This will keep the geometries simple enough to be tractable without losing sight of any of the interesting physics. The reader should keep in mind that any of the 4 generic classes above can be reconstructed by considering a sequence of the model spacetimes of Eq. (9.7) in which M, ℓ_1 and ℓ_2 become functions of time. We will discuss some features of such dynamical spacetimes below.

9.2.1 Regularization in Simply Connected Topologies

Within the family of geometries we are considering as an example, see Eq. (9.7), the simply connected topologies are given by $\ell_2 = 0$, so that we have the metric

$$F(r) = 1 - \frac{2M\rho^2(r)}{\rho^3(r) + 2M\ell_1^2}, \qquad \rho^2(r) = r^2. \tag{9.8}$$

This metric can describe three distinct kinds of objects, depending on the relative values of the parameters M and ℓ_1 controlling the roots of the polynomial in the numerator of $F(r)$:

$$F(r) = \frac{r^3 - 2Mr^2 + 2M\ell_1^2}{r^3 + 2M\ell_1^2}. \tag{9.9}$$

The properties of these different objects are described in the sections below. We will keep M fixed and explore this family of geometries as ℓ_1 takes values in the interval $(0, +\infty)$.

Regular Black Holes

For ℓ_1 sufficiently small, more precisely $\ell_1 \in (0, \ell_1^\star \equiv 4M/[3\sqrt{3}])$, the function $F(r)$ has two distinct positive roots $r_\pm$ which correspond to outer and inner horizons as defined above. (The third negative root is in this situation unphysical.) See Fig. 9.2 for the relevant Penrose diagram. The derivative of $F(r)$ at each of the horizons provides their surface gravities $\kappa_\pm$,

$$\kappa_\pm = \frac{1}{2}\frac{dF}{dr}\bigg|_{r=r_\pm}, \tag{9.10}$$

and, according to the definition of outer and inner horizons, we have

$$\kappa_- < 0, \qquad \kappa_+ > 0. \tag{9.11}$$

This implies that at the inner horizon there is a exponential focusing of null rays [37, 38, 40, 51]. Within general relativity, it is well known that such focusing leads to an exponential instability of the inner horizon [97] with a characteristic timescale fixed by the inner horizon surface gravity κ_-.

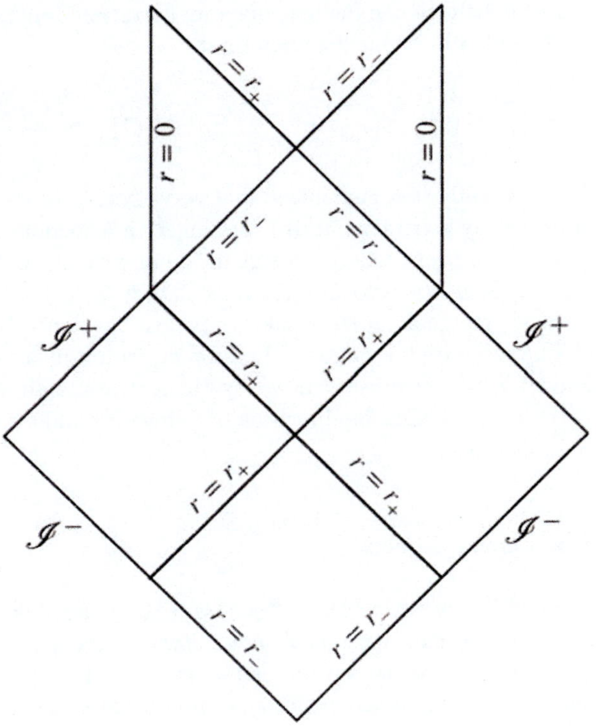

Fig. 9.2 Two-horizon RBH, of the type presented in Eq. (9.8), corresponding to $0 < \ell_1 < \ell_1^\star \equiv 4M/(3\sqrt{3})$. For the maximally extended Penrose diagram one should repeat the construction infinitely many times in the vertical direction. Note that the Penrose diagram is qualitatively similar to that for Reissner–Nordström—except that the timelike curve $r = 0$ is now carefully arranged to be regular, not singular

In the context of regular black hole geometries, we do not know the dynamics of the theory. However, it is still possible to prove a linear instability on purely geometrical grounds, once a small perturbation is added to the background [37, 38, 40, 51]. While it is impossible determine the endpoint of such instability without knowing the field equations of the theory, the presence of a linear instability constitutes a very general result—one that should be taken as a strong cautionary note regarding the viability, as a stable final state, of any geometry with a generic inner horizon.

The geometry around the inner horizon can be deformed so that $\kappa_- = 0$, which removes the unstable behavior [39, 60]. These *inner-extremal* regular black holes may therefore represent a suitable end state towards which the dynamical evolution triggered by the unstable nature of generic inner horizons could tend to. However, this problem has not been analyzed in detail yet, and other possibilities remain open.

Extremal Regular Black Holes

As seen in the Penrose diagram presented in Fig. 9.4, if we increase the value of the parameter ℓ_1, the outer and inner horizon move towards each other. In particular, for $\ell_1 \to \ell_1^\star \equiv 4M/(3\sqrt{3})$, the two horizons merge into a single extremal horizon, located at $r_E = 4M/3$, and the corresponding spacetime describes an extremal regular black hole. (There is also a third unphysical root at $r_{\text{unphysical}} = -2M/3$.) Figure 9.3 shows the Penrose diagram corresponding to this configuration

Specifically in this extremal limit

$$r^3 - 2M(r^2 - \ell_1^2) \to r^3 - 2M(r^2 - [\ell_1^\star]^2) = \frac{(3r + 2M)(3r - 4M)^2}{27}. \tag{9.12}$$

Geometrically it is guaranteed that very special things happen at all extremal horizons; a fully general analysis is presented in Appendix 1. For now let us just observe that in the current context outside the outer horizon, and inside the inner horizon, the Misner–Sharp quasilocal mass must satisfy $2m(r) < r$. Thence as inner and outer horizons merge at an extremal r_E we must have both $2m'(r_E) = 1$ and $m''(r_E) < 0$. Furthermore on the extremal-horizon curvature invariants are functions only of r_E and $m''(r_e)$. These observations survive even in situations much more general than the current 1-parameter modification of Schwarzschild ($\ell_1 \neq 0$, $\ell_2 = 0$). See Appendix 1 for details.

Horizonless Objects

Geometries for which $\ell_1 \in (\ell_1^\star = 4M/[3\sqrt{3}], +\infty)$ do not contain any type of horizon. We will call these *horizonless stars*, as these geometries describe a spherical distribution of matter surrounded by vacuum. These represent another possible end state for the dynamical evolution of regular black holes.

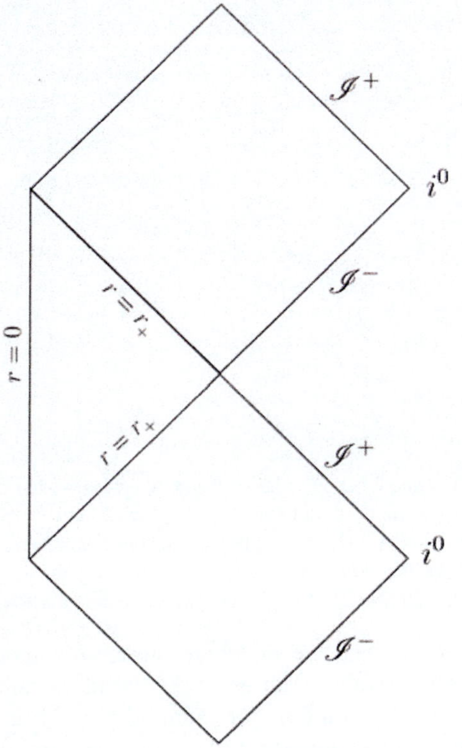

Fig. 9.3 Extremal regular black hole, of type presented in Eq. (9.8), corresponding to $\ell_1 = \ell_1^\star$. As for the non-extremal case, the maximally extended Penrose diagram is obtained repeating the construction infinitely many times in the vertical direction, and it is qualitatively similar to that for an extremal Reissner–Nordström black hole—except that the timelike curve $r = 0$ is now carefully arranged to be regular, not singular

Depending on the value of ℓ_1, these stars are more or less compact. Indeed for $\ell_1 - \ell_1^\star \ll \ell_1^\star$, the corresponding stars are ultracompact, while in the opposite limit $\ell_1 \to +\infty$ the stars become more and more dilute until eventually becoming indistinguishable from flat spacetime. Regarding the causal structure of these spacetimes, the Penrose diagram for any of these geometries is equivalent to that of Minkowski spacetime.

Concerning the formation of such horizonless stars, it appears far from obvious that they can be realized without an intermediates state involving a trapped region. Indeed, without exotic physics entering at energy densities beyond nuclear density, it is difficult to avoid the conclusion that the gravitational collapse will occur almost in free fall [22–24]. In this case, it was shown that even quantum effects cannot change the expected behaviour and prevent the formation of a trapping horizon [22–24].[3]

[3] This can be somewhat expected based on the heuristic argument that, if ones starts at early times with a dilute star in quasi-Minkoswki vacuum then, in the absence of a Cauchy horizon [66], a

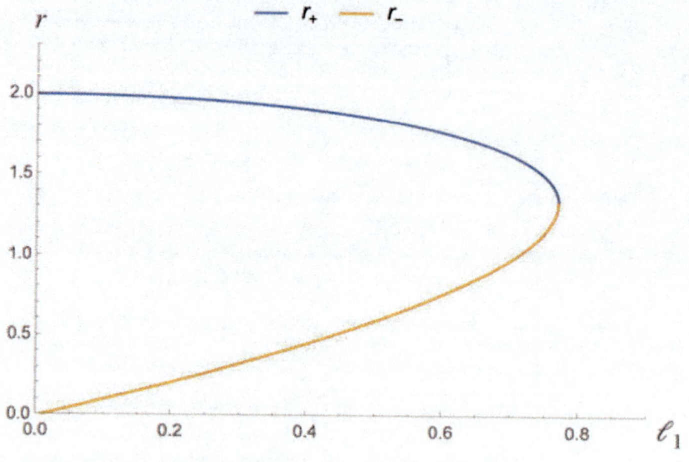

Fig. 9.4 Location of outer and inner horizons of the geometry specified in (9.8), as determined by the roots of $F(r) = 0$. This is equivalent to determining the positive roots of $r^3 - 2Mr^2 + 2M\ell_1^2$. Note that the two horizons merge, to form a single extremal horizon, when $\ell_1 \to \ell_1^\star \equiv 4M/[3\sqrt{3}] \approx 0.77M$. When this happens one has $r_E = 4M/3 \approx 1.333M$

However, the above mentioned, generic inner horizon instability of regular black holes seems to entail a possible evolution of regular black holes towards some sort of stable configuration. This could be for example an inner extremal regular black hole with $\kappa_- = 0$ at the inner horizon [39, 60], or possibly the ultracompact but horizonless limit of the very same geometric family [44].

Concerning the stability, not very much can be safely said due to the lack of a proper understanding of the dynamics associated to these objects. Nonetheless it is interesting to note that horizonless stars are generically endowed with pairs of light rings (closed photon orbits): an outer, unstable, one corresponding to the usual structure present e.g. in Schwarzschild spacetime, and an inner, stable one (at least for static configurations) [26, 48, 78]. This novel feature is potentially dangerous as it might lead to a nonlinear instability due to the accumulation of energy (e.g. photons and/or gravitons) which might bring the horizonless star back within its gravitational radius and hence lead to the formation of a trapping horizon (see e.g. the discussion of this instability for the case of boson stars [48]). While this is surely a feature worth investigating in greater detail (see e.g. the caveats raised in [120]), for now we just want to point out that such inner light ring necessarily has to appear in a region where the metric has order one deviations from the Schwarzschild one, which is tantamount to saying that they will lie in non vacuum regions. (Alternatively one

free-fall collapse would allow one to keep renormalizing the stress energy tensor at different radii, obtaining small deviations from the initial vacuum in the local inertial frame. Hence, this would prevent the build up of large quantum effects able to slow down the collapse before the formation of a trapping horizon. Of course, one might consider the possibility of different initial conditions, e.g. concerning the vacuum state at past null infinity, something that has so far received quite limited attention (see e.g. [61]).

might say that inner light rings will always lie below the stellar surface if this is defined as containing almost all of the ADM mass.) In this sense any study concerning the potential instability at these inner light rings cannot avoid the need to model the matter interaction with the massless field—such as for example its reflectivity or absorption properties—which would generally dampen this possible unstable behavior.

9.2.2 *Regularization in Non-simply Connected Topologies*

Within the family of geometries we are considering as an example, the non-simply connected topologies are given by $\ell_1 = 0$, so that we have the Simpson–Visser metric

$$F(r) = 1 - \frac{2M}{\rho(r)}, \qquad \rho^2(r) = r^2 + \ell_2^2. \tag{9.13}$$

The geometries in this class have the characteristic of possessing a minimum areal radius $\rho_{\min} = \rho(0) = \ell_2$. This metric can describe three distinct kinds of objects, depending on the relative values of the parameters M and ℓ_2 controlling the roots of the polynomial in the numerator of $F(r)$:

$$F(r) = \frac{\sqrt{r^2 + \ell_2^2} - 2M}{\rho(r)}. \tag{9.14}$$

The properties of these different objects are described in the sections below. We will keep M fixed and explore this family of geometries as ℓ_2 takes values in the interval $(0, +\infty)$.

Hidden Wormholes

For ℓ_2 sufficiently small, more precisely $\ell_2 \in (0, \ell_2^\star \equiv 2M)$, the function $F(r)$ has a single (positive) root r_+ which corresponds to an outer horizon, similarly to the cases discussed above. The location of this root is given by the condition

$$r^2 = 4M^2 - \ell_2^2. \tag{9.15}$$

That is

$$r_\pm = \pm\sqrt{4M^2 - \ell_2^2}. \tag{9.16}$$

The regularization is achieved in this case, not by the introduction of an inner horizon, but by the introduction of a minimum length ℓ_2 so that spheres in this spacetime cannot have an area below $4\pi(\ell_2)^2$. This implies that the topology of these

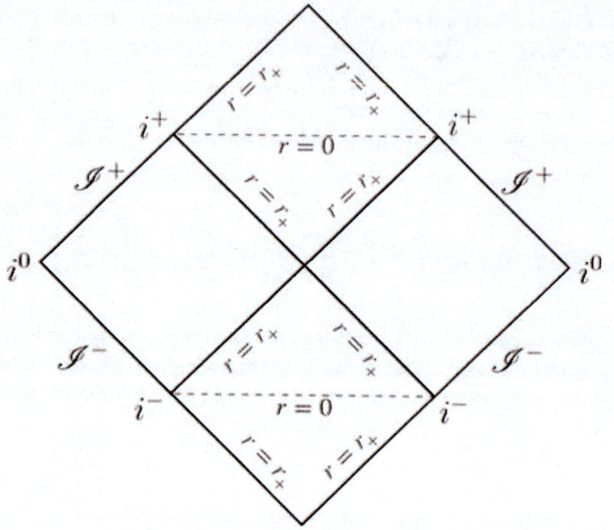

Fig. 9.5 "Black bounce": In this situation the (spacelike) wormhole throat at $r = 0$ is a regular part of the spacetime and is hidden behind a horizon. For the maximally extended Penrose diagram one should repeat the construction infinitely many times in the vertical direction (One could attempt to "simplify" the Penrose diagram by identifying past and future wormhole throats; this would indeed make the maximally extended Penrose diagram simpler, at the cost of introducing closed timelike curves (CTCs), which in particular would destroy global hyperbolicity)

spacetimes is $\mathbb{R}^2 \times S^2$, which are therefore non-simply connected. The minimum radius hypersurface is within the trapped region and it is spacelike. Hence, it can only by traversed in one direction.

From a physical perspective, these spacetimes describe black holes that contain a wormhole throat in their interior. The wormhole throat in this particular situation is a spacelike hypersurface which is "hidden" for observers that remain outside the black hole. This is typically referred to as a "black bounce" [104, 105]. One horizon is a black hole horizon in "our" universe, the other is a white hole horizon in the "future" universe. See Fig. 9.5 for a suitable Penrose diagram. The maximally extended Penrose diagram for the "black bounce" spacetime is qualitatively similar to an infinite vertical stack of Schwarzschild spacetime Penrose diagrams—except that what was the spacelike singularity at $r = 0$ has been replaced by a regular spacelike bounce into the next part of the Penrose diagram. The main issue with these geometries is that, for these to arise in gravitational collapse with standard initial conditions, there must be a change of topology of Cauchy hypersurfaces from $\mathbb{R}^3$ to $\mathbb{R} \times S^2$. This is incompatible with global hyperbolicity [27].

Null Wormholes

As seen in the Penrose diagram of Fig. 9.6, if we now increase the value of the parameter ℓ_2, the wormhole throat at $r = 0$ and the horizons at $r = \pm([2M]^2 - \ell_2^2)^{1/2}$ move towards each other. In particular, for $\ell_2 \to \ell_2^\star \equiv 2M$, both structures merge into a null wormhole throat. Because the wormhole throat is null, it is at best one-way traversable [104, 105]. These null throat wormholes represent a new paradigm that lies well outside the traditional realm of two-way traversable wormholes such as those discussed in [32, 76, 77, 82, 91, 92, 104, 105, 111, 112, 114].

Naked Wormholes

Geometries for which $\ell_2 \in (\ell_2^\star \equiv 2M, +\infty)$ do not contain any type of horizon. See Fig. 9.7 for a suitable Penrose digram. In these geometries, the wormhole throat at $r = 0$ is now timelike, and is no longer "hidden". Such wormholes would be (at least in principle) globally traversable, and have been the subject of intensive investigation for quite different reasons. See for instance [91, 92, 111, 112, 114]

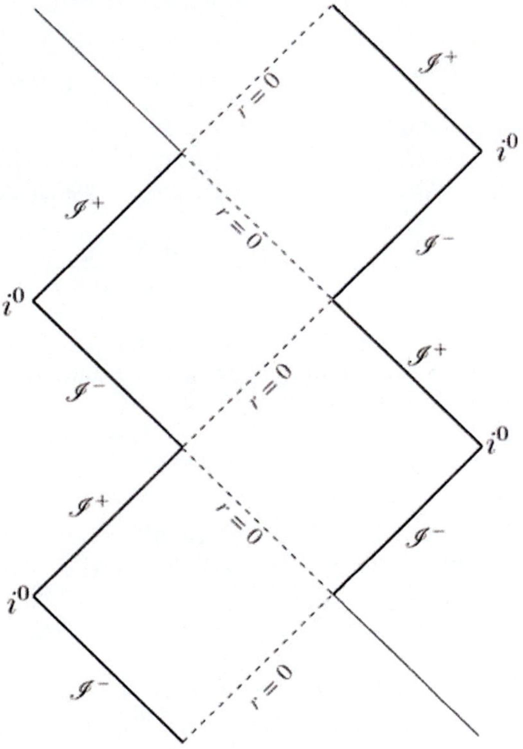

Fig. 9.6 Null wormhole throat. For the maximally extended Penrose diagram one should again repeat the construction infinitely many times in the vertical direction (One could again attempt to "simplify" the Penrose diagram by identifying past and future null wormhole throats; this would indeed make the maximally extended Penrose diagram simpler, at the cost of introducing closed timelike curves (CTCs), which in particular would destroy global hyperbolicity)

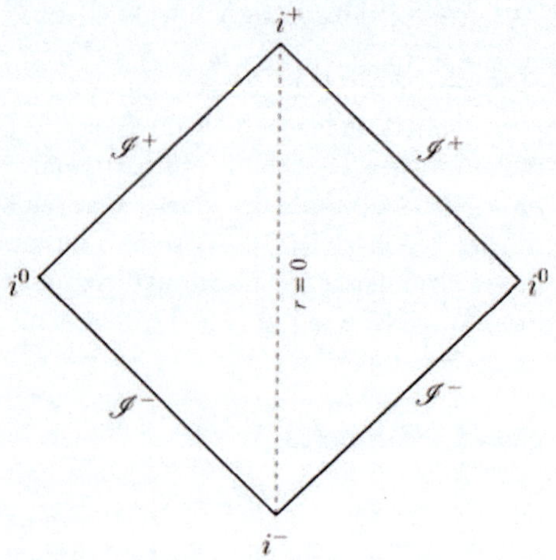

Fig. 9.7 Timelike wormhole throat at $r = 0$. This implies a "naked", in principle globally traversable, wormhole. To determine practical traversability one would additionally need to ensure that tidal forces could be kept suitably small

and [32, 76, 77, 82, 104, 105]. The qualitative analysis presented herein does not extend to quantitatively estimating the tidal forces, which would be necessary for verifying practical traversability.

9.2.3 Comment on "Mixed" Geometries

For completeness, let us consider the case in which both ℓ_1 and ℓ_2 are different from zero. Within the family of geometries we are considering as an example, we have the metric

$$F(r) = 1 - \frac{2M\rho^2(r)}{\rho^3(r) + 2M\ell_1^2}, \qquad \rho^2(r) = r^2 + \ell_2^2. \tag{9.17}$$

This metric can describe three distinct kinds of objects, depending on the relative values of the parameters M and ℓ_1 controlling the roots of the polynomial in the numerator of $F(r)$:

$$F(r) = \frac{\rho^3(r) - 2M(\rho^2(r) - \ell_1^2)}{\rho^3(r) + 2M\ell_1^2}. \tag{9.18}$$

The properties of these different objects are sketched below.

Similarly to what we saw with previous examples, horizons are located at the roots of the polynomial

$$P(\rho) \equiv \rho^3 - 2M(\rho^2 - \ell_1^2). \tag{9.19}$$

This is a cubic in ρ, so mathematically has three roots $\rho_i(M, \ell_1)$ with $i \in \{1, 2, 3\}$, not all of which need be physical. This then corresponds to six roots in terms of r, symmetrically placed around $r = 0$, at locations:

$$r_i(M, \ell_1, \ell_2) = \pm\sqrt{\rho_i^2(M, \ell_1) - \ell_2^2}. \tag{9.20}$$

Not all of these roots need be real, which limits the number of actual horizons around the wormhole throat.

To find the three roots $\rho_i(M, \ell_1)$ the analysis is then very similar to that for the $\ell_2 = 0$ special case analyzed above. We again define

$$\ell_1^\star = \frac{4M}{3\sqrt{3}}. \tag{9.21}$$

For $\ell < \ell_1^\star$ there are three real roots $\rho_i(M, \ell_1)$, two of them positive and one negative. For $\ell = \ell_1^\star$ there are two real roots, a repeated positive root at $\rho = 4M/3$, and a singleton negative root at $\rho = -2M/3$. For $\ell > \ell_1^\star$ there is only one real root, a singleton negative root. Note that, in order to keep the physical interpretation of ρ as an areal radius, as well as for recovering the standard Hayward regular black hole solution in the limit $\ell_2 \to 0$, one needs to take $\rho > 0$ and hence keep only the positive roots in the above analysis.

- Let us now set $\ell_1 < \ell_1^\star \equiv 4M/(3\sqrt{3})$ and study the situation for different values of ℓ_2. We recall that in this situation (Fig. 9.8)

$$r_i(M, \ell_1, \ell_2) = \pm\sqrt{\rho_i^2(M, \ell_1) - \ell_2^2}. \tag{9.22}$$

  with two distinct roots for $\rho_i(M, \ell_1)$ (remember that we are discarding negative roots). With reference to Fig. 9.9, the wormhole throat is always located at $r = 0$, so $\rho_{\text{throat}} \equiv \ell_2$. For sufficiently small values of ℓ_2 there will be four horizons located at

$$r_i(M, \ell_1, \ell_2) \approx \pm|\rho_i(M, \ell_1)|. \tag{9.23}$$

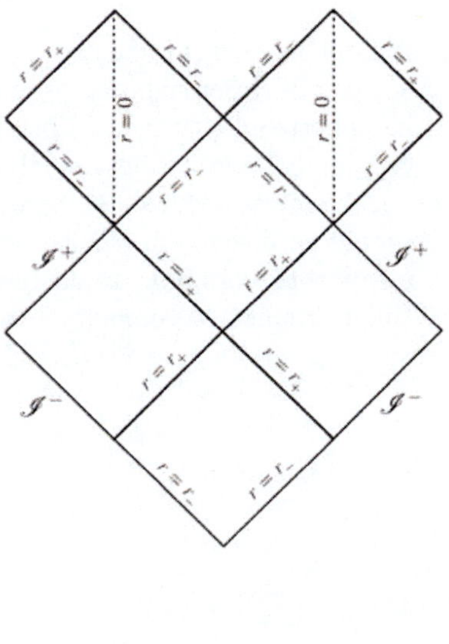

Fig. 9.8 Timelike wormhole throat inside the inner horizon obtained for $\ell_1 \neq 0$ and ℓ_2 sufficiently small. The full Penrose diagram is obtained by repeating the construction both vertically and horizontally

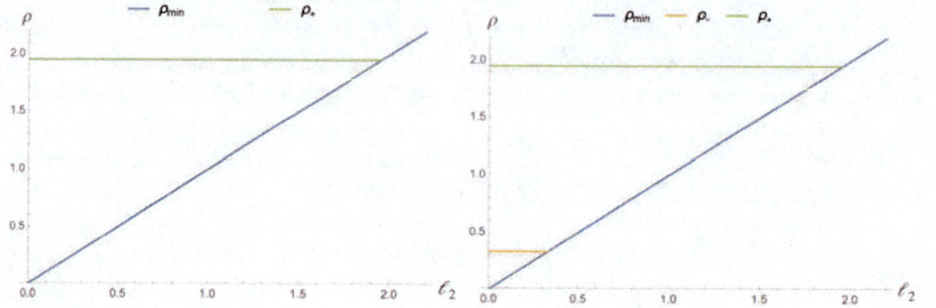

Fig. 9.9 Location of the inner and outer horizons $\rho_\pm$ and the wormhole throat, ($r_{\rm throat} = 0$ corresponding to $\rho_{\rm throat} = \ell_2$), for $M = 1$ and $\ell_1 = 0$ (left) and $\ell_1 = 0.3$ (right) as a function of ℓ_2

This configuration describes, for each side of the universe, a pair of outer and inner horizons shielding a timelike wormhole throat. A Penrose diagram of this spacetime is provided in Fig. 9.8. Increasing the value of ℓ_2, the throat at $r = 0$ moves toward both the inner and outer horizon. For $\ell_2 > \min|\rho_i(M, \ell_1)|$ one pair of horizons disappears: there is now only a pair of outer horizons (one for each side of the universe) and the wormhole throat is spacelike, corresponding to a "black bounce". This spacetime is qualitatively equivalent to the hidden wormhole described in Sect. 9.2.2. Finally, for $\ell_2 > \max|\rho_i(M, \ell_1)|$, no horizons are present and the geometry describes a naked wormhole, a globally traversable wormhole. Note that for $\ell_2 \to 0$ the standard branches of the Hayward geometry described in Sect. 9.2.1 are recovered.

- Let us now set $\ell_1 \to \ell_1^\star \equiv 4M/(3\sqrt{3})$ and study the situation for different values of ℓ_2. There are now two value of r corresponding to the degenerate positive root of ρ

$$r_E = \pm\sqrt{\left(\frac{4M}{3}\right)^2 - \ell_2^2}, \tag{9.24}$$

As long as ℓ_2 is sufficiently small ($\ell_2 < 4M/3$) these will be real, corresponding to a pair of extremal horizons (one for each side of the throat). For $\ell_2 > 4M/3$ one has a horizonless object with a timelike throat at $r = 0$. For $\ell_2 \to 0$ one recovers the simply connected extremal geometries discussed in Sect. 9.2.1.

- For $\ell > \ell_1^\star \equiv 4M/(3\sqrt{3})$ there is only one real root for ρ which being negative should be discarded. Hence, no horizons are present and one recovers a naked traversable wormhole. In the limit $\ell_2 \to 0$ one recovers as expected a horizonless simply connected geometry as described in Sect. 9.2.1.

9.3 Dynamical Geometries

The static geometries described above are non-physical, in the sense that these cannot describe situations in which black holes are formed in a gravitational collapse process. The static nature of these solutions causes further problems, such as the existence of Cauchy horizons, that are generally not present in time-dependent situations. Hence, it is important to keep in mind that we will use these static geometries carefully, as useful building blocks of time-dependent situations, but do not take seriously some of the issues that appear only in the (unphysical) static limit.

With this in mind, the geometries we will be working with in this section can be written in Eddington–Finkelstein form as

$$ds^2 = -F(v,r)\,dv^2 + 2\,dv\,dr + \rho^2(v,r)\,d\Omega^2\,, \tag{9.25}$$

where $F(v,r)$ and $\rho(v,r)$ are obtained from the expressions in Eq. (9.7) when implicit v-dependence through the functions $M(v)$, $\ell_1(v)$ and $\ell_2(v)$ is allowed:

$$F(v,r) = 1 - \frac{2M(v)\rho^2(r)}{\rho^3(r) + 2M\ell_1^2(v)}\,, \qquad \rho^2(v,r) = r^2 + \ell_2^2(v). \tag{9.26}$$

The three functions $M(v)$, $\ell_1(v)$ and $\ell_2(v)$ play three very different roles:

- $M(v)$ is the Bondi mass, and thus determines the total amount of mass contained in the corresponding spacetime at a given moment v. It has a pronounced effect on the location of the outer horizon, if existing.
- $\ell_1(v)$ has a pronounced effect on the location of the inner horizon, if existing.
- $\ell_2(v)$ has a pronounced effect on the location of the wormhole throat.

Hence, these geometries can describe the formation of outer/inner horizons and wormhole throats that evolve dynamically. Let us discuss the possible dynamical behaviors that can arise, starting from the topological classification that was also useful in static situations.

9.3.1 Dynamical Regularization in Simply Connected Topologies

Within the family of dynamical geometries we are considering as an example, the simply connected topologies are given by $\ell_2 = 0$, so that we have the metric

$$F(v,r) = 1 - \frac{2M(v)\rho^2(r)}{\rho^3(r) + 2M\ell_1^2(v)}\,, \qquad \rho^2(r) = r^2. \tag{9.27}$$

This general metric contains metrics previously analyzed in the literature. In particular, if we drop the time dependence in $\ell_1(v)$, writing it as a constant quantity, we recover the metric used by Hayward in his analysis of evaporating regular black holes [71]. In fact, it is interesting to note that the assumption of varying $M(v)$ with $\ell_1 \neq 0$ held fixed has been routinely used in the analysis of the dynamics of regular black holes (see e.g. [63–65, 71, 104–106]).

That the possible time evolution of ℓ_1 has been ignored is likely both due to an implicit assumption that this quantity must remain constant, as well as technical limitations in the analysis of semiclassical backreaction:

- The assumption that ℓ_1 must remain constant comes from its interpretation as a fundamental scale set by quantum gravity (e.g. the Planck scale) [34, 67]. However, this quantity is just a dynamical scalar contained in the metric and, therefore, while it may be reasonable that its initial value can be fixed according to these considerations, there is no strong reason to discard a subsequent dynamical evolution towards different values.
- The analysis of the backreaction of quantum fields on black holes have so far typically been limited to the region around and outside outer horizons [30, 62], with relatively little work on investigating the interior, although some analyses using toy models [89, 90], as well as the actual renormalized stress-energy tensor [5, 20, 21] have been carried out. Focusing on the region around and outside the outer horizon leads to the standard result of evaporation due to the emission of Hawking radiation [72], which is described as the time dependence in $M(v)$. The analysis of backreaction inside the black hole, and in particular around inner horizons, presents some technical limitations, though recent attempts at analyzing these issues have shown [20, 21] that backreaction turns into a time dependence for $\ell_1(v)$.

In summary, given our current knowledge, it is reasonable to expect that semiclassical backreaction on regular black holes will lead to time dependence in both scales $M(v)$ and $\ell_1(v)$. The former scale controls the dynamical evolution of the geometry around the outer horizon, while the latter scale controls the dynamical evolution of the geometry around the inner horizon. The precise dynamical evolution of regular black holes cannot be determined without a better understanding of the underlying dynamical laws.

Regardless of this uncertainty, we do know that the resulting geometry must belong to the classes $\{1, 3\}$. We provide below, in Fig. 9.10, a Penrose diagram for the physically more relevant class 1. The diagram for class 3 can be seen as the everlasting limit of this one. Hence, while not all details are fixed (such as the duration of the horizon structure in class 1), the qualitative behavior associated with this regularization mechanism is well understood. Among qualitative universal features, stands out the lack of Cauchy horizons. The existence of Cauchy horizons are associated with the unphysical restriction to static geometries, but these are replaced by dynamical inner horizons in dynamical situations.

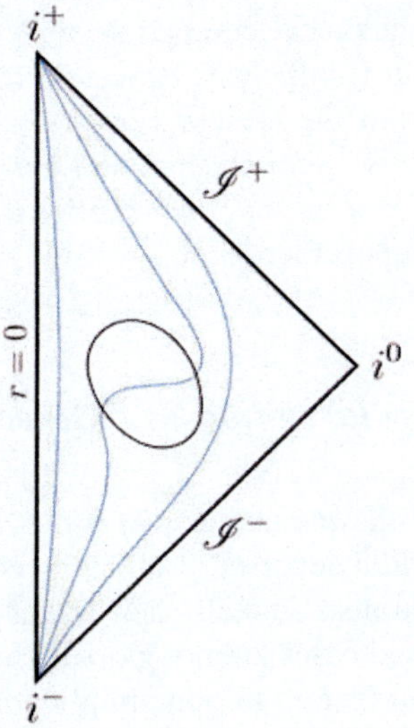

Fig. 9.10 Dynamical regular black hole formed by gravitational collapse. Here we allow $M(v)$ and $\ell_1(v)$ to be time dependent, but keep $\ell_2 = 0$. Once formed, the trapped region disappears in finite time, albeit the latter is undetermined (missing a specific dynamics) and hence can be also very large, formally even infinite. The blue lines correspond to lines of constant radius and are timelike outside the trapped region and spacelike inside it. Contrary to the eternal case, this spacetime is globally hyperbolic

9.3.2 Dynamical Regularization in Non-simply Connected Topologies

Within the family of geometries we are considering as an example, the non-simply connected topologies are given by $\ell_1 = 0$, so that we have the metric

$$F(v,r) = 1 - \frac{2M(v)}{\rho(v,r)}, \qquad \rho^2(v,r) = r^2 + \ell_2^2(v). \tag{9.28}$$

These geometries can describe the creation of a wormhole at a finite time, by including either non-smooth functions or non-analytic smooth functions in the definition of $\ell_2(v)$. As mentioned above, this leads to the breakdown of global hyperbolicity. (The special case of varying $M(v)$ with $\ell_2 \neq 0$ held fixed is discussed extensively in Ref. [105].)

In principle, the lack of global hyperbolicity makes these geometries less appealing. However, the creation of wormholes (and thus, the associated violation) can happen at arbitrarily small scales, perhaps associated with quantum gravity. In a similar way as we discussed with the function $\ell_1(v)$ for regular black holes, in the present case $\ell_2(v)$ can start taking Planckian values and then flowing dynamically to macroscopic values.

Whether or not this behavior is compatible with known laws of physics (e.g. semiclassical physics) is still unknown. However, if any geometries of this kind is realized, it must belong to the classes {2, 4}. Penrose diagrams for this class of spacetimes are generically multi-sheeted and hence difficult to draw (this is a straightforward consequence of the topology-change associated to these geometries). For this reason we do not report them here.

9.3.3 Dynamical Regularization in "Mixed" Cases

One could let all three of the parameters, $M(v)$, $\ell_1(v)$, and $\ell_2(v)$, become time dependent. This would yield a superset of all cases considered above. While technically somewhat more complex, no really new issues of principle are involved. All non-trivial ($\ell_2(v) \neq 0$) mixed cases are not globally hyperbolic. Hence, as with the models of dynamical regularization in non-simply connected topologies, topology change must occur in gravitational collapse for these spacetimes to provide a viable description.

9.4 Discussion and Wrap-up

In wrapping up this chapter let us reflect on the major points we have considered. Firstly, a purely kinematic "geometrographic" analysis—inspired by the desire to avoid singularities—already places significant constraints on just how singularity avoidance might be achieved. To merely appeal to a generic "quantum smoothing" and hiding all of the details into the "too hard basket" does not make a good contribution to knowledge; there is a reasonably well fleshed out taxonomy of kinematically acceptable scenarios that can plausibly be justified from first principles.

We have found good reason to focus on spacetime geometries of the form

$$ds^2 = -F(v,r)\,dv^2 + 2\,dv\,dr + \rho^2(v,r)\,d\Omega^2\,. \tag{9.29}$$

Here $F(v,r)$ and $\rho(v,r)$ are required to satisfy certain conditions (discussed above) to keep the spacetime regular. A sufficiently broad class of models is obtained by letting the 2-variable functions $F(v,r)$ and $\rho(v,r)$ depend implicitly on v through the three 1-parameter functions $M(v)$, $\ell_1(v)$ and $\ell_2(v)$ by setting:

$$F(v,r) = 1 - \frac{2M(v)\rho^2(r)}{\rho^3(r) + 2M\ell_1(v)^2}\,, \qquad \rho^2(v,r) = r^2 + \ell_2^2(v). \tag{9.30}$$

This explicit class of models is still good enough to cover all of the general classes of spacetimes determined to be of interest from our "geometrographic" arguments.

Focusing on these specific models allows us to say quite a bit about candidate regular black holes—and their extremal limits and horizonless counterparts. In the longer run, such considerations are also of use for further phenomenological analyses and for planning future observational projects [17, 18, 29, 46].

Acknowledgements RCR acknowledges financial support through a research grant (29405) from VILLUM fonden. FDF acknowledges financial support by Japan Society for the Promotion of Science Grants-in-Aid for international research fellow No. 21P21318. SL acknowledges funding from the Italian Ministry of Education and Scientific Research (MIUR) under the grant PRIN MIUR 2017-MB8AEZ. CP acknowledges the financial support provided under the European Union's H2020 ERC, Starting Grant agreement no. DarkGRA–757480 and support under the MIUR PRIN and FARE programmes (GW- NEXT, CUP: B84I20000100001). MV was supported by the Marsden Fund, via a grant administered by the Royal Society of New Zealand.

Appendix 1: Extremal Horizons

A general feature implicit in the discussion above is that special things happen to the spacetime geometry at horizons, and that even more special things happen at extremal horizons. However different special things might happen for inner *versus* outer horizons. In this appendix we shall seek to present a coherent overview of this topic. (In a somewhat similar vein, it has been known for some time that special things happen at wormhole throats [76].)

We find it convenient to work with static spacetimes in area coordinates:

$$ds^2 = -e^{-2\Phi(r)}\left(1 - \frac{2m(r)}{r}\right)dt^2 + \frac{dr^2}{1 - \frac{2m(r)}{r}} + r^2\, d\Omega_2^2. \tag{9.31}$$

The horizons are located at solutions (if any) of the equation $r_H = 2m(r_H)$. If we are dealing with a wormhole throat, we will need two coordinate systems of this type, carefully matched at the throat [114].

A purely geometrical result is that in a suitable orthonormal basis [114]

$$G_{\hat{t}\hat{t}} = \frac{2m'(r)}{r^2}; \qquad G_{\hat{r}\hat{r}} = -\frac{2m'(r)}{r^2} + \left(1 - \frac{2m(r)}{r}\right)\frac{2\Phi'(r)}{r}; \tag{9.32}$$

Thence at any horizon (inner or outer, extremal or non-extremal) one has

$$G_{\hat{t}\hat{t}} + G_{\hat{r}\hat{r}} \quad \to \quad 0. \tag{9.33}$$

This on-horizon "enhanced symmetry" for the Einstein (and Ricci) tensors is a recurring theme in near horizon physics [85]. Another useful and very general result is that the surface gravity is [113]:

$$\kappa_H = e^{-\Phi(r_H)} \, \frac{1 - 2m'(r_H)}{2r_H}. \tag{9.34}$$

An extremal horizon is characterized by the vanishing of the surface gravity which we see requires $2m'(r_H) = 1$.

At any extremal horizon (outer or inner) the Einstein tensor (and Ricci tensor) become particularly simple. Specifically

$$G_{\hat{t}\hat{t}|r_E} = -G_{\hat{r}\hat{r}|r_E} = \frac{1}{r_E^2}; \qquad G_{\hat{\theta}\hat{\theta}|r_E} = G_{\hat{\phi}\hat{\phi}|r_E} = -\frac{m''(r_E)}{r_E}. \tag{9.35}$$

In fact at any extremal horizon all orthonormal components of the Riemann tensor are proportional to either $1/r_E^2$, or $m''(r_E)/r_E$, or are zero. Thence at any extremal horizon all of the polynomial curvature invariants are simply multi-nomial functions $f(1/r_E^2, m''(r_E)/r_E)$ of these two quantities. Furthermore at any extremal horizon all nonzero orthonormal components of the Weyl tensor are simply miltiples the single quantity $(1 + r_E m''(r_E))/r_E^2$. So the spacetime geometry simplifies quite drastically on any extremal horizon (either outer or inner).

Finally, what can we say about $m''(r_E)$? This will depend on how many non-extremal horizons merge to yield the extremal horizon of interest. If two non-extremal horizons merge then $m''(r_E) \neq 0$; if three (or more) non-extremal horizons merge then $m''(r_E) = 0$. For instance in the extremal Reissner–Nordström geometry (where two horizons merge) we have $m(r) = m - \frac{1}{2}q^2/r$ and at extremality we obtain $m''(r_E) = -q^2/r_E^3 = -1/r_E < 0$. In contrast for the extremal inner horizons explored in Ref. [39] we have three merging horizons, $m(r)$ is a rational quartic and it is easy to check that $m''(r_E) = 0$.

In short, although it is perhaps not all that well appreciated, geometrically it is guaranteed that very special things happen at all extremal horizons; and these special properties will have a role to play in both phenomenology and in stability analyses for RBHs.

Appendix 2: Light Rings

From the discussion above we have seen that interesting things happen for light rings in extremal, near-extremal, and super-extremal objects. See also [26, 78, 120]. That something unusual happens with light rings in the extremal limit can already be deduced from the very simple and explicit example of Reissner–Nordström spacetime. Since this situation already captures the key features of the discussion with an absolute minimum of fuss, we present some brief pedagogical comments here, before looking at the general situation.

Reissner–Nordström Light-Rings

The Reissner–Nordström spacetime in area coordinates is

$$ds^2 = -(1 - 2m/r + q^2/r^2)dt^2 + \frac{dr^2}{1 - 2m/r + q^2/r^2} + r^2 d\Omega^2 \tag{9.36}$$

It is a quite standard result that in Reissner–Nordström spacetime the light rings can be found by inspecting the effective potential

$$V(r) = \left(1 - \frac{2m}{r} + \frac{q^2}{r^2}\right)\left(\frac{L^2}{r^2}\right) \tag{9.37}$$

The circular photon orbits are located at r_c such that $V'(r_c) = 0$, and stability depends on the sign of $V''(r_c)$. If $V''(r_c) > 0$ then the light ring is stable; if $V''(r_c) < 0$ then the light ring is stable; if $V''(r_c) = 0$ then the light ring exhibits neutral (marginal) stability.

Unfortunately, the potential $V(r)$ *is not unique*, a circumstance which can sometimes cause confusion. Indeed, let $F(x)$ be any monotone increasing function and define $\tilde{V}(r) = F(V(r))$. Then

$$\tilde{V}(r)' = F'(V(r))\, V'(r); \qquad \tilde{V}(r)'' = F''(V(r))\, [V'(r)]^2 + F'(V(r))\, V''(r). \tag{9.38}$$

So the extrema r_c of $V(r)$ coincide with extrema of $\tilde{V}(r)$. Furthermore, at these extrema one has $sign\{\tilde{V}''(r_c)\} = sign\{V''(r_c)\}$.

To locate the light rings we note

$$V'(r) = \frac{2L^2}{r^5}(3mr - 2q^2 - r^2), \tag{9.39}$$

and

$$V''(r) = \frac{2L^2}{r^6}(3r^2 + 10q^2 - 12mr). \tag{9.40}$$

The outer and inner horizons are located at

$$r_H = m \pm \sqrt{m^2 - q^2}. \tag{9.41}$$

The outer and inner light rings are located at

$$r_c = \frac{3m}{2} \pm \frac{\sqrt{9m^2 - 8q^2}}{2}. \tag{9.42}$$

Distinct inner and outer light rings exist for $9m^2 > 8q^2$, and merge at $9m^2 = 8q^2$. that is, beyond extremality. The light rings merge at $r_c = \frac{3m}{2}$.

- At the outer light ring

$$V''(r_c) = -\frac{64L^2\sqrt{9m^2 - 8q^2}}{(3m + \sqrt{9m^2 - 8q^2})^5} < 0, \tag{9.43}$$

so the outer light ring is always unstable.
- At the inner light ring

$$V''(r_c) = \frac{64L^2\sqrt{9m^2 - 8q^2}}{(3m - \sqrt{9m^2 - 8q^2})^5} > 0, \tag{9.44}$$

so the inner light ring is always stable.
- At $r_c = \frac{3m}{2}$, where the light rings merge, $9m^2 = 8q^2$ so $V''(r_c) = 0$, and the merged light ring exhibits neutral stability.

At extremality ($m = |q|$) the light rings are formally located at

$$r_c = \frac{3m}{2} \pm \frac{m}{2} = \{m, 2m\}. \tag{9.45}$$

Here $r_c = 2m$ corresponds to a true light ring, while $r_c = m$ represents the light sheet defining the extremal horizon. (There is now no angular motion, so this is not a "ring".) At extremality ($m = |q|$) for the outer light ring

$$r_c = 2m = 2r_H; \qquad V''(r_c) \to -\frac{L^2}{16m^4} < 0. \tag{9.46}$$

At extremality ($m = |q|$) for the inner light sheet

$$r_c = m = r_H; \qquad V''(r_c) \to +\frac{2L^2}{m^4} > 0. \tag{9.47}$$

So the extremal horizon is a stable light sheet. Perhaps counter-intuitively, the fact that the light sheet is stable will destabilize the spacetime—since the light sheet is stable, massless particles cam pile up there; eventually back-reaction will become large, and the spacetime detabilizes. The situation is summarized in Fig. 9.11.

The key observation here is that the Reissner–Nordström spacetime is already subtle enough to exhibit a stable light ring at extremality, and multiple light rings in a small region beyond extremality. This does have implications for more general RBHs, since one can always cut off the core of the Reissner–Nordström spacetime at some $r_{core} < |q|$ and replace it with a Reissner–Nordström-inspired RBH that would then (by construction) exhibit exactly the same light rings as the Reissner–Nordström spacetime itself. In short, the existence of unstable light rings exterior to generic extremal black holes should not really come as a surprise.

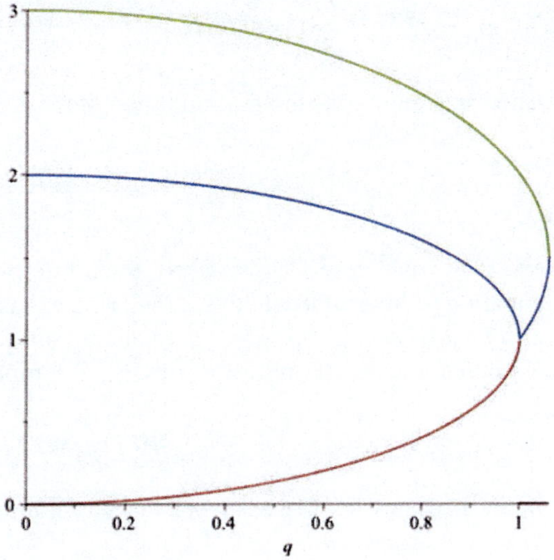

Fig. 9.11 Reissner–Nordström inner and outer horizons, and inner and outer light rings

Generic Light-Rings

What can we say about light rings in the generic case? We already have rather good intuition based on what we saw happening for Reissner–Nordström spacetime.

For any geometry of the form given in Eq. (9.31) it is easy to check that the effective potential governing the light rings is

$$V(r) = e^{-2\Phi(r)} \left(1 - \frac{2m(r)}{r}\right) \frac{L^2}{r^2}. \tag{9.48}$$

It is then easy to check that

$$V'(r) = e^{-2\Phi(r)} \frac{2L^2}{r^4} \left(\{3m(r) - rm'(r) - r\} - \Phi'(r) r^2 (1 - 2m(r)/r)\right). \tag{9.49}$$

Purely geometrically this leads to

$$V'(r) = e^{-2\Phi(r)} \frac{2L^2}{r^4} \{3m(r) - r + r^3 G_{\hat{r}\hat{r}}(r)\}. \tag{9.50}$$

This can also be written as

$$V'(r) = e^{-2\Phi(r)} \frac{2L^2}{r^4} \{3m(r) - r - rm'(r) + r^3 [G_{\hat{t}\hat{t}}(r) + G_{\hat{r}\hat{r}}(r)]\}. \tag{9.51}$$

Furthermore one can easily verify that

$$V''(r) = e^{-2\Phi(r)}\frac{2L^2}{r^5}\{3r - 12m(r) + 6rm'(r) - r^2m''(r)\} - 2\Phi'(r)V'(r) - 2\Phi''(r)V(r). \tag{9.52}$$

Thence at any light ring that might be present the condition $V'(r_c) = 0$ implies

$$r_c = \frac{3m(r_c) + r_c^3[G_{\hat{t}\hat{t}}(r_c) + G_{\hat{r}\hat{r}}(r_c)]}{1 - m'(r_c)} \tag{9.53}$$

We now wish to self-consistently bound the location of possible solutions to this equation to determine whether a light ring exists for $r_c > r_H$. The purely geometrical *null convergence condition* (guaranteeing the convergence of null geodesics), when applied to the radial null geodesics, would imply

$$[G_{\hat{t}\hat{t}}(r) + G_{\hat{r}\hat{r}}(r)] \geq 0. \tag{9.54}$$

If the Misner-Sharp quasi-local mass is non-decreasing outside the horizon then this would imply $m'(r) \geq 0$. Combining, if we assume the light ring exists, then its location is bounded below by:

$$r_c \geq 3m(r_c) \geq 3m(r_H) = \frac{3}{2}\, r_H. \tag{9.55}$$

To establish an upper bound we start from Eq. (9.50). Setting $V'(r_c) \to 0$ yields

$$r_c = 3m(r_c) + r_c^3 G_{\hat{r}\hat{r}}(r_c). \tag{9.56}$$

Then from the DCC (dominant convergence condition): $|G_{\hat{r}\hat{r}}| \leq |G_{\hat{t}\hat{t}}| = 2m'/r^2$ we see

$$r_c \leq 3m(r_c) + 2r_c m'(r_c). \tag{9.57}$$

We need one more condition to get a useful bound: $(m(r)/r^3)' < 0$, implying $m'(r) < 3m(r)/r$. (This condition corresponds to the volume-averaged density decreasing as one moves outwards, and is a very popular condition used in building relativistic and Newtonian models.) Then

$$r_c \leq 9m(r_c) \leq 9m_\infty. \tag{9.58}$$

Overall, under plausible structural conditions, and assuming existence of the light ring, we have

$$\frac{3}{2}r_H \leq r_c \leq 9m_\infty. \tag{9.59}$$

However, proving actual existence of the light rings is slightly more subtle, and requires slightly different arguments for outer-non-extremal and outer-extremal horizons.

Outer Non-extremal Horizons

For an outer non-extremal horizon in terms of the surface gravity we can calculate

$$V'(r_H) = \frac{2L^2 e^{-\Phi(r_H)}\,\kappa_H}{r_H^2} > 0. \tag{9.60}$$

On the other hand for an asymptotically flat geometry, at large r we will have $m(r) = m_\infty + O(1/r)$ and $\Phi(r) = O(1/r)$ whence asymptotically[4]

$$V'(r) = -\frac{2L^2}{r^3} + O\left(\frac{1}{r^4}\right) < 0. \tag{9.61}$$

The sign flip guarantees that there will be at least one light ring somewhere between the outer horizon and spatial infinity.

Outer Extremal Horizons

For any extremal horizon we can calculate

$$V'(r_H) = 0; \qquad V''(r_H) = -\frac{2L^2 e^{-2\Phi(r_H)} m''(r_H)}{r_H^3}. \tag{9.62}$$

If this is to be an outer extremal horizon then we must have $m''(r_H) < 0$ and so $V''(r_H) > 0$. But then $V'(r) > 0$ in the region immediately above the horizon. On the other hand, for any asymptotically flat geometry we still have

$$V'(r) = -\frac{2L^2}{r^3} + O\left(\frac{1}{r^4}\right) < 0. \tag{9.63}$$

The sign flip again guarantees that there will be at least one light ring somewhere between the outer horizon and spatial infinity.

Regular Horizonless Objects

For a regular horizonless object, (cf. a super-extremal Reissner–Nordström geometry with a regularized core at $r_{\text{core}} < |q|$), at short distances we would demand $m(r) = O(r^3)$ and $\Phi(r) = O(r^2)$. Consequently

$$V'(r) = -\frac{2L^2}{r^3} + O(1) < 0, \tag{9.64}$$

[4] There are additional complications if one abandons asymptotic flatness. For instance in asymptotically de Sitter spacetimes one also encounters OSCOs, outermost stable circular orbits [28, 33].

while at large distances

$$V'(r) = -\frac{2L^2}{r^3} + O\left(\frac{1}{r^4}\right) < 0. \quad (9.65)$$

There is now no sign flip and consequently there must be an even number (possibly zero) of light rings.

References

1. B.P. Abbott et al., [LIGO Scientific and Virgo], Observation of gravitational waves from a binary black hole merger. Phys. Rev. Lett. **116**(6), 061102 (2016). https://doi.org/10.1103/PhysRevLett.116.061102. arXiv:1602.03837 [gr-qc]
2. B.P. Abbott et al., [LIGO Scientific and Virgo], GW170817: observation of gravitational waves from a binary neutron star inspiral. Phys. Rev. Lett. **119**(16), 161101 (2017). https://doi.org/10.1103/PhysRevLett.119.161101. arXiv:1710.05832 [gr-qc]
3. B.P. Abbott et al., [LIGO Scientific and Virgo], GW151226: observation of gravitational waves from a 22-solar-mass binary black hole coalescence. Phys. Rev. Lett. **116**(24), 241103 (2016). https://doi.org/10.1103/PhysRevLett.116.241103. arXiv:1606.04855 [gr-qc]
4. B.P. Abbott et al., [LIGO Scientific, Virgo, et al.], Multi-messenger observations of a binary neutron star merger. Astrophys. J. Lett. **848**(2), L12 (2017). https://doi.org/10.3847/2041-8213/aa91c9. arXiv:1710.05833 [astro-ph.HE]
5. J. Abedi, H. Arfaei, Obstruction of black hole singularity by quantum field theory effects. JHEP **03**, 135 (2016). https://doi.org/10.1007/JHEP03(2016)135. arXiv:1506.05844 [gr-qc]
6. R. Abuter et al., [GRAVITY], Detection of the gravitational redshift in the orbit of the star S2 near the Galactic centre massive black hole. Astron. Astrophys. **615**, L15 (2018). https://doi.org/10.1051/0004-6361/201833718. arXiv:1807.09409 [astro-ph.GA]
7. K. Akiyama et al., [Event Horizon Telescope], First M87 event horizon telescope results. I. The shadow of the supermassive black hole. Astrophys. J. Lett. **875**, L1 (2019). https://doi.org/10.3847/2041-8213/ab0ec7. arXiv:1906.11238 [astro-ph.GA]
8. K. Akiyama et al., [Event Horizon Telescope], First M87 event horizon telescope results. IV. Imaging the central supermassive black hole. Astrophys. J. Lett. **875**(1), L4 (2019). https://doi.org/10.3847/2041-8213/ab0e85. arXiv:1906.11241 [astro-ph.GA]
9. K. Akiyama et al., [Event Horizon Telescope], First M87 event horizon telescope results. VI. The shadow and mass of the central black hole. Astrophys. J. Lett. **875**(1), L6 (2019). https://doi.org/10.3847/2041-8213/ab1141. arXiv:1906.11243 [astro-ph.GA]
10. K. Akiyama et al., [Event Horizon Telescope], First sagittarius A* event horizon telescope results. I. The shadow of the supermassive black hole in the center of the Milky Way. Astrophys. J. Lett. **930**(2), L12 (2022). https://doi.org/10.3847/2041-8213/ac6674
11. J. Ambjorn, J. Jurkiewicz, R. Loll, A Nonperturbative Lorentzian path integral for gravity. Phys. Rev. Lett. **85**, 924–927 (2000). https://doi.org/10.1103/PhysRevLett.85.924. arXiv:hep-th/0002050 [hep-th]
12. J. Ambjorn, J. Jurkiewicz, R. Loll, Dynamically triangulating Lorentzian quantum gravity. Nucl. Phys. B **610**, 347–382 (2001). https://doi.org/10.1016/S0550-3213(01)00297-8. arXiv:hep-th/0105267 [hep-th]
13. L. Andersson, M. Mars, W. Simon, Local existence of dynamical and trapping horizons. Phys. Rev. Lett. **95**, 111102 (2005). https://doi.org/10.1103/PhysRevLett.95.111102. arXiv:gr-qc/0506013 [gr-qc]
14. S. Ansoldi, Spherical black holes with regular center: a review of existing models including a recent realization with Gaussian sources. arXiv:0802.0330 [gr-qc]

15. A. Ashtekar, M. Bojowald, Quantum geometry and the Schwarzschild singularity. Class. Quant. Grav. **23**, 391–411 (2006). https://doi.org/10.1088/0264-9381/23/2/008. arXiv:gr-qc/0509075 [gr-qc]
16. C. Bambi, Testing black hole candidates with electromagnetic radiation. Rev. Mod. Phys. **89**(2), 025001 (2017). https://doi.org/10.1103/RevModPhys.89.025001. arXiv:1509.03884 [gr-qc]
17. L. Barack, C. Cutler, LISA capture sources: approximate waveforms, signal-to-noise ratios, and parameter estimation accuracy. Phys. Rev. D **69**, 082005 (2004). https://doi.org/10.1103/PhysRevD.69.082005. arXiv:gr-qc/0310125 [gr-qc]
18. E. Barausse, E. Berti, T. Hertog, S.A. Hughes, P. Jetzer, P. Pani, T.P. Sotiriou, N. Tamanini, H. Witek, K. Yagi et al., Prospects for fundamental physics with LISA. Gen. Rel. Grav. **52**(8), 81 (2020). https://doi.org/10.1007/s10714-020-02691-1. arXiv:2001.09793 [gr-qc]
19. J.M. Bardeen, Non-singular general relativistic gravitational collapse, in *Abstracts of the International Conference GR5* (Georgia, former U.S.S.R, Tbilisi, 1968)
20. C. Barceló, V. Boyanov, R. Carballo-Rubio, L.J. Garay, Black hole inner horizon evaporation in semiclassical gravity. Class. Quant. Grav. **38**(12), 125003 (2021). https://doi.org/10.1088/1361-6382/abf89c. arXiv:2011.07331 [gr-qc]
21. C. Barceló, V. Boyanov, R. Carballo-Rubio, L.J. Garay, Classical mass inflation versus semiclassical inner horizon inflation. Phys. Rev. D **106**(12), 124006 (2022). https://doi.org/10.1103/PhysRevD.106.124006. arXiv:2203.13539 [gr-qc]
22. C. Barceló, S. Liberati, S. Sonego, M. Visser, Fate of gravitational collapse in semiclassical gravity. Phys. Rev. D **77**, 044032 (2008). https://doi.org/10.1103/PhysRevD.77.044032. arXiv:0712.1130 [gr-qc]
23. C. Barceló, S. Liberati, S. Sonego, M. Visser, Revisiting the semiclassical gravity scenario for gravitational collapse. AIP Conf. Proc. **1122**(1), 99–106 (2009). https://doi.org/10.1063/1.3141347. arXiv:0909.4157 [gr-qc]
24. C. Barceló, S. Liberati, S. Sonego, M. Visser, Black Stars. Not Holes. Sci. Am. **301**(4), 38–45 (2009). https://doi.org/10.1038/scientificamerican1009-38
25. C. Barceló, M. Visser, Twilight for the energy conditions? Int. J. Mod. Phys. D **11**, 1553–1560 (2002). https://doi.org/10.1142/S0218271802002888. arXiv:gr-qc/0205066 [gr-qc]
26. P. Bargueno, Light rings in static and extremal black holes. arXiv:2211.16899 [gr-qc]
27. A.N. Bernal, M. Sanchez, On smooth Cauchy hypersurfaces and Geroch's splitting theorem. Commun. Math. Phys. **243**, 461–470 (2003). https://doi.org/10.1007/s00220-003-0982-6. arXiv:gr-qc/0306108 [gr-qc]
28. T. Berry, A. Simpson, M. Visser, Photon spheres, ISCOs, and OSCOs: astrophysical observables for regular black holes with asymptotically Minkowski cores. Universe **7**(1), 2 (2020). https://doi.org/10.3390/universe7010002. arXiv:2008.13308 [gr-qc]
29. E. Berti, V. Cardoso, C.M. Will, On gravitational-wave spectroscopy of massive black holes with the space interferometer LISA. Phys. Rev. D **73**, 064030 (2006). https://doi.org/10.1103/PhysRevD.73.064030. arXiv:gr-qc/0512160 [gr-qc]
30. N.D. Birrell, P.C.W. Davies, *Quantum Fields in Curved Space* (Cambridge University Press, Cambridge, 1984). ISBN 978-0-521-27858-4. https://doi.org/10.1017/CBO9780511622632
31. P. Boonserm, M. Visser, Buchdahl-like transformations for perfect fluid spheres. Int. J. Mod. Phys. D **17**, 135–163 (2008). https://doi.org/10.1142/S0218271808011912. arXiv:0707.0146 [gr-qc]
32. P. Boonserm, T. Ngampitipan, A. Simpson, M. Visser, Exponential metric represents a traversable wormhole. Phys. Rev. D **98**(8), 084048 (2018). https://doi.org/10.1103/PhysRevD.98.084048. arXiv:1805.03781 [gr-qc]
33. P. Boonserm, T. Ngampitipan, A. Simpson, M. Visser, Innermost and outermost stable circular orbits in the presence of a positive cosmological constant. Phys. Rev. D **101**(2), 024050 (2020). https://doi.org/10.1103/PhysRevD.101.024050. arXiv:1909.06755 [gr-qc]
34. I. Booth, B. Creelman, J. Santiago, M. Visser, Evading the Trans-Planckian problem with Vaidya spacetimes. JCAP **09**, 067 (2019). https://doi.org/10.1088/1475-7516/2019/09/067. arXiv:1809.10412 [gr-qc]

35. A. Borde, Geodesic focusing, energy conditions and singularities. Class. Quant. Grav. **4**, 343–356 (1987). https://doi.org/10.1088/0264-9381/4/2/015
36. A.E. Broderick, T. Johannsen, A. Loeb, D. Psaltis, Testing the no-hair theorem with event horizon telescope observations of Sagittarius A*. Astrophys. J. **784**, 7 (2014). https://doi.org/10.1088/0004-637X/784/1/7. arXiv:1311.5564 [astro-ph.HE]
37. R. Carballo-Rubio, F. Di Filippo, S. Liberati, C. Pacilio, M. Visser, On the viability of regular black holes. JHEP **07**, 023 (2018). https://doi.org/10.1007/JHEP07(2018)023. arXiv:1805.02675 [gr-qc]
38. R. Carballo-Rubio, F. Di Filippo, S. Liberati, C. Pacilio, M. Visser, Inner horizon instability and the unstable cores of regular black holes. JHEP **05**, 132 (2021). https://doi.org/10.1007/JHEP05(2021)132. arXiv:2101.05006 [gr-qc]
39. R. Carballo-Rubio, F. Di Filippo, S. Liberati, C. Pacilio, M. Visser, Regular black holes without mass inflation instability. JHEP **09**, 118 (2022). https://doi.org/10.1007/JHEP09(2022)118. arXiv:2205.13556 [gr-qc]
40. R. Carballo-Rubio, F. Di Filippo, S. Liberati, C. Pacilio, M. Visser, Comment on "Stability properties of Regular Black Holes". arXiv:2212.07458 [gr-qc]
41. R. Carballo-Rubio, F. Di Filippo, S. Liberati, M. Visser, Phenomenological aspects of black holes beyond general relativity. Phys. Rev. D **98**(12), 124009 (2018). https://doi.org/10.1103/PhysRevD.98.124009. arXiv:1809.08238 [gr-qc]
42. R. Carballo-Rubio, F. Di Filippo, S. Liberati, M. Visser, Geodesically complete black holes. Phys. Rev. D **101**, 084047 (2020). https://doi.org/10.1103/PhysRevD.101.084047. arXiv:1911.11200 [gr-qc]
43. R. Carballo-Rubio, F. Di Filippo, S. Liberati, M. Visser, Constraints on horizonless objects after the EHT observation of Sagittarius A*. JCAP **08**(08), 055 (2022). https://doi.org/10.1088/1475-7516/2022/08/055. arXiv:2205.13555 [astro-ph.HE]
44. R. Carballo-Rubio, F. Di Filippo, S. Liberati, M. Visser, A connection between regular black holes and horizonless ultracompact stars. arXiv:2211.05817 [gr-qc]
45. V. Cardoso, L. Gualtieri, Testing the black hole 'no-hair' hypothesis. Class. Quant. Grav. **33**(17), 174001 (2016). https://doi.org/10.1088/0264-9381/33/17/174001. arXiv:1607.03133 [gr-qc]
46. V. Cardoso, P. Pani, Testing the nature of dark compact objects: a status report. Living Rev. Rel. **22**(1), 4 (2019). https://doi.org/10.1007/s41114-019-0020-4. arXiv:1904.05363 [gr-qc]
47. K. Crowther, S. De Haro, Four attitudes towards singularities in the search for a theory of quantum gravity. arXiv:2112.08531 [gr-qc]
48. P.V.P. Cunha, C. Herdeiro, E. Radu, N. Sanchis-Gual, The fate of the light-ring instability. arXiv:2207.13713 [gr-qc]
49. E. Curiel, A primer on energy conditions. Einstein Stud. **13**, 43–104 (2017). https://doi.org/10.1007/978-1-4939-3210-8_3. arXiv:1405.0403 [physics.hist-ph]
50. M. Dafermos, Spherically symmetric space-times with a trapped surface. Class. Quant. Grav. **22**, 2221–2232 (2005). https://doi.org/10.1088/0264-9381/22/11/019. arXiv:gr-qc/0403032 [gr-qc]
51. F. Di Filippo, R. Carballo-Rubio, S. Liberati, C. Pacilio, M. Visser, On the inner horizon instability of non-singular black holes. Universe **8**(4), 204 (2022). https://doi.org/10.3390/universe8040204. arXiv:2203.14516 [gr-qc]
52. I. Dymnikova, Int. J. Mod. Phys. D **12**, 1015–1034 (2003). https://doi.org/10.1142/S021827180300358X. arXiv:gr-qc/0304110 [gr-qc]
53. I. Dymnikova, Class. Quant. Grav. **21**, 4417–4429 (2004). https://doi.org/10.1088/0264-9381/21/18/009. arXiv:gr-qc/0407072 [gr-qc]
54. A. Einstein, The foundation of the general theory of relativity. Ann. Phys. **49**(7), 769–822 (1916). https://doi.org/10.1002/andp.19163540702
55. C.J. Fewster, G.J. Galloway, Singularity theorems from weakened energy conditions. Class. Quant. Grav. **28**, 125009 (2011). https://doi.org/10.1088/0264-9381/28/12/125009. arXiv:1012.6038 [gr-qc]
56. C.J. Fewster, Lectures on quantum energy inequalities. arXiv:1208.5399 [gr-qc]

57. L.H. Ford, T.A. Roman, Averaged energy conditions and quantum inequalities. Phys. Rev. D **51**, 4277–4286 (1995). https://doi.org/10.1103/PhysRevD.51.4277. arXiv:gr-qc/9410043 [gr-qc]
58. L.H. Ford, The classical singularity theorems and their quantum loop holes. Int. J. Theor. Phys. **42**, 1219–1227 (2003). https://doi.org/10.1023/A:1025754515197. arXiv:gr-qc/0301045 [gr-qc]
59. E. Franzin, S. Liberati, J. Mazza, A. Simpson, M. Visser, Charged black-bounce spacetimes. JCAP **07**, 036 (2021). https://doi.org/10.1088/1475-7516/2021/07/036. arXiv:2104.11376 [gr-qc]
60. E. Franzin, S. Liberati, J. Mazza, V. Vellucci, Stable rotating regular black holes. arXiv:2207.08864 [gr-qc]
61. K. Fredenhagen, R. Haag, On the derivation of Hawking radiation associated with the formation of a black hole. Commun. Math. Phys. **127**, 273 (1990). https://doi.org/10.1007/BF02096757
62. V.P. Frolov, I.D. Novikov, *Black Hole Physics: Basic Concepts and New Developments*. https://doi.org/10.1007/978-94-011-5139-9
63. V.P. Frolov, Information loss problem and a 'black hole' model with a closed apparent horizon. JHEP **05**, 049 (2014). https://doi.org/10.1007/JHEP05(2014)049. arXiv:1402.5446 [hep-th]
64. V.P. Frolov, A. Zelnikov, Quantum radiation from a sandwich black hole. Phys. Rev. D **95**(4), 044042 (2017). https://doi.org/10.1103/PhysRevD.95.044042. arXiv:1612.05319 [hep-th]
65. V.P. Frolov, A. Zelnikov, Quantum radiation from an evaporating nonsingular black hole. Phys. Rev. D **95**(12), 124028 (2017). https://doi.org/10.1103/PhysRevD.95.124028. arXiv:1704.03043 [hep-th]
66. S.A. Fulling, M. Sweeny, R.M. Wald, Singularity structure of the two point function in quantum field theory in curved space-time. Commun. Math. Phys. **63**, 257–264 (1978). https://doi.org/10.1007/BF01196934
67. L.J. Garay, Quantum gravity and minimum length. Int. J. Mod. Phys. A **10**, 145–166 (1995). https://doi.org/10.1142/S0217751X95000085. arXiv:gr-qc/9403008 [gr-qc]
68. R.P. Geroch, What is a singularity in general relativity? Ann. Phys. **48**, 526–540 (1968). https://doi.org/10.1016/0003-4916(68)90144-9
69. S.E. Gralla, Can the EHT M87 results be used to test general relativity? Phys. Rev. D **103**(2), 024023 (2021). https://doi.org/10.1103/PhysRevD.103.024023. arXiv:2010.08557 [astro-ph.HE]
70. S.A. Hayward, General laws of black hole dynamics. Phys. Rev. D **49**, 6467–6474 (1994). https://doi.org/10.1103/PhysRevD.49.6467
71. S.A. Hayward, Formation and evaporation of regular black holes. Phys. Rev. Lett. **96**, 031103 (2006). https://doi.org/10.1103/PhysRevLett.96.031103. arXiv:gr-qc/0506126 [gr-qc]
72. S.W. Hawking, Particle creation by black holes. Commun. Math. Phys. **43**, 199–220 (1975). [erratum: Commun. Math. Phys. **46**, 206 (1976)]. https://doi.org/10.1007/BF02345020
73. S.W. Hawking, G.F.R. Ellis, *The Large Scale Structure of Space-Time* (Cambridge University Press, Cambridge, 2011). ISBN 978-0-521-20016-5, 978-0-521-09906-6, 978-0-511-82630-6, 978-0-521-09906-6. https://doi.org/10.1017/CBO9780511524646
74. S.W. Hawking, R. Penrose, The singularities of gravitational collapse and cosmology. Proc. Roy. Soc. Lond. A **314**, 529–548 (1970). https://doi.org/10.1098/rspa.1970.0021
75. A. Held, R. Gold, A. Eichhorn, Asymptotic safety casts its shadow. JCAP **06**, 029 (2019). https://doi.org/10.1088/1475-7516/2019/06/029. arXiv:1904.07133 [gr-qc]
76. D. Hochberg, M. Visser, Geometric structure of the generic static traversable wormhole throat. Phys. Rev. D **56**, 4745–4755 (1997). https://doi.org/10.1103/PhysRevD.56.4745. arXiv:gr-qc/9704082 [gr-qc]
77. D. Hochberg, M. Visser, Dynamic wormholes, anti-trapped surfaces, and energy conditions. Phys. Rev. D **58**, 044021 (1998). https://doi.org/10.1103/PhysRevD.58.044021. arXiv:gr-qc/9802046 [gr-qc]
78. S. Hod, Extremal black holes have external light rings. arXiv:2211.15983 [gr-qc]

79. T. Johannsen, A.E. Broderick, P.M. Plewa, S. Chatzopoulos, S.S. Doeleman, F. Eisenhauer, V.L. Fish, R. Genzel, O. Gerhard, M.D. Johnson, Testing general relativity with the shadow size of Sgr A*. Phys. Rev. Lett. **116**(3), 031101 (2016). https://doi.org/10.1103/PhysRevLett.116.031101. arXiv:1512.02640 [astro-ph.GA]
80. T. Johannsen, Sgr A* and general relativity. Class. Quant. Grav. **33**(11), 113001 (2016). https://doi.org/10.1088/0264-9381/33/11/113001. arXiv:1512.03818 [astro-ph.GA]
81. E.A. Kontou, K. Sanders, Energy conditions in general relativity and quantum field theory. Class. Quant. Grav. **37**(19), 193001 (2020). https://doi.org/10.1088/1361-6382/ab8fcf. arXiv:2003.01815 [gr-qc]
82. F.S.N. Lobo, M.E. Rodrigues, M.V. de Sousa Silva, A. Simpson, M. Visser, Novel black-bounce spacetimes: wormholes, regularity, energy conditions, and causal structure. Phys. Rev. D **103**(8), 084052 (2021). https://doi.org/10.1103/PhysRevD.103.084052. arXiv:2009.12057 [gr-qc]
83. M. Mars, M.M. Martín-Prats, J. Senovilla, Models of regular Schwarzschild black holes satisfying weak energy conditions. Class. Quant. Grav. **13**(5), L51–L58 (1996). https://doi.org/10.1088/0264-9381/13/5/003
84. P. Martín-Moruno, M. Visser, Classical and semi-classical energy conditions. Fund. Theor. Phys. **189**, 193–213 (2017). https://doi.org/10.1007/978-3-319-55182-1_9. arXiv:1702.05915 [gr-qc]
85. P. Martín-Moruno, M. Visser, Hawking-Ellis classification of stress-energy tensors: test fields versus backreaction. Phys. Rev. D **103**(12), 124003 (2021). https://doi.org/10.1103/PhysRevD.103.124003. arXiv:2102.13551 [gr-qc]
86. J. Mazza, E. Franzin, S. Liberati, A novel family of rotating black hole mimickers. JCAP **04**, 082 (2021). https://doi.org/10.1088/1475-7516/2021/04/082. arXiv:2102.01105 [gr-qc]
87. P.O. Mazur, E. Mottola, Gravitational condensate stars: an alternative to black holes. arXiv:gr-qc/0109035 [gr-qc]
88. P.O. Mazur, E. Mottola, Gravitational vacuum condensate stars. Proc. Nat. Acad. Sci. **101**, 9545–9550 (2004). https://doi.org/10.1073/pnas.0402717101. arXiv:gr-qc/0407075 [gr-qc]
89. L. Mersini-Houghton, Backreaction of Hawking radiation on a gravitationally collapsing star I: black holes? Phys. Lett. B **738**, 61–67 (2014). https://doi.org/10.1016/j.physletb.2014.09.018. arXiv:1406.1525 [hep-th]
90. L. Mersini-Houghton, H.P. Pfeiffer, Back-reaction of the Hawking radiation flux on a gravitationally collapsing star II. arXiv:1409.1837 [hep-th]
91. M.S. Morris, K.S. Thorne, Wormholes in space-time and their use for interstellar travel: a tool for teaching general relativity. Amer. J. Phys. **56**, 395–412 (1988). https://doi.org/10.1119/1.15620
92. M.S. Morris, K.S. Thorne, U. Yurtsever, Wormholes, Time machines, and the weak energy condition. Phys. Rev. Lett. **61**, 1446–1449 (1988). https://doi.org/10.1103/PhysRevLett.61.1446
93. L. Parker, S.A. Fulling, Quantized matter fields and the avoidance of singularities in general relativity. Phys. Rev. D **7**, 2357–2374 (1973). https://doi.org/10.1103/PhysRevD.7.2357
94. R. Penrose, Gravitational collapse and space-time singularities. Phys. Rev. Lett. **14**, 57–59 (1965). https://doi.org/10.1103/PhysRevLett.14.57
95. R. Penrose, Gravitational collapse: the role of general relativity. Riv. Nuovo Cim. **1**, 252–276 (1969). https://doi.org/10.1023/A:1016578408204
96. R. Penrose, *Techniques of Differential Topology in Relativity* (SIAM Press, Philadelpha, 1972)
97. E. Poisson, W. Israel, Internal structure of black holes. Phys. Rev. D **41**, 1796–1809 (1990). https://doi.org/10.1103/PhysRevD.41.1796
98. D. Psaltis, Testing general relativity with the event horizon telescope. Gen. Rel. Grav. **51**(10), 137 (2019). https://doi.org/10.1007/s10714-019-2611-5. arXiv:1806.09740 [astro-ph.HE]
99. T.A. Roman, P.G. Bergmann, Stellar collapse without singularities? Phys. Rev. D **28**, 1265–1277 (1983). https://doi.org/10.1103/PhysRevD.28.1265
100. C. Rovelli, L. Smolin, Discreteness of area and volume in quantum gravity. Nucl. Phys. B **442**, 593–622 (1995). [erratum: Nucl. Phys. B **456**, 753–754 (1995)]. https://doi.org/10.1016/0550-3213(95)00150-Q. arXiv:gr-qc/9411005 [gr-qc]

101. M. Rummel, C.P. Burgess, Constraining fundamental physics with the event horizon telescope. JCAP **05**, 051 (2020). https://doi.org/10.1088/1475-7516/2020/05/051. arXiv:2001.00041 [gr-qc]
102. E. Schnetter, B. Krishnan, Non-symmetric trapped surfaces in the Schwarzschild and Vaidya spacetimes. Phys. Rev. D **73**, 021502 (2006). https://doi.org/10.1103/PhysRevD.73.021502. arXiv:gr-qc/0511017 [gr-qc]
103. J.M.M. Senovilla, D. Garfinkle, The 1965 Penrose singularity theorem. Class. Quant. Grav. **32**(12), 124008 (2015). https://doi.org/10.1088/0264-9381/32/12/124008. arXiv:1410.5226 [gr-qc]
104. A. Simpson, M. Visser, Black-bounce to traversable wormhole. JCAP **02**, 042 (2019). https://doi.org/10.1088/1475-7516/2019/02/042. arXiv:1812.07114 [gr-qc]
105. A. Simpson, P. Martín-Moruno, M. Visser, Vaidya spacetimes, black-bounces, and traversable wormholes. Class. Quant. Grav. **36**(14), 145007 (2019). https://doi.org/10.1088/1361-6382/ab28a5. arXiv:1902.04232 [gr-qc]
106. A. Simpson, M. Visser, Regular black holes with asymptotically Minkowski cores. Universe **6**(1), 8 (2019). https://doi.org/10.3390/universe6010008. arXiv:1911.01020 [gr-qc]
107. A. Simpson, M. Visser, The eye of the storm: a regular Kerr black hole. JCAP **03**(03), 011 (2022). https://doi.org/10.1088/1475-7516/2022/03/011. arXiv:2111.12329 [gr-qc]
108. A. Simpson, M. Visser, Astrophysically viable Kerr-like spacetime. Phys. Rev. D **105**(6), 064065 (2022). https://doi.org/10.1103/PhysRevD.105.064065. arXiv:2112.04647 [gr-qc]
109. R.D. Sorkin, Causal sets: discrete gravity. https://doi.org/10.1007/0-387-24992-3_7. arXiv:gr-qc/0309009 [gr-qc]
110. T.P. Sotiriou, M. Visser, S. Weinfurtner, Spectral dimension as a probe of the ultraviolet continuum regime of causal dynamical triangulations. Phys. Rev. Lett. **107**, 131303 (2011). https://doi.org/10.1103/PhysRevLett.107.131303. arXiv:1105.5646 [gr-qc]
111. M. Visser, Traversable wormholes: some simple examples. Phys. Rev. D **39**, 3182–3184 (1989). https://doi.org/10.1103/PhysRevD.39.3182. arXiv:0809.0907 [gr-qc]
112. M. Visser, Traversable wormholes from surgically modified Schwarzschild space-times. Nucl. Phys. B **328**, 203–212 (1989). https://doi.org/10.1016/0550-3213(89)90100-4. arXiv:0809.0927 [gr-qc]
113. M. Visser, Dirty black holes: thermodynamics and horizon structure. Phys. Rev. D **46**, 2445–2451 (1992). https://doi.org/10.1103/PhysRevD.46.2445. arXiv:hep-th/9203057 [hep-th]
114. M. Visser, *Lorentzian Wormholes: From Einstein to Hawking* (AIP Press — now Springer, New York, 1995)
115. M. Visser, C. Barceló, S. Liberati, S. Sonego, Small, dark, and heavy: but is it a black hole? PoS **BHGRS**, 010 (2008). https://doi.org/10.22323/1.075.0010. arXiv:0902.0346 [gr-qc]
116. M. Visser, Which number system is "Best" for describing empirical reality? Mathematics **10**(18), 3340 (2022). https://doi.org/10.3390/math10183340. arXiv:1212.6274 [math-ph]
117. S.H. Völkel, E. Barausse, N. Franchini, A.E. Broderick, EHT tests of the strong-field regime of general relativity. Class. Quant. Grav. **38**(21), 21LT01 (2021). https://doi.org/10.1088/1361-6382/ac27ed. arXiv:2011.06812 [gr-qc]
118. U. Yurtsever, Does quantum field theory enforce the averaged weak energy condition? Class. Quant. Grav. **7**, L251–L258 (1990). https://doi.org/10.1088/0264-9381/7/11/005
119. O.B. Zaslavskii, Regular black holes and energy conditions. Phys. Lett. B **688**, 278–280 (2010). https://doi.org/10.1016/j.physletb.2010.04.031. arXiv:1004.2362 [gr-qc]
120. Z. Zhong, V. Cardoso, E. Maggio, On the instability of ultracompact horizonless spacetimes. arXiv:2211.16526 [gr-qc]

Chapter 10
Stability Properties of Regular Black Holes

Alfio Bonanno and Frank Saueressig

10.1 Introduction

To an external observer black holes are extremely simple objects. In the wake of a gravitational collapse quadrupole moments and all the deformations produced by the star either get swallowed inside the event horizon or are carried away by gravitational radiation. At late times, the external field settles into a Kerr-Newman geometry and is completely described by its mass, charge, and angular momentum.

The stationary exterior field hides a rather complicated dynamics in the black hole interior which drives spacetime towards its final classical fate, determined by Penrose's celebrated 1965 theorem [1]. The key property of the black hole interior is that the Schwarzschild radial coordinate r becomes time-like within the event horizon. Thus *a descent into a black hole is a progression in time*. The inner layers do not only enclose but actually *precede* the core. This peculiar causal structure allows us to theoretically explore the interior by boring in "layer-by-layer" starting from shells situated at larger radii. The causal structure guarantees that what we learn about the outer zone, where the classical description of the geometry is still possible, cannot be affected by our ignorance about quantum gravity effects potentially operating in the innermost regions close to the spacetime singularity. From this point of view, it is possible to speak of an "evolution" with increasing advanced time and subsequent relaxation to its final state. The final "big-crunch" can then be imagined as an abrupt

A. Bonanno (✉)
INAF, Osservatorio Astrofisico di Catania, via S.Sofia 78, 95123 Catania, Italy
e-mail: alfio.bonanno@inaf.it

INFN, Sezione di Catania, via S.Sofia 64, 95123 Catania, Italy

F. Saueressig (✉)
Institute for Mathematics, Astrophysics and Particle Physics (IMAPP), Radboud University, Heyendaalseweg 135, 6525 AJ Nijmegen, The Netherlands
e-mail: f.saueressig@science.ru.nl

C. Bambi (ed.), *Regular Black Holes*, Springer Series in Astrophysics and Cosmology,
https://doi.org/10.1007/978-981-99-1596-5_10

stop or a rupture in the classical geometry as, accordingly to the strong cosmic censorship, the inner singularity is space-like [2].

In the case of no angular momentum the static configuration approached outside the horizon is mirrored in the interior by an almost spherically symmetric configuration where the internal disturbances are exponentially damped. As the radial coordinate plays the role of time, it is conceivable that, before the singularity, there exists a layer where the dynamics admits a semiclassical description in terms of an effective Einstein equation originating from quantum field theory in a curved spacetime [3]

$$G_{\mu\nu} = 8\pi G \langle T_{\mu\nu} \rangle . \tag{10.1}$$

Here $G_{\mu\nu}$ is the (classical) Einstein tensor, G is Newton's coupling, and $\langle T_{\mu\nu} \rangle$ is the (suitably regularized) expectation value of the stress-energy tensor. Using Schwarzschild coordinates, the sign of $\langle T_t{}^t \rangle$ dictates the fate of the singularity. If it is negative, quantum polarization effects have a self-regulatory effect. The Misner-Sharp quasi-local mass, formally introduced in Eq. (10.4), behaves like $M(r) \sim r^3$ near Planckian distances. This in turn implies a de Sitter core inside the black hole [4].

The idea of replacing the spacetime singularity by a patch of de Sitter space actually comes with a long history. It was proposed in [5], introduced in [6] in the context of the limiting-curvature conjecture and further developed in [7, 8], also see [9] for a review. The formation and evaporation of these regular black holes has been first discussed in [10]. From a fundamental perspective these types of regular black holes have been motivated based on the gravitational asymptotic safety program [11, 12] (also see [13–21] for selected follow-up works) and as Planck stars [22] (further explored in [23–25] and reviewed in [26]) inspired by loop quantum gravity.

The de Sitter core shared by these models has important consequences for their causal structure, since it implies the presence of an inner horizon, a so-called Cauchy horizon. This feature may be crucial when considering the regularity of the geometry in the presence of perturbations. Already in 1968 Penrose noted [27] that the ingoing sheet of the inner horizon, corresponding to infinite advanced time, is a surface of infinite blueshift for a wavelike disturbance propagating inwards. In particular, the radiative tail of a generic collapse experiences an exponential blueshift close to the Cauchy horizon. Most likely, this property signals the presence of an instability.

This possibility was then explored in several perturbative studies. The first investigation of the backreaction of the blueshifted influx onto the geometry near the Cauchy horizon was performed by Hiscock [28], but it was Poisson and Israel [29] and Ori [30] who demonstrated that the combined effect of influx and outflux produces an exponential growth of the Coulomb component of the Weyl curvature as a function of the advanced time coordinate v. This has been dubbed the "mass-inflation effect". For an astrophysical (non-zero angular momentum) black hole this effect has dramatic consequences: at variance with the strong singularity at the center, the curvature reaches planckian levels even if the radius of the inner horizon is macroscopically large. Semiclassical quantum corrections on a dynamically inflating geometry cannot really halt the build-up of this singularity. It turns out though that the new singularity is weak in the Tipler sense. For this reason, some authors have speculated that a C^1-extension of the geometry is not ruled out [31]. The arguments suggesting a possible continuation of spacetime beyond the Cauchy horizon carry over to the case of a regular black hole. In this case, the Cauchy horizon may be located at planckian distances from the center and only a complete and consistent theory of quantum gravity may be able to deliver a final answer on the possibility of determining plausible extensions of the geometry though.

The relevant question is if it is possible to identify a *dynamical mechanism* damping of the mass-inflation singularity which operates near the central nucleus, despite our ignorance of the details concerning the structure of the spacetime near the center. In fact, the late advanced time geometry can be described in clearer physical terms because the initial data is well known in this case. It is composed of two elements: the influx of gravitational waves transmitted by the outer potential barrier inside the black hole and the Hawking flux. The decay of the mass associated with the former is well-described by an inverse power law of the type $1/v^{p-1}$ [32], the so-called Price's tail ($p = 12$ for a quadrupole moment). The latter contribution has been conjectured in [33] and it was argued that this should stop the growths of the mass function at *early* advanced times already for mini black holes of mass <100 kg, approximately.

In the case of regular black holes the Hawking flux has a dramatic impact on the final state of the geometry: it is given by an extremal configuration which is reached in an infinite amount of advanced time. The problem of the stability of the inner structure of a regular black hole then turns into a delicate dynamical question with two competing time scales, the one associated with the mass-inflation instability and the characteristic time-scale of the evaporation. If the latter is much longer than the dynamical time-scale of the instability, one can completely neglect the effect of the Hawking radiation unless the black hole is of very low mass. On the other hand recent calculations showed [34] that for specific types of regular black holes the exponential instability is turned into a much milder power-law growth of the curvature near the Cauchy horizon. In this case one has to asses the impact of the Hawking flux on the instability, using a dynamically evolving background which is running towards its critical state in the $v \to \infty$ limit. This analysis was recently performed in [35] and discovered two new classes of late-time attractors governing the dynamics of mass-inflation at asymptotically late times.

The goal of this chapter is to review the mass-inflation effect for regular black hole geometries in case of a static background as well as in the presence of the Hawking flux. In order to make the discussion self-contained we review the key properties of spherically symmetric spacetimes and regular black holes in Sects. 10.2 and 10.3 respectively. Section 10.4 covers the "classical" analysis of the mass-inflation effect for static Reissner–Nordström black holes. Section 10.5 contrasts this situation to the mass-inflation effect encountered for regular black holes. Our conclusion that *regular black holes come with stable cores* is detailed in Sect. 10.6. Throughout this chapter we work in the probe-approximation, neglecting the back-reaction of the geometry onto perturbations.

10.2 Spherically Symmetric Spacetimes: A Primer

We start by summarizing the relevant properties of spherically symmetric spacetimes. Throughout the discussion we adopt geometric units $G = c = 1$. A line-element exhibiting spherical symmetry can be cast into the form

$$ds^2 = g_{ab}dx^a dx^b + r^2 d\Omega^2 \,, \qquad a, b = 0, 1 \,. \tag{10.2}$$

Here, x^a is any pair of coordinates that labels the set of two-spheres, $d\Omega^2 \equiv d\theta^2 + \sin^2(\theta) d\varphi^2$ is the metric on a unit sphere and $r(x^a)$ is a function of x^a defined by the geometric condition that the area A of the two-spheres is given by $A = 4\pi r^2$. We then define $f(x^a)$ through the gradient

$$f(x^a) \equiv g^{ab}(\partial_a r)(\partial_b r) \,. \tag{10.3}$$

The function $f(x^a)$ is related to the Misner-Sharp mass $M(x^a)$ through the definition

$$f(x^a) = 1 - \frac{2M(x^a)}{r} \,. \tag{10.4}$$

A convenient choice for the coordinates x^a is ingoing Eddington-Finkelstein coordinates $x^a = (v, r)$ where (10.2) takes the form

$$ds^2 = -f(v, r)\, dv^2 + 2dr dv + r^2 d\Omega^2 \,. \tag{10.5}$$

We then use a dot to denote a derivative with respect to v, $\dot{f} \equiv \frac{\partial f(v,r)}{\partial v}$, while derivatives with respect to r are indicated by primes, $f' \equiv \frac{\partial f(v,r)}{\partial r}$.

Starting from (10.5), it is instructive to express the most common curvature scalars in terms of $M(v, r)$ and its derivatives. For the Ricci scalar R one readily finds

$$R = \frac{2rM'' + 4M'}{r^2} \,. \tag{10.6}$$

In addition, we define the two curvature scalars appearing at second order of the spacetime curvature. The Kretschmann scalar $K \equiv R_{\alpha\beta\gamma\delta}R^{\alpha\beta\gamma\delta}$ and the square of the Weyl-tensor $C^2 \equiv C_{\alpha\beta\gamma\delta}C^{\alpha\beta\gamma\delta}$ evaluate to

$$\begin{aligned} K &= \frac{16MM''}{r^4} + \frac{4(M'')^2}{r^2} - \frac{64MM'}{r^5} + \frac{32(M')^2}{r^4} - \frac{16M'M''}{r^3} + \frac{48M}{r^6} \,, \\ C^2 &= \frac{4\left(r^2M'' - 4rM' + 6M\right)^2}{3r^6} \,. \end{aligned} \tag{10.7}$$

Since the geometries considered in this work are in general not Ricci flat, $R_{\mu\nu} \neq 0$, K and C^2 are not identical and capture different curvature properties. For later reference we also define the Coulomb-component $\Psi_2 \equiv -\frac{1}{2}C^{\theta\varphi}{}_{\theta\varphi}$ of the Weyl-tensor which evaluates to

$$\Psi_2 = -\frac{M}{r^3} + \frac{2M'}{3r^2} - \frac{M''}{6r} \,. \tag{10.8}$$

For static geometries, where $f(v, r)$ is a function of r only, the location of the horizons is determined by the condition

$$f(r) = 0 \,, \qquad r > 0 \,. \tag{10.9}$$

For the generic situation discussed in this review, this equation has two solutions. The event horizon (EH) is located at r_+. In addition, there is a Cauchy horizon (CH) at $r_- < r_+$. The surface gravity $\kappa_\pm$ at these points is defined as

$$\kappa_\pm \equiv \pm \left. \frac{1}{2} \frac{\partial f(r)}{\partial r} \right|_{r=r_\pm} \,. \tag{10.10}$$

The choice of sign ensures that $\kappa_\pm > 0$. In non-static spacetimes the solutions of (10.9) may depend on v and thus constitute apparent horizons.

We exemplify the general setting for the static Reissner–Nordström (RN)-geometry. In this case, Eq. (10.3) is given by

$$f(r) = 1 - \frac{2m}{r} + \frac{Q^2}{r^2} \,, \tag{10.11}$$

where m is the asymptotic mass of the configuration and Q denotes the charge of the black hole. In this case the EH and CH are situated at

$$r_\pm = m \pm (m^2 - Q^2)^{1/2} \,. \tag{10.12}$$

The corresponding surface gravity is given by

$$\kappa_\pm = \frac{m^2 - Q^2}{r_\pm^2} \,. \tag{10.13}$$

For $Q = m$ the position of the horizons coincide and one has the extremal RN black hole. For this configuration, the surface gravity (10.13) vanishes.

10.3 Regular Black Holes

Upon completing our review of spherically symmetric geometries, we briefly summarize the key properties of regular black holes. The idea of a regular de Sitter core is introduced in Sect. 10.3.1 and various models implementing this idea are reviewed in Sect. 10.3.2. We close with a short discourse on the thermodynamical properties of regular black holes in Sect. 10.3.3.

10.3.1 de Sitter Cores

The idea of a de Sitter core regularizing the interior of a Schwarzschild black hole has been implemented in various settings. It is therefore instructive to start by reviewing the general idea, also see [36] for a general discussion. Let us consider a static, spherically symmetric geometry characterized by the Misner-Sharp mass $M(r)$ introduced in Eq. (10.4). The general expressions for the Ricci scalar R and the Kretschmann scalar K for this case have been given in Eqs. (10.6) and (10.7). Demanding that the curvature scalars remain finite as $r \to 0$ implies that $M(r)$ cannot be constant. Regularity demands that

$$\lim_{r \to 0} r^{-3} M(r) = \text{const}\,. \tag{10.14}$$

Assuming a polynomial expansion around $r = 0$, regularity then implies

$$M(r) \simeq a_3\, r^3 + a_4\, r^4 + a_5\, r^5 + \mathcal{O}(r^5)\,, \tag{10.15}$$

where $a_i, i = 3, 4, \ldots$ are real coefficients. Substituting this expansion into Eq. (10.4) gives

$$f(r) \simeq 1 - 2a_3\, r^2 - 2a_4\, r^3 - 2a_5\, r^4 + \mathcal{O}(r^5)\,. \tag{10.16}$$

For $a_3 > 0$ the local geometry at $r = 0$ is the one of de Sitter space.[1] Combining the asymptotics (10.16) with the condition of asymptotic flatness shows that a black hole with a de Sitter core must come with an even number of horizons. This result is independent on the adopted field equations. It follows from the analytical properties of $M(r)$ near the origin. In static spacetimes with only two horizons, the second, inner horizon is a Cauchy horizon. For non-static spacetimes, the regularity condition implies that the apparent horizon cannot cross $r = 0$.

10.3.2 Examples of Regular Black Hole Geometries

Regular black holes with a de Sitter core have been constructed by several authors. Prominent examples are the Bardeen black hole [5, 39], regular black holes constructed by Dymnikova [7, 40], the Hayward black hole [10], regular black holes constructed within the gravitational asymptotic safety program [11], and Planck stars

---

[1] For $a_3 < 0$ one encounters a regular anti-de Sitter core. Regular geometries of this type have recently been discussed in the context of Gauss black holes [37]. The loop black hole suggested in [38] also falls into this classification having $a_3 = 0$.

motivated from loop quantum gravity [22, 41]. For the sake of conciseness, we will limit the discussion to the Hayward (H) geometry and the regular black holes found in the gravitational asymptotic safety program (RG-improved black holes).

Hayward black holes. Perhaps the simplest implementation of the above requirements has been proposed by Hayward in [10]. In this case, the function f is given by

$$f = 1 - \frac{2M(r)}{r}, \qquad M(r) = \frac{mr^3}{r^3 + 2ml^2}. \tag{10.17}$$

Here l is a characteristic scale of the order of Planck length. Interestingly, (10.17) was actually first derived by Poisson and Israel [4] assuming a simple relation between vacuum energy and curvature via the effective Einstein field Eq. (10.1). In the "evolutionary" picture of the interior described in the introduction one would then expect that a consistent theory of quantum gravity should be able reproduce a similar behavior at smaller radii.

Following the general discussion of Sect. 10.2, the explicit form of the horizon condition (10.9) for the static Hayward geometry is

$$2l^2m - 2mr^2 + r^3 = 0. \tag{10.18}$$

For $m > m_{\rm cr}$ this equation has two real, positive roots $r_+ > r_-$, i.e., one encounters an event and one Cauchy horizon. For the critical mass $m_{\rm cr} = \frac{3\sqrt{3}}{4}l$, the two horizons coincide and the black hole is extremal. For $m < m_{\rm cr}$ one has a regular geometry without horizons. This characteristic structure of $f(r)$ is illustrated in Fig. 10.1. It is universal in the sense that it is essentially identical for all regular black holes building on a de Sitter core.

It is also straightforward to evaluate the curvature scalars (10.7) for the Misner-Sharp mass (10.17). The result is shown in Fig. 10.2 for masses m given by multiples of $m_{\rm cr}$. The Kretschmann scalar reaches its maximum at $r = 0$ and decreases monotonically with increasing r. The square of the Weyl tensor exhibits zeros at the origin

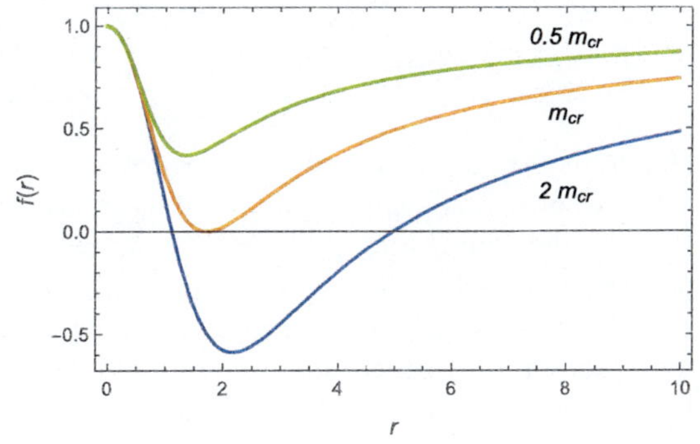

Fig. 10.1 Illustration of the horizon structure of the Hayward black hole (10.17). For $m > m_{cr}$ one encounters an outer event horizon and an inner Cauchy horizon. These horizons merge for $m = m_{cr}$. For $m < m_{cr}$ no horizons appear. The graph has been obtained for a Hayward geometry with $l = 1$

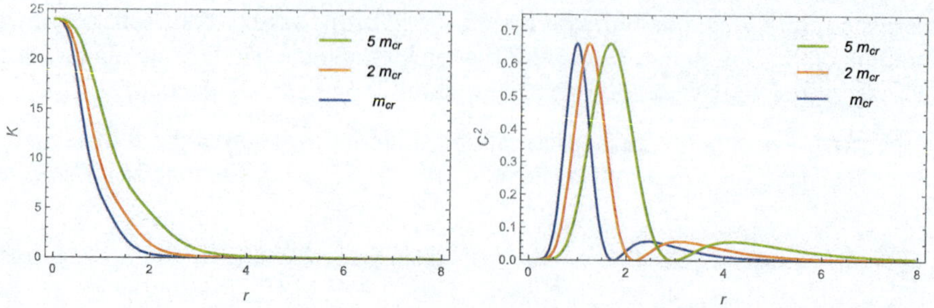

Fig. 10.2 Illustration of the Kretschmann scalar K (left) and the squared Weyl tensor C^2 (right) evaluated for the Hayward geometry (10.17) with $m = m_{\rm cr}$ (blue line), $m = 2m_{\rm cr}$ (orange line), and $m = 5m_{\rm cr}$ (green line). The curvature invariants are finite everywhere and the geometry obeys the limiting curvature hypothesis. The graphs have been obtained for $l = 1$

and one specific point located between the horizons r_- and r_+. Thus, the regular *static* black hole geometry is compatible with the limiting curvature hypothesis [42–44] for all values m.

RG-improved black holes. Interestingly, the gravitational asymptotic safety program [45, 46] has provided non-trivial hints that the theory supports regular black holes with a de Sitter core. Reference [11] applied the method of renormalization group (RG)-improvement to a Schwarzschild black hole in order to obtain an effective geometry taking quantum gravity corrections into account. This leads to a Misner sharp mass exhibiting singularity resolution. In this case, the effective running of the Newton constant at high energies produces an effective mass $M(r)$ which vanishes as r^3 at small distances. The essential elements of the construction, leading to the geometry (10.20), can be summarized as follows.

Investigating asymptotic safety based on solutions of the Wetterich equation adapted to gravity [47] leads to a scale-dependent Newton's coupling $G(k)$ depending on a coarse graining scale k. Using the arguably simplest approximate solution of the Wetterich equation based on the Einstein-Hilbert truncation [47–49], this scale-dependence can be approximated by [11]

$$G(k) = \frac{G}{1 + \omega\, G\, k^2} \,. \tag{10.19}$$

Here G is the laboratory value of Newton's coupling measured at $k = 0$ and ω is a *positive* constant. At large distances, $k \to 0$, $G(k)$ approaches G. At short distances, $k \to \infty$, the scaling $\lim_{k\to\infty} k^2 G(k) = \omega^{-1} = g_* > 0$ is dictated by the Reuter fixed point providing the UV-completion of the theory.

Following the original derivation [11], the RG-improvement process starts from the function $f(r)$ describing a classical Schwarzschild black hole. Subsequently $G \mapsto G(k)$ is promoted to the scale-dependent coupling (10.19). The RG-improved geometry is then obtained by identifying the coarse-graining scale k with the inverse radial proper distance $d(r)$ between the origin and a point located at radius r, $k^2 = \xi^2/d(r)^2$. This procedure leads to the geometry [11]

$$f(r) = 1 - \frac{2Gmr^2}{r^3 + \widetilde{\omega}G(r + \gamma mG)}, \qquad M(r) = \frac{mr^3}{r^3 + \widetilde{\omega}G(r + \gamma mG)}. \tag{10.20}$$

Here $\widetilde{\omega} \equiv \omega\xi^2$ is a positive constant and $\gamma \approx 9/2$ but the qualitative features of the model are independent of its precise value. We now return to geometric units $G = 1$. Moreover, we will set $\gamma = 0$. While this does not strictly correspond to a regular black hole with a de Sitter core in the sense of Eq. (10.14), this limit still captures all essential features of the horizons exhibited by the regular model while significantly simplifying the discussion. The case $\gamma = 9/2$ can be found in [11].

Qualitatively, the features of the function $f(r)$ describing the RG-improved black hole are identical to the ones shown in Fig. 10.1. The horizon condition resulting from (10.20) is ($\gamma = 0$)

$$r\left(r^2 - 2mr + \widetilde{\omega}\right) = 0\,. \tag{10.21}$$

Analogous to the Hayward case, there exists a critical mass value

$$m_{\rm cr} = \sqrt{\widetilde{\omega}}\,. \tag{10.22}$$

For $m > m_{\rm cr}$, $f(r)$ has two simple zeros at

$$r_{\pm} = m \pm \sqrt{m^2 - m_{\rm cr}^2}\,. \tag{10.23}$$

Hence, the spacetime has an outer event horizon at r_+ and an inner (Cauchy) horizon at r_-. For $m = m_{\rm cr}$ the two horizons coincide and there is one double zero at $r_+ = r_- = \sqrt{\widetilde{\omega}}$ and the black hole is extremal. If $m < m_{\rm cr}$, the spacetime is free from any horizon.

10.3.3 Hawking Effect and Final State

So far, our exposition has been limited to static geometries. Following Hawking's seminal work [50], it is expected that black holes emit thermal radiation in form of a Hawking flux. This radiation comes with a perfect black body spectrum with temperature

$$T_{\rm BH} = \frac{\kappa_+}{2\pi}\,, \tag{10.24}$$

where κ_+ is the surface gravity at the event horizon (10.10). The resulting energy loss leads to a decrease of the black hole's mass. For a Schwarzschild black hole with just an event horizon this entails that the black hole evaporates completely within a

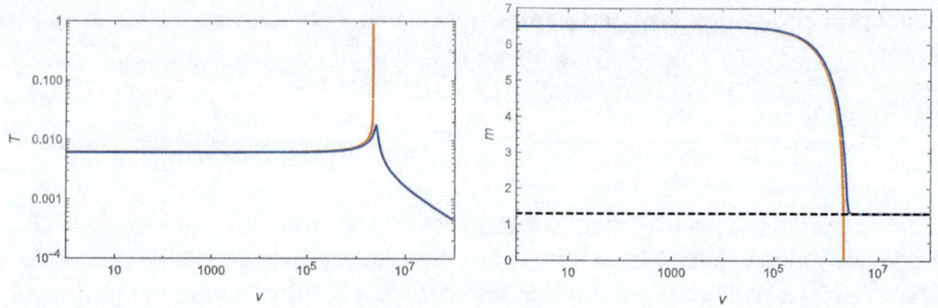

Fig. 10.3 Illustration of the thermodynamical properties of a regular black hole. Its temperature (left) remains finite and the mass (right) approaches the critical mass m_{cr} (dashed line) asymptotically. The graphs have been obtained for a Hayward geometry with $l = 1$. The Schwarzschild geometry ($l = 0$) has been added as the orange line for comparison

finite proper time interval. This can be traced back to the temperature being inversely proportional to the mass m, so that the black hole turns hotter the lighter it gets.

The presence of a Cauchy horizon changes this picture drastically. While the black hole is still expected to experience mass loss due to the Hawking flux, the temperature of the radiation remains finite since the two horizons approach each other as the black hole becomes lighter, c.f. Fig. 10.1. As a consequence, the Hawking temperature decreases in the final stage of the evaporation process and the geometry asymptotes to a cold remnant given by the extremal black hole. Figure 10.3 contrasts these two situations. The black hole evaporation of a Schwarzschild black hole is shown as orange lines indicating that the process terminates in a finite time-span. The generic situation encountered for regular black holes with a Cauchy horizon is shown by the blue curves, illustrating that one obtains a remnant with mass $m_{\rm cr}$.

We illustrate these properties using the technically simplest setting of the RG-improved geometry (10.20) with $\gamma = 0$. The analysis of other regular black hole geometries featuring a Cauchy horizon follows along the same lines but results in expressions which are significantly more complex. Thus we opt for this example for pedagogical reasons. In order to stress the universal features of the construction, we replace the model parameter $\widetilde{\omega}$ by $m_{\rm cr}$ using (10.22).

Starting from (10.20) and evaluating (10.24) gives the black hole temperature as a function of its mass

$$T_{\rm BH}(m) = \frac{1}{4\pi m} \frac{\sqrt{1-\Omega^2}}{1+\sqrt{1-\Omega^2}} \,, \qquad \Omega \equiv \frac{m_{cr}}{m} \,. \tag{10.25}$$

The temperature vanishes for $m \searrow m_{\rm cr}$, *i.e.*, $\Omega \nearrow 1$. This feature underlies the interpretation of an extremal black hole as a "cold" remnant.

Given the temperature (10.25), the luminosity of the black hole can be estimated via the Stefan-Boltzmann law $L = \sigma T^4$, where $\sigma = \pi^2/60$ for a single, massless degree of freedom. The emitted power is obtained by multiplying L with the area of the event horizon $A \equiv 4\pi r_+^2$,

$$P(m) = \sigma\left(4\pi r_+^2\right) T^4\,. \tag{10.26}$$

Here T and r_+ are understood as functions of m. The mass-loss is then determined by solving the mass-loss formula

$$\dot{m} = -P(m)\,. \tag{10.27}$$

For the concrete example of the RG-improved black hole

$$P(m) = \frac{\sigma}{(4\pi)^3\, m^2}\, \frac{\left(1-\Omega^2\right)^2}{\left(1+\sqrt{1-\Omega^2}\right)^2}\,. \tag{10.28}$$

The final part of the evaporation process is described by those terms in the above expressions which are dominant for $m \searrow m_{\rm cr}$ ($\Omega \nearrow 1$). Expanding (10.25) gives the asymptotic form of the temperature

$$T(m) \simeq \frac{\sqrt{m-m_{\rm cr}}}{2\sqrt{2}\pi\, m_{\rm cr}^{3/2}}\,, \tag{10.29}$$

and the emitted power

$$P(m) \simeq \frac{\sigma}{16\pi^3}\, \frac{(m-m_{\rm cr})^2}{m_{\rm cr}^4}\,. \tag{10.30}$$

Integrating (10.27) in the asymptotic regime then yields

$$m(v) \simeq m_{\rm cr} + \frac{m_0 - m_{\rm cr}}{1+\alpha(m_0-m_{\rm cr})(v-v_0)} \tag{10.31}$$

Here $\alpha \equiv \sigma/(16\pi^3\tilde{\omega}^2)$, and v_0 is a time, already in the late-time regime, where $m(v_0) = m_0$ is imposed. The key result is that for $v \to \infty$

$$m(v) - m_{\rm cr} \propto 1/v\,. \tag{10.32}$$

This result is *universal*, i.e. it does not depend on the details of the geometry of the regular black hole [35]. Equation (10.75) shows that it carries over to the Reissner–Nordström geometry and the Hayward regular black hole as well. The profound consequence of (10.32) arises from the comparison with Price's law

$$m_-(v) \simeq m_0 - \frac{\beta}{(v/v_0)^{p-1}}\,, \qquad p \geq 12\,, \tag{10.33}$$

where m_0 is the asymptotic mass of the black hole, $\beta > 0$ is a quantity with the dimension of a mass, and v_0 is the initial time which we set to one in the sequel. This shows that the Hawking flux dominates over the gravitational waves contribution which vanishes as $1/v^p$. Hence the Hawking effect may play an important role for the stability of a regular black hole at asymptotically late times.

In order to judge the phenomenological viability of regular black holes, understanding whether these geometries are robust once perturbations are included is highly relevant. This applies in particular to perturbations of the asymptotically extreme geometry with respect to the combined influx of Hawking radiation and outflux of the star. At this point we have all the prerequisites to review the current understanding of this situation in the next sections.

10.4 The Elementary Mechanics of Mass-Inflation

We start by reviewing the basic mechanisms underlying the mass-inflation effect. Our discussion focuses on the historical analysis carried out in the context of the spherically symmetric Reissner–Nordström geometry. The key result of the analysis is that the mass function $m_+(v)$ in the vicinity of the Cauchy horizon grows exponentially once perturbations which naturally appear in a gravitational collapse of a star into a black hole are included,

$$m_+(v) \sim v^{-(p-1)} e^{\kappa_- v}\,, \qquad \text{as} \quad v \to \infty\,, \tag{10.34}$$

with p being the exponent appearing in Price's law (10.33). We start by giving a qualitative discussion of the mass-inflation mechanism in Sect. 10.4.1. Section 10.4.2 then derives (10.34) based on the two-dimensional wave equation for the generalized mass function in the presence of a cross-flow of outgoing and ingoing streams of radiation. We close with a brief discussion on the strength of the resulting singularity in Sect. 10.4.3.

10.4.1 Colliding Mass-Shells and DTR-Relations

Qualitatively, the divergence of the local mass in the interior of a realistic black hole can be understood with the following example (see [51]). Let us consider a time-like moving shell of radius $R(\tau)$ where τ is the proper time. The shell divides spacetime into a region $\mathscr{M}_+$ inside and $\mathscr{M}_-$ outside the shell and we assume that the mass functions m_+ and m_- in each sector are constant. The proper mass of the shell, m_{shell}, satisfies

$$dm_{\text{shell}} + P\, d(4\pi R^2) = 0 \tag{10.35}$$

which implies mass conservation if $P = 0$. The equation of motion of the shell is ruled by [52]

$$m_- - m_+ = m_{\text{shell}}\left(1 - \frac{2m_+}{R} + \left(\frac{dR}{d\tau}\right)^2\right)^{1/2} - \frac{m_{\text{shell}}^2}{2R} \tag{10.36}$$

It expresses the total conserved gravitating mass $m_- - m_+$ of the shell as a sum of four terms (expanding the square root to first order): the rest-mass m_{shell}, the kinetic energy $\frac{1}{2}m_{\text{shell}}\dot{R}^2$, the mutual potential energy $-m_{\text{shell}}m_+/R$, and a self-potential energy $-\frac{1}{2}m_{\text{shell}}^2/R$. Note that the "potential energy" contributes to the total mass of the *outer* body, it is a "binding energy" [53]. If the outer body can be released from its gravitational binding, its gravitating mass can increase.

This can be easily illustrated in the light-like limit. In this case, we may gain intuition from the DTR-relations, first discovered by Dray, 't Hooft, and Redmount [54, 55], and subsequently generalized in [56]. In the latter work the (generalized) DTR-relations are derived as geometrical consistency conditions on the function f, beyond spherical symmetry, for an energy-momentum tensor which is vacuum except for a distributional delta function source representing the shells, also see [53] for an instructive discussion.

The idea behind the DTR-relations is to consider the collision of two infinitely thin, light-like, pressureless shells describing an ingoing and outgoing perturbation, see Fig. 10.4.[2] The shells provide a highly idealized model of matter perturbations in the black hole interior which are naturally expected from the collapse of a star into a black hole. They capture the essential features underlying the mass-inflation effect, the counter-streaming of matter located between the inner and outer horizon of the black hole in the optical geometric limit [58].

As depicted in Fig. 10.5, the ingoing and outgoing shells separate spacetime in four regions $\mathcal{M}_A, \mathcal{M}_B, \mathcal{M}_C, \mathcal{M}_D$. The generalized DTR-relations then encode the fact that the spacetime metric in each region must agree at the collision point r_0. We then assume that in each region the metric takes the form (10.5) with $f_i(v, r) = f_i(r)$, $i = A, B, C, D$, being static. Following the pedagogical derivation given in [59], this entails that

$$f_A\, f_B = f_C\, f_D\,, \quad \text{at} \quad r_0\,. \tag{10.37}$$

For the sake of simplicity, let us assume that the metric in each sector has the form of the Schwarzschild metric. Assuming that the region $\mathcal{M}_C$ corresponds to flat space, the functions f_i in each region take the form

$$f_C = 1\,, \qquad f_A = 1 - \frac{2m_1}{r_0}\,, \qquad f_B = 1 - \frac{2m_4}{r_0}\,, \tag{10.38}$$

---

[2] For more details on the infinitely thin-shell formalism, see [57].

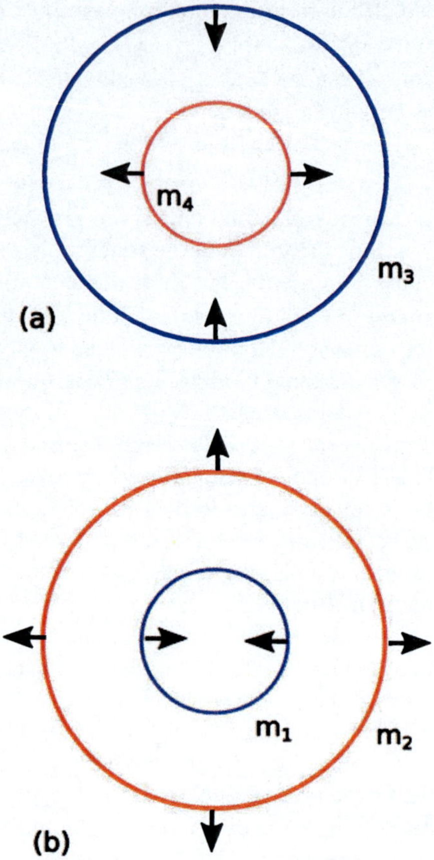

Fig. 10.4 Two spherically symmetric, transparent, concentric shells colliding at the speed of light: **a** before collision, **b** after the collision. The blue circle represent ingoing radiation close to the Cauchy horizon, while the red shell represents the outgoing radiation

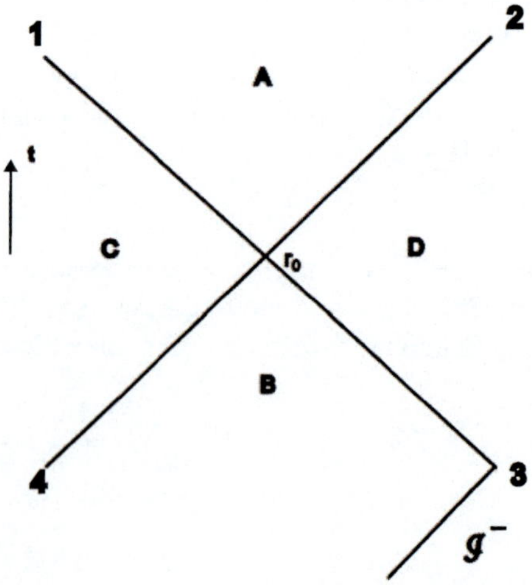

Fig. 10.5 Figure 10.4 recast as a spacetime diagram: r_0 is the colliding radius and sectors B and C represent the future and past evolution of the spacetime, respectively

together with

$$f_D = 1 - \frac{2(m_3 + m_4)}{r_0} = 1 - \frac{2(m_1 + m_2)}{r_0}\,. \tag{10.39}$$

The relation (10.39) implies the conservation law

$$m_1 + m_2 = m_3 + m_4\,. \tag{10.40}$$

Moreover, evaluating (10.37) gives the additional conditions

$$m_1 = m_3(1 - 2m_4/r_0)^{-1}, \quad m_2 = m_4(1 - 2m_1/r_0)\,. \tag{10.41}$$

We now consider a collision just outside the horizon of the interior field, $r_0 = 2m_4 + \delta$ with $\delta \to 0^+$. The first identity in (10.41) then entails that, close to the inner horizon, the mass increase $m_1 - m_3$ diverges as

$$m_1 - m_3 = \frac{2m_3 m_4}{\delta}\,. \tag{10.42}$$

This divergence implies that the new Schwarzschild mass of the outgoing shell would then become negative because of its potential energy.

This example, albeit very simple, illustrates the basic physical mechanism underling the mass-inflation phenomenon: the imploding shell can mimic the fallout from the radiative tail of the collapse while the outgoing shell models the outflow from the collapsing star as depicted in Fig. 10.4. In a spacetime diagram the Cauchy horizon would develop beyond the history of region A, extending region B and the collision point r_0 is near the actual Cauchy horizon. The fact that the DTR-relations are not limited to spherical symmetry then suggests that the same phenomenon also appears in less symmetric situations including a Kerr black hole [33].

This is the best that can be concluded on the basis of DTR-relations. In particular, promoting $m_1 - m_3$ to a function of v is beyond the framework and should not be used to draw conclusions on the strength of the singularity, albeit this point is sometimes not appreciated in the literature [60, 61]. Determining the strength of the divergence requires a more detailed analysis based on dynamical models. We cover this in the next subsection.

10.4.2 Dynamical Models of Mass Inflation

The simple analogy discussed in the previous section can be useful to understand the interior of realistic black holes. In this case, it is useful to make some simplifying assumptions. A realistic, rotating (non-spherical) black hole may be schematized as a spherically symmetric, charged black hole because their horizon structures are similar. Moreover, the tail of gravitational quadrupolar waves may be idealized as spherical waves. Since the perturbations get blue-shifted near the Cauchy horizon, one can use an "optical" approximation and describe the matter infalling onto the Cauchy horizon as a stream of light-like particles.

In 1981, Hiscock [28] treated the dynamics of this situation by considering the charged Vaidya metric,

$$ds^2 = 2drdv - \left(1 - \frac{2m(v)}{r} + \frac{Q^2}{r^2}\right) dv^2 + r^2 d\Omega^2\,, \tag{10.43}$$

promoting the mass function of the Reissner–Nordström metric to a v-dependent function. Then $m(v)$ is the externally measured mass of the black hole which varies with advanced time v because of the inflow. The associated stress-energy tensor (the electrostatic component is irrelevant in this discussion) corresponds to radially infalling light-like dust:

$$T_{\alpha\beta} = \frac{\dot{m}(v)}{4\pi r^2} l_\alpha l_\beta, \quad l_\alpha = -\partial_\alpha v\,, \quad l_\alpha l^\alpha = 0\,. \tag{10.44}$$

The ansatz

$$m(v) = m_0 - \frac{\beta}{(v/v_0)^{p-1}} \tag{10.45}$$

reproduces the power-law decay of the radiative tail. (For the generic case of the "quadrupole" waves $p = 12$.) It is easy to show from (10.44) and (10.45) that an observer approaching the Cauchy horizon ($v \to \infty$) reaches the horizon in a finite proper time and measures an energy flux diverging like $e^{2\kappa_- v}$, with κ_- being the surface gravity (10.13) for the asymptotic, stationary Reissner–Nordström black hole of mass m_0 and charge Q.

In [62], Poisson and Israel extended this model by including a crossflow of radially ingoing and outgoing radiation. This analysis shows that the geometry near the Cauchy horizon changes dramatically. Approaching the Cauchy horizon, the Coulomb component of the Weyl curvature

$$\Psi_2 = -\frac{m(v)}{2r^3} + \frac{Q^2}{r^4}\,, \tag{10.46}$$

and hence the mass parameter $m(v)$ diverges like

$$m(v) \simeq v^{-(p-1)} e^{\kappa_- v}, \quad v \to \infty\,. \tag{10.47}$$

This behavior has to be contrasted with the situation outside the event horizon, where m approaches its ADM value at infinity.

It is interesting to see how this effect originates from the point of view of the field equations. A spherically geometry can be described by the metric (10.2). The Einstein equations can be reformulated as a two-dimensional, covariant wave equation for m,

$$\Box m = -16\pi^2 r^3 T^{ab} T_{ab} \,. \tag{10.48}$$

Going to Kruskal coordinates, the energy momentum tensor for a null-crossflowing radiation can be written as

$$T_{ab} = \frac{L_{\text{in}}(V)}{4\pi r^2}\partial_a V \partial_b V + \frac{L_{\text{out}}(U)}{4\pi r^2}\partial_a U \partial_b U \,. \tag{10.49}$$

Due to $\partial_a V$ and $\partial_a U$ being null vectors, the square of the energy momentum tensor vanishes if only one component T_{ab}^{out} or T_{ab}^{in} is considered. In this case the source on the right-hand side of Eq. (10.48) vanishes and there is no mass-inflation. It is only when *both components* T_{ab}^{out} or T_{ab}^{in} are present that the wave equation contains a non-trivial source term triggering the growth of m.

At this stage, one has to adopt assumptions for L_{in} and L_{out}. L_{in} uses Price's law (10.45), which in Kruskal coordinates V reads

$$L_{in}(V) = \frac{\beta}{(-\kappa_- V)^2}(-\log(-\kappa_- V))^{-p}. \tag{10.50}$$

$L_{\text{out}}(U)$ receives two contributions, one from the collapsing star and a second one from gravitational waves reflected by the inner potential barrier. At late advanced time the first one is negligible. The second one is basically following Price's law, but written in terms of the coordinate U. Thus, we have control over $L_{\text{out}}(U)$ at asymptotically late times only and every analysis going beyond this regime has to track the full dynamics numerically.

Near the Cauchy horizon,[3] assuming that $r \approx r_0$, Eq. (10.48) then reads

$$\partial_U \partial_V m = -\frac{2}{r_0} L_{\text{in}}(V) L_{\text{out}}(U) \,. \tag{10.51}$$

In the limit $V \to 0_-$ this relation is readily integrated, producing a divergence of the type

$$m \simeq \frac{1}{-\kappa_- V(\log(-\kappa_- V))^{(p-1)}} \,. \tag{10.52}$$

In terms of the original advanced time coordinate v, (10.52) entails the exponential growths of the mass function underlying the mass-inflation effect (10.34).

[3] This makes the crucial assumption that the Cauchy horizon actually exists. While there is good evidence for forming such a horizon for the black holes encountered in general relativity, the question whether such a horizon actually forms in the collapse to a regular black hole is a largely open question, see [63, 64] for some recent work in this direction. For a first self-consistent dynamical calculation of mass-inflation without assuming the existence of a Cauchy horizon we refer to [65].

10.4.3 Mass-Inflation Instability and Singularities of Spacetime

At this stage it is interesting to discuss the implications of the growths of the mass function (10.34) for the singularity structure of spacetime. We recall that a singularity for which all algebraic scalars of curvature are finite and yet some component of the curvature diverge in some frame is known as a "whimper" singularity. In most of the cases these singularities are unstable and tend to evolve in a strong, "scalar" singularity. As demonstrated in the previous section, it turns out that if the Hiscock model is perturbed with an additional *outflow* of radiation, which in a realistic collapse would always be present either for the presence of the collapsing star or for the scattering gravitational waves from the inner potential barrier, the Weyl curvature diverges without limit at the Cauchy horizon.

The strength of the resulting singularity can then be understood as follows. We focus on the future sector of the shell, denoting the coordinates and mass function in this sector with the subscript "+". Using the advanced coordinate, the asymptotic form of the metric near the Cauchy horizon reads

$$ds^2 \simeq 2\frac{dv_+}{r}\left(r dr + m_+(v_+)dv_+\right) + r^2 d\Omega^2 \,. \tag{10.53}$$

We then define a new coordinate u through the relation $du = (r dr + m_+(v_+)dv_+)$. This coordinate is regular at the Cauchy horizon. The line-element (10.53) then becomes [65]

$$ds^2 \simeq 2\frac{dv_+ du}{r} + r^2 d\Omega^2 \,. \tag{10.54}$$

Again, this expression is manifestly regular at the Cauchy horizon. Since it is possible to find a coordinate system where the metric is regular, the singularity building up at the Cauchy horizon is rather weak. This fact has profound consequences: as already realized by Ori [30] and further investigated by Burko [66], the mass-inflation singularity does not satisfy the necessary conditions to be strong in the Tipler sense [67].[4] A measure of the tidal distortion experienced by an observer is obtained by integrating the square of the Weyl curvature twice. In the case of the standard mass-inflation scenario one finds

$$(\Psi_2)^2 \simeq C_{\mu\nu\rho\sigma}C^{\mu\nu\rho\sigma} \simeq \frac{1}{(\kappa_- V)^2(\log(-\kappa_- V))^{2(p-1)}} \,. \tag{10.55}$$

Here Ψ_2 is the Coulomb-component of the Weyl curvature (10.46) and $V \propto \tau$ is proportional to the proper time of an observer impacting on the horizon. The tidal

[4] According to Tipler, a null singularity is called "strong" if there exists at least one component of the Riemann tensor (in a parallelly propagated frame) which does not converge when integrated with respect to the affine parameter τ twice. The physical meaning of this requirement is that the tidal distortion is not finite as an observer crosses the singularity.

distortion is obtained by twice integrating (10.55) and is therefore finite. It has further been argued by Ori that this behavior could be sufficient to determine a C^1-extension of the spacetime beyond the Cauchy horizon [30]. However, according to Królak, Eq. (10.55) still signals a strong singularity, as the expansion of the congruence is divergent [68]: if the components of the Riemann tensor are integrated only once, the integral does not converge on the singularity.

10.5 Mass-Inflation for Regular Black Holes

We now embark on the central theme of this chapter, contrasting the mass-inflation scenario for Reissner–Nordström and regular black hole geometries. In the latter case we use the Hayward geometry as an explicit representative. The results are more general though, since the Reissner–Nordström and Hayward geometry constitute representatives for the two classes of universal late-time behavior encountered in the literature [34, 35]. The discussion will be based on the Ori-model introduced in Sect. 10.5.1. The case of a static background is covered in Sect. 10.5.2 while the modifications due to the mass-loss generated by the Hawking effect are highlighted in Sect. 10.5.3. In order to contrast the structural differences of the mass-inflation effect for the Reissner–Nordström case and regular black holes, we discuss the two geometries in parallel.

10.5.1 The Ori-Model—General Setup and Dynamics

The Ori-model constitutes a simplification of the original Poisson-Israel model geared towards making the mass-inflation effect accessible by analytic methods. In this case the outgoing energy flux is modelled by a spherically symmetric, pressureless null shell Σ placed between the (apparent) inner and outer horizon of the black hole geometry. In the language of Eq. (10.49) it corresponds to taking $L_{\text{out}}(U)$ to be a delta-function. The shell then acts as a catalyst triggering the mass-inflation instability. In this section, we derive the equations capturing the dynamics of the system for a generic, spherically symmetric black hole spacetime exhibiting an event and a Cauchy horizon.

The shell Σ modeling the ingoing perturbation, divides spacetime into a region $\mathcal{M}_+$ inside and $\mathcal{M}_-$ outside the shell. Denoting the coordinates on $\mathcal{M}_\pm$ with subscripts $\pm$, the metric in each sector can be written as

$$ds^2 = -f_\pm(r, v_\pm)dv_\pm + 2dr dv_\pm + r^2 d\Omega^2 . \tag{10.56}$$

Insisting that both regions lead to the same induced metric on Σ yields that the radial coordinate r can be taken the same in both regions. Equation (10.56) already anticipated this result. In general, the relation between v_+ and v_- is non-trivial though.

Their dependence is fixed by noting that the position of Σ in the two coordinate systems is

$$f_+ \, dv_+ = f_- \, dv_- \,. \tag{10.57}$$

In practice, we use this relation to express all quantities in terms of the coordinate in the outer sector of the shell, setting $v \equiv v_-$. Equations (10.56) and (10.57) are independent of the dynamics and completely fixed by the geometrical setup of the model.

Our next task is to find the relation between the Misner-Sharp mass $M_\pm$ in the two sectors. Evaluating Einstein's equations in each sector implies

$$\frac{\partial M}{\partial r} = -4\pi r^2 T_v{}^v \,, \qquad \frac{\partial M}{\partial v} = 4\pi r^2 T_v{}^r \,, \tag{10.58}$$

combined with $T_{rr} = 0$. We then introduce the null generators of Σ

$$s_\pm^\mu \equiv \frac{dx_\pm^\mu}{dr} = (2/f_\pm, 1, 0, 0) \,, \tag{10.59}$$

where it is convenient to use r as a parameter. Continuity of the flux across Σ requires

$$\left[T_{\mu\nu} \, s^\mu s^\nu \right] = 0 \tag{10.60}$$

where the square brackets indicate the discontinuity of a scalar quantity across the position of the shell. In terms of the lapse- and the mass-functions, Eq. (10.60) implies

$$\left. \frac{1}{f_+^2} \frac{\partial M_+}{\partial v_+} \right|_\Sigma = \left. \frac{1}{f_-^2} \frac{\partial M_-}{\partial v_-} \right|_\Sigma \,. \tag{10.61}$$

We then recast this equation in terms of the coordinate v. Using, Eq. (10.57), it is convenient to write

$$\left. \frac{1}{f_+} \frac{\partial M_+}{\partial v} \right|_\Sigma = F(v) \,, \tag{10.62}$$

where

$$F(v) \equiv \left. \frac{1}{f_-} \frac{\partial M_-}{\partial v} \right|_\Sigma \,. \tag{10.63}$$

The introduction of $F(v)$ will facilitate the analysis of the late-time dynamics of the model later on.

Finally, we need an equation determining the dynamics of the shell. Exploiting that Σ moves light-like, Eq. (10.56) gives the relation $f_- \, dv_- = 2dr$. Thus the position of the shell $R(v)$ follows from

$$\frac{dR}{dv} = \left. \frac{1}{2} f_- \right|_\Sigma \,. \tag{10.64}$$

Equations (10.62) and (10.64) then form a coupled dynamical system determining the position of the shell and the Misner-Sharp mass in its interior in terms of the mass-function m_- in the outer sector of the shell. Since the dynamics of $R(v)$ is independent of M_+, one can first study the motion of the shell based on (10.64) before substituting the solution into (10.62).

The Ori-model applied to a static black hole background fixes m_- by imposing the Price's tail behavior (10.33). The Price tail governs the decay of a perturbation at asymptotically late times. We stress that the relation (10.33) applies only asymptotically. At intermediate times one expects that the time-dependence of $m_-(v)$ may be significantly more complicated than indicated by the asymptotic relation.

10.5.2 The Ori-Model on Static Backgrounds

The Misner-Sharp mass for the Reissner–Nordström (RN) and Hayward (H) black hole is

$$\begin{aligned} \text{RN}: \quad & M(r) = m - \frac{Q^2}{2r}\,, \\ \text{H}: \quad & M(r) = \frac{mr^3}{r^3 + 2ml^2}\,. \end{aligned} \tag{10.65}$$

Here Q is the charge of the black hole and $l > 0$ is a parameter with the dimension of mass which ensures the regularity of the Hayward geometry. The static background analysis includes the effect of the perturbation by identifying the mass function $m_-(v)$ in the outer sector of the shell with the Price-tail behavior given in Eq. (10.33).

Specifying the general Eqs. (10.62) and (10.64) to our exemplary geometries, one finds that the dynamics of the shell is given by

$$\begin{aligned} \text{RN}: \quad & \dot{R}(v) = \frac{R^2 - 2m\,R + Q^2}{2R^2}\,, \\ \text{H}: \quad & \dot{R}(v) = \frac{R^3 - 2m\,R^2 + 2l^2\,m}{2\left(R^3 + 2l^2 m\right)}\,. \end{aligned} \tag{10.66}$$

The mass function $m_+(v)$ in the interior is determined from

$$\dot{m}_+(v) = p(m_+)\,F(v)\,. \tag{10.67}$$

The explicit forms of the functions $F(v)$ are

$$\begin{aligned} \text{RN}: \quad & F(v) = \frac{\dot{m}_-}{R^2 - 2R\,m_- + Q^2}\,, \\ \text{H}: \quad & F(v) = \frac{\dot{m}_-}{\left(R^3 + 2l^2 m_-\right)\left(R^3 - 2(R^2 - l^2)m_-\right)}\,, \end{aligned} \tag{10.68}$$

while the dependence of the right-hand side on m_+ is captured by the polynomials

$$\begin{aligned} \text{RN}: &\quad p(m_+) = R^2 - 2m_+ R + Q^2\,, \\ \text{H}: &\quad p(m_+) = \left(R^3 + 2l^2 m_+\right)\left(R^3 - 2(R^2 - l^2)m_+\right)\,. \end{aligned} \tag{10.69}$$

The polynomials (10.69) encode the key difference of the two geometries: in the Reissner–Nordström case $p(m_+)$ is linear in the mass while for Hayward it is quadratic. As it turns out, this makes a decisive difference in the stability of the two models.

We proceed by analyzing the late-time dynamics of the two systems. At the analytic level, this is conveniently done by employing the Frobenius method. In the specific case at hand, the v-dependent functions are expanded in a generalized power series in $1/v$ which take the general form

$$f(v) = \frac{1}{v^s}\sum_{k=0}^{\infty}\frac{a_k}{v^{k/2}}\,. \tag{10.70}$$

The parameter s and the coefficients a_k are obtained from substituting this ansatz in the corresponding differential equation, performing an expansion for large v, and extracting a hierarchy of equations given by the coefficients appearing at each order in the large-v-expansion. This hierarchy is then solved recursively for s (lowest order equation) and the coefficients a_k. Note that Eq. (10.70) already anticipates that consistent solutions require the inclusion of non-integer powers of $1/v$.

Based on this strategy, one finds the following late-time behavior. Starting with the dynamics of the shell, one first establishes that r_- is a fixed point of (10.66) since $f_-|_{r=r_-} = 0$ by definition of the Cauchy horizon. The first correction term obtained from the Frobenius analysis shows that r_- is indeed a late-time attractor,

$$R(v) \simeq r_- + c v^{-(p-1)}\,, \tag{10.71}$$

The constant c is a positive coefficient depending on m_0 and β and again $\simeq$ denotes that the expression holds asymptotically for large v. Integrating (10.66) numerically confirms this property. Some sample solutions arising from imposing initial conditions at different values v are shown in Fig. 10.6. This confirms that the shell impacts on r_- rather quickly from $R(v) > r_-$ and then essentially keeps its position close to the Cauchy horizon.

Based on the asymptotic solutions for $m_-(v)$ and $R(v)$ one then readily deduces the asymptotic behavior of $F(v)$,

$$F(v) \simeq -\frac{1}{2} r_- \kappa_-\,, \tag{10.72}$$

with $\kappa_- > 0$ the surface gravity at the Cauchy horizon. Remarkably, the negative constant appearing in this relation is universal in the sense that it is the same for

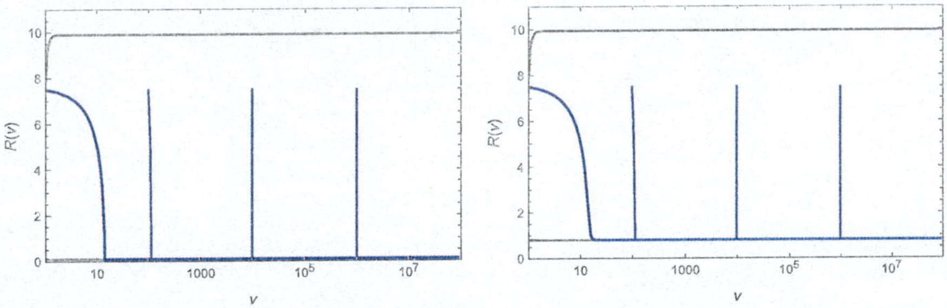

Fig. 10.6 Illustration of the shell dynamics (blue curves) for the background spacetime given by a static Reissner–Nordström (left) and Hayward black hole (right). The gray lines depict the positions of the apparent event (top) and Cauchy horizons (bottom). The shells reach r_- at asymptotically late times. The model parameters Q and l have been chosen such that the critical mass of the remnant is $m_{cr} = 1$

both geometries. Equation (10.72) then allows to conclude the asymptotic behavior of $m_+(v)$. It is at this point, where the different structures of $p(m_+)$ enter, yielding

$$\begin{aligned} \text{RN}: \quad & m_+(v) \simeq c_1\, e^{\kappa_- v}\, v^{-(p-1)}\,, \\ \text{H}: \quad & m_+(v) \simeq -\frac{r_-^3}{2l^2}\,. \end{aligned} \tag{10.73}$$

In the case of $p(m_+)$ being linear, $m_+(v)$ grows exponentially in v, justifying the terminology "mass-inflation". For quadratic polynomials $p(m_+)$ admits a second class of solutions where $m_+(v)$ takes a finite, negative value asymptotically. This value is given by the negative root of $p(m_+)$ and thus constitutes a fixed point of (10.67). The role of this attractor is illustrated at the level of numerical solutions in the top row of Fig. 10.7. Imposing initial conditions at $v = 1$ (which may be outside of the validity of the Ori-model) one finds that solutions for the Reissner–Nordström geometry terminate at finite value v (blue curves). In addition to this behavior, the Hayward geometry admits a second class of solutions (green curves): these extend to asymptotically late times and approach the attractor (10.73).

In order to understand the physics consequences entailed by the late-time behavior of the mass-function, it is instructive to study the growth of the Kretschmann scalar (10.7) evaluated at the position of the shell. For the asymptotic solutions (10.73) this yields

$$\begin{aligned} \text{RN}: \quad & K|_\Sigma \propto e^{2\kappa_- v}\, v^{-2(p-1)}\,, \\ \text{H}: \quad & K|_\Sigma \propto v^{6(p-1)}\,. \end{aligned} \tag{10.74}$$

Thus the structure of $p(m_+)$ *can lead to a significant weakening of the mass-inflation effect for regular black holes*. The origin of the growths (10.74) in the two cases is quite different though: while the exponential growths in the Reissner–Nordström case is directly related to the growths of the mass-function, the polynomial growths

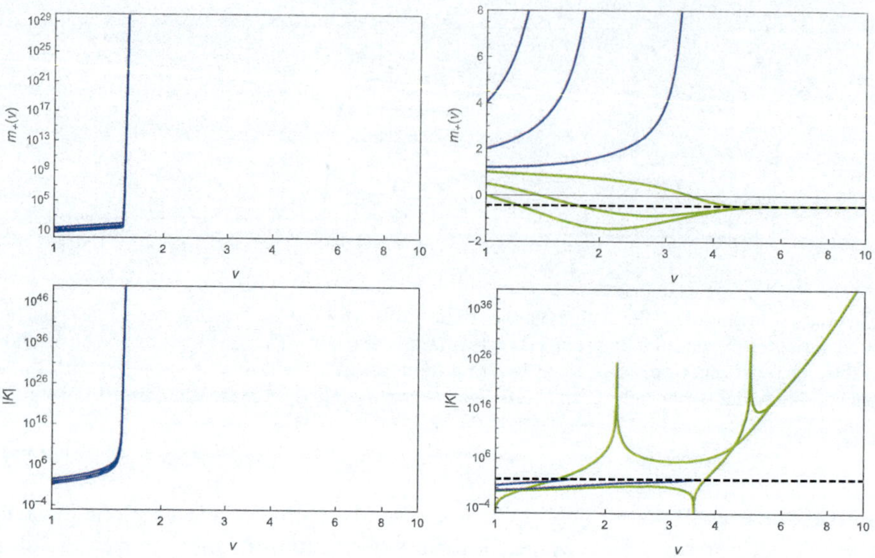

Fig. 10.7 Top: Illustration of the mass function $m_+(v)$ inside the shell for a static Reissner–Nordström (left) and Hayward black hole (right). For the blue lines the numerical integration terminates at finite value v while the green lines reach the late-time attractor of the static Hayward geometry where $m_+^\infty \equiv \lim_{v\to\infty} m_+(v)$ remains finite (black dashed line). Botton: Illustration of the Kretschmann scalar evaluated at the position of the shell. For the green solutions, the curvature scalar grows polynomially in v. The dashed line added in the Hayward case gives the value of $|K|$ at the point where the blue solutions terminate, indicating that the curvature scalar remains finite. The model parameters Q and l have been chosen such that the critical mass of the remnant is $m_{cr} = 1$

in the Hayward case is tracked back to the fact that the quantity $(R^3 + 2l^2 m_+)$, which vanishes at the late-time attractor, also appears in the denominator of the Kretschmann scalar. The scaling laws (10.74) are illustrated in the bottom row of Fig. 10.7. For the blue solutions $K|_\Sigma$ diverges at finite values v without reaching the late-time attractor. This is different for the green solutions where the numerical integration confirms the polynomial growths of $K|_\Sigma$ in the Hayward case.

The weakening of the curvature singularity (10.74) may have profound consequences for the geodesic completeness of spacetime. Solving the geodesic equation for a massive, radially free-falling observer in the static, non-critical black hole spacetime shows that $v = \infty$ can be reached in a finite amount of the observer's proper time. Technically, the new late-time attractor turns the mass-inflation singularity into a weak singularity with respect to both the Tipler and the Królak definition. This may open the possibility to extend geodesics beyond the singularity. While this is certainly an exciting possibility, this point is currently still awaiting its final clarification.

10.5.3 The Ori-Model Including Hawking Radiation

When discussing the thermodynamics of regular black holes in Sect. 10.3.2, we argued that the final state of the evaporation process is a cold remnant with finite mass m_{cr}. The mass-inflation effect associated with the Cauchy horizon then raises the question whether this picture is robust against perturbations. The analysis for static, regular black holes showed that there are mechanisms which tame the growth of the curvature singularity, delaying the growths of the Misner-Sharp mass to asymptotically late times. This suggests that the Hawking effect can influence the dynamics. The analysis based on the Ori-model [35], summarized in this section, indeed confirms this expectation.

Upon including the Hawking effect, the mass function of the black hole is no longer constant but turns into a v-dependent function. The time-dependence can be determined from the mass-loss formula (10.27). Following the strategy of the previous section, focusing on the late-time behavior of the solutions, it is straightforward to determine the leading terms from the Frobenius method

$$\begin{aligned} \text{RN}: \quad & m_{\text{Hawking}}(v) \simeq m_{cr}\left(1 + 2\,\frac{15\pi\, m_{cr}^3}{v} + 48\left(\frac{15\pi m_{cr}^3}{v}\right)^{3/2} + \cdots\right), \\ \text{H}: \quad & m_{\text{Hawking}}(v) \simeq m_{cr}\left(1 + 6\,\frac{80\pi\, m_{cr}^3}{v} + 80\left(\frac{80\pi\, m_{cr}^3}{v}\right)^{3/2} + \cdots\right). \end{aligned} \tag{10.75}$$

Here the dependence of $m(v)$ on Q and l are encoded in m_{cr}. The expansion shows that $\lim_{v\to\infty} m(v) = m_{cr}$, giving the expected mass for the cold remnant. Remarkably, the late-time behavior is universal in the sense that there are no free parameters entering the first three terms of the expansion.

Starting from Price's law (10.33), we then include the dynamics of the background by $m_0 \mapsto m_{\text{Hawking}}(v)$

$$m_-(v) = m_{\text{Hawking}}(v) - \frac{\beta}{(v/v_0)^{p-1}}\,. \tag{10.76}$$

The expansion (10.75) then entails that the late-time behavior of $m_-(v)$ is actually fixed by the Hawking effect, the contribution of the Price tail being subleading compared to the time-dependence of $m_{\text{Hawking}}(v)$. Moreover, the position of the roots of the lapse function $f(r, v)$ is no longer static: one inherits an apparent event horizon $r_+(v)$ and an apparent Cauchy horizon $r_-(v)$ which both approach the critical radius r_{cr} at asymptotically late times.

Having $m_-(v)$ at our disposal, we can again use (10.64) to find the dynamics of the shell. The result obtained from numerical integration is illustrated in Fig. 10.8. Here the position of the apparent horizons is indicated by the opaque gray lines. In analogy to the static case, a shell starting between the apparent horizons quickly falls

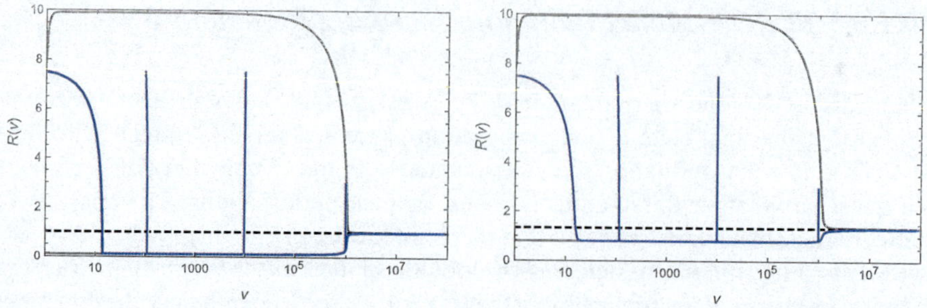

Fig. 10.8 Illustration of the shell dynamics for the background spacetime given by an evaporating Reissner–Nordström (left) and Hayward black hole (right) (blue lines). The gray lines depict the positions of the apparent event (top) and Cauchy horizons (bottom). In both cases the dashed line depicts r_{cr}. The model parameters Q and l have been chosen such that the critical mass of the remnant is $m_{cr} = 1$

towards the apparent Cauchy horizon and subsequently trails the dynamics of $r_-(v)$. This behavior is independent of the initial conditions and details of the geometry.

The crucial difference with the static background then occurs at the level of (10.72). Including the mass loss one has

$$\begin{aligned} \text{RN}: \quad & F(v) \simeq -\left(\frac{15\pi\, m_{cr}^3}{v}\right)^{1/2}, \\ \text{H}: \quad & F(v) \simeq -\left(\frac{80\pi\, m_{cr}^3}{v}\right)^{1/2}. \end{aligned} \tag{10.77}$$

Thus the function $F(v)$ appearing on the right-hand side of the dynamical equation for $m_+(v)$ vanishes asymptotically. The power-law governing the decay of $F(v)$ is again independent of the geometry under consideration. Heuristically, this can be understood from the fact that the asymptotic geometry is a remnant with vanishing surface gravity. Hence the leading terms in (10.72) vanish and the dynamics of $F(v)$ starts with the subleading order as compared to the static case.

The modification (10.77) has profound consequences for the late-time behavior of $m_+(v)$. In contrast to the static case, the mass function in the interior of the shell approaches a constant value

$$\begin{aligned} \text{RN}: \quad & m_+(v) \simeq m_{cr} + \frac{30\pi\, m_{cr}^4}{v}, \\ \text{H}: \quad & m_+(v) \simeq -2m_{cr} + 12 m_{cr} \left(\frac{80\pi\, m_{cr}^3}{v}\right)^{1/2}. \end{aligned} \tag{10.78}$$

Again this new attractor appears in both geometries. The different fall-off behavior in the subleading term for the Reissner–Nordström case is due to the cancellation of the $v^{-1/2}$-contributions in the series. The approach of numerical solutions to this

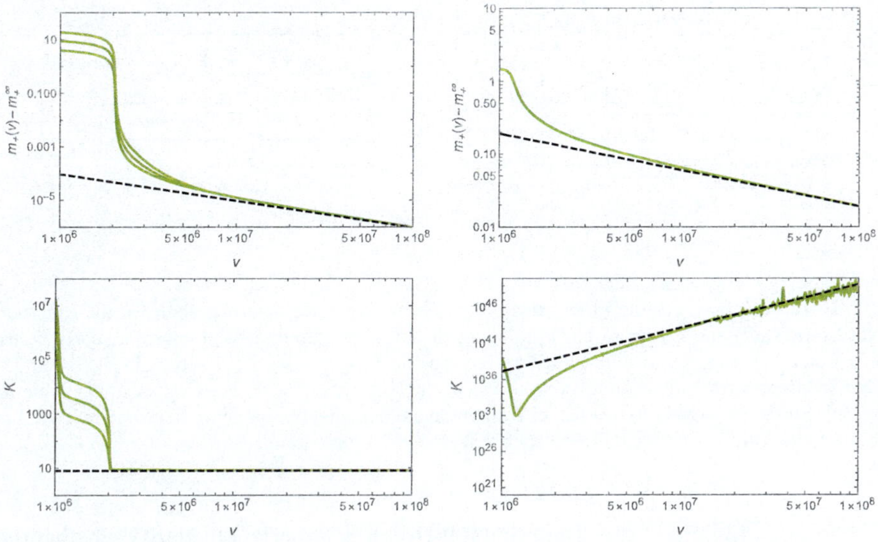

Fig. 10.9 Top: Illustration of the mass function $m_+(v)$ inside the shell for a Reissner–Nordström (left) and Hayward black hole (right) undergoing mass-loss due to Hawking radiation. Initial conditions are imposed at late times. All green lines are attracted to the late-time attractor where $m_+(v)$ approaches the constant $m_+^\infty \equiv \lim_{v \to \infty} m_+(v)$. Botton: Illustration of the Kretschmann scalar evaluated at the position of the shell. For the Reissner–Nordström attractor $K|_\Sigma$ remains finite. In the Hayward case, it grows polynomially in v. The model parameters Q and l have been chosen such that $m_{cr} = 1$

late-time attractor is illustrated in Fig. 10.9, top row. The new attractor behavior then also propagates into the Kretschmann scalar evaluated at the position of the shell

$$\begin{aligned} \text{RN}: \quad & K|_\Sigma \simeq \frac{8}{m_{cr}^4}, \\ \text{H}: \quad & K|_\Sigma \simeq \frac{59049\, v^6}{4096\, m_{cr}^{10}}. \end{aligned} \tag{10.79}$$

The approach of numerical solutions to this attractor is illustrated in the bottom row of Fig. 10.9. The remarkable feature of this result is that *there is no curvature singularity building up in the Reissner–Nordström case*. The mechanism underlying the polynomial growths in the Hayward case is identical to the one observed in the static case. A consistent solution for $m_+(v)$ requires the vanishing of the first factor in $p(m_+)$ which then leads to cancellations among leading terms in the denominator of the curvature scalar. As a result, one again experiences a power-law growth of $K|_\Sigma$.

An important prerequisite for solutions following the attractor (10.78) is that the initial conditions for the shell are imposed at sufficiently late times, close to the regime where the black hole has almost reached the final phase of its evaporation process,

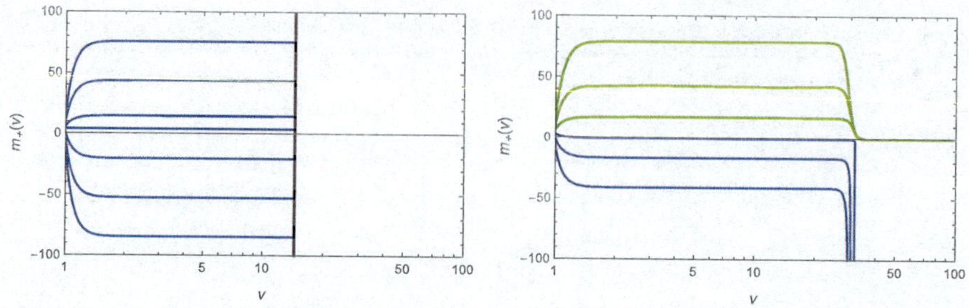

Fig. 10.10 Solutions for the mass function m_+ in the inner sector of the shell for the Reissner–Nordström (left) and Hayward black hole (right). Initial conditions are imposed at "early times" $v = 1$ and the background is provided by the most left trajectories $R(v)$ shown in Fig. 10.8. The blue solutions terminate at finite value of v. For the Hayward case there exists a critical value for $m_+(v)$. Solutions starting above this critical value are depicted by the green lines and extended up to $v = \infty$. Again Q and l are chosen such that $m_{cr} = 1$

cf. Fig. 10.3. In light of the discussion [69, 70], it is important to clarify whether this attractor can also be reached from initial conditions imposed at early times, $v = 1$ say. While the corresponding analysis is most likely outside of the validity range of the Ori-model (the Price tail and the approximation of the perturbation by a thin shell being reasonable at asymptotically late times only), we give a tentative answer in Fig. 10.10. This reveals that the static and dynamical background models share the same early-time behavior: for the Reissner–Nordström black hole all solutions again terminate at finite v. For the Hayward case there is again a range of initial conditions where the solutions connect to the asympotic late-time attractors. These are highlighted by the green lines in the right diagram of Fig. 10.10. The existence of these solutions can again be traced back to the quadratic nature of $p(m_+)$ which induces a plateau for m_+ even before one enters into the regime where $F(v)$ decays.

In order to conclude our discussion, it is important to highlight the relevance of the attractor (10.78) with respect to the geodesic completeness of spacetime. In this context, we stress that the asymptotic geometry is an extremal black hole. Investigating the geodesics of radially free-falling massive observers in such a background one finds that the curvature singularity building up at $v = \infty$ can no longer be reached in a finite amount of the observer's proper time. Hence these observers will not encounter the singularity and questions related to its strength may actually become academic. Whether this result extends to all observers and the full dynamical background is currently still an open question. Nevertheless, the exposition in this section shows that the dynamics of the black hole mass function induced by the emission of Hawking radiation is a crucial element when analyzing the late-time stability of regular black holes.

10.6 Conclusions

Static black hole solutions in general relativity are characterized by curvature singularities hidden behind an event horizon. The singularities are often taken as a signal for the breakdown of the classical theory which should be removed by quantum (gravity) effects. A common strategy anticipating such an effect replaces the singular part of the black hole spacetime by a regular patch of de Sitter space. In this way one arrives at regular black holes satisfying the limited curvature hypothesis [9]. Examples include Bardeen-type black holes [5, 39], the Hayward geometry [10], RG-improved black holes [11] and Planck stars [22].

A direct consequence of the de Sitter core is that asymptotically flat regular black holes must have (at least) two horizons, an outer event horizon and an inner Cauchy horizon. Thus, in their simplest incarnation they have the same horizon structure as a charged Reissner–Nordström black hole. The appearance of the Cauchy horizon entails drastic consequences for the black hole evaporation process due to the emission of Hawking radiation: instead of evaporating completely within a finite time-span, the final state of a spherically symmetric regular black hole is a cold, regular remnant corresponding to an extremal black hole.

The similarity to the Reissner–Nordström geometry suggests that regular black holes may suffer from a dynamical instability, the so-called mass-inflation effect. In brief the effect states that a tiny perturbation crossing the event horizon and impacting on the Cauchy horizon induces a curvature singularity in its interior. Extrapolating this effect from a static, non-extremal Reissner–Nordström black hole to regular black holes then suggests that the instability can reintroduce spacetime singularities dynamically.

The analysis of the mass-inflation effect based on the Ori-model reveals that this analogy comes with significant limitations though [34, 35]. At the level of static, non-extremal black holes the equation controlling the dynamics of the mass-inflation effect *is structurally different* for the Reissner–Nordström geometry and regular black holes of the Hayward and RG-improved type. As a consequence, the latter admit solutions where the mass-function remains constant and the curvature in the inner sector of the shell grows polynomial in the ingoing Eddington-Finkelstein coordinate v only.[5]

[5] This conclusion has been challenged in [69, 70] (also see [60] for an earlier analysis based on the DTR-relations), claiming that the onset of the attractor behavior is preceded by a "fatal" phase of exponential growth. This conclusion is flawed for three reasons though. Firstly, it builds on an analysis of the system at "early times" outside the region of validity of the underlying assumptions: Price's law holds at asymptotically late times only where the optical geometric limit is valid and one is allowed to neglect the "finite size" effects of the inner potential barrier. In a realistic collapse the early-time dynamics is significantly more complicated because of the presence of the flux from the collapsing star. Secondly, closing eyes to this difficulty and extrapolating the model to early times, Fig. 10.7 establishes that the analysis of [69, 70] is incomplete. There are initial conditions which actually reach the salient late-time attractor also from early times. Determining the precise initial data requires the analysis of the full dynamical process and is beyond the scope of a simplified model whose fundamental limitation is the assumption that the Cauchy Horizon always exist at

The static analysis is modified significantly once the mass-loss due to Hawking radiation is taken into account [35]. This leads to two novel types late-time attractors where the curvature either grows polynomially in v (Hayward, RG-improved) or remains finite (Reissner–Nordström, Bardeen). Moreover, the fact that the final state is an extremal black hole suggests that no observer can actually reach these singularities in a finite proper time. These features are important theoretical prerequisits when trying to establish regular black holes as valid alternatives to the black holes from general relativity.

The existence of late-time attractors taming or even expelling the mass-inflation effect for regular black holes is highly encouraging. This raises the crucial question whether the full dynamics actually reaches these salient regimes. Settling this question may require a full-fledged numerical analysis beyond the analytic models describing the mass-inflation effect at late times. Making this connection will again be an important step towards establishing regular black holes as valid alternatives to the black holes described by general relativity.

Acknowledgements We thank our collaborators N. Alkhofer, J. Daas, M. Galis, G. d'Odorico, I. van der Pas, A. Platania, A. Khosravi, B. Koch,, S. Silveravalle, F. Vidotto, M. Wondrak, and, foremost, M. Reuter for many inspiring discussions developing our understanding of spacetime singularities, regular black holes, and the stability properties of Cauchy horizons. The work of F.S. is supported by the Dutch Black Hole Consortium.

References

1. R. Penrose, Phys. Rev. Lett. **14**, 57 (1965). https://doi.org/10.1103/PhysRevLett.14.57
2. R. Penrose, Riv. Nuovo Cim. **1**, 252 (1969). https://doi.org/10.1023/A:1016578408204
3. N.D. Birrell, P.C.W. Davies, *Quantum Fields in Curved Space*. Cambridge Monographs on Mathematical Physics (Cambridge University Press, Cambridge, UK, 1984). https://doi.org/10.1017/CBO9780511622632
4. E. Poisson, W. Israel, Class. Quant. Grav. **5**, L201 (1988). https://doi.org/10.1088/0264-9381/5/12/002
5. J.M. Bardeen, in *Proceedings of the International Conference GR5 Tbilisi*, vol. 174 (1968)
6. M.A. Markov, ZhETF Pisma Redaktsiiu **36**, 214 (1982)
7. I. Dymnikova, Gen. Rel. Grav. **24**, 235 (1992). https://doi.org/10.1007/BF00760226
8. V.P. Frolov, M. Markov, V.F. Mukhanov, Phys. Rev. D **41**, 383 (1990). https://doi.org/10.1103/PhysRevD.41.383
9. S. Ansoldi, Spherical black holes with regular center: a Review of existing models including a recent realization with Gaussian sources
10. S.A. Hayward, Phys. Rev. Lett. **96**, 031103 (2006). https://doi.org/10.1103/PhysRevLett.96.031103

$v = \infty$. Thirdly, a divergence in the asymptotic mass m_+ may just not be fatal for the geometry as the Misner-Sharp mass controlling the curvature of spacetime may remain finite. This is clear from Eq. (10.65) which shows that the limit $m \to \infty$ has different physical implications for the Reissner–Nordström and Hayward geometry. Most remarkably, very similar conclusions have already been reached when analyzing the mass-inflation effect for loop black holes [59]. For a recent discussion of this viewpoint by the authors of [69, 70] also see [71].

11. A. Bonanno, M. Reuter, Phys. Rev. D **62**, 043008 (2000). https://doi.org/10.1103/PhysRevD.62.043008
12. A. Bonanno, M. Reuter, Phys. Rev. D **73**, 083005 (2006). https://doi.org/10.1103/PhysRevD.73.083005
13. Y.F. Cai, D.A. Easson, JCAP **09**, 002 (2010). https://doi.org/10.1088/1475-7516/2010/09/002
14. M. Reuter, E. Tuiran, Phys. Rev. D **83**, 044041 (2011). https://doi.org/10.1103/PhysRevD.83.044041
15. B. Koch, F. Saueressig, Class. Quant. Grav. **31**, 015006 (2014). https://doi.org/10.1088/0264-9381/31/1/015006
16. B. Koch, F. Saueressig, Int. J. Mod. Phys. A **29**(8), 1430011 (2014). https://doi.org/10.1142/S0217751X14300117
17. R. Torres, Phys. Lett. B **733**, 21 (2014). https://doi.org/10.1016/j.physletb.2014.04.010
18. A. Bonanno, B. Koch, A. Platania, Found. Phys. **48**(10), 1393 (2018). https://doi.org/10.1007/s10701-018-0195-7
19. J.M. Pawlowski, D. Stock, Phys. Rev. D **98**(10), 106008 (2018). https://doi.org/10.1103/PhysRevD.98.106008
20. A. Adeifeoba, A. Eichhorn, A. Platania, Class. Quant. Grav. **35**(22), 225007 (2018). https://doi.org/10.1088/1361-6382/aae6ef
21. A. Platania, Eur. Phys. J. C **79**(6), 470 (2019). https://doi.org/10.1140/epjc/s10052-019-6990-2
22. C. Rovelli, F. Vidotto, Int. J. Mod. Phys. D **23**(12), 1442026 (2014). https://doi.org/10.1142/S0218271814420267
23. A. Barrau, C. Rovelli, Phys. Lett. B **739**, 405 (2014). https://doi.org/10.1016/j.physletb.2014.11.020
24. T. De Lorenzo, C. Pacilio, C. Rovelli, S. Speziale, Gen. Rel. Grav. **47**(4), 41 (2015). https://doi.org/10.1007/s10714-015-1882-8
25. M. Christodoulou, C. Rovelli, S. Speziale, I. Vilensky, Phys. Rev. D **94**(8), 084035 (2016). https://doi.org/10.1103/PhysRevD.94.084035
26. A. Perez, Rept. Prog. Phys. **80**(12), 126901 (2017). https://doi.org/10.1088/1361-6633/aa7e14
27. R. Penrose, in *Battelle Rencontres - 1967 Lectures in Mathematics and Physics*, Seattle, WA, USA, 16–31 July 1967 (1968), pp. 121–235
28. W.A. Hiscock, Physics Letters A **83**(3), 110 (1981). https://doi.org/10.1016/0375-9601(81)90508-9
29. E. Poisson, W. Israel, Phys. Rev. Lett. **63**, 1663 (1989). https://doi.org/10.1103/PhysRevLett.63.1663
30. A. Ori, Phys. Rev. Lett. **67**, 789 (1991). https://doi.org/10.1103/PhysRevLett.67.789
31. L.M. Burko, A. Ori, Phys. Rev. Lett. **74**, 1064 (1995). https://doi.org/10.1103/PhysRevLett.74.1064
32. R.H. Price, Phys. Rev. D **5**, 2419 (1972). https://doi.org/10.1103/PhysRevD.5.2419
33. R. Balbinot, P.R. Brady, W. Israel, E. Poisson, Phys. Lett. A **161**(3), 223 (1991). https://doi.org/10.1016/0375-9601(91)90007-U
34. A. Bonanno, A.P. Khosravi, F. Saueressig, Phys. Rev. D **103**(12), 124027 (2021). https://doi.org/10.1103/PhysRevD.103.124027
35. A. Bonanno, A.P. Khosravi, F. Saueressig, Phys. Rev. D **107**(2), 024005 (2023). https://doi.org/10.1103/PhysRevD.107.024005
36. V.P. Frolov, Phys. Rev. D **94**(10), 104056 (2016). https://doi.org/10.1103/PhysRevD.94.104056
37. J. Boos, (2021)
38. L. Modesto, Int. J. Theor. Phys. **49**, 1649 (2010). https://doi.org/10.1007/s10773-010-0346-x
39. E. Ayon-Beato, A. Garcia, Phys. Lett. B **493**, 149 (2000). https://doi.org/10.1016/S0370-2693(00)01125-4
40. I. Dymnikova, Class. Quant. Grav. **19**, 725 (2002). https://doi.org/10.1088/0264-9381/19/4/306
41. F. Saueressig, N. Alkofer, G. D'Odorico, F. Vidotto, PoS **FFP14**, 174 (2016). https://doi.org/10.22323/1.224.0174

42. M. Markov, Pis'ma v ZHETF **36**, 214 (1982)
43. V.P. Frolov, M. Markov, V.F. Mukhanov, Phys. Lett. B **216**(3–4), 272 (1989)
44. V.P. Frolov, M. Markov, V.F. Mukhanov, Phys. Rev. D **41**(2), 383 (1990)
45. R. Percacci, *An Introduction to Covariant Quantum Gravity and Asymptotic Safety*, *100 Years of General Relativity*, vol. 3 (World Scientific, 2017). https://doi.org/10.1142/10369
46. M. Reuter, F. Saueressig, *Quantum Gravity and the Functional Renormalization Group: The Road Towards Asymptotic Safety* (Cambridge University Press, 2019)
47. M. Reuter, Phys. Rev. D **57**, 971 (1998). https://doi.org/10.1103/PhysRevD.57.971
48. O. Lauscher, M. Reuter, Phys. Rev. D **65**, 025013 (2002). https://doi.org/10.1103/PhysRevD.65.025013
49. M. Reuter, F. Saueressig, Phys. Rev. D **65**, 065016 (2002). https://doi.org/10.1103/PhysRevD.65.065016
50. S.W. Hawking, Commun. Math. Phys. **43**, 199 (1975). https://doi.org/10.1007/BF02345020. [Erratum: Commun. Math. Phys. **46**, 206 (1976)]
51. B.L. Hu, T.A. Jacobson, M.P. Ryan, C.V. Vishveshwara (eds.). *Directions in general relativity. Proceedings, 1993 International Symposium, College Park, USA, May 27–29, 1993. Vol. 1: Papers in honor of Charles Misner. Vol. 2: Papers in honor of Dieter Brill* (1995)
52. J.E. Chase, Nuovo Cimento Serie B **67**(2), 136 (1970). https://doi.org/10.1007/BF02711502
53. S.K. Blau, Phys. Rev. D **39**, 2901 (1989). https://doi.org/10.1103/PhysRevD.39.2901
54. T. Dray, G. 't Hooft, Commun. Math. Phys. **99**, 613 (1985). https://doi.org/10.1007/BF01215912
55. I.H. Redmount, Progr. Theor. Phys. **73**(6), 1401 (1985). https://doi.org/10.1143/PTP.73.1401
56. C. Barrabes, W. Israel, Phys. Rev. D **43**, 1129 (1991). https://doi.org/10.1103/PhysRevD.43.1129
57. W. Israel, Nuovo Cim. B **44S10**, 1 (1966). https://doi.org/10.1007/BF02710419. [Erratum: Nuovo Cim. B **48**, 463 (1967)]
58. A.J.S. Hamilton, P.P. Avelino, Phys. Rept. **495**, 1 (2010). https://doi.org/10.1016/j.physrep.2010.06.002
59. E.G. Brown, R.B. Mann, L. Modesto, Phys. Rev. D **84**, 104041 (2011). https://doi.org/10.1103/PhysRevD.84.104041
60. R. Carballo-Rubio, F. Di Filippo, S. Liberati, C. Pacilio, M. Visser, JHEP **07**, 023 (2018). https://doi.org/10.1007/JHEP07(2018)023
61. M. Bertipagani, M. Rinaldi, L. Sebastiani, S. Zerbini, Phys. Dark Univ. **33**, 100853 (2021). https://doi.org/10.1016/j.dark.2021.100853
62. E. Poisson, W. Israel, Phys. Rev. D **41**, 1796 (1990). https://doi.org/10.1103/PhysRevD.41.1796
63. J. Ziprick, G. Kunstatter, Phys. Rev. D **82**, 044031 (2010). https://doi.org/10.1103/PhysRevD.82.044031
64. A. Biasi, O. Evnin, S. Sypsas, (2022)
65. A. Bonanno, S. Droz, W. Israel, S.M. Morsink, Proc. Roy. Soc. Lond. A **450**, 553 (1995). https://doi.org/10.1098/rspa.1995.0100
66. L.M. Burko, Phys. Rev. D **60**, 104033 (1999). https://doi.org/10.1103/PhysRevD.60.104033
67. F.J. Tipler, Phys. Lett. A **64**, 8 (1977). https://doi.org/10.1016/0375-9601(77)90508-4
68. A. Krolak, J. Math. Phys. **28**(11), 2685 (1987). https://doi.org/10.1063/1.527829
69. R. Carballo-Rubio, F. Di Filippo, S. Liberati, C. Pacilio, M. Visser, JHEP **05**, 132 (2021). https://doi.org/10.1007/JHEP05(2021)132
70. F. Di Filippo, R. Carballo-Rubio, S. Liberati, C. Pacilio, M. Visser, Universe **8**(4), 204 (2022). https://doi.org/10.3390/universe8040204
71. R. Carballo-Rubio, F. Di Filippo, S. Liberati, C. Pacilio, M. Visser, arXiv:2212.07458 [gr-qc]

Chapter 11
Regular Rotating Black Holes

Ramón Torres

11.1 A Glance at Classical Rotating Black Holes

Most astrophysically significant bodies are rotating. If a rotating body collapses, the rate of rotation will speed up, maintaining constant angular momentum. Through a rather complicated process, the body could finally generate a black hole which would be a *rotating black hole* (*RBH*). From a classical point of view (*no-hair conjecture* [64]) the resulting spacetime will be described by a Kerr solution (or a Kerr-Newman solution, in the charged case). This implies that an idealized classical model of a lonely rotating body eventually generates an axially symmetric, stationary and asymptotically flat spacetime with certain horizons, a specific causal structure and a curvature singularity.

In order to compare these characteristics with those of regular RBH, let us now briefly summarize them for the classical uncharged RBH solution. (The reader can consult, for example, [45, 64] and references therein for more information). In Boyer-Lidquist (B-L) coordinates $\{t, r, \theta, \phi\}$, the Kerr metric takes the form

$$ds^2 = -\frac{\Delta}{\Sigma}(dt - a\sin^2\theta d\phi)^2 + \frac{\Sigma}{\Delta}dr^2 + \Sigma d\theta^2 + \frac{\sin^2\theta}{\Sigma}(adt - (r^2 + a^2)d\phi)^2, \tag{11.1}$$

where

$$\Sigma = r^2 + a^2\cos^2\theta, \qquad \Delta = r^2 - 2mr + a^2,$$

[1] In case the RBH is also charged, then it is described by using the Kerr-Newman solution in which m should be replaced by $m - e^2/(2r)$, where e is the total charge of the RBH.

R. Torres (✉)
Department of Physics, Universitat Politècnica de Catalunya, Barcelona, Spain
e-mail: ramon.torres-herrera@upc.edu

C. Bambi (ed.), *Regular Black Holes*, Springer Series in Astrophysics and Cosmology,
https://doi.org/10.1007/978-981-99-1596-5_11

m is the black hole mass[1] and a is a *rotation parameter* that measures the (Komar) angular momentum per unit of mass [64]. The spacetime is type D whenever $m \neq 0$.

If $m \neq 0$ there is a curvature singularity at $(r = 0,\ \theta = \pi/2)$, as can be shown by the divergence of the curvature invariant $R_{\alpha\beta\gamma\delta}R^{\alpha\beta\gamma\delta}$. Remarkably, for $a \neq 0$ and $\theta \neq \pi/2$, a surface defined by t =constant and $r = 0$ is singularity-free and has metric

$$ds_2^2 = a^2 \cos^2\theta d\theta^2 + a^2 \sin^2\theta d\phi^2 = dx^2 + dy^2,$$

where the coordinate change $x \equiv a \sin\theta \cos\phi$, $y \equiv a \sin\theta \sin\phi$ has been made to make explicit that the surface is flat. The curvature singularity corresponds to the *ring* $x^2 + y^2 = a^2$, while the flat surface corresponds to $x^2 + y^2 < a^2$. As it is customary, we will call this flat surface *the disk*.

In this way, the curves that reach $r = 0$ with $\theta \neq \pi/2$ are reaching a regular point in the disk. In order to continue the curves it is usually argued that an analytic extension of the spacetime has to be obtained through $r = 0$. The procedure requires letting the coordinate r to take negative values [49]. The $r < 0$ extended spacetime can be seen as a negative mass spacetime. Causality violations occur in the extended spacetime [29].

The metric has a coordinate singularity at $\Delta = 0$, which can be easily removed by a coordinate change [22]. At $\Delta = 0$ the hypersurface r =constant becomes light-like and no observer can remain at the specific value for r, thus the hypersurface is called a *null horizon*. In the Kerr case, if $m^2 > a^2$ there are two roots: $r_{\pm}^{Kerr} = m \pm \sqrt{m^2 - a^2}$, where the null horizon r_{+}^{Kerr} is an event horizon, while the null horizon r_{-}^{Kerr} is a Cauchy horizon. The limiting case $m^2 = a^2$ has a degenerate null horizon and the spacetime is called the *extreme* Kerr black hole. If $m^2 < a^2$ there are no roots for $\Delta = 0$ and the curvature singularity is naked. This is the so called *hyperextreme* case.

11.2 Kerr-like Rotating Black Holes

Several authors have suggested that the existence of singularities in the solutions of General Relativity has to be considered as a weakness of the theory rather than as a real physical prediction. The problem of obtaining singularity-free models for black holes was first approached for spherically symmetric black holes. In this context, some authors introduced non-standard energy-momentum tensors mainly acting in the core of the black hole (see, for example, [10, 11, 13, 17]). However, most authors expect that the inclusion of quantum theory in the description of black holes could avoid the existence of their singularities (see, for example, [9, 21, 39, 40, 47, 50, 77] and references therein).

We do not yet have a mature and reliable candidate for a quantum theory of gravity so that it is difficult to accurately describe even non-rotating (quantum) black holes. On the other hand, a glimpse back into classical black hole history indicates that finding an accurate description of a (quantum) RBH could be even much more difficult: Kerr solution was only discovered following 48 years of struggle after the

Einstein field equations were first developed. There are some works on approximated solutions, only valid in the slow-rotation limit [66, 84]. Unfortunately, they are not adequate enough to be used in astrophysical observations. In this way, it is necessary to try phenomenological approaches to check for possible models of regular RBHs and their implications, including the possibility of observable astrophysical predictions.

Even if a regular RBH model comes from an approach to Quantum Gravity Theory, we will assume in this chapter that it can be reasonably well described by a manifold endowed with its corresponding metric. Nevertheless, it should be taken into account that, in the absence of a full Quantum Gravity Theory, probably one can only guarantee this to be a good description of the RBH up to the high curvature planckian regime.

Recently, there have appeared different proposals for *regular* RBHs spacetimes with their corresponding metrics (see Sect. 11.9). While they have been obtained by different approaches, most of them share a common *Kerr-like form*. The general metric corresponding to this kind of RBH, was found by Gürses-Gürsey [46] as a particular rotating case of the algebraically special Kerr-Schild metric:

$$ds^2 = (\eta_{\alpha\beta} + 2\,H k_\alpha k_\beta) dx^\alpha dx^\beta, \tag{11.2}$$

where η is the metric of Minkowski, H is a scalar function and $\mathbf{k}$ is a light-like vector both with respect to the spacetime metric and to Minkowski's metric. Specifically, in Kerr-Schild coordinates $\{\tilde{t}, x, y, z\}$ the Gürses-Gürsey metric (11.2) corresponds with the choices

$$H = \frac{\mathcal{M}(r) r^3}{r^4 + a^2 z^2}$$

and

$$k_\alpha dx^\alpha = -\frac{r(xdx + ydy) - a(xdy - ydx)}{r^2 + a^2} - \frac{zdz}{r} - d\tilde{t},$$

where r is a function of the Kerr-Schild coordinates implicitly defined by

$$r^4 - r^2(x^2 + y^2 + z^2 - a^2) - a^2 z^2 = 0, \tag{11.3}$$

$\mathcal{M}(r)$ is known as the *mass function* and the constant a is a rotation parameter.

This metric can be written in Boyer-Lindquist-like coordinates by using the coordinate change defined by

$$\begin{aligned} x + iy &= (r + ia) \sin\theta \exp\left[i \int (d\phi + \frac{a}{\Delta} dr) \right] \\ z &= r \cos\theta \\ \tilde{t} &= t + \int \frac{r^2 + a^2}{\Delta} dr - r. \end{aligned}$$

where now $\Delta = r^2 - 2\mathcal{M}(r) r + a^2$. The resulting metric takes the form

$$ds^2 = -\frac{\Delta}{\Sigma}(dt - a\sin^2\theta d\phi)^2 + \frac{\Sigma}{\Delta}dr^2 + \Sigma d\theta^2 + \frac{\sin^2\theta}{\Sigma}(adt - (r^2+a^2)d\phi)^2, \tag{11.4}$$

where, again, $\Sigma = r^2 + a^2\cos^2\theta$. Note that this metric reduces to Kerr's solution in B-L coordinates if $\mathcal{M}(r) = m$=constant and that it reduces to the (charged) Kerr-Newman solution if $\mathcal{M}(r) = m - e^2/(2r)$, where e is the charge.

In order to analyze the general properties of the RBH spacetime we will use the following null tetrad-frame:

$$\mathbf{l} = \frac{1}{\Delta}\left((r^2+a^2)\frac{\partial}{\partial t} + \Delta\frac{\partial}{\partial r} + a\frac{\partial}{\partial\phi}\right),$$
$$\mathbf{k} = \frac{1}{2\rho^2}\left((r^2+a^2)\frac{\partial}{\partial t} - \Delta\frac{\partial}{\partial r} + a\frac{\partial}{\partial\phi}\right),$$
$$\mathbf{m} = \frac{1}{\sqrt{2}\varrho}\left(ia\sin\theta\frac{\partial}{\partial t} + \frac{\partial}{\partial\theta} + i\csc\theta\frac{\partial}{\partial\phi}\right),$$
$$\bar{\mathbf{m}} = \frac{1}{\sqrt{2}\bar{\varrho}}\left(-ia\sin\theta\frac{\partial}{\partial t} + \frac{\partial}{\partial\theta} - i\csc\theta\frac{\partial}{\partial\phi}\right),$$

where $\varrho \equiv r + ia\cos\theta$, $\bar{\varrho} \equiv r - ia\cos\theta$ and the tetrad is normalized as follows $\mathbf{l}^2 = \mathbf{k}^2 = \mathbf{m}^2 = \bar{\mathbf{m}}^2 = 0$ and $\mathbf{l}\cdot\mathbf{k} = -1 = -\mathbf{m}\cdot\bar{\mathbf{m}}$.

Theorem 11.1 ([79]) *The RBH metric (11.4) is Petrov type D and the two double principal null directions are* **l** *and* **k**.

We can also define a real orthonormal basis $\{\mathbf{t}, \mathbf{x}, \mathbf{y}, \mathbf{z}\}$ formed by a timelike vector $\mathbf{t} \equiv (\mathbf{l}+\mathbf{k})/\sqrt{2}$ and three spacelike vectors: $\mathbf{z} \equiv (\mathbf{l}-\mathbf{k})/\sqrt{2}$, $\mathbf{x} = (\mathbf{m}+\bar{\mathbf{m}})/\sqrt{2}$ and $\mathbf{y} = (\mathbf{m}-\bar{\mathbf{m}})i/\sqrt{2}$. Then, $\mathbf{t}$ and $\mathbf{z}$ are two eigenvectors of the Ricci tensor with eigenvalue [79]

$$\lambda_1 = \frac{2a^2\cos^2\theta\mathcal{M}' + r\Sigma\mathcal{M}''}{\Sigma^2}. \tag{11.5}$$

$\mathbf{x}$ and $\mathbf{y}$ are two eigenvectors of the Ricci tensor with eigenvalue

$$\lambda_2 = \frac{2r^2\mathcal{M}'}{\Sigma^2}. \tag{11.6}$$

In this way, the Ricci tensor can be written as

$$R_{\mu\nu} = \lambda_1(-t_\mu t_\nu + z_\mu z_\nu) + \lambda_2(x_\mu x_\nu + y_\mu y_\nu), \tag{11.7}$$

what shows the following

Theorem 11.2 ([79]) *The metric (11.4) with $\mathcal{M}$ ≠constant is Segre type [(1,1) (1 1)].*

Note that the $\mathcal{M}$ ≠constant case is precisely the case we are interested in for our regular RBHs, since the $\mathcal{M}$ =constant (i.e., Kerr's case) is singular.

11.3 Regularity in Kerr-like Rotating Black Holes

In order for the model of a RBH to be regular it should be devoid of curvature singularities. Let us now specifically analyze the absence of *scalar* curvature singularities. We say that there is a *scalar curvature singularity* in the spacetime if any scalar invariant polynomial in the Riemann tensor diverges when approaching it along any incomplete curve.

It is well-known [83] that an arbitrary spacetime possesses at most 14 s order algebraically independent invariants. The finiteness of *all* the invariants is a necessary and sufficient condition for the absence of scalar curvature singularities.

A minimum set of reliable independent invariants for the RBH spacetime exists. This can be shown thanks to the following result by Zakhary and McIntosh [85]

Theorem 11.3 *The algebraically complete set of second order invariants for a Petrov type D spacetime and Segre type [(1,1) (1 1)] is* $\{\mathscr{R}, I, I_6, K\}$.

Apart form the well-known curvature scalar $\mathscr{R}$, the rest of the invariants are defined as[2]

$$I_6 \equiv \frac{1}{12} S_\alpha{}^\beta S_\beta{}^\alpha ,$$
$$I \equiv \frac{1}{24} \bar{C}_{\alpha\beta\gamma\delta} \bar{C}^{\alpha\beta\gamma\delta} ,$$
$$K \equiv \frac{1}{4} \bar{C}_{\alpha\gamma\delta\beta} S^{\gamma\delta} S^{\alpha\beta} ,$$

where $S_\alpha{}^\beta \equiv R_\alpha{}^\beta - \delta_\alpha{}^\beta \mathscr{R}/4$ and $\bar{C}_{\alpha\beta\gamma\delta} \equiv (C_{\alpha\beta\gamma\delta} + i \ * C_{\alpha\beta\gamma\delta})/2$ is the complex conjugate of the selfdual Weyl tensor being $*C_{\alpha\beta\gamma\delta} \equiv \epsilon_{\alpha\beta\mu\nu} C^{\mu\nu}{}_{\gamma\delta}/2$ the dual of the Weyl tensor. Note that $\mathscr{R}$ and I_6 are real, while I and K are complex. Therefore for this type of spacetimes there are only 6 independent real scalars.

It trivially follows from our previous propositions

Corollary 11.1 ([79]) *The algebraically complete set of second order invariants for the RBH metric (11.4) is* $\{\mathscr{R}, I, I_6, K\}$.

Similarly to Kerr's case, a straightforward inspection of the metric (11.4) tell us that it is singular if there are values of r such that $\Delta = 0$ and if $\Sigma = 0$. However, $\Delta = 0$ is not a scalar curvature singularity since the curvature scalars do not diverge there. for the values of r ($\neq 0$) where $\Delta = 0$. It is simply a coordinate singularity that can be removed through a coordinate change. (See Sect. 11.6).

Scalar curvature singularities do may appear if $\Sigma = 0$ or, in other words, in ($r = 0, \theta = \pi/2$). (We already confirmed this possibility in Sect. 11.1 for the particular case of Kerr's solution). Now, by explicitly computing the complete set of scalars in our case, one directly gets a necessary and sufficient condition for the absence of scalar curvature singularities:

[2] Here the invariants are written in tensorial form. See [85] for their spinorial form.

Theorem 11.4 ([79]) *Assuming a RBH metric (11.4) possessing a C^3 function $\mathscr{M}(r)$, all its second order curvature invariants will be finite at* $(r = 0, \theta = \pi/2)$ *if, and only if,*

$$\mathscr{M}(0) = \mathscr{M}'(0) = \mathscr{M}''(0) = 0. \tag{11.8}$$

The absence of curvature singularities is a necessary condition in order to have a regular RBH. The theorem allows to control the specific case of *scalar curvature singularities*, which arguably are the most serious type of curvature singularities. However, since scalar polynomials do not fully characterize the Riemann tensor, it does not cover the possibility of the existence of curvature singularities with respect to a parallelly propagated basis (*p.p. curvature singularity*) [49]. This possibility has not yet been fully analyzed in the literature.

11.4 Violation of the Energy Conditions

The energy conditions were first developed in the framework of Einstein's General Relativity. These are conditions imposed on the energy-momentum tensor of the spacetime as a means of ensuring plausible matter-energy contents [49]. Even if here we are not confined to General Relativity we can take profit of the energy conditions by considering the existence of an *effective energy-momentum tensor* defined through

$$T_{\mu\nu} \equiv R_{\mu\nu} - \frac{1}{2}\mathscr{R} g_{\mu\nu}.$$

In our more general context, it is usually argued that it seems reasonable to demand the spacetime describing a realistic isolated RBH to fulfill the standard energy conditions in asymptotically flat regions (thus, imitating the classical RBH solutions at large distances/low curvatures). Nevertheless, probably it would be more accurate to say that one should expect extremely small violations of the energy conditions in the asymptotically flat regions. This is due to the fact that, as pointed out by Donoghue [31], the standard perturbative quantization of Einstein gravity leads to a well-defined, finite prediction for the leading large distance correction to Newton's potential. Specifically, it is shown that quantum effects produce deviations in the gravitational field for spherically symmetric fields of the order Gml_p^2/r^3 whenever $r \gg 2m$, being l_p Planck's length. This implies an extremely small reduction of the classically expected (negative) gravitational field. In the weak field approximation this leads to an effective energy-momentum tensor that (*slightly*) violates the dominant energy conditions [21, 75].

Let us now treat the behaviour of the energy conditions in the region around $r = 0$ for regular RBH. If we take the expression obtained for the Ricci tensor (11.7) one can explicit $\mathbf{T}$ for a RBH as

$$T_{\mu\nu} = -\lambda_2(-t_\mu t_\nu + z_\mu z_\nu) - \lambda_1(x_\mu x_\nu + y_\mu y_\nu).$$

Since **T** diagonalizes in the orthonormal basis $\{\mathbf{t}, \mathbf{x}, \mathbf{y}, \mathbf{z}\}$, the RBH spacetime possesses an (effective) energy-momentum tensor of type I [49]. The (effective) density being $\mu = \lambda_2$ and the (effective) pressures being $p_x = p_y = -\lambda_1$ and $p_z = -\lambda_2$. The weak energy conditions [49] require $\mu \geq 0$ and $\mu + p_i \geq 0$. In other words, in this case they require

$$\lambda_2 \geq 0 \qquad \text{and} \qquad \lambda_2 - \lambda_1 \geq 0.$$

By using this and expressions (11.5) and (11.6) it is easy to show the following

Theorem 11.5 ([79]) *Assume that a* regular *RBH has a function $\mathcal{M}(r)$ that can be approximated by a Taylor polynomial around $r = 0$, then the weak energy conditions should be violated around $r = 0$.*

Note that for this type I effective energy-momentum the violation of the weak energy condition also implies the violation of the *dominant* and the *strong* energy conditions.

In this way, no model with *normal* matter (matter satisfying the energy conditions) can produce a *regular* RBH of the type (11.4). However, the violation of the WEC around $r = 0$ is not problematic since it is well-known that quantum effects can violate the WEC (Casimir effect). Moreover, singularity theorems require the spacetime to fulfill some energy condition in order to predict the existence of singularities. In this sense, the violation of energy conditions just *helps* to avoid the existence of singularities.[3]

11.5 Extensions Beyond $r = 0$

As stated in Sect. 11.1, for Kerr's solution one could considered the possibility of extending the spacetime through the disk. Now, in order to analyze the general situation for regular RBHs with metric (11.2), let us proceed with an analysis similar to the one usually carried out for the classical RBH case. Consider the metric component

$$g_{tt} = -1 + \frac{2\mathcal{M}(r)r^3}{r^4 + a^2 z^2}. \tag{11.9}$$

Let us imagine and observer crossing $r = 0$ moving in the z *axis* ($x = y = 0$). If we persist in considering r as non-negative, then (11.3) implies that $r = |z|$ along the trajectory of the observer, so that along it

$$g_{tt} = -1 + \frac{2\mathcal{M}(|z|)|z|}{z^2 + a^2}. \tag{11.10}$$

[3] Of course, regularity can also be obtained by violating other assumptions in the singularity theorems.

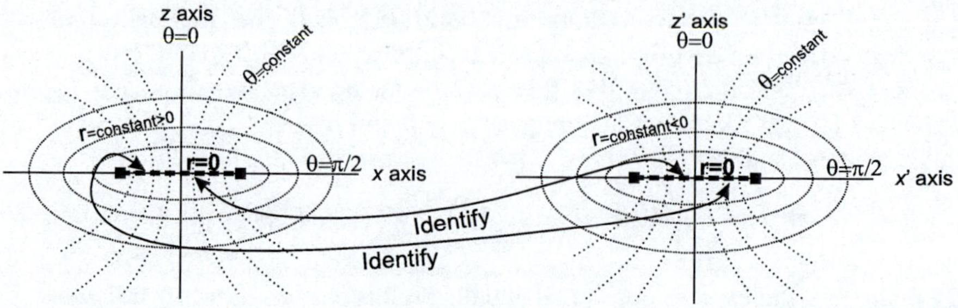

Fig. 11.1 In Kerr's case and approach (**a**), the extension through $r = 0$ is obtained by identifying the top of the surface ($r = 0, t =$constant) in the hypersurface described by coordinates $\{x, y, z\}$ with the bottom of the surface ($r = 0, t =$constant) in the hypersurface described by coordinates $\{x', y', z'\}$, and vice versa. Only the $y = 0$, $y' = 0$ sections of theses hypersurfaces are represented here

The numerator in the fraction indicates that the derivative of this metric component along the axis, as well the Christoffel symbols and the extrinsic curvature of the surface can be discontinuous across the disk depending on the chosen mass function. In particular, this discontinuity occurs if the mass function is constant (Kerr's case).

Historically, the differentiability problems in Kerr's RBH have been approached:

(a) By analytically extending the spacetime through $r = 0$ with negative values for r. This requires considering two spacetimes, one with positive r and another with negative r and properly identifying points in their $r = 0$ surfaces by a standard procedure which is illustrated in Fig. 11.1 (see, for example, [49]).[4]
(b) By considering the discontinuity of the derivatives of the metric components in the disk and, thus, the discontinuities in the second fundamental form, as indicating the presence of a thin shell in the surface [53].

It was soon noticed that by following approach (a), and extending through $r = 0$ with negative values of r, can produce closed causal curves and, thus, causality problems [49]. This will be treated, for the general case of RBH, in Sect. 11.7,

With regard to approach (b), it not only leads to the concentration of mass-energy in a infinitesimally thin surface, but it also needs its matter to move faster than light [48, 54].

At first sight, the situation for general regular RBH looks much better. Assuming that the regular RBH has a mass function $\mathcal{M}(r) \sim r^n$ with $n \geq 3$ around $r = 0$, the metric component along the trajectory (11.10) will not have differentiability problems in $z = 0$ ($\partial_z g_{tt}(z = 0) = 0$) (and, in fact, it will be at least C^n). This suggests that the extension through the disk could not be necessary for regular RBH. In order to show it, one has to go beyond a particular trajectory intersecting the disk and beyond the analysis of a single metric component. Let's start by noticing that while approaching a point in the disk ($x^2 + y^2 < a^2$), according to (11.3), the function r approaches

[4] Note also that this approach has been criticized in [41].

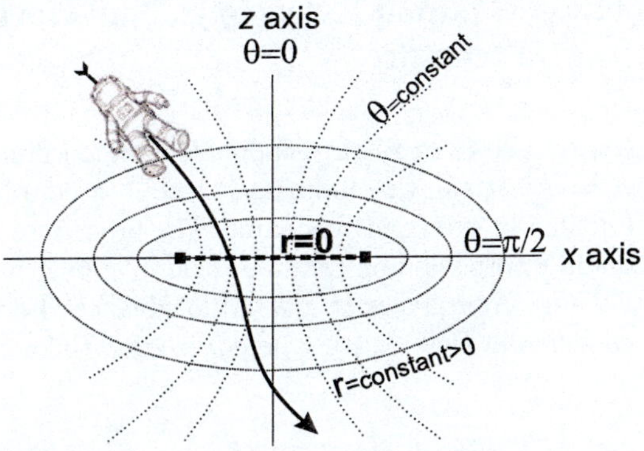

Fig. 11.2 For regular RBH no extension through $r = 0$ is required. An observer crossing the surface $(r = 0, t = \text{constant})$ from positive z to negative z following a non-geodesic time-like curve can stay in its original spacetime. Along the trajectory of the observer r just decreases until reaching the surface $r = 0$ (disk), where it increases again

zero whenever z approaches zero and vice versa. If we insist in having a positive r, we get, solving for r in (11.3), that around $z = 0$

$$r \simeq \frac{|a|}{\sqrt{a^2 - (x^2 + y^2)}}|z|$$

If we introduce this into the metric component (11.9) and considering a mass function $\mathcal{M}(r) \sim r^n$ with $n \geq 3$ around $r = 0$, we see that the metric component takes the form

$$g_{tt} \simeq -1 + \frac{f(x, y)|z|^{n+1}}{g(x, y)z^2 + a^2},$$

where f and g are finite differentiable functions in the disk. In this way, g_{tt} is differentiable at the disk. (In particular, again $(\partial_z g_{tt}(z = 0) = 0)$). The reader can check that the same situation is found for the rest of metric components. Let us only remark that the metric will not be analytic at the disk. Not all metric components will be infinitely differentiable. For example, even if the particular metric component (11.9) and for odd n is C^∞ other metric components like

$$g_{tz} = \frac{2\mathcal{M}(r)r^2 z(ay + xr)}{(a^2 + r^2)(a^2 z^2 + r^4)}$$

are not. Nevertheless, such a degree of differentiability is not required at all[5]. In this way, regular RBH do not have differentiability problems and an extension through the $r = 0$ is not needed[6]. An observer could cross through $r = 0$ while remaining in the $(r \geq 0)$ spacetime (See Fig. 11.2). Furthermore, most of the problems found in Kerr's RBH would be nonexistent.

[5] Usually the metric is required to be at least C^2 [49]. However, many authors consider this degree of differentiability too restrictive.

[6] Let us comment that, even if not mathematically needed, the possibility of extending through $r = 0$ with negative values of r exists, in principle, for all regular RBH.

11.6 Maximal Extensions, Null Horizons and Global Structure

Metric (11.4) in Boyer-Lindquist-like coordinates has a coordinate singularity at $\Delta = 0$ that can be eliminated through a coordinate change in order to obtain the maximally extended spacetime. The procedure is similar to the one usually carried out in Kerr's solution [22]. For example, one can perform a coordinate change from B-L-like coordinates $\{t, r, \theta, \phi\}$ to advanced Eddington-Finkelstein-like or *Kerr-like coordinates* $\{u, r, \theta, \varphi\}$, where u is a light-like coordinate, through[7]

$$u \equiv t + \int \frac{r^2 + a^2}{\Delta} dr \qquad ; \qquad \varphi = \phi + \int \frac{a}{\Delta} dr. \tag{11.11}$$

In these Kerr-like coordinates the metric takes the form

$$ds^2 = -\left(1 - \frac{2\mathscr{M}(r)r}{\Sigma}\right) du^2 + 2dudr + \Sigma d\theta^2 - 2a \sin^2\theta dr d\varphi$$
$$+ \left(r^2 + a^2 + \frac{2a^2\mathscr{M}(r)r \sin^2\theta}{\Sigma}\right) d\varphi^2 - \frac{4\mathscr{M}(r)ra}{\Sigma} \sin^2\theta du d\varphi$$

and the problems with $\Delta = 0$ disappear.

The causal character of the r =constant hypersurfaces is defined by the sign of $g^{rr} = \Delta/\Sigma$. Since $\Sigma > 0$ (except at the ring $r = 0, \theta = \pi/2$), the causal character of the r =constant hypersurfaces depends on the sign of Δ. In particular, this hypersurface will be light-like if $\Delta = 0$, so that observers will not be able to remain at $r =$ constant at these particular hypersurfaces, thus called *null horizons*. We have already treated the null horizons in Kerr's solution in Sect. 11.1. Now, in order to get the null horizons in the general RBH case we should solve

$$\Delta = r^2 - 2\mathscr{M}(r)r + a^2 = 0. \tag{11.12}$$

Without the knowledge of a specific $\mathscr{M}(r)$ it is not possible to know the *exact* position of the horizons. Nevertheless, one can analyze the general behaviour of the horizons by taking into account the following considerations:

- If we assume an asymptotically flat spacetime, at large distances $\mathscr{M}(r) \simeq m =$ constant, so that one (approximately) recovers the behaviour for the Kerr solution. Then $\Delta > 0$ and r will be a spacelike coordinate.
- For $r \simeq 0$ ($a \neq 0$) a regular RBH has $\Delta > 0$ thanks to the effect of the rotation and, again, r will be a spacelike coordinate. (Note that this already happens in the classical Kerr solution).

---

[7] By means of these kind of coordinate changes -advanced and retarded- the *maximal* extension is obtained by following the procedure in [22].

- If we assume the existence of a RBH and, thus, the existence of an exterior horizon r_+ (solution of $\Delta = 0$) then the continuity of Δ and the two previous items imply either a single horizon (*extreme* RBH), two horizons r_- and r_+ ($> r_-$) or, in general, an even number of horizons.
- If no solutions of (11.12) exist, then no null horizons exist and we are in a *hyper-extreme* case. The regular rotating astrophysical object without an event horizon is not properly a black hole. The regularity implies that, contrary to the classical case, there is not a naked singularity.

In practice, the usual regular RBH in the literature has one or two null horizons, as in the classical case. This is not surprising if one considers deviations from General Relativity as coming from Quantum Gravity effects. Then, based on a simple dimensional analysis, one could expect the Planck scale to be the most natural scale in which to expect the departure from General Relativity to occur, what would imply only strong deviations from the classical solution around $r \sim r_{Planck}$ and, thus, only small corrections to the horizons (at least for RBH with masses much larger than the planckian mass). One also expects that associated with non-singular RBH there would be a *weakening of gravity*. An effect which should be very important at high curvature scales. In this way, comparing with the classical case, it is usual to obtain bigger inner horizons and smaller outer horizons. Of course, the Planck scale approach could turn out to be too naive and bigger deviations from the classical solutions could be possible, what would be good news for the observational aspects of RBH (see Sect. 11.10).

Nevertheless, in order to illustrate the global causal structure of regular RBH let us follow the approach of small perturbations with respect to the classical horizons. We will compare this regular RBH causal structure with the usual one for Kerr's RBH, where we extend the spacetime through $r = 0$ into negative values for r. There are three possible qualitatively different causal structures for Kerr's RBH spacetime which are represented in the Penrose diagrams of Fig. 11.3 (for the case with two null horizons) and of Fig. 11.4 (for the *extreme* case and the *hyperextreme* case).

If we are in the regular RBH case then there is no need for an extension through $r = 0$. We can have three possible qualitatively different causal structures for the BH spacetime which are represented in the Penrose diagrams of Fig. 11.5 (for the case with two null horizons) and of Fig. 11.6 (for the *extreme* case and the *hyperextreme* case).

The absence of an event horizon in the hyperextreme case is interesting, since this implies that an observer could receive information from the inner high curvature regions near $r = 0$. In principle, this could be used to observationally test the different approaches to Quantum Gravity. The problem is whether such RBHs are feasible. In the framework of General Relativity, it does not seem possible to obtain such high speed RBH ($a^2 > m^2$) from a collapsing star and any attempt to overspin an existing black hole destroying its event horizon has failed, in agreement with the weak cosmic censorship conjecture. However, for regular RBHs it has been suggested that it could be possible to destroy the event horizon [58].

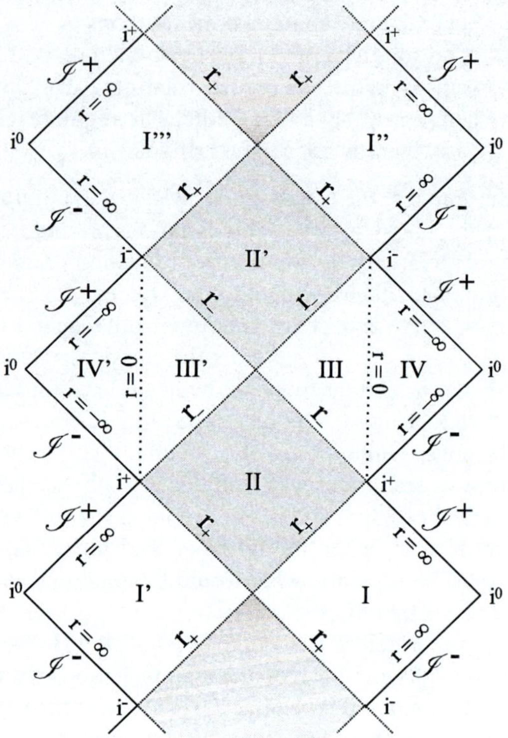

Fig. 11.3 Penrose diagram for Kerr's RBH with two horizons. The spacetime has been extended through $r = 0$ to asymptotically flat regions with negative values for r (IV or IV'). The grey regions are the regions where the coordinate r is timelike. Starting from the asymptotically flat region I, one could enter region II by traversing the *event horizon* r_+. Region III could next be reached by traversing the *Cauchy horizon* r_-. Then, the asymptotically flat region IV could be reached by passing through the regular $r = 0$. Note that the diagram is valid for $\theta \neq \pi/2$. The diagram with $\theta = \pi/2$ will require to draw the ring singularity

A warning is relevant here: A RBH solution should be stable in the region outside the event horizon and also inside. In the previous considerations (and figures) we have not taken into account the effects that instabilities could have in the global structure of the spacetime. Precisely, a non-trivial problem for RBH is the stability of their inner horizon. The first works in instability of inner horizons come from the study of the classical charged Reissner-Nordström black hole which suffers the so called *mass-inflation instability* [68]. Studies of the instability of Kerr's horizons were developed in [20, 67]. The consideration of regular black holes coming from different approaches to quantum gravity does not seem to alleviate the problem since, first, even in the non-rotating case they seem to require the existence of an (usually unstable) inner horizon and, second, even the backscattered flux of Hawking radiation coming from the black hole itself could be enough to destabilize its inner horizon [76]. Other studies on the stability of regular black holes can be found in [26–28]. Even a fine-tuned stable regular RBH can be found in [38].

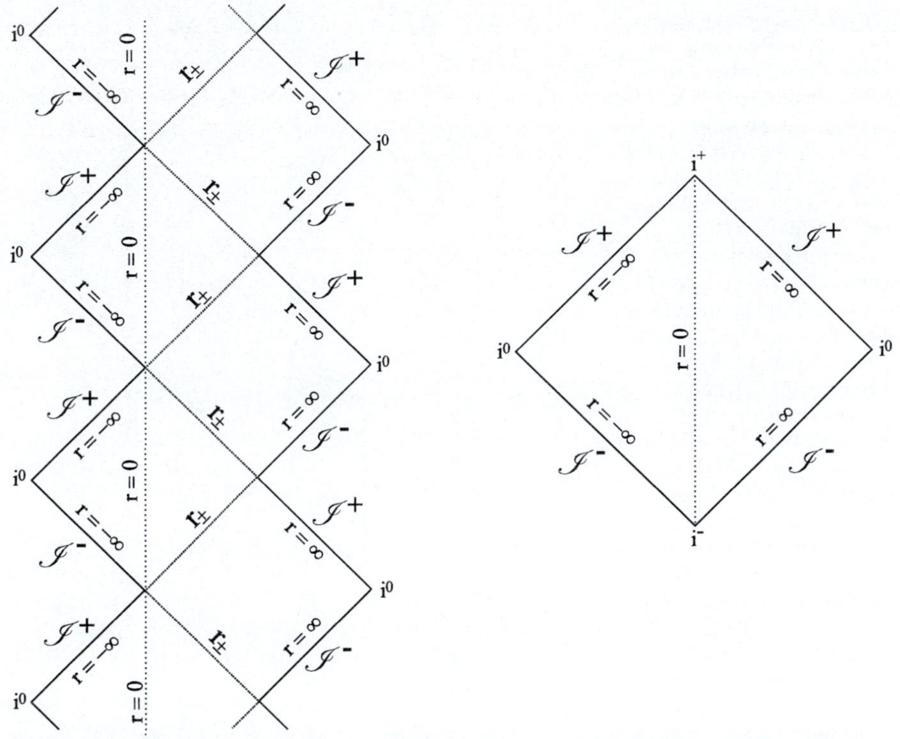

Fig. 11.4 Penrose diagrams for Kerr's extreme RBH (to the left) and for the hyperextreme case (to the right). In the extreme case there is only one horizon denoted by $r_\pm$ in which the coordinate r is lightlike. r is never timelike. $r_\pm$ acts both as an event and as a Cauchy horizon. In the hyperextreme case there are no horizons and r is always spacelike. In both cases, the spacetime has been extended through $r = 0$ to an asymptotically flat region with negative values for r. (Note that, again, the diagrams are valid for $\theta \neq \pi/2$). The diagram with $\theta = \pi/2$ will require to draw the ring singularity

11.7 Causality

In general, it seems reasonable to ask a time orientable spacetime to be absent of closed causal curves. The existence of such curves would seem to lead to logical paradoxes: One could travel following these curves and arrive back before one's departure, so that one could prevent oneself from setting out in the first place. A spacetime absent of closed causal curves is said to be *causal* [49]. If, in addition, no closed causal curve appears even under any small perturbation of the metric the spacetime is called *stably causal*. It is well-known that the usual analytical extension of the Kerr metric is non-causal. Since Kerr metric is a particular case of the metric (11.4), it is natural to ask whether the maximal extensions of regular RBH should also be non-causal.

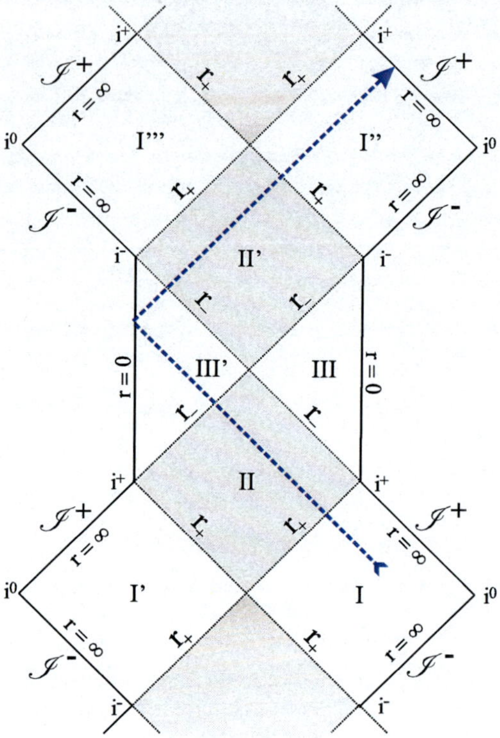

Fig. 11.5 Penrose diagram for a regular RBH with two null horizons. In this case an extension through $r = 0$ is not required. The grey regions are the regions where the coordinate r is timelike. We have depicted a light-like geodesic (dashed blue line) that, starting from the asymptotically flat region I, enters region II by traversing the *event horizon* r_+. Then it reaches region III' by traversing the null horizon r_-. The value of r first decreases along the geodesic until reaching $r = 0$, where it increases again. It makes it to another null horizon r_-, enters region II', traverses another event horizon r_+ to enter the asymptotically flat region I'' where it travels towards the future null infinity. (Note that, since there are not singularities, the diagram is valid for all θ)

Along the lines in [60], in order to examine this issue we will use proposition 6.4.9 in [49] that states that when a time function f exists in the spacetime such that its normal $\mathbf{n} \equiv \nabla_\mu f\, dx^\mu$ is timelike, then the spacetime is stably causal. (f can be thought as *the* time in the sense that it increases along every future-directed causal curve). Let us choose the time coordinate $\tilde{t}$ in Kerr-Schild coordinates as our time function f. The timelike character of $\mathbf{n}$ can be checked as follows:

$$\mathbf{n}^2 = g^{\mu\nu}\nabla_\mu \tilde{t}\nabla_\nu \tilde{t} = g^{\tilde{t}\tilde{t}} = -1 - \frac{2\mathcal{M}(r)r^3}{r^4 + a^2 z^2}. \tag{11.13}$$

Since we would like this to be negative, it trivially follows

Theorem 11.6 ([60]) *If $r\mathcal{M}(r) \geq 0$ for all r, then the model of RBH with metric (11.2) [or (11.4)] will be stably causal.*

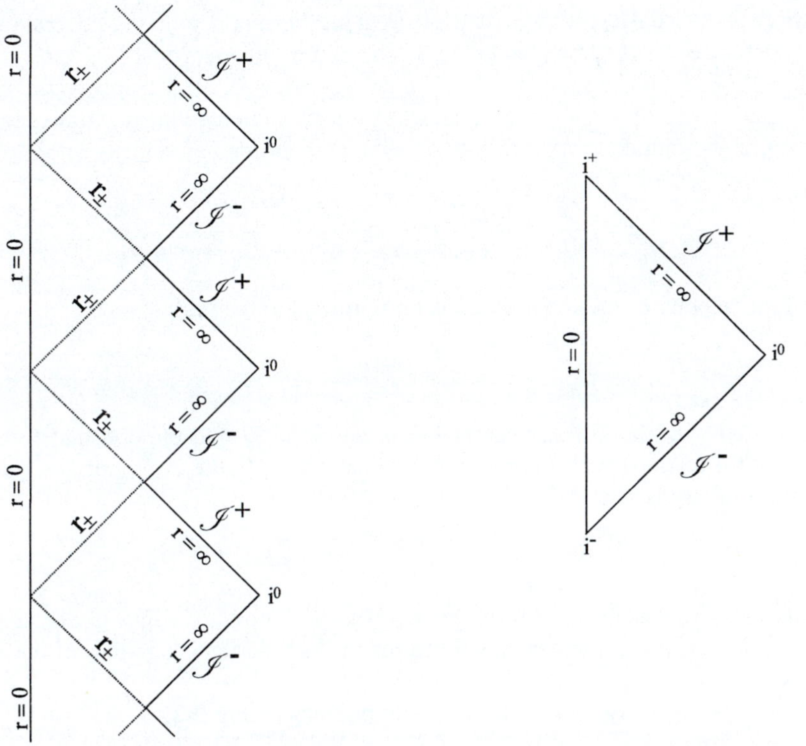

Fig. 11.6 Penrose diagrams for an extreme regular RBH (to the left) and for a hyperextreme case (to the right). In both cases, an extension through $r = 0$ is not required. In the extreme case there is only one horizon denoted by $r_\pm$ in which the coordinate r is lightlike. r is never timelike. $r_\pm$ acts as an event horizon. In the hyperextreme case there are no horizons and r is always spacelike. (Note that, again, since there are not singularities, the diagrams are valid for all θ)

Note that for a regular RBH (unextended through $r = 0$), a non-negative mass function suffices to guarantee a stably causal spacetime.

11.8 Thermodynamics

The consideration of the thermodynamics of black holes started in the 1970's with a series of articles with fundamental contributions by Bekenstein and Hawking [18, 19]. In 1975, Hawking proved that quantum mechanical effects cause Schwarzschild black holes to create and emit particles as if they were black bodies with a temperature proportional to their surface gravity. Since then, the thermodynamics of many different black holes coming from General Relativity and from alternative theories has been analyzed. Here we would like to treat the thermodynamics of general regular RBH described by the line element (11.4) at an introductory level.

In a RBH there is a family of *stationary observers*, i.e., those observers moving with constant angular velocity at fixed r and θ without perceiving any time variation of the gravitational field. It is easy to check that their angular velocity is

$$\Omega = \frac{d\phi}{dt} = -\frac{g_{t\phi}}{g_{\phi\phi}} = \frac{a(a^2 + r^2 - \Delta)}{(r^2 + a^2)^2 - \Delta a^2 \sin^2\theta}.$$

The four velocity of these observers is proportional to the killing vector

$$\boldsymbol{\xi} \equiv \mathbf{t} + \Omega \boldsymbol{\phi},$$

constructed with the killing vectors $\mathbf{t} = \partial_t$ and $\boldsymbol{\phi} = \partial_\phi$ and the constant (r and θ fixed for the stationary observers) angular velocity Ω. In the event horizon (r_+) the angular velocity is just

$$\Omega(r_+) = \Omega_+ = \frac{a}{r_+^2 + a^2}$$

and it can be checked that the kiliing vector on the horizon $\boldsymbol{\xi}\rfloor_{r_+}$ is light-like ($\boldsymbol{\xi} \cdot \boldsymbol{\xi}\rfloor_{r_+} = 0$). In this way, this killing is tangent to the null geodesics generators of the event horizon.

The surface gravity κ in the horizon can be found using [82]

$$\kappa^2 = -\frac{1}{2}\nabla_\mu \xi_\nu \nabla^\mu \xi^\nu.$$

In our regular RBH it is just

$$\kappa = \frac{\Delta'(r_+)}{2(r_+^2 + a^2)},$$

where the prime in Δ' stands for derivative with respect to r. Note that the surface gravity is constant in the event horizon, what links it to the temperature of the RBH. In fact, it is usually assumed that the temperature is just $T_+ = \kappa/2\pi$ [82] and, thus,

$$T_+ = \frac{\Delta'(r_+)}{4\pi(r_+^2 + a^2)}.$$

In this way, whenever $\Delta'(r_+) \neq 0$ the black hole will have a non-zero temperature and will emit Hawking radiation. Note also that a differentiable Δ requires, in the extremal case, that $\Delta'(r_\pm) = 0$. In this way, in case it exists, the temperature of an extremal RBH would be zero and it would not emit Hawking radiation.

In order to check the correctness of the result, one can compute the temperature of the particular case $\mathscr{M}(r) = m =$ constant (i.e., the well-known Kerr's RBH) obtaining the expected result

$$T_+^{Kerr} = \frac{r_+^2 - a^2}{4\pi r_+(r_+^2 + a^2)}.$$

Further discussions of the thermodynamic properties and Hawking radiation for particular regular RBHs can be found in [6, 52, 69, 74].

11.9 Obtaining Regular Rotating Black Hole Models

Different articles dealing with regular RBHs propose different forms for the function $\mathscr{M}(r)$. Its exact expression depends on the procedure used to obtain the RBH. In many cases the authors just propose heuristic forms for $\mathscr{M}$. The idea behind this heuristic approach is to try to ascertain the main characteristics that a regular RBH should have. Thus, for instance, the possible differences between the event horizons in the proposed models and the event horizon in Kerr's solution can be analyzed and maybe observationally tested (see Sect. 11.10). Of course, for a regular RBH the specific mass function $\mathscr{M}$ is chosen to avoid the existence of singularities. It is usually also demanded that the spacetime should be asymptotically flat. Other goals may include the (approximated) fulfillment of energy conditions beyond the event horizon, the stability of the model [38] or a good causal behaviour of the model. (See, for instance, [12, 14, 30, 60, 62, 71]). Even d-dimensional ($d > 4$) regular RBH have been studied heuristically. (See, for instance, [5, 8]).

In other cases a physical approach provides a specific $\mathscr{M}(r)$. Let us just mention a few of them. Some authors, inspired by the work of Bardeen [17], have taken the path of nonlinear electrodynamics, which provides the necessary modifications in the energy-momentum tensor in order to avoid singularities in the RBH [35, 42, 80]. Yet, another way of addressing the problem of singularities is to take into account that quantum gravity effects should play an important role in the core of black holes, so that it would seem convenient to directly derive the black hole behaviour from an approach to quantum gravity. In this way, regular RBHs deduced in the Quantum Einstein Gravity approach can be found in [69, 78], in the framework of Conformal Gravity in [15], in the framework of Shape Dynamics in [44], inspired by Supergravity in [24], by Loop Quantum Gravity in [25] and by non-commutative gravity in [72].

In the case of non-heuristic models, theoretically one *obtains* a specific expression for the mass function and then one has to check for the avoidance of singularities and for the rest of desirable properties cited above. The study of their event horizon is particularly important here since it may be observationally tested in the future (see Sect. 11.10), what could, for instance, help selecting among the different candidates to a Quantum Gravity Theory.

11.9.1 Generalized Newman-Janis Algorithms

In 1965, Newman and Janis [65] discovered that it was possible to obtain Kerr's solution by applying an algorithm to a spherically symmetric and static *seed* metric: Schwarzschild's solution. A great step towards understanding the algorithm, its possibilities, generalizations and limitations was carried out in [32]. The generalized algorithm allows to take any static spherically symmetric seed metric and obtain a rotating axially symmetric offspring from it[8]. The application to regular RBH followed [14]: One starts with a regular static and spherically symmetric black hole from a specific framework. Then, one applies the generalized N-J algorithm to try to construct a regular RBH.

The generalized Newman-Janis algorithm is a five-step procedure [32]:

1. Take a static spherically symmetric line element and write it in advanced null coordinates.
2. Express the contravariant form of the metric in terms of a null tetrad Z_a^μ.
3. Extend the coordinates x^ρ to a new set of complex coordinates

$$x^\rho \rightarrow \tilde{x}^\rho = x^\rho + i y^\rho(x^\sigma)$$

and let the null tetrad vectors Z_a^μ undergo a transformation

$$Z_a^\mu \rightarrow \tilde{Z}_a^\mu(\tilde{x}^\rho, \bar{\tilde{x}}^\rho).$$

Require that the transformation recovers the old tetrad and metric when $\tilde{x}^\rho = \bar{\tilde{x}}^\rho$.

4. Obtain a new metric by making a complex coordinate transformation

$$\tilde{x}^\rho = x^\rho + i\gamma^\rho(x^\sigma)$$

5. Apply a coordinate transformation $u = t + \mathscr{F}(r)$, $\phi = \varphi + \mathscr{H}(r)$ to transform the metric to Boyer-Lindquist-type coordinates.

While the seed spacetime can be a general spherically symmetric static spacetime, we will restrict ourselves here to a specific family of seed spacetimes that will allow us to connect with our family of regular RBH (11.4). Lets say that one has found a line element for a static regular spherically symmetric black hole (in a determined framework or just in a heuristic manner). Assume that the found line element can be written in coordinates $\{t, r, \theta, \varphi\}$ as a member of the family of static regular spherically symmetric black holes with metric:

$$ds^2 = -f(r)dt^2 + f^{-1}(r)dr^2 + r^2 d\Omega^2, \tag{11.14}$$

[8] The reader should be aware that the *offspring* has different geometrical properties and also different physical properties. For example, the seed metric can be a perfect fluid, but the offspring will never be another perfect fluid [32].

where $d\Omega^2 = d\theta^2 + \sin^2\theta d\varphi^2$. The function $f(r)$ can be rewritten as

$$f(r) = 1 - 2\frac{M(r)}{r},$$

by using the *mass function* $M(r)$ defined for general spherically symmetric space-times [63].

Let us now see that *the generalized N-J algorithm provides us with a means of obtaining the corresponding RBH line element* (11.4) *from the static spherically symmetric seed metric* (11.14).

The specific five steps for this case would be:

1. The coordinate change $du = dt + dr/f(r)$ allows us to rewrite the metric in advanced null coordinates[9]

$$ds^2 = -f(r)du^2 + 2dudr + r^2 d\Omega^2.$$

2. The null tetrad $Z_a^\mu = (l^\mu, n^\mu, m^\mu, \bar{m}^\mu)$ satisfying $l_\mu n^\mu = -m_\mu \bar{m}^\mu = -1$ and $l_\mu m^\mu = n_\mu m^\mu = 0$ can be chosen as

$$l^\mu = -\delta_r^\mu, \qquad n^\mu = \delta_u^\mu + \frac{f(r)}{2}\delta_r^\mu, \qquad m^\mu = \frac{1}{\sqrt{2}r}\left(\delta_\theta^\mu + \frac{i}{\sin\theta}\delta_\varphi^\mu\right)$$

so that $g^{\mu\nu} = -l^\mu n^\nu - l^\nu n^\mu + m^\mu \bar{m}^\nu + m^\nu \bar{m}^\mu$. (Note that both **l** and **n** are future directed).

3. We perform the coordinate change

$$r' = r - i\,a\cos\theta, \qquad u' = u - i\,a\cos\theta.$$

and demand r' and u' to be real. In this way the null tetrad transforms into ($Z_a'^\mu = Z_a^\nu \partial x^{\mu'}/\partial x^\nu$)

$$l'^\mu = -\delta_r^\mu, \qquad n'^\mu = \delta_u^\mu + \frac{\bar{f}(r')}{2}\delta_r^\mu, \qquad m'^\mu = \frac{1}{\sqrt{2}r'}\left(\delta_\theta^\mu + \frac{i}{\sin\theta}\delta_\varphi^\mu + i\,a\sin\theta(\delta_u^\mu + \delta_r^\mu)\right)$$

The function $\bar{f}$ comes from the complexification of f and, for the moment, we only know that it must be real and that it must reproduce Kerr solution if the complexified mass function is just a constant. This is possible if, as usual [14, 32, 65], one uses the complexification

$$\frac{1}{r} \to \frac{1}{2}\left(\frac{1}{r'} + \frac{1}{\bar{r}'}\right),$$

[9] Note that in the literature on the NJ algorithm there is some confusion between the advanced and the retarded ($dw = dt - dr/f(r)$) null coordinates. The first is suitable for describing black holes, the second for white holes.

that provides us with

$$\bar{f} = 1 - \frac{2\bar{M}(r,\theta)r}{\Sigma}, \tag{11.15}$$

where there is still some freedom in choosing the function $\bar{M}(r,\theta)$.

4. The new non-zero metric coefficients can be computed to be

$$\begin{aligned} g_{uu} &= -\bar{f}(r,\theta), \quad & g_{ur} &= +1, \quad & g_{u\varphi} &= -a\sin^2\theta[1-\bar{f}(r,\theta)] \\ r\varphi &= -a\sin^2\theta, \quad & g_{\theta\theta} &= \Sigma, \quad & g_{\varphi\varphi} &= \sin^2\theta[\Sigma + a^2\sin^2\theta(2-\bar{f})] \end{aligned} \tag{11.16}$$

5. In order to get the metric in Boyer-Lindquist type coordinates $\{u, r, \theta, \phi\}$ we perform the coordinate change $u = t + \int F(r)dr$, $\varphi = \phi + \int H(r)dr$, where

$$F(r) = \frac{r^2 + a^2}{\bar{f}(r,\theta)\Sigma + a^2\sin^2\theta} \quad \text{and} \quad H(r) = \frac{a}{\bar{f}(r,\theta)\Sigma + a^2\sin^2\theta}. \tag{11.17}$$

Thus, $\bar{f}$ should be chosen in such a way that F and H must be functions of r alone. Note that (11.17) implies

$$\Sigma\bar{f}(r,\theta) + a^2\sin^2\theta = D(r),$$

and substituting $\bar{f}$ using (11.15) one immediately sees that this step requires $\bar{M} = \bar{M}(r)$, i.e., $\bar{M}$ cannot depend on θ. Thus, we arrive at the natural choice $\bar{M}(r) = M(r)$. In effect, in this case F and H are really functions of r alone since

$$F(r) = \frac{r^2 + a^2}{r^2 + a^2 - 2\mathscr{M}(r)r} \quad \text{and} \quad H(r) = \frac{a}{r^2 + a^2 - 2\mathscr{M}(r)r}$$

(see (11.11)). Therefore, in this way it is possible to write the solution in Boyern-Lindquist type coordinates as

$$ds^2 = -\frac{\Delta}{\Sigma}(dt - a\sin^2\theta d\phi)^2 + \frac{\Sigma}{\Delta}dr^2 + \Sigma d\theta^2 + \frac{\sin^2\theta}{\Sigma}(adt - (r^2 + a^2)d\phi)^2. \tag{11.18}$$

where the mass function $M(r)$ (appearing in Δ) should just be relabelled as $\mathscr{M}(r)$ in order to be exactly (11.4).

Since we started with a *regular* spherically symmetric seed spacetime, let us assume that its mass function satisfies $M(r) = O(r^n)$ with $n \geq 3$ around $r = 0$. Thus, its offspring also has a mass function $\mathscr{M}(r) = M(r) = O(r^n)$ and, therefore, is also devoid of scalar polynomial curvature singularities (according to Theorem 11.4).

The moral of the procedure is that, if one wants a regular RBH in B-L-like coordinates, then one can simply take the mass function obtained in a spherically symmetric framework with metric (11.14) and use it as the mass function in the metric (11.4).

Note that, if one considers step 5 (obtaining the line element in B-L-like coordinates) as non-compulsory, then one could analyze the complexification for two different cases [14]:

1. *Type I* in which we impose $\mathscr{M} = \mathscr{M}(r)$. This is the case that we have just considered and the usual approach in the literature.
2. *Type II* in which we allow $\mathscr{M} = \mathscr{M}(r, \theta)$. The new rotating metric can be written in Kerr form (with a null coordinate in the style of Eddington-Finkelstein coordinates), but the N-J algorithm cannot be completed since it is not possible to write the rotating metric in the final Boyer-Lindquist form. Specific models of this type have been proposed and explored in [14, 36, 37].

11.10 Phenomenology

Recent developments have greatly enhanced our ability to probe theoretical predictions concerning black holes. These include, in the one hand, the direct observation of gravitational waves emanating from astrophysical sources by LIGO-Virgo collaboration [70] and, in the other hand, the images of black holes taken by the Event Horizon Telescope (EHT) [3, 4]. Moreover, a considerable enhancement is expected in the near future thanks to the LISA project [16] and the new planned ground-based observatories [55]. In this way, the physics in strong gravitational fields near black holes is becoming an important topic not only in theoretical physics but also in astrophysical phenomenology. There is now a need for compiling the maximum amount of theoretical results about realistic RBHs. It is hoped that the phenomenological evidence will help us to choose among the different proposals for RBH models and, as a consequence, among the alternative approaches to gravitational theories.

11.10.1 Shadows

A defining characteristic of a black hole is the event horizon. To a distant observer, the event horizon casts a relatively large "shadow" with, according to General Relativity, an apparent diameter of ~ 10 gravitational radii that is due to the bending of light by the black hole. Of course, the specific theoretical characteristics of this shadow depend on the alternative gravitational theory chosen and the properties of the modeled RBH. Currently, there are numerous studies about RBH shadows in different frameworks for alternative theories. Just to mention a few: heuristic approaches can be found in [1, 2, 7, 36, 57, 59, 73], results from different approaches to Quantum Gravity can be found in [23, 51].

Assuming a metric of the Gürses-Gürsey type (11.4), a general formula for the contour of the shadow can be easilly obtained [81]: The photons that would define the shadow of the regular RBH are described by an action $S = S(x^\alpha)$. The momentum of the photons is

$$p_\mu \equiv \frac{\partial S}{\partial x^\mu}$$

and satisfy

$$g^{\alpha\beta} p_\alpha p_\beta = 0. \tag{11.19}$$

The stationary and axisymmetric of the spacetime described by the metric (11.4) imply two conserved quantities in the trajectory of the photon: the energy $E \equiv -p_t$ and the angular momentum $L \equiv p_\phi$. If there is a separable solution for S, by using the definition of the momentum, we could rewrite it as

$$S = -Et + L\phi + S_r + S_\theta,$$

where we have introduced the new functions $S_r = S_r(r)$ and $S_\theta = S_\theta(\theta)$. In this way, (11.19) can now be written as

$$-\Delta\left(\frac{dS_r}{dr}\right)^2 + \frac{[(r^2+a^2)E - aL]^2}{\Delta} = \left(\frac{dS_\theta}{d\theta}\right)^2 + \frac{(L - aE\sin^2\theta)^2}{\sin^2\theta}. \tag{11.20}$$

In this equation, the left-hand side depends only on r, while the right-hand side depends only on θ. In this way, they define a constant which we will denote by

$$K = \left(\frac{dS_\theta}{d\theta}\right)^2 + \frac{(L - aE\sin^2\theta)^2}{\sin^2\theta}. \tag{11.21}$$

From $dx^\mu/d\lambda = p^\mu = g^{\mu\nu}p_\nu$ and using (11.20) and (11.21) one gets

$$\Sigma\frac{dr}{d\lambda} = \pm\sqrt{R(r)}, \tag{11.22}$$

where $R(r) \equiv P(r)^2 - \Delta[(L - aE)^2 + \mathcal{Q}]$, $P(r) \equiv E(r^2 + a^2) - aL$ and $\mathcal{Q} \equiv K - (L - aE)^2$ is the Carter constant.

Equation (11.22) implies that there would be unstable circular orbits at a certain $r = r_0$ whenever $R(r_0) = R'(r_0) = 0$ and $R''(r_0) > 0$. To exploit this, note that the definition of R can also be rewritten as

$$R/E^2 = r^4 + (a^2 - \xi^2 - \eta)r^2 + 2\mathcal{M}(r)[(\xi - a)^2 + \eta]r - a^2\eta,$$

where $\eta \equiv \mathcal{Q}/E^2$ and $\xi \equiv L/E$. The derivative of this expression with respect to r provides

$$R'/E^2 = 4r^3 + 2(a^2 - \xi^2 - \eta)r + 2\mathcal{M}(r)[(\xi - a)^2 + \eta]f(r),$$

where

$$f(r) \equiv 1 + \frac{r\mathcal{M}'}{\mathcal{M}}.$$

Using the conditions for the orbit one gets the quadratic equation with respect to ξ:

$$\begin{aligned} &a^2(r_0 - f_0\mathcal{M}_0)\xi^2 - 2a\mathcal{M}_0[(2 - f_0)r_0^2 - f_0a^2]\xi - r_0^5 + \\ &+ (4 - f_0)\mathcal{M}_0 r_0^4 - 2a^2r_0^3 + 2a^2\mathcal{M}_0(2 - f_0)r_0^2 - a^4r_0 - a^4\mathcal{M}_0 f_0 = 0, \end{aligned}$$

where $\mathcal{M}_0 \equiv \mathcal{M}(r_0)$ and $f_0 \equiv f(r_0)$. In order to describe the black hole shadow, we must choose the solution

$$\xi_- \equiv \frac{4\mathcal{M}_0 r_0^2 - (r_0 + f_0\mathcal{M}_0)(r_0^2 + a^2)}{a(r_0 - f_0\mathcal{M}_0)}.$$

that implies

$$\eta = \eta_- \equiv \frac{r_0^3[4(2 - f_0)a^2\mathcal{M}_0 - r_0[r_0 - (4 - f_0)\mathcal{M}_0]^2]}{a^2(r_0 - f_0\mathcal{M}_0)^2}.$$

We consider an observer at a large distance from the RBH in the asymptotically flat spacetime thats observes the RBH with an inclination θ_o. The contour of the shadow of the black hole can be expressed by celestial coordinates α and β [81] as

$$\alpha = \frac{\xi_-}{\sin\theta_0} \quad ; \quad \beta = \pm\sqrt{\eta_- + (a - \xi_-)^2 - \left(a\sin\theta_o - \frac{\xi_-}{\sin\theta_o}\right)^2}$$

In this way, we have finally arrived at the expressions that link the parameters describing the RBH with the observer's celestial coordinates. Some applications in particular models can be found in [81].

The current observations of black hole shadows by the Event Horizon Telescope are so far consistent with the shadow predicted for Kerr's RBHs. However, this classical solution of Einstein's equations cannot provide a complete understanding of black holes since, for instance, it implies the existence of an inner singularity. An analysis of the regular RBH studied so far shows that their shadows are usually also compatible with the observed shadows. In fact, the shadows of the regular RBH are indistinguishable from Kerr black holes shadows within the current observational uncertainties [56, 57, 59]. Future mm/sub-mm VLBI facilities will be able to greatly increase the current observational resolution. Even so, it will be challenging to test these metrics in the near future. A primary reason for this is (as explained in Sect. 11.6) that only slightly more compact event horizons and smaller shadows are usually expected. In fact, if the deviation from General Relativity comes from Quantum Gravity effects and it is Planck's scale which provides us with the scale in which to expect the departure, then it would be practically impossible to observe these effects in the shadows of a massive RBH. A better prospect for future observations would be expected if, on the contrary, the scale could be much bigger, as pointed out by some authors [33, 61], or the resolution of singularities were not related to Quantum Gravity effects.

11.11 Summary

Assuming that a manifold endowed with its corresponding metric is a fairly good approximation for describing a regular RBH, most of the models in the literature are of the Gürses-Gürsey type, whose general properties have been discussed in this text. We have seen that the regularity condition for these models translates into a condition for their mass function. We showed that the requirement of regularity leads to the violation of the energy conditions. Remarkably, regular RBH do not seem to require an extension through their disk. In this way, causality problems could be avoided simply if their mass function could remain non-negative. With regard to the choice of their mass function, in the literature it has been either chosen heuristically or derived from some gravitational theory. In either case, the generalized Newman-Janis algorithm provides us with an alibi to just use the mass function from regular spherically symmetric static black hole models.

The recent observational developments (LIGO-Virgo-KAGRA collaboration, the Event Horizon Telescope or, in the near future, the LISA project) opened the possibility to probe our theoretical predictions on rotating black holes. With our current observational resolution, so far the observations are consistent both with General Relativistic black holes and with alternative regular RBH models. Nevertheless, there is hope that a future increase in observational resolution could lead to discern among the different theoretical predictions.

There are some open questions concerning regular RBHs. For example, the analysis of the possible existence of parallely propagated curvature singularities, the treatment and resolution of their inner horizon instabilities, their future evolution due to the emission of Hawking radiation (including the problem of the possible formation of remnants) and a deeper generalization of the models beyond the Gürses-Gürsey type.

References

1. A. Abdujabbarov , M. Amir, B. Ahmedov, S.G. Ghosh, Phys. Rev. D **93**, 104004 (2016). (arXiv:1604.03809 [gr-qc])
2. F. Ahmed, D.V. Singh, S.G. Ghosh, Gen. Rel. Grav. **54**, 21 (2022). (arXiv:2002.12031 [gr-qc])
3. K. Akiyama et al., [Event Horizon Telescope] Astrophys. J. Lett. **875** L1, L2, L3, L4, L5, L6 (2019)
4. K. Akiyama et al., [Event Horizon Telescope] Astrophys. J. Lett. **930** L12, L13, L14, L15, L16, L17 (2022)
5. M.S. Ali, S.G. Ghosh, Phys. Rev. D **98**, 084025 (2018)
6. M.S. Ali, S.G. Ghosh, Phys. Rev. D **99**, 024015 (2019)
7. M. Amir, S.G. Ghosh, Phys. Rev. D **94**, 024054 (2016). (arXiv:1603.06382 [gr-qc])
8. M. Amir, M.S. Ali, S.D. Maharaj, Class. Quantum Grav. **37**, 145014 (2020)
9. A. Ashtekar, M. Bojowald, Class. Quantum Grav. **22**, 3349 (2005)
10. E. Ayon-Beato, A. Garcia, Phys. Lett. B **493**, 149 (2000)
11. E. Ayon-Beato, A. Garcia, Gen. Rel. Grav. **37**, 635 (2005)
12. M. Azreg-Aïnou, Phys. Rev. D **90**, 064041 (2014). (arXiv:1405.2569 [gr-qc])
13. L. Balart, E.C. Vagenas, Phys. Rev. D **90**, 124045 (2014)
14. C. Bambi, L. Modesto, Phys. Lett. B **721**, 329 (2013). (arXiv:1302.6075 [gr-qc])
15. C. Bambi, L. Modesto, L. Rachwal, JCAP **05**, 003 (2017). (arXiv:1611.00865 [gr-qc])
16. E. Barausse et al., Gen. Rel. Grav. **52**, 81 (2020)
17. J.M. Bardeen, in *Conference Proceedings of GR5*, p. 174 (1968)
18. J.M. Bardeen, B. Carter, S.W. Hawking, Commun. Math. Phys. **31**, 161 (1973)
19. J.D. Bekenstein, Lett. Nuovo Cimento **4**, 737 (1972)
20. F.W. Bernard, J. Math. Phys. **30**, 1301 (1989)
21. A. Bonanno, M. Reuter, Phys. Rev. D **62**, 043008 (2000). (arXiv:0002196 [hep-th])
22. R.H. Boyer, R.W. Lindquist, J. Math. Phys. **8**, 265 (1967)
23. S. Brahma, C.-Y. Chen, D.-H. Yeom, Phys. Rev. Lett. **126**, 181301 (2021). (arXiv:2012.08785 [gr-qc])
24. A. Burinskii, Czech. J. Phys. **52**, C471 (2002)
25. F. Caravelli, L. Modesto, Class. Quant. Grav. **27**, 245022 (2010). (arXiv:1006.0232 [gr-qc])
26. R. Carballo-Rubio, F. Di Filippo, S. Liberati, C. Pacilio, M. Visser, JHEP **07**, 023 (2018). (arXiv:1805.02675 [gr-qc])
27. R. Carballo-Rubio, F. Di Filippo, S. Liberati, C. Pacilio, M. Visser (2021). arXiv: 2101.05006 [gr-qc]
28. R. Carballo-Rubio, F. Di Filippo, S. Liberati, C. Pacilio, M. Visser (2022). arXiv: 2205.13556 [gr-qc]
29. B. Carter, Phys. Rev. **174**, 1559 (1968)
30. T. De Lorenzo, A. Giusti, S. Speziale, Gen. Rel. Grav. **48**, 31 (2016). Corrigendum at Gen. Rel. and Grav. **48** 111
31. J.F. Donoghue, Phys. Rev. Lett. **72**, 2996 (1994). Phys. Rev. D **50**, 3874 (1994)
32. S.P. Drake, P. Szekeres, Gen. Rel. Grav. **32**, 445 (2000)
33. G. Dvali, C. Gomez (2011). arXiv:1112.3359 [hep-th]
34. I. Dymnikova, Class. Quantum Grav. **21**, 4417 (2004). (arXiv:0407072 [gr-qc])
35. I. Dymnikova, E. Galaktionov, Class. Quant. Grav. **32**, 165015 (2015). (arXiv:1510.01353 [gr-qc])
36. A. Eichhorn, A. Held, JCAP **05**, 073 (2021). (arXiv:2103.13163 [gr-qc])
37. A. Eichhorn, A. Held, Eur. Phys. J. C **81**, 933 (2021). (arXiv:2103.07473 [gr-qc])
38. E. Franzin, S. Liberati, J. Mazza, V. Vellucci (2022). arXiv:2207.08864 [gr-qc]
39. V.P. Frolov, JHEP **5**, 049 (2014). (arXiv:1402.5446 [hep-th])
40. R. Gambini, J. Pullin, Proc. Sci. **14**, 038 (2015). (arXiv:1408.3050 [gr-qc])
41. H. García-Compeán, V.S. Manko, Prog. Theor. Exp. Phys. **2015**, 043E02 (2015). (arXiv:1205.5848 [gr-qc])

42. S.G. Ghosh, Eur. Phys. J. C **75**, 532 (2015). (arXiv:1408.5668 [gr-qc])
43. G.W. Gibbons, M.S. Volkov, Phys. Rev. D **96**, 024053 (2017). (arxiv:1705.07787 [hep-th])
44. H. Gomes, G. Herczeg, Class. Quant. Grav. **31**, 175014 (2014). (arXiv:1310.6095 [gr-qc])
45. J.B. Griffths, J. Podolský *Exact Space-Times in Einstein's General Relativity* (Cambridge University Press, Cambridge, 2009)
46. M. Gürses, F. Gürsey, Phys. Rev. D **11**, 967 (1975)
47. H.M. Haggard, C. Rovelli, Phys. Rev. D **92**, 104020 (2015). (arXiv:1407.0989 [gr-qc])
48. V. Hamity, Phys. Lett. A **56**, 77 (1976)
49. S.W. Hawking, G.F.R. Ellis, *The Large Scale Structure of Space-Time* (Cambridge University Press, Cambridge, 1973)
50. S.A. Hayward, Phys. Rev. Lett. **96**, 031103 (2006)
51. A. Held, R. Gold, A. Eichhorn, JCAP **1906**, 029 (2019). (arXiv:1904.07133 [gr-qc])
52. S.H. Hendi, S.N. Sajadi, M. Khademi, Phys. Rev. D **103**, 064016 (2021). (arXiv:2006.11575 [gr-qc])
53. W. Israel, Nuovo Cimento B **44**, 1 (1967). Nuovo Cimento B **48**, 463
54. W. Israel, Phys. Rev. D **2**, 641 (1970)
55. V. Kalogera et al. arXiv:2111.06990 [gr-qc]
56. R. Kumar, A. Kumar, S.G. Ghosh, Astrophys. J **896**, 89 (2020)
57. F. Lamy, E. Gourgoulhon, T. Paumard, F.H. Sand Vincent, Class. Quantum Grav. **35**, 115009 (2018). (arXiv:1802.01635 [gr-qc])
58. Z. Li, C. Bambi, Phys. Rev. D **87**, 124022 (2013). (arXiv:1304.6592 [gr-qc])
59. Z. Li, C. Bambi, JCAP **1401**, 041 (2014). (arXiv:1309.1606 [gr-qc])
60. H. Maeda (2021). (arXiv:2107.04791 [gr-qc])
61. S.D. Mathur, Fortsch. Phys. **53**, 793 (2005). (arXiv:0502050 [hep-th])
62. J. Mazza, E. Franzin, S. Liberati, JCAP **04**, 082 (2021)
63. C.W. Misner, D.H. Sharp, Phys. Rev. **136**, B571 (1964)
64. C.W. Misner, K.S. Thorne, J.A. Wheeler, *Gravitation* (Freeman and Company, New York, WH, 1970)
65. E.T. Newman, A.I. Janis, J. Math. Phys. **6**, 915 (1965)
66. P. Pani, V. Cardoso, Phys. Rev. D **79**, 084031 (2009). (arXiv:0902.1569 [gr-qc])
67. E. Poisson, W. Israel, Phys. Rev. Lett. **63**, 1663 (1989)
68. E. Poisson, W. Israel, Phys. Rev. D **41**, 1796 (1990)
69. M. Reuter, E. Tuiran, Phys. Rev. D **83**, 044041 (2011)
70. See, for example, https://en.wikipedia.org/wiki/List_of_gravitational_wave_observations for a list of gravitational wave observations
71. A. Simpson, M. Visser, JCAP **03**, 011 (2022). (arXiv:2111.12329 [gr-qc])
72. A. Smailagic, E. Spallucci, Phys. Lett. B **688**, 82 (2010). (arxiv:1003.3918 [hep-th])
73. Z. Stuchlík, J. Schee, Eur. Phys, J. **79**, 44 (2019)
74. S. Takeuchi, Class. Quant. Grav. **33**, 225016 (2016). (arXiv:1603.04159 [gr-qc])
75. R. Torres, Class. Quantum Grav. **30**, 065014 (2013)
76. R. Torres, Phys. Lett. B **724**, 338 (2013). (arXiv:1309.1083 [gr-qc])
77. R. Torres, Phys. Lett. B **733**, 21 (2014)
78. R. Torres, Gen. Rel. Grav. **49**, 74 (2017). (arXiv:1702.03567 [gr-qc])
79. R. Torres, F. Fayos, Gen. Rel. Grav. **49**, 2 (2017). (arXiv:1611.03654 [gr-qc])
80. B. Toshmatov, B. Ahmedov, A. Abdujabbarov , Z. Stuchlik, Phys. Rev. D **89**, 104017 (2014). (arXiv:1404.6443 [gr-qc])
81. N. Tsukamoto, Phys. Rev. D **97**, 064021 (2018). (arXiv:1708.07427 [gr-qc])
82. R.M. Wald, *General Relativity* (University of Chicago Press, Chicago, 1984)
83. S. Weinberg, *Gravitation and Cosmology: Principles and applications of the general theory of relativity* (Wiley, New York, 1972)
84. K. Yagi, N. Yunes, T. Tanaka, Phys. Rev. D **86**, 044037 (2012). (arXiv:1206.6130 [gr-qc]
85. E. Zakhary, C.B.G. McIntosh, Gen. Rel. Grav. **29**, 539 (1997)

Chapter 12
Semi-classical Dust Collapse and Regular Black Holes

Daniele Malafarina

12.1 Introduction

In 1939 Oppenheimer and Snyder [58] and independently Datt [23] developed the first mathematical model describing complete gravitational collapse in General Relativity (GR). The model, usually referred to as OSD, is given by an exact solution of the field equations for a dynamical spherically symmetric cloud of homogeneous collisionless matter, usually called 'dust', collapsing under its own gravity. Homogeneous dust collapse results in the formation of a Schwarzschild black hole where the central spacetime singularity, that develops as the endstate of collapse, is covered by the horizon at all times.

We now know, from the singularity theorems, that singularities must inevitably appear as the endstate of collapse once a series of conditions are met. These conditions are (i) the validity of GR during collapse, (ii) the validity of some energy condition, (iii) global hyperbolicity of the spacetime and (iv) the formation of trapped surfaces at some point during collapse. Taken as hypotheses, these assumption lead to a series of theorems that show that singularities are an inevitable outcome of collapse [32, 34, 60, 66].

One could conjecture that singularities must always be hidden behind horizons in order to preserve the causal structure of the spacetime [61]. However, even before the formulation of the singularity theorems, many researchers speculated that singularities should not form in the real universe and therefore one or more of the above hypotheses must be violated at some stage during collapse. Intuitively speaking this means that some kind of repulsive effect must appear to halt the attractive force of gravity before the formation of the singularity. We know that for objects that are sufficiently massive and sufficiently compact the known forces of nature are not able to halt collapse [57]. Therefore the repulsive effects must come from some physics

D. Malafarina (✉)
Department of Physics, Nazarbayev University, Kabanbay Batyr 53, 010000 Astana, Kazakhstan
e-mail: daniele.malafarina@nu.edu.kz

C. Bambi (ed.), *Regular Black Holes*, Springer Series in Astrophysics and Cosmology,
https://doi.org/10.1007/978-981-99-1596-5_12

that is yet unknown to us and dominates when the spacetime curvature becomes large enough. This may be due to the effects of a new theory of gravity, thus modifying hypothesis (i) [30], and/or to a modification of the other forces and thus the averaged properties of matter, thus modifying hypothesis (ii) [64].

In order to investigate the implications of such modifications for black hole physics several toy models have been developed over the past few decades. These include modifications of black hole geometries such as [6, 10, 14, 18, 29, 31, 37, 41, 69] as well as modified collapse models such as [7, 9, 11, 13, 15, 33, 65]. For a recent review see [52]. In turn, these modifications often present interesting features that may bear significant consequences for astrophysical black hole candidates. One of the most interesting class of modified black hole geometries is that of regular black hole solutions obtained within some theory of non linear electrodynamics (NLED) [3, 4, 19–21, 25, 70]. Then one is naturally led to consider under what circumstances such NLED regular black holes can develop from collapse [53]. In the present chapter we provide a detailed construction of semi-classical models for dust collapse and investigate the conditions under which they may lead to black holes, bounces or regular black holes as final states.

The chapter is organized as follows: In Sect. 12.2 we review the field equations for spherical collapse, the most important features of collapse solutions and the general procedure to formulate semi-classical models. Section 12.3 is devoted to the spacetimes describing the exterior of the collapsing sphere with particular attention to regular black holes in NLED. In Sect. 12.4 we review the formalism for matching the interior and exterior geometries across a collapsing time-like surface, while in Sect. 12.5 we discuss in detail dust collapse with semi-classical corrections and the conditions for the formation of a regular black hole. Finally Sect. 12.6 provides a brief summary and conclusions.

Throughout the chapter we shall adopt the convention of absorbing the factor 8π in Einstein's equations into the definition of the energy-momentum tensor and we will use geometrized units taking $G = c = 1$.

12.2 Interior: Gravitational Collapse

Let us first review the general formalism to describe relativistic collapse [43]. We shall consider the line-element for the spherical collapsing interior in co-moving coordinates $\{t, r, \theta, \phi\}$ as [47, 51]

$$ds^2 = -e^{2\nu}dt^2 + e^{2\psi}dr^2 + C^2 d\Omega^2 \,, \tag{12.1}$$

where $d\Omega^2 = d\theta^2 + \sin^2\theta d\phi^2$ is the line element on the unit 2-sphere and the metric functions depend only on t and r, namely $\nu(t, r)$, $\psi(t, r)$, $C(t, r)$. Notice that r is a coordinate radius 'attached' to the infalling particles while C is the so-called area-radius function related to the area of collapsing spheres of constant r. Then collapse is described by C decreasing in time. For the energy momentum tensor we shall start with an anisotropic inhomogeneous fluid with

$$T^{\mu\nu} = (\varepsilon + p_\theta)u^\mu u^\nu + p_\theta g^{\mu\nu} + (p_r - p_\theta)\xi^\mu \xi^\nu \,, \tag{12.2}$$

where $\varepsilon(t, r)$ is the energy density, $p_r(t, r)$ and $p_\theta(t, r)$ are the radial and tangential pressures, u^μ is the fluid's four velocity and ξ^μ is a space-like unit vector orthogonal to u^α. We can define a mass function F known as Misner-Sharp mass [56] as

$$F = C(1 - e^{-2\psi}C'^2 + e^{-2\nu}\dot{C}^2) = C(1 - G + H) \,, \tag{12.3}$$

where we used primed quantities for partial derivatives with respect to r and dotted quantities for derivatives with respect to t and we have introduced two new functions defined as

$$G = e^{-2\psi}C'^2, \;\; H = e^{-2\nu}\dot{C}^2 \,. \tag{12.4}$$

The Misner-Sharp mass may intuitively be understood as describing the amount of matter contained within the shell r at the time t. Additionally we need to consider conservation of energy momentum, i.e. the Bianchi identities, given by $\nabla_\mu T^{\mu\nu} = 0$ which for the zero component gives

$$\nu' = -\frac{p_r'}{\varepsilon + p_r} + 2\frac{p_\theta - p_r}{\varepsilon + p_r}\frac{C'}{C} \,. \tag{12.5}$$

In the case of a perfect isotropic fluid, given by $p_\theta = p_r = p$, the Bianchi identity becomes

$$\nu' = -\frac{p'}{\varepsilon + p} \,, \tag{12.6}$$

and the field equations then become

$$\varepsilon = \frac{F'}{C^2 C'} \,, \tag{12.7}$$

$$p_r = p = -\frac{\dot{F}}{C^2 \dot{C}} \,, \tag{12.8}$$

$$2\dot{C}' = C'\frac{\dot{G}}{G} + \dot{C}\frac{H'}{H} \,. \tag{12.9}$$

Using Eq. (12.6) we can also rewrite Eq. (12.9) as

$$\frac{\dot{G}}{G} = 2\nu'\frac{\dot{C}}{C'} \,. \tag{12.10}$$

We are left with a set of five equations, namely three field equations, one Bianchi identity and the definition of the Misner-Sharp mass, for six unknown quantities, i.e. F, p, ε and three metric functions ν, ψ and C. Therefore in order to close the system we need one additional relation. This is usually given in the form of an equation of state relating density and pressure $p = p(\varepsilon)$. In the following we will consider the simplest case of non interacting particles, also called 'dust', for which $p = 0$. However, for completeness, it is worth mentioning the most commonly used equations of state in astrophysics and cosmology, which are the linear barotropic equation

$$p = \omega\varepsilon \ , \tag{12.11}$$

with $\omega \in [-1, 1]$ and the polytropic equation

$$p = K\varepsilon^{(n+1)/n} \ , \tag{12.12}$$

with the polytropic index n usually taken between 0.5 and 1 for compact objects such as neutron stars [22, 68].

12.2.1 Regularity and Scaling

To ensure that the density and pressures are regular and the mass function F is well behaved at the center at the initial time t_i some additional conditions are necessary [46]. In fact it is immediately clear from Eq. (12.7) that there exist mass functions F for which the density diverges at $C = 0$ at all times. If we wish for collapse to start from a regular configuration we must restrict the allowed functions F to those that give a finite density everywhere at t_i. Since we still have some gauge freedom in specifying the initial value of the area-radius function C we can impose

$$C(t_i, r) = r \ , \tag{12.13}$$

and thus we see that we can define an adimensional function $a(t, r)$, usually called 'scale factor', in such a way that

$$C(t, r) = ra(t, r) \ , \tag{12.14}$$

with the initial condition for a set as

$$a(t_i, r) = 1 \ . \tag{12.15}$$

Then the condition for collapse is given by $\dot{a} < 0$ and collapse ends in a singularity if $a \to 0$ in a finite time. Using the scaling (12.14) and Eq. (12.15) in the field equation for the energy density (12.7) we get the initial condition for the density as

$$\varepsilon(t_i, r) = \frac{F'(t_i, r)}{r^2(1 + ra'(t_i, r))} \ , \tag{12.16}$$

which diverges for $r \to 0$ unless $F \sim r^n$ with $n \geq 3$ for r close to zero. Then we are led to impose the following rescaling for the Misner-Sharp mass

$$F(t, r) = r^3 m(t, r) , \tag{12.17}$$

with $m(t_i, 0) = \text{const.} \neq 0$, which ensures that the mass function is sufficiently regular at the center at the initial time and the singularity may develop only at a later time. Also, from the definition of the Misner-Sharp mass (12.3) we see that the introduced scaling leads to

$$m = a\left(\frac{1-G}{r^2} + e^{-2\nu}\dot{a}^2\right) . \tag{12.18}$$

Imposing regularity of m at the center then imposes the additional condition

$$G(t, r) = 1 + r^2 b(t, r) . \tag{12.19}$$

with $b(t_i, 0) = \text{const.} \neq 0$. To summarize, we have imposed the following rescaling

$$C(t, r) = r a(t, r) , \tag{12.20}$$

$$F(t, r) = r^3 m(t, r) , \tag{12.21}$$

$$G(t, r) = 1 + r^2 b(t, r) , \tag{12.22}$$

and thus we can rewrite the field equations, the Misner-Sharp mass and the Bianchi identity for an isotropic perfect fluid as

$$\varepsilon = \frac{3m + rm'}{a^2(a + ra')} , \tag{12.23}$$

$$p = -\frac{\dot{m}}{a^2\dot{a}} , \tag{12.24}$$

$$\frac{r\dot{b}}{1 + r^2 b} = 2\nu' \frac{\dot{a}}{a + ra'} , \tag{12.25}$$

$$\nu' = -\frac{p'}{\varepsilon + p} , \tag{12.26}$$

$$m = a(e^{-2\nu}\dot{a}^2 - b) . \tag{12.27}$$

If $a \to 0$ in a finite co-moving time the density ε diverges and it can be shown that the Kretschmann scalar, which is the invariant scalar obtained from the Riemann tensor as $\mathscr{K} = R_{\alpha\beta\mu\nu}R^{\alpha\beta\mu\nu}$, also diverges, thus giving rise to a true curvature singularity.

Notice that if the pressure does not vanish there is one more condition to impose to ensure that the initial pressure $p_i = p(t_i, r)$ is finite. In fact if $\dot{a}(t_i, r) = 0$ for some value of r from Eq. (12.24) we see that p_i might diverge. From Eq. (12.27) we get

$$\frac{\dot{m}}{\dot{a}} = e^{-2\nu}\left(\dot{a}^2 + 2a\ddot{a} - 2\dot{\nu}a\dot{a}\right) - b - \frac{\dot{b}a}{\dot{a}}, \tag{12.28}$$

from which we see that if $b = b(r)$ then p_i is finite at t_i even if $\dot{a}(t_i, r) = 0$ provided that ν, b and a are well behaved. On the other hand if $b = b(t, r)$ then an additional condition on $b(t_i, r)$ must be imposed ensuring that $\dot{b}/\dot{a}$ is finite at t_i. Of course this condition is not necessary in the case of dust.

12.2.2 *Trapped Surfaces, Singularities and Energy Conditions*

Gravitational collapse produces a black hole when some kind of trapped surface appears as matter collapses from an initially non trapped configuration [17, 28, 36, 73]. For spherical collapse models we can define the apparent horizon as the curve $t_{\rm ah}(r)$ for which the surface $C(t_{\rm ah}(r), r)$ becomes null. Namely from the metric this condition can be written as

$$X(t, r) = g^{\mu\nu}(\partial_\mu C)(\partial_\nu C) = 0\,. \tag{12.29}$$

Then according to the implicit function theorem $X(t, r) = 0$ describes implicitly the curve $t_{ah}(r)$ (or $r_{ah}(t)$) which gives the time at which the shell r becomes trapped. Applying the definition of the Misner-Sharp mass from Eq. (12.3) to the above equation we obtain the condition for the formation of trapped surfaces as

$$1 - \frac{F}{C} = 1 - \frac{r^2 m(t, r)}{a(t, r)} = 0\,. \tag{12.30}$$

Then, looking at Eq. (12.27) the apparent horizon curve $t_{ah}(r)$ is given by the condition

$$\frac{r^2 e^{-2\nu}}{1 + r^2 b} = \frac{1}{\dot{a}^2}\,. \tag{12.31}$$

Keep in mind that in general we may have $b = b(t, r)$ and $\nu = \nu(t, r)$, however for dust collapse, as we shall see later, we have $b = b(r)$ and $\nu = 0$ making the left-hand side a function of r only.

In constructing a collapse model one wants to start with a configuration that has no trapped surfaces. This can be done by imposing that the at the initial time the solutions of Eq. (12.31), if any, are located outside the boundary of the cloud r_b, i.e. Eq. (12.31) has no solutions for $r \le r_b$ at $t = t_i$, or equivalently $r_{\rm ah}(t_i) > r_b$.

As mentioned collapse ends in a spacetime singularity if $a \to 0$. This condition can also be described via a curve $t_s(r)$ denoting the time at which the shell r becomes singular. In this case, all geodesics located inside the trapped region must terminate at the singularity. Again the curve $t_s(r)$ can be given implicitly by

$$a(r, t_s(r)) = 0\,. \tag{12.32}$$

Notice that in the OSD model, with the energy density being homogeneous, we get that $t_s(r) = \text{const.}$ and $t_{\rm ah}(r)$ is monotonically decreasing, which means that the singularity at the end of collapse is always covered by the trapped surface. However, even in simple inhomogeneous collapse models such as the Lemaître-Tolman-Bondi dust case [16, 49, 67] this is not obvious. In fact there exist models with $\varepsilon(t_i, r)$ decreasing outwards in r where both $t_s(r)$ and $t_{\rm ah}(r)$ are monotonically increasing outwards and $t_s(0) = t_{\rm ah}(0)$, thus leaving the first point of the singularity curve not necessarily covered by the trapped surface [45]. In both cases $t_{\rm ah}(r) \leq t_s$ for all r (with $t_{\rm ah}(0) = t_s(0)$) and therefore for $r > 0$ the singularity is covered [51].

The energy momentum tensor in Einstein's equations describes the averaged properties of matter at macroscopic scales. Therefore conditions must be imposed to ensure that it describes physically viable matter fields. To this aim there are three inequalities that can be imposed for $T_{\mu\nu}$ to be considered physically valid [35]:

1. The weak energy condition (w.e.c.) states that $T_{\mu\nu}$ must satisfy $T_{\mu\nu}V^\mu V^\nu \geq 0$ for any time-like (and null) vector V^μ. This implies that the energy density must be non negative in any reference frame. Then the weak energy conditions in the co-moving frame can be written as

$$\varepsilon \geq 0\,,\ \varepsilon + p \geq 0\,. \tag{12.33}$$

2. The additional requirement that the total mass is conserved leads to the dominant energy condition (d.e.c.). This implies $T_{\mu\nu}V^\mu V^\nu \geq 0$ for every time-like vector V^μ and $T_{\mu\nu}V^\mu$ must be null or time-like. This is a more stringent condition with respect to the w.e.c. because it also requires that the flow of ε must be locally non space-like. In the co-moving frame used here this translates to the additional requirement that the energy density must be greater than the pressures. Namely

$$\varepsilon \geq 0\,,\ -\varepsilon \leq p \leq \varepsilon\,. \tag{12.34}$$

3. Finally the strong energy condition (s.e.c.) requires that $(T_{\mu\nu} - g_{\mu\nu}T/2)V^\mu V^\nu \geq 0$ for every time-like vector V^μ. In the co-moving frame the s.e.c. requires

$$\varepsilon \geq 0\,,\ \varepsilon + p \geq 0\,,\ \varepsilon + 3p \geq 0\,. \tag{12.35}$$

Notice that the w.e.c. does not require the conservation of the baryon number of $T_{\mu\nu}$ and therefore new particles can be created if one does not impose other energy conditions. The d.o.c. is more stringent than the w.e.c. as it requires mass conservation and also it does not allow for faster than light speed of sound in the medium. Finally the s.e.c. is more stringent than the other two and it may be violated by physically valid matter models such as scalar fields. It is important to note that the energy conditions refer to the behavior of matter at macroscopic scales and they need not apply to matter fields in the strong curvature regime, close to the formation of the singularity, where quantum effects and some corresponding quantum energy conditions may dominate [8, 54]. Therefore models allowing for violations of the energy conditions towards the end of collapse may be considered even within GR.

The singularity theorems tell us that in GR if the energy conditions are satisfied and a trapped surface appears during collapse then a singularity must form. Therefore in order to avoid the formation of the singularity while retaining the formation of trapped surfaces, we must require that either GR does not hold for the whole collapse and/or that energy conditions are violated at some point.

12.2.3 Semi-classical Collapse

As mentioned earlier in order to avoid the formation of a singularity at the end of collapse one or more of the hypothesis of the singularity theorems must be violated. Keeping the assumption that the spacetime be globally hyperbolic and assuming that trapped surfaces do form during collapse (after all we do observe black hole candidates in the universe) we may look for violations of the energy conditions or a breakdown of GR in the last stages of collapse. We may then describe both scenarios in a unified formalism if we further assume that the breakdown of GR takes place in a way that can be written in the form of Einstein's equations with semi-classical corrections, namely assuming that the field equations become

$$G_{\mu\nu} + < G_{\mu\nu} >= T_{\mu\nu} \,, \tag{12.36}$$

where $T_{\mu\nu} = T_{\mu\nu}^{\text{matter}}$ accounts for the matter content including the part of the energy momentum tensor violating energy conditions, if any, while $< G_{\mu\nu} >= G_{\mu\nu}^{\text{corr}}$ is obtained from averaging the effects of the modifications to the geometry in such a way that the new theory still obeys Einstein equations for an effective geometry $g_{\mu\nu}^{\text{eff}} = g_{\mu\nu} + < g_{\mu\nu} >$ [9]. Then we can bring $G_{\mu\nu}^{\text{corr}}$ on the right-hand side of Eq. (12.36) and treat it as an additional, non physical, component of the energy-momentum tensor. From the above considerations we obtain the effective energy momentum tensor as

$$T_{\mu\nu}^{\text{eff}} = T_{\mu\nu}^{\text{matter}} + T_{\mu\nu}^{\text{corr}} \,, \tag{12.37}$$

with the strong field corrections to GR now described by $T_{\mu\nu}^{\text{corr}} = -G_{\mu\nu}^{\text{corr}}$. Keep in mind that $T_{\mu\nu}^{\text{corr}}$ does not describe a matter source. Therefore even if $T_{\mu\nu}^{\text{matter}}$ obeys the energy conditions we may have that $T_{\mu\nu}^{\text{eff}}$ violates them as a consequence of the modifications to the theory. We can also write the action for the semi-classical collapse model as

$$\mathscr{A} = \frac{1}{2} \int d^4x \sqrt{|g|} \left(\mathrm{R} + \mathscr{L}_{\text{matter}} + \mathscr{L}_{\text{corr}} \right) \,, \tag{12.38}$$

where $\mathscr{L}_{\text{corr}}$ is the Lagrangian density describing the strong curvature corrections to GR.

In the case of a perfect fluid $T^{\mu\nu}_{\text{matter}}$ is given by

$$T^{\mu\nu}_{\text{matter}} = (\varepsilon + p)u^{\mu}u^{\nu} + pg^{\mu\nu} , \tag{12.39}$$

with u^{κ} being the 4-velocity of the fluid. In typical scenarios we may define a critical density ε_{cr} for which the deviations from GR become non negligible and write $T^{\mu\nu}_{\text{corr}}$ as an expansion in $\varepsilon/\varepsilon_{\text{cr}}$ close to zero, i.e. for $\varepsilon \ll \varepsilon_{\text{cr}}$. Then the effective density be written as

$$\varepsilon_{\text{eff}} = \varepsilon + \alpha_1\varepsilon^2 + \alpha_2\varepsilon^3 + \dots , \tag{12.40}$$

where the parameters α_i depend on ε_{cr} and are obtained from the expansion of $T^{\mu\nu}_{\text{corr}}$.

In Einstein's equations the geometry side of the equations remains unchanged while the matter fields in the field Eqs. (12.23) and (12.24) can still be written in the same form with the effective quantities replacing the classical ones. Namely we get

$$\varepsilon_{\text{eff}} = \frac{3m_{\text{eff}} + rm'_{\text{eff}}}{a^2(a + ra')} , \tag{12.41}$$

$$p_{\text{eff}} = -\frac{\dot{m}_{\text{eff}}}{a^2\dot{a}} , \tag{12.42}$$

with the effective Misner-Sharp mass being

$$m_{\text{eff}} = a(e^{-2\nu}\dot{a}^2 - b) . \tag{12.43}$$

As expected, depending on the specific choice of the effective energy momentum tensor the final outcome of collapse, need not necessarily be a singularity. Besides black holes one may obtain models that bounce, models that 'evaporate' and models that settle to massive compact remnants. One notable example of a bouncing model was considered in [7] and it is given by the choice $\alpha_1 = -1/\varepsilon_{\text{cr}}$ and $\alpha_i = 0$ for $i > 1$. This case, while being the simplest possible, is also well motivated as it arises from the effective description proposed in Loop Quantum Cosmology [1, 2, 13]. The aim then is to construct a physically well motivated $T^{\text{eff}}_{\mu\nu}$, solve Einstein's equations to obtain $a(t, r)$ and then investigate whether a singularity occurs and the behavior of the trapped region delimited by $t_{\text{ah}}(r)$.

12.3 Exterior: Regular Black Holes

To have a global geometry for the model we need to provide a line-element for the exterior spacetime to match to the line element (12.1), which describes the collapsing interior, across a suitable boundary r_b. The natural choice to match the OSD model is a Schwarzschild exterior, which describes a static black hole once the boundary of the collapsing cloud crosses the horizon. However, modified models may require for the exterior to be modified accordingly. Similarly to what has been discussed for semi-classical collapse one can consider a semi-classical description of the black hole geometry. Starting from the Schwarzschild solution describing a classical static black hole one then can devise new non vacuum solutions, where the non vanishing energy momentum tensor is interpreted as the effective correction $T_{\mu\nu}^{\rm corr}$. In turn these solutions may not present a central singularity. For example one may consider the exterior line element in coordinates $\{T, R, \theta, \phi\}$ written as

$$ds^2 = -f(R)dT^2 + \frac{dR^2}{f(R)} + R^2 d\Omega^2 \,, \tag{12.44}$$

where, in analogy with the Schwarzschild case, we can take f defined in terms of a mass function $M(R)$ as

$$f(R) = 1 - \frac{2M(R)}{R} \,, \tag{12.45}$$

with $M(R) \to M_0$ for R large in order to retrieve the Schwarzschild solution in the weak field where semi-classical corrections are negligible. Obviously these are not vacuum solutions, and, as said, the energy momentum can be understood as a semi-classical correction to the Schwazrschild vacuum. The energy momentum tensor for the metric (12.44) is

$$T_0^0 = T_1^1 = -\frac{2M'(R)}{R^2} \,, \tag{12.46}$$

$$T_2^2 = T_3^3 = -\frac{M''(R)}{R} \,. \tag{12.47}$$

from which we see that one must choose $M(R)$ in such a way that $T^{\mu\nu}$ goes to zero at large distances. The Kretschmann scalar for this spacetime is

$$\mathcal{K} = \frac{48M^2}{R^6} - \frac{16M}{R^3}\left(\frac{4M_{,R}}{R^2} - \frac{M_{,RR}}{R}\right) + 4\left(\frac{8M_{,R}^2}{R^4} - \frac{4M_{,R}M_{,RR}}{R^3} + \frac{M_{,RR}^2}{R^2}\right) , \tag{12.48}$$

and the condition for avoidance of the central singularity is then given by $M(R)/R^3$ being finite for $R \to 0$. Then noting that in this case $F(R) \to 1$ both for $R \to 0$ and $R \to +\infty$ we see that there must be either zero or an even number of roots of $F(R) = 0$, corresponding to no horizons or an even number of horizons. The simplest case where horizons are present is that of two horizons, namely an outer one, corresponding to the black hole event horizon and an inner one, which is a Cauchy horizon. In fact if the dominant energy condition holds then it can be shown that the number of horizons must be exactly two [24].

Of course there are other possibilities that may be considered for the exterior geometry, depending on the properties of the interior one. For example, radiating solutions may require to be matched to an exterior Vaidya [72] or generalized Vaidya metric [74].

12.3.1 Regular Black Holes in Non-linear Electrodynamics

An interesting class of regular black hole in the form (12.44) can be obtained from GR coupled to a theory of non linear electrodynamics (NLED) [3, 59, 62]. The action for GR coupled to NLED is given by

$$\mathscr{A} = \frac{1}{16\pi}\int d^4x\sqrt{|\mathrm{g}|}\,(\mathrm{R} - \mathscr{L}_{\mathrm{NLED}}(\mathrm{F}))\ , \tag{12.49}$$

where $|\mathrm{g}|$ is the determinant of the metric and the Lagrangian for NLED is

$$\mathscr{L}_{\mathrm{NLED}}(\mathrm{F}) = \frac{4\lambda}{\alpha}\frac{(\alpha\mathrm{F})^{(\kappa+3)/4}}{[1+(\alpha\mathrm{F})^{\kappa/4}]^{1+\lambda/\kappa}}\ , \tag{12.50}$$

with α the coupling parameter. The Faraday tensor $\mathrm{F}_{\mu\nu}$ of Maxwell's electrodynamics gives

$$\mathrm{F} = \mathrm{F}_{\mu\nu}\mathrm{F}^{\mu\nu}\ . \tag{12.51}$$

To consider a vacuum solution for the exterior we must take the energy momentum tensor as due only to $\mathscr{L}_{\mathrm{NLED}}$. For a spherically symmetric black hole coupled to NLED we have

$$T_{\mu\nu} = \frac{1}{4\pi}\left(\partial_{\mathrm{F}}\mathscr{L}_{\mathrm{NLED}}\mathrm{F}_{\mu}^{\sigma}\mathrm{F}_{\nu\sigma} - \frac{1}{4}g_{\mu\nu}\mathscr{L}_{\mathrm{NLED}}\right)\ . \tag{12.52}$$

Then taking the NLED source as a magnetic charge q_* and using Schwarzschild coordinates we obtain the line element in the form (12.44) with

$$M(R) = \frac{M_0 R^\lambda}{(R^\kappa + q_*^\kappa)^{\lambda/\kappa}}\ . \tag{12.53}$$

It is obvious that the case $\lambda = 0$ reduces to the Schwarzschild solution as does the case of vanishing NLED charge $q_* = 0$. Also in order for the solution to be regular at $R = 0$ we must evaluate the Kretschmann scalar $\mathscr{K}$ which is given by Eq. (12.48) from which we see that the condition of regularity at the center is $\lambda \geq 3$ [25].

Simple examples of regular black holes belonging to the above class may be obtained for $\lambda = 3$ for different values of κ [71]:

1. For $\kappa = 1$ we obtain the so-called Maxwellian black hole with

$$f = 1 - \frac{2M_0}{R}\left(1 + \frac{q_*}{R}\right)^{-3} . \tag{12.54}$$

2. For $\kappa = 2$ we obtain the so-called Bardeen black hole [10] with

$$f = 1 - \frac{2M_0}{R}\left(1 + \frac{q_*^2}{R^2}\right)^{-3/2} . \tag{12.55}$$

3. For $\kappa = 3$ we obtain the so-called Hayward black hole [37] with

$$f = 1 - \frac{2M_0}{R}\left(1 + \frac{q_*^3}{R^3}\right)^{-1} . \tag{12.56}$$

12.4 Matching

We shall now discuss the matching of the collapsing interior to a given exterior geometry [63]. The collapsing cloud is separated from the exterior spacetime by a boundary hypersurface, which in the following we will assume to follow a time-like trajectory. Junction conditions at the boundary describe the change of the matter field from the interior to the exterior, such as, for example, the separation between a dust interior from a vacuum exterior in the OSD model. The junction conditions are obtained by assuming that the manifold $\mathscr{M}$ is divided into two distinct regions $\mathscr{M}^+$ and $\mathscr{M}^-$ separated by a three-dimensional hypersurface Σ and requiring that the metric be continuous across Σ while discontinuities on Σ may be interpreted as a matter distribution concentrated on Σ [26, 27, 42].

Einstein's field equations hold in both regions and the line element in $\mathscr{M}^\pm$ can be written as

$$ds_\pm^2 = g_{\mu\nu}^\pm dx_\pm^\mu dx_\pm^\nu , \tag{12.57}$$

with $\{x^\mu\}^\pm$ being the coordinates in $\mathscr{M}^\pm$ ($\mu, \nu = 0, 1, 2, 3$). The line element on the three-dimensional boundary surface Σ can be written as

$$ds_\Sigma^2 = \gamma_{ab} dy^a dy^b , \tag{12.58}$$

where $\{y^a\}$ are the coordinates on the Σ (with latin indices a, b taking three values). The hypersurface Σ can be written in parametric form on both sides as

$$\Phi^\pm(x_\pm^\mu(y^a)) = 0 . \tag{12.59}$$

The first junction conditions then are given by the requirement that the induced metric γ_{ab} must be the same on both sides. Since the induced metric is

$$\gamma^{\pm}_{ab} = \frac{\partial x^{\mu}_{\pm}}{\partial y^a}\frac{\partial x^{\nu}_{\pm}}{\partial y^b} g^{\pm}_{\mu\nu} = e^{\mu}_{(a)} e^{\nu}_{(b)} g^{\pm}_{\mu\nu} , \tag{12.60}$$

with $e^{\mu}_{(a)} = \partial x^{\mu}_{\pm}/\partial y^a$ the basis vectors tangent to Σ, in order for $\gamma^{\pm}_{ab}$ to be the same on both sides there must exist a coordinate transformation on Σ for which $\gamma^{\pm}_{ab} = \gamma_{ab}$ or

$$[\gamma_{ab}] = \gamma^{+}_{ab} - \gamma^{-}_{ab} = 0 , \tag{12.61}$$

where $[\mathscr{A}] = \mathscr{A}_+ - \mathscr{A}_-$ defines the jump of a quantity $\mathscr{A}$ across Σ. Then the metric is continuous everywhere on $\mathscr{M}$, even though its first derivatives might still be discontinuous across Σ.

The second junction conditions must be imposed on the first derivatives of the metric. If they are also continuous, then the hypersurface Σ is truly a boundary. To evaluate these junction conditions one needs to evaluate the extrinsic curvature, also known as second fundamental form, $K^{\pm}_{ab}$ on both sides. Given the unit vector n^{μ} normal to Σ

$$n_{\mu} = \frac{\partial \Phi/\partial x^{\mu}}{\sqrt{g^{\alpha\beta}\frac{\partial \Phi}{\partial x^{\alpha}}\frac{\partial \Phi}{\partial x^{\beta}}}} , \tag{12.62}$$

the induced metric can be found from

$$e^{\mu}_{(a)} e^{\nu}_{(b)} \gamma^{ab} = g^{\mu\nu} - \varepsilon n^{\mu} n^{\nu} , \tag{12.63}$$

with $\varepsilon = 0$ for a null surface, $\varepsilon = 1$ for a spacelike surface and $\varepsilon = -1$ for a timelike surface. Then the extrinsic curvature is defined as

$$K_{ab} = g_{\mu\nu} n^{\mu} \nabla_a e^{\nu}_{(b)} , \tag{12.64}$$

or, expressed in coordinates,

$$K^{\pm}_{ab} = \frac{\partial x^{\mu}_{\pm}}{\partial y^a}\frac{\partial x^{\nu}_{\pm}}{\partial y^b} \nabla_{\mu} n_{\nu} = -n_{\sigma}\left(\frac{\partial^2 x^{\sigma}}{\partial y^a \partial y^b} + \Gamma^{\sigma}_{\mu\nu}\frac{\partial x^{\mu}}{\partial y^a}\frac{\partial x^{\nu}}{\partial y^b}\right) . \tag{12.65}$$

The Einstein tensor contains second derivatives of the metric and since, as we have seen, the first derivatives may be discontinuous across Σ this means that the second derivatives can be written as a Dirac delta on Σ. For simplicity, let us consider a coordinate system such that Σ is given by $x = x^3 = 0$. Such a coordinate always exists and the change of coordinates implies only a gauge fixing. Then the energy-momentum tensor can be written as

$$T_{\mu\nu} = T^{+}_{\mu\nu}\theta(x) + T^{-}_{\mu\nu}\theta(-x) + S_{\mu\nu}\delta(x) , \tag{12.66}$$

where the function $\theta(x)$ is the step function

$$\theta(x) = \begin{cases} 0, \ x < 0 \, , \\ 1, \ x > 0 \, , \end{cases} \tag{12.67}$$

for which $d\theta/dx = \delta(x)$ with δ being the Dirac delta. Then $S_{\mu\nu}$ describes the part of the energy-momentum tensor concentrated on Σ, which means that the components of $S_{\mu\nu}$ outside the shell $x = 0$ must vanish, i.e. in the gauge used here this implies that we must have $S_{33} = S_{a3} = 0$ and

$$S^{\mu\nu} = S^{ab} e^{\mu}_{(a)} e^{\nu}_{(b)} \, , \tag{12.68}$$

with $a, b = 0, 1, 2$. From Einstein's equations we then obtain the so-called Lanczos equation as

$$S_{ab} = [K_{ab}] - \gamma_{ab}[K] \, , \tag{12.69}$$

or inversely

$$[K_{ab}] = S_{ab} - \frac{1}{2}\gamma_{ab} S \, . \tag{12.70}$$

A boundary surface is defined by $S_{ab} = 0$, which means that the energy momentum tensor has a discontinuity only across the surface. This is reflected in the extrinsic curvature for which

$$[K_{ab}] = 0 \, , \tag{12.71}$$

implying that the first derivatives of the metric are continuous on Σ.

In the following we will consider the boundary surface for the collapsing cloud to be spherical, and assume that it will follow a time-like trajectory, although the considerations can be extended to null boundary surfaces in a rather straightforward way [48].

12.4.1 Spherical Time-Like Matching

We will now derive the first and second fundamental forms and the junction conditions for a spherical time-like shell in an arbitrary dynamical spacetime. Let's consider a generic spherical line element in the coordinates $\{x^{\mu}\} = \{t, r, \theta, \phi\}$ (with $\mu = 0, 1, 2, 3$) given by

$$ds^2 = -A^2 dt^2 + B^2 dr^2 + C^2 d\Omega^2 \, . \tag{12.72}$$

This is the same line element as in Eq. (12.1) with $A = e^{\nu}$ and $B = e^{\psi}$. The hypersurface Σ of a spherical time-like boundary can be given in parametric form as

$$\Phi(x^{\mu}) = r - R_b(t) = 0 \, , \tag{12.73}$$

so that the metric restricted on Σ becomes

$$ds_\Sigma^2 = -\left[A^2 - B^2\left(\frac{\partial R_b}{\partial t}\right)^2\right]dt^2 + C^2 d\Omega^2 \,, \tag{12.74}$$

where we understand that a generic function $X(t, r)$ on Σ becomes $X_b(t) = X(t, R_b(t))$, and we will omit the subscript 'b' to avoid making the notation too cumbersome.

The metric on Σ in coordinates $\{y^a\} = \{\tau, \theta, \phi\}$ (with $a = 0, 2, 3$) is also

$$ds_\Sigma^2 = -d\tau^2 + C_b(\tau)^2 d\Omega^2 \,. \tag{12.75}$$

Since the two line elements (12.74) and (12.75) must be the same we get

$$\left(\frac{d\tau}{dt}\right)^2 = A^2 - B^2\left(\frac{\partial R_b}{\partial t}\right)^2 \,, \tag{12.76}$$

$$C_b(\tau) = C\left(t(\tau), R_b(t(\tau))\right) \,. \tag{12.77}$$

Since we shall use the proper time on the shell τ as the trajectory's affine parameter, it is useful to invert (12.76) to get

$$\left(\frac{dt}{d\tau}\right)^2 = \frac{1}{A^2}\left(1 + B^2 \dot{R}_b^2\right) \,, \tag{12.78}$$

with $\dot{R}_b = dR_b/d\tau$. Notice that in this section we use 'dot' to denote derivatives with respect to the co-moving time on Σ, i.e. τ, while in Sect. 12.2 we used the same notation to denote derivatives with respect the time coordinate t. The two will be shown to be the same for homogeneous collapse models, but one should keep in mind that they need not be the same in general.

To have a physically viable matching we need to consider also the continuity of the extrinsic curvature K_{ab} on the surface which is defined by Eq. (12.65). In the case of a spherical time-like shell we get

$$n_t = -\dot{R}_b A B \,, \tag{12.79}$$

$$n_r = AB\frac{\partial t}{\partial \tau} \,, \tag{12.80}$$

$$n_\theta = n_\varphi = 0 \,. \tag{12.81}$$

A somewhat tedious calculation for the extrinsic curvature then gives

$$K_{\tau\tau} = -\frac{B\ddot{R}_b + B_{,r}\dot{R}_b^2}{\sqrt{1 + B^2\dot{R}_b^2}} - 2\frac{\dot{R}_b B_{,t}}{A} - \frac{A_{,r}}{AB}\sqrt{1 + B^2\dot{R}_b^2} \,, \tag{12.82}$$

$$K_{\theta\theta} = C\left(\frac{B}{A}\dot{R}_b C_{,t} + \frac{\sqrt{1 + B^2\dot{R}_b^2}}{B}C_{,r}\right) \,, \tag{12.83}$$

$$K_{\phi\phi} = K_{\theta\theta} \sin^2\theta \,, \tag{12.84}$$

where one should remember that for the purpose of the matching A, B and C must be evaluated on Σ, i.e. they are A_b, B_b and C_b.

The above formalism can be used to evaluate the first and second fundamental forms for both interior and exterior by making suitable choices for A, B and C. The continuous matching between an interior and a given exterior is then obtained by imposing the following junction conditions for the first and second fundamental forms

$$\gamma^+_{\tau\tau} = \gamma^-_{\tau\tau} \,, \tag{12.85}$$

$$\gamma^+_{\theta\theta} = \gamma^-_{\theta\theta} \,, \tag{12.86}$$

$$K^+_{\tau\tau} = K^-_{\tau\tau} \,, \tag{12.87}$$

$$K^+_{\theta\theta} = K^-_{\theta\theta} \,. \tag{12.88}$$

Of course, the conditions for $\gamma_{\phi\phi}$ and $K_{\phi\phi}$ are immediately obtained from those for $\gamma_{\theta\theta}$ and $K_{\theta\theta}$ due to spherical symmetry since we have $\gamma_{\phi\phi} = \gamma_{\theta\theta} \sin^2\theta$ and $K_{\phi\phi} = K_{\theta\theta} \sin^2\theta$.

12.4.2 Interior Geometry: Collapse

We shall now specialise the above treatment to some special cases. First we consider the interior $\mathcal{M}^-$ with line element (12.1) in coordinates $\{x^\mu\}^- = \{t, r, \theta, \phi\}$. The boundary surface Σ given by $\Phi^- = r - r_b(t) = 0$. The metric on Σ is given by (12.74) with $A^2 = e^{2\nu}$ and $B^2 = e^{2\psi}$. Then for the continuity of the metric we get

$$\frac{dt}{d\tau} = e^{-\nu}\sqrt{1 + e^{2\psi}\dot{r}_b^2} \,, \tag{12.89}$$

$$C_b(\tau) = C(t, r_b(t)) \,, \tag{12.90}$$

with $t = t(\tau)$. Similarly for the extrinsic curvature we get the normal unit vector as

$$n_t = -\dot{r}_b e^{\nu+\psi} \,, \tag{12.91}$$

$$n_r = e^{2\psi}\sqrt{e^{-2\psi} + \dot{r}_b^2} \,, \tag{12.92}$$

$$n_\theta = n_\phi = 0 \,, \tag{12.93}$$

so that

$$K^-_{\tau\tau} = -\frac{\ddot{r}_b + \psi' \dot{r}_b^2}{\sqrt{e^{-2\psi} + \dot{r}_b^2}} - 2\dot{r}_b \psi_{,t} e^{\psi-\nu} - \nu' \sqrt{e^{-2\psi} + \dot{r}_b^2} \,, \tag{12.94}$$

$$K^-_{\vartheta\theta} = C \left(C_{,t} \, \dot{r}_b e^{\psi-\nu} + C' \sqrt{e^{-2\psi} + \dot{r}_b^2} \right) , \tag{12.95}$$

where we used primed quantities for derivatives with respect to r but kept the subscript $X_{,t}$ for derivatives with respect to t and dotted quantities for derivatives with respect to τ. In the case of a co-moving boundary $r_b =$ const. the above equations reduce to

$$\frac{dt}{d\tau} = e^{-\nu} \,, \tag{12.96}$$

$$C(t, r_b) = C_b(\tau) \,, \tag{12.97}$$

$$K^-_{\tau\tau} = \nu' e^{-\psi} \,, \tag{12.98}$$

$$K^-_{\vartheta\theta} = C C' e^{-\psi} \,. \tag{12.99}$$

As we shall see later, for homogeneous dust collapse we may take $\nu = 0$ and therefore identify t with τ.

12.4.3 Exterior Geometry: Regular Black Holes

If we consider the exterior geometry $\mathscr{M}^+$ with line element (12.44) in Schwarzschild coordinates $\{x^\mu\}^+ = \{T, R, \theta, \phi\}$, thus setting $A^2 = f$ and $B^2 = 1/f$, the boundary surface Σ can be given by $\Phi_+ = R - R_b(T) = 0$, with $R_b(T)$ being the trajectory of a radial infalling particle in the spacetime. Then for the continuity of the metric we get

$$\frac{dT}{d\tau} = \frac{\sqrt{f + \dot{R}_b^2}}{f} \,, \tag{12.100}$$

$$C_b(\tau) = R_b(T(\tau)) \,, \tag{12.101}$$

with $T = T(\tau)$. For the extrinsic curvature we have

$$K^+_{\tau\tau} = -\frac{1}{\sqrt{f + \dot{R}_b^2}} \left(\ddot{R}_b + \frac{f_{,R}}{2} \right) , \tag{12.102}$$

$$K^+_{\vartheta\theta} = R_b \sqrt{f + \dot{R}_b^2} \,. \tag{12.103}$$

In the case of a regular black hole with $M(R)$ given by Eq. (12.53) we then have

$$f,_R = \frac{2M(R)}{R^2} - \frac{2M,_R}{R} = \frac{2M_0 R^{\lambda-2}}{(R^\kappa + q_*^\kappa)^{\lambda/\kappa}} \frac{1 + (1-\lambda) q_*^\kappa / R^\kappa}{1 + q_*^\kappa / R^\kappa} . \tag{12.104}$$

The Schwarzschild case is immediately obtained from the regular black hole case if we impose $M = M_0 =$ const. or $\lambda = 0$. We then get

$$\frac{dT}{d\tau} = \frac{\sqrt{1 - 2M_0/R_b + \dot{R}_b^2}}{1 - 2M_0/R_b} , \tag{12.105}$$

$$C_b(\tau) = R_b(T(\tau)) , \tag{12.106}$$

$$K_{\tau\tau}^+ = -\frac{\ddot{R}_b + M_0/R_b^2}{\sqrt{1 - 2M_0/R_b + \dot{R}_b^2}} , \tag{12.107}$$

$$K_{\vartheta\theta}^+ = R_b \sqrt{1 - 2M_0/R_b + \dot{R}_b^2} . \tag{12.108}$$

In the following we will consider semi-classical homogeneous dust collapse models for the interior and investigate under what conditions they may be matched to exterior solutions given by regular black holes in GR coupled to NLED.

12.5 Dust Collapse

Dust collapse can be obtained from Eqs. (12.23) to (12.27) by setting $p = 0$. Then Eq. (12.24) becomes $\dot{m} = 0$ and implies that $m = m(r)$ while Eq. (12.26) gives $\nu = \nu(t)$ which can be set to $\nu = 0$ by a suitable rescaling of the co-moving time [51]. Equation (12.9) becomes $\dot{G} = 0$ from which we get $b = b(r)$ in Eq. (12.25). Then from Eq. (12.27) we finally get

$$\dot{a}^2 = \frac{m}{a} + b , \tag{12.109}$$

from which we see that in the case of inhomogeneous dust we must have $a = a(t, r)$. The Kretschmann scalar for dust collapse is

$$\mathscr{K} = 48\frac{m^2}{a^6} - 32\frac{m(3m + rm')}{a^5(a + ra')} + 12\frac{(3m + rm')^2}{a^4(a + ra')^2} , \tag{12.110}$$

which diverges for $a = 0$ signalling the occurrence of the singularity. The complete set of Einstein's equations for inhomogeneous dust collapse may be complicated to solve analytically [44–46] and can require the aid of numerical methods [40, 55]. For example, in inhomogeneous models one has to consider the relative trajectories of different shells, which may overlap leading to 'shell crossing' singularities [38, 39, 76, 77]. Therefore applying semi-classical corrections to inhomogeneous dust may be complicated [12, 50] and one may look at the homogeneous case as a more manageable toy model.

12.5.1 Homogeneous Dust

The OSD model describes homogeneous dust collapse and is obtained by further requiring that $\varepsilon = \varepsilon(t)$, i.e. the density is homogeneous. Imposing homogeneity in Eq. (12.23) implies that $m' = 0$ and $a' = 0$ and therefore $m = m_0 =$ const. and $a = a(t)$. The energy density is then simply given by

$$\varepsilon(t) = \frac{3m_0}{a^3} . \tag{12.111}$$

From Eq. (12.109) with $a = a(t)$ and $m = m_0$ we then get the additional condition $b = k =$ const. We can then restrict the allowed values of k to $k = 0, \pm 1$ via the additional rescaling $r \to r/\sqrt{k}$, $t \to t/\sqrt{k}$. The case $k = 0$ is called 'marginally bound' and it corresponds to infalling particles having zero velocity at spatial infinity. The case $k = 1$ is called 'unbound' and it corresponds to infalling particles having positive velocity at spatial infinity. Finally the case $k = -1$ is called 'bound' collapse and it corresponds to infalling particles reaching zero velocity at a finite radius. The system is fully solved once we find the solution of the Eq. (12.109), written in the form

$$\dot{a} = -\sqrt{\frac{m_0}{a} + k} , \tag{12.112}$$

with the minus sign chosen in order to describe collapse.

In the marginally bound case, given by $k = 0$, with initial condition $a(0) = 1$, the above equation is immediately integrated to give

$$a(t) = \left(1 - \frac{3}{2}\sqrt{m_0}t\right)^{2/3} , \tag{12.113}$$

and the singularity is reached for $a = 0$ in a finite comoving time $t_s = 2/(3\sqrt{m_0})$. In the bound and unbound cases we obtain the solution in parametric form as

$$a(t) = \frac{m_0}{2}(1 - \cos\eta) , \tag{12.114}$$

with

$$\eta - \sin\eta = \frac{2}{m_0}(t - t_s) , \tag{12.115}$$

if $k = -1$ and

$$a(t) = \frac{m_0}{2}(\cosh\eta - 1) , \tag{12.116}$$

with

$$\sinh\eta - \eta = \frac{2}{m_0}(t - t_s) , \tag{12.117}$$

if $k = +1$. The line element then takes the simple form

$$ds^2 = -dt^2 + a^2\left(\frac{dr^2}{1 + kr^2} + r^2 d\Omega^2\right) , \tag{12.118}$$

and the Krestchmann scalar reduces to

$$\mathscr{K} = 60\frac{m_0^2}{a^6} . \tag{12.119}$$

The matching must be done with the Schwarzschild metric in the exterior since there is no inflow or outflow of matter through any shell r of the interior all the way up to the boundary $r_b =$ const.. In order to perform the matching with the interior for any value of $k = 0, \pm 1$ we need to rewrite the line element with the following change of coordinates

$$r = \begin{cases} \sin\zeta \;\text{ for }\; k = -1 , \\ \zeta \;\text{ for }\; k = 0 , \\ \sinh\zeta \;\text{ for }\; k = +1 , \end{cases} \tag{12.120}$$

which gives the line element (12.118) as

$$ds^2 = -dt^2 + a^2\left(d\zeta^2 + r(\zeta)^2 d\Omega^2\right) , \tag{12.121}$$

and consider ζ as the radial coordinate with the boundary at $r_b = \sin\zeta_b$ for $k = 1$ and $r_b = \sinh\zeta_b$ for $k = -1$.

In the case of homogeneous dust collapse the first and second fundamental forms are immediately obtained from Eqs. (12.96) to (12.99) and give

$$\frac{dt}{d\tau} = 1 , \tag{12.122}$$

$$C(t, r_b) = r_b a(t) , \tag{12.123}$$

$$K^-_{\tau\tau} = 0 , \tag{12.124}$$

$$K^-_{\vartheta\theta} = r_b a(t) . \tag{12.125}$$

The first and second fundamental forms for the Schwarzschild case are given by Eqs. (12.105)–(12.108) so that continuity of the metric gives the Schwarzschild time T as a function of the co-moving time t from

$$\frac{dT}{dt} = \frac{\sqrt{1 - 2M_0/R_b + \dot{R}_b^2}}{1 - 2M_0/R_b} , \tag{12.126}$$

and the trajectory of the collapsing boundary as

$$R_b(T(t)) = r_b a(t) \, . \tag{12.127}$$

The matching of the extrinsic curvature then gives the two additional relations

$$0 = -\frac{\ddot{R}_b + M_0/R_b^2}{\sqrt{1 - 2M_0/R_b + \dot{R}_b^2}} \, , \tag{12.128}$$

$$r_b a(t) = R_b\sqrt{1 - 2M_0/R_b + \dot{R}_b^2} \, , \tag{12.129}$$

which can be rewritten as

$$\ddot{R}_b = -\frac{M_0}{R_b^2} \, , \tag{12.130}$$

$$1 = \sqrt{1 - 2M_0/R_b + \dot{R}_b^2} \, . \tag{12.131}$$

It is easy to see that these two equations are equivalent, since by squaring the second equation and deriving with respect to t we obtain the first one. Also Eq. (12.128) is equivalent to the equation of motion (12.112) evaluated at the boundary if we make use of the matching condition (12.127) and impose that

$$2M_0 = F(r_b) = m_0 r_b^3 \, . \tag{12.132}$$

The above discussion shows that for homogeneous dust the trajectory of a particle at the boundary of the cloud is a geodesic determined by the amount of matter contained within it. We should then be able to write the same equation of motion for a particle on the boundary using the interior or the exterior metric. This can be done by using Lemaître coordinates for the exterior.

12.5.2 *Schwarzschild in Lemaître Coordinates*

In the case of dust collapse the absence of pressures implies that each particle at a co-moving radius r_0 must follow the geodesic determined by the matter content present within $r \leq r_0$. This is the same geodesic followed by a particle in radial free fall in the Schwarzschild geometry with the mass parameter given by the amount of mass contained within $r \leq r_0$. One easy way to illustrate the above idea is to consider for the exterior a set of coordinates used by an observer in free fall. These coordinates, known as Lemaître coordinates $\{\tau, \rho\}$ [49], can be defined for a general exterior of the form (12.44), which includes Schwarzschild. Lemaître coordinates are obtained from the two transformations $R = R(\tau, \rho)$, $T = T(\tau, \rho)$ given by

$$d\tau = dT + \frac{g(R)}{f(R)}dR\,, \tag{12.133}$$

$$d\rho = dT + \frac{1}{g(R)f(R)}dR\,, \tag{12.134}$$

with $g = \sqrt{1-f}$ and f given by (12.45), so that for Schwarzschild we get the line element as

$$ds^2 = -d\tau^2 + \frac{2M_0}{R}d\rho^2 + R(\tau,\rho)^2 d\Omega^2\,. \tag{12.135}$$

A particle in radial free fall in the Schwarzschild geometry in Lemaître coordinates is then located at $\rho = \rho_0$, $\theta = \theta_0$ and $\phi = \phi_0$. Then we can find the particle's trajectory as $R_0(\tau) = R(\tau, \rho_0)$ from the change of coordinates, since from

$$d\rho - d\tau = \frac{1}{g}dR = \sqrt{\frac{R}{2M_0}}dR\,, \tag{12.136}$$

evaluated at $\rho = \rho_0$, i.e. $d\rho_0 = 0$, we get

$$\frac{dR_0}{d\tau} = -\sqrt{\frac{2M_0}{R_0}}\,. \tag{12.137}$$

Integrating the above equation with the initial condition $R_0(0) = \rho_0$ gives the trajectory as

$$R_0(\tau) = \rho_0\left(1 - \frac{3}{2}\sqrt{\frac{2M_0}{\rho_0^3}}\tau\right)^{2/3} = \rho_0 a(\tau)\,, \tag{12.138}$$

where the adimensional function $a(\tau)$ is the scale factor and we can define $2M_0 = m_0\rho_0^3$ which then relates to the matching with the interior. Notice that, as expected, Eq. (12.137) is identical to the equation of motion for marginally bound homogeneous dust collapse. Notice that the cases of bound and unbound collapse may also be obtained in a similar manner by considering a particle in radial free fall with an energy per unit mass such at its initial velocity zero at a finite radius R or is positive at spatial infinity. In fact if we call E the energy per unit mass of the test particle its velocity is

$$\frac{dR_0}{d\tau} = -\sqrt{\frac{2M_0}{R_0} + E^2 - 1}\,. \tag{12.139}$$

which gives the bound (unbound) case for $k = E^2 - 1 < 0$ ($E^2 > 1$, respectively).

12.5.3 Semi-classical Homogeneous Dust

Semi-classical corrections to the energy momentum tensor can be introduced to model repulsive effects at large curvature which may lead to the singularity resolution. In general one may consider an effective density given in the form (12.40). However, the choice of $\varepsilon_{\rm eff}$ must be well motivated by some approach to gravity at large curvatures. Also, in order to avoid an unnecessary proliferation of arbitrary parameters, it may be wise to impose that α_i depend upon one single free parameter for all i. One simple model for a modification of the OSD scenario, inspired by Loop Quantum Gravity (LQG) was proposed in [7]. The effective energy density in this model is given by

$$\varepsilon_{\rm eff} = \varepsilon\left(1 - \frac{\varepsilon}{\varepsilon_{\rm cr}}\right) , \tag{12.140}$$

where $\varepsilon_{\rm cr} = 3m_0/a_{\rm cr}^3$ is a critical density scale related to α_1 in Eq. (12.40) via $\alpha_1 = -1/\varepsilon_{\rm cr}$. Notice that $\alpha_i = 0$ for $i > 1$. The corresponding equation of motion obtained from Eq. (12.27) is

$$\dot{a} = -\sqrt{\frac{m_0}{a}\left(1 - \frac{a_{\rm cr}^3}{a^3}\right) + k} \; . \tag{12.141}$$

The above model can be easily generalised if we consider

$$\varepsilon_{\rm eff} = \varepsilon\left[1 \pm \left(\frac{\varepsilon}{\varepsilon_{\rm cr}}\right)^{\beta}\right]^{\gamma} , \tag{12.142}$$

for which the equation of motion becomes

$$\dot{a} = -\sqrt{\frac{m_0}{a}\left(1 \pm \frac{a_{\rm cr}^{3\beta}}{a^{3\beta}}\right)^{\gamma} + k} \; . \tag{12.143}$$

Since we wish to retain terms in ε^2 in the expansion of the effective energy momentum, namely keeping $\alpha_1 \neq 0$, we must choose $\beta = 1$.

Different scenarios are possible depending on the sign chosen and the values of γ and k. In brief, a singularity will develop if $a \to 0$ while $\dot{a}$ remains negative. On the other hand, collapse will halt if $\dot{a} \to 0$ while a remains finite. In this case collapse will turn into expansion if $\ddot{a} \neq 0$ when $\dot{a} = 0$. Therefore it is clear that in order to understand the behavior of such models one has to solve Eq. (12.143) and study a, $\dot{a}$ and $\ddot{a}$.

The question now is whether this kind of models can be obtained from some approach to modify GR at large curvatures. If we describe the semi-classical corrections to dust collapse in the form of an effective energy momentum tensor, then such energy momentum must carry to the exterior and affect the geometry outside the collapsing sphere. We must then investigate the conditions under which the collapsing interior described by the equation of motion (12.143) can be matched to a suitable exterior. In the following we will take a regular black hole exterior obtained from GR coupled to NLED of the form (12.44) with $M(R)$ in the form given in Eq. (12.53).

To perform the matching of the above models we need to apply the junction conditions developed in Sect. 12.4. Homogeneity of collapse implies again that we can identify the co-moving time t with the proper time on Σ. Then the relation between T in the exterior and t is given by

$$\frac{dT}{dt} = \frac{\sqrt{f + \dot{R}_b^2}}{f} , \tag{12.144}$$

and junction conditions for the metric in the case of a constant co-moving boundary $r = r_b$ imply simply

$$r_b a(t) = R_b(T(t)) . \tag{12.145}$$

The remaining junction conditions for the extrinsic curvature become

$$0 = -\frac{1}{\sqrt{f + \dot{R}_b^2}} \left(\ddot{R}_b + \frac{f_{,R}}{2} \right) , \tag{12.146}$$

$$r_b a(t) = R_b \sqrt{f + \dot{R}_b^2} . \tag{12.147}$$

which again reduce to two equivalent equations, namely

$$\ddot{R}_b = -\frac{f_{,R}}{2} , \tag{12.148}$$

$$1 = f + \dot{R}_b^2 . \tag{12.149}$$

as it can easily be seen by differentiating the second one with respect to t. Equation (12.149) with f given by Eq. (12.44) is formally identical to Eq. (12.143) and so we see that the junction conditions are fully satisfied once we identify

$$2M_0 = r_b^3 m_0 , \tag{12.150}$$

$$q_*^\kappa = (r_b a_{\rm cr})^{3\beta} , \tag{12.151}$$

$$\frac{\lambda}{\kappa} = -\gamma . \tag{12.152}$$

12.5.4 NLED Black Holes in Lemaitre Coordinates

Similarly to what we did for homogeneous dust and Schwarzschild in Lemaître coordinates we can follow the same steps for semi-classical homogeneous dust and regular black holes. To move to Lemaître coordinates $\{\tau, \rho\}$ we define

$$g(R) = \sqrt{\frac{2M(R)}{R}} = \sqrt{\frac{2M_0 R^{\lambda-1}}{(R^\kappa + q_*^\kappa)^{\lambda/\kappa}}} . \tag{12.153}$$

The line element can then be written in the form

$$ds^2 = -d\tau^2 + \frac{2M(R)}{R} d\rho^2 + R(\tau, \rho)^2 d\Omega^2 , \tag{12.154}$$

and the change of coordinates gives

$$d\rho - d\tau = \frac{1}{g} dR = \sqrt{\frac{R}{2M(R)}} dR . \tag{12.155}$$

A free falling observer at $\rho = \rho_0 =$ const. with constant values for θ and ϕ follows the trajectory $R_0(\tau) = R(\tau, \rho_0)$ and must satisfy the equation of motion

$$\frac{dR_0}{d\tau} = -\sqrt{\frac{2M_0}{R_0}\left(1 + \frac{q_*^\kappa}{R_0^\kappa}\right)^{-\lambda/\kappa}} . \tag{12.156}$$

With the scaling $R_0(\tau) = \rho_0 a(\tau)$, $2M_0 = m_0 \rho_0^3$ and defining $q_* = \rho_0 q$ the above equation becomes

$$\frac{da}{d\tau} = -\sqrt{\frac{m_0}{a}\left(1 + \frac{q^\kappa}{a^\kappa}\right)^{-\lambda/\kappa}} , \tag{12.157}$$

which resembles the equation of motion of semi-classical dust collapse (12.143) for $k = 0$. In fact the two equations coincide if we make use of the junction conditions and identify $\gamma = -\lambda/\kappa$ and $\kappa = 3\beta$. Also notice that in the collapse model $a_{\text{cr}} > 0$ while in principle q may be positive or negative. Having fixed $\beta = 1$ implies that we can take $q > 0$ and consider the case of negative q by changing the sign in front of q^κ / a^κ, thus retrieving exactly Eq. (12.143) in the marginally bound case. The bound and unbound cases are again obtained by writing Eq. (12.156) with a different energy per unit mass for the test particle, which corresponds to a different velocity, similarly to Eq. (12.139).

We thus have established a correspondence between a class of regular black holes in NLED and a class of semi-classical homogeneous dust collapse models. However, we know that not all values of λ give a regular black hole solution and thus we need now to investigate in detail the possible endstates of collapse depending on the sign of the NLED charge q and the parameter λ. To keep the analysis as general as possible we will also allow for collapse to be bound, unbound or marginally bound, thus reintroducing the parameter k in the equation of motion.

12.5.5 Final Fates

To study the qualitative behavior of the class of semi-classical dust collapse models obtained from GR coupled to NLED let's consider the one dimensional dynamical system given by Eq. (12.143) with $\beta = 1$, namely

$$\dot{a} = j(a) = -\sqrt{\frac{1}{a}\left(1 \pm \frac{q^3}{a^3}\right)^\gamma + k}\,, \tag{12.158}$$

with $a > 0$, and where we have normalized the scale factor by substituting $a \to m_0 a$ and $q \to m_0 q$. Therefore the assumption that $q << m_0$ now becomes $q << 1$. The rescaling also implies that for negative k we must take $k > -1$ in order for j to be real at initial time. Since $j \leq 0$ the system moves towards smaller values of a from the initial state which may be taken as $a(t_i) = 1$, for which $j(1) = -\sqrt{(1 \pm q^3)^\gamma + k}$. Since the system is one-dimensional the only possible outcome is that a goes to a fixed point in a finite or infinite amount of time. In the marginally bound case, the fixed point being zero in the case of a plus sign and q in the case of minus sign. For simplicity, in the following we shall denote as (+) the case with plus sign and (−) the case with minus sign in Eq. (12.158). Accordingly, the only possible outcomes are that j goes to $-\infty$, zero or a finite, non zero, value and such outcomes may be reached as a goes to zero or as a goes to a finite value.

For each value of k we can distinguish scenarios based on the sign in Eq. (12.158) and the value of γ. Notice that for $k < 0$ we will have a zero of $j(a)$ at large a, which could be taken as an initial condition, and the dynamics will occur for smaller a where j could potentially have another zero. The results are summarised as follows

1. Sign (+) and $k > 0$:
   - $\gamma < -1/3 \Rightarrow j \xrightarrow[a\to 0]{} -\sqrt{k}$.
   - $\gamma = -1/3 \Rightarrow j \xrightarrow[a\to 0]{} -\sqrt{k + 1/q}$.
   - $\gamma > -1/3 \Rightarrow j \xrightarrow[a\to 0]{} -\infty$.
2. Sign (−) and $k > 0$:
   - $\gamma < 0 \Rightarrow j \xrightarrow[a\to q]{} -\infty$.
   - $\gamma = 0 \Rightarrow j \xrightarrow[a\to 0]{} -\infty$.
   - $\gamma > 0$ not even $\Rightarrow j \xrightarrow[a\to a^*]{} 0$ with $a^* < q$.
   - $\gamma > 0$ and even $\Rightarrow j \xrightarrow[a\to 0]{} -\infty$.

3. Sign (+) and $k = 0$:
   - $\gamma < -1/3 \Rightarrow j \xrightarrow[a\to 0]{} 0$.
   - $\gamma = -1/3 \Rightarrow j \xrightarrow[a\to 0]{} -1/\sqrt{q}$.
   - $\gamma > -1/3 \Rightarrow j \xrightarrow[a\to 0]{} -\infty$.
4. Sign (−) and $k = 0$:
   - $\gamma < 0 \Rightarrow j \xrightarrow[a\to q]{} -\infty$.
   - $\gamma = 0 \Rightarrow j \xrightarrow[a\to 0]{} -\infty$.
   - $\gamma > 0$ not even $\Rightarrow j \xrightarrow[a\to q]{} 0$.
   - $\gamma > 0$ and even $\Rightarrow j \xrightarrow[a\to 0]{} -\infty$[1].
5. Sign (+) and $k < 0$:
   - $\gamma < -1/3 \Rightarrow j \xrightarrow[a\to a^*]{} 0$ with $a^* < q$.
   - $\gamma = -1/3 \Rightarrow j \xrightarrow[a\to 0]{} -\sqrt{k + 1/q}$.
   - $\gamma > -1/3 \Rightarrow j \xrightarrow[a\to 0]{} -\infty$.
6. Sign (−) and $k < 0$:
   - $\gamma < 0 \Rightarrow j \xrightarrow[a\to q]{} -\infty$.
   - $\gamma = 0 \Rightarrow j \xrightarrow[a\to 0]{} -\infty$.
   - $\gamma > 0 \Rightarrow j \xrightarrow[a\to a^*]{} 0$ with $a^* < q$.

The acceleration is another important element to determine the final outcome. Differentiating Eq. (12.158) with respect to t we find

$$\ddot{a} = -\frac{1}{2}\frac{(a^3 \pm q^3)^\gamma}{a^{3\gamma+2}}\left(1 \pm \frac{3\gamma q^3}{a^3 \pm q^3}\right), \tag{12.159}$$

from which we can easily see that for (+) we have

- $\ddot{a} \xrightarrow[a\to 0]{} 0$ if $\gamma < -2/3$,
- $\ddot{a} \xrightarrow[a\to 0]{} q^2/2$ if $\gamma = -2/3$,
- $\ddot{a} \xrightarrow[a\to 0]{} -\infty$ if $\gamma > -2/3$,

while for (−) we have

- $\ddot{a} \xrightarrow[a\to q]{} -\infty$ if $\gamma < 1$,
- $\ddot{a} \xrightarrow[a\to q]{} 3/(2q^2)$ if $\gamma = 1$,
- $\ddot{a} \xrightarrow[a\to q]{} 0$ if $\gamma > 1$.

[1] Notice that $j(q) = 0$ in this case.

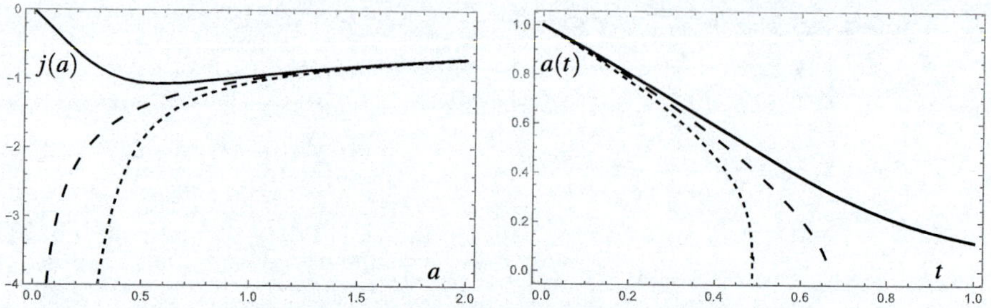

Fig. 12.1 Semi-classical marginally bound collapse, i.e. $k = 0$, with (+) in Eq. (12.158). Left panel: The function $j(a) = \dot{a}$ is plotted for $\gamma = 0$ (large dashed), $\gamma = 1$ (short dashed) and $\gamma = -1$ (solid). For the sake of clarity the range of the horizontal axis is (0, 2) even if the initial condition is taken as $a(t_i) = 1$. Right panel: The scale factor $a(t)$ is plotted for $\gamma = 0$ (large dashed), $\gamma = 1$ (short dashed) and $\gamma = -1$ (solid). The case $\gamma = 0$ corresponds to OSD collapse, while $\gamma = -1$ corresponds to collapse leading to the Hayward black hole. The plots are obtained for $q^3 = 0.1$

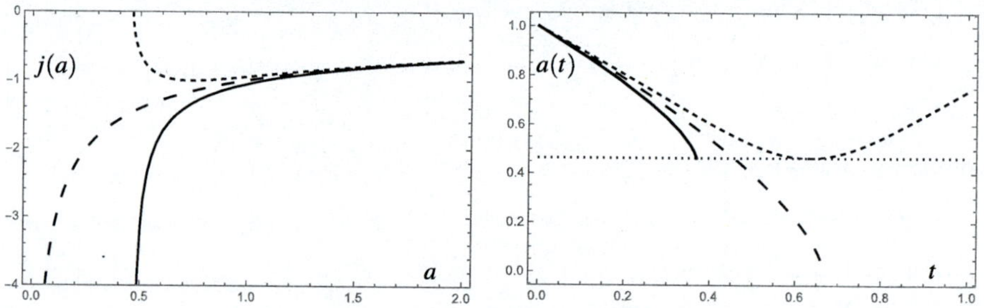

Fig. 12.2 Semi-classical marginally bound collapse, i.e. $k = 0$, with (−) in Eq. (12.158). Left panel: The function $j(a) = \dot{a}$ is plotted for $\gamma = 0$ (large dashed), $\gamma = 1$ (short dashed) and $\gamma = -1$ (solid). For the sake of clarity the range of the horizontal axis is (0, 2) even if the initial condition is taken as $a(t_i) = 1$. Right panel: The scale factor $a(t)$ is plotted for $\gamma = 0$ (large dashed), $\gamma = 1$ (short dashed) and $\gamma = -1$ (solid). The case $\gamma = 0$ corresponds to OSD collapse, while $\gamma = 1$ corresponds to the LQG inspired bounce model described in [7]. The plots are obtained for $q^3 = 0.1$

For example, the OSD case, i.e. $\gamma = 0$, we have j going to minus infinity as a goes to zero and $\ddot{a}$ also goes to minus infinity. On the other hand for $\gamma = 1$ with (−) we have that j goes to zero as a goes to a finite value and $\ddot{a}$ goes to a finite value. Also, for $\gamma = -1$ in the (+) case we have that j goes to zero if $k \leq 0$ and $\ddot{a}$ goes to zero leading asymptotically to an equilibrium configuration.

These possibilities are illustrated for three possible values of $k = 0, \pm 0.1$ in Figs. 12.1, 12.2, 12.3, 12.4, 12.5 and 12.6. In Fig. 12.1 are shown $j(a)$ (left panel) and $a(t)$ (right panel) for marginally bound collapse models with (+) for three values of $\gamma = 0, \pm 1$. The scale factor a goes to zero in all three cases but j goes to zero only in the case $\gamma = -1$, which corresponds to the Hayward black hole [53]. The corresponding models with $k = 0.1$ and $k = -0.1$ are shown in Figs. 12.3 and 12.5 respectively.

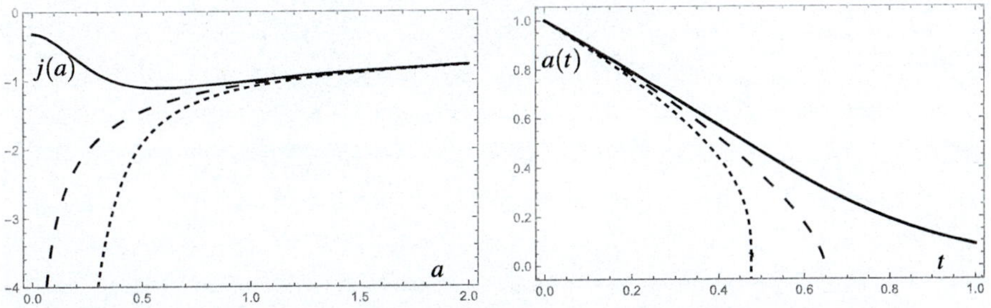

Fig. 12.3 Semi-classical unbound collapse, i.e. $k > 0$, with (+) in Eq. (12.158). Left panel: The function $j(a) = \dot{a}$ is plotted for $\gamma = 0$ (large dashed), $\gamma = 1$ (short dashed) and $\gamma = -1$ (solid). For the sake of clarity the range of the horizontal axis is (0, 2) even if the initial condition is taken as $a(t_i) = 1$. Right panel: The scale factor $a(t)$ is plotted for $\gamma = 0$ (large dashed), $\gamma = 1$ (short dashed) and $\gamma = -1$ (solid). The plots are obtained for $q^3 = 0.1$ and $k = 0.1$

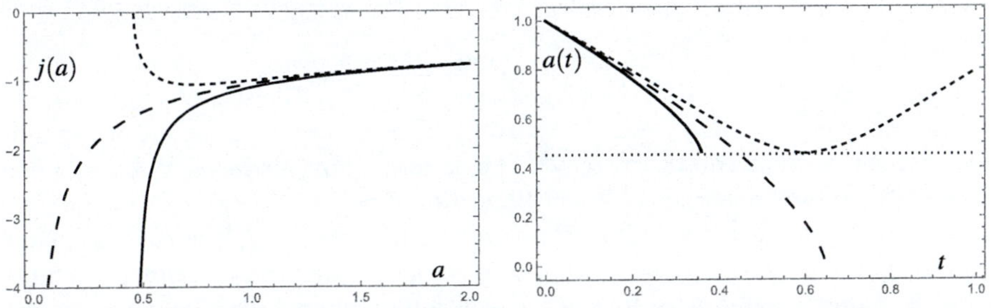

Fig. 12.4 Semi-classical unbound collapse, i.e. $k > 0$, with (−) in Eq. (12.158). Left panel: The function $j(a) = \dot{a}$ is plotted for $\gamma = 0$ (large dashed), $\gamma = 1$ (short dashed) and $\gamma = -1$ (solid). For the sake of clarity the range of the horizontal axis is (0, 2) even if the initial condition is taken as $a(t_i) = 1$. Right panel: The scale factor $a(t)$ is plotted for $\gamma = 0$ (large dashed), $\gamma = 1$ (short dashed) and $\gamma = -1$ (solid). The plots are obtained for $q^3 = 0.1$ and $k = 0.1$

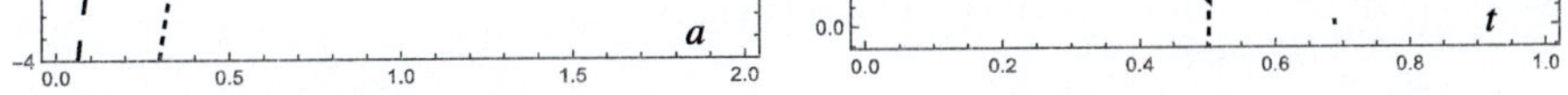

Fig. 12.5 Semi-classical bound collapse, i.e. $k < 0$, with (+) in Eq. (12.158). Left panel: The function $j(a) = \dot{a}$ is plotted for $\gamma = 0$ (large dashed), $\gamma = 1$ (short dashed) and $\gamma = -1$ (solid). For the sake of clarity the range of the horizontal axis is (0, 2) even if the initial condition is taken as $a(t_i) = 1$. Right panel: The scale factor $a(t)$ is plotted for $\gamma = 0$ (large dashed), $\gamma = 1$ (short dashed) and $\gamma = -1$ (solid). The plots are obtained for $q^3 = 0.1$ and $k = -0.1$

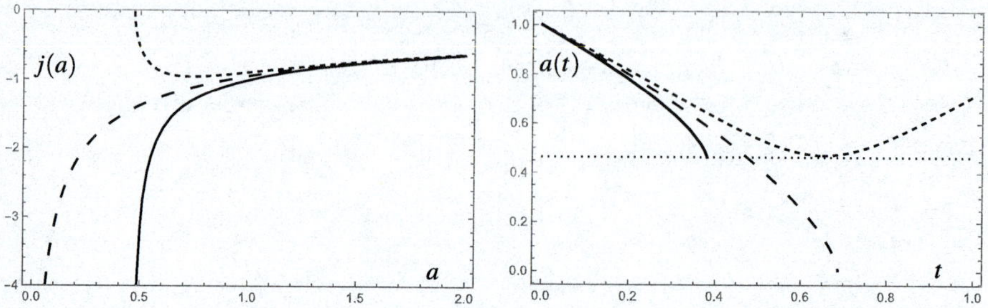

Fig. 12.6 Semi-classical bound collapse, i.e. $k < 0$, with $(-)$ in Eq. (12.158). Left panel: The function $j(a) = \dot{a}$ is plotted for $\gamma = 0$ (large dashed), $\gamma = 1$ (short dashed) and $\gamma = -1$ (solid). For the sake of clarity the range of the horizontal axis is (0, 2) even if the initial condition is taken as $a(t_i) = 1$. Right panel: The scale factor $a(t)$ is plotted for $\gamma = 0$ (large dashed), $\gamma = 1$ (short dashed) and $\gamma = -1$ (solid). The plots are obtained for $q^3 = 0.1$ and $k = -0.1$

In Fig. 12.2 are shown $j(a)$ (left panel) and $a(t)$ (right panel) for marginally bound collapse models with $(-)$ for three values of $\gamma = 0, \pm 1$. The scale factor a goes to a finite value if $\gamma \neq 0$ and j goes to zero only in the case $\gamma = 1$, which corresponds to the bouncing model discussed in [7]. The corresponding models with $k = 0.1$ and $k = -0.1$ are shown in Figs. 12.4 and 12.6 respectively.

Physically, in order to avoid the formation of the singularity, it seems reasonable to look for those models for which j goes to zero, i.e. collapse halts, either in a finite time or asymptotically. Also for collapse to settle to an equilibrium configuration we should require that $\ddot{a}$ goes to zero. Finally we wish the endstate of collapse to be a regular black hole. Notice that $\gamma > 0$, for which j goes to zero in the $(-)$ case, implies $\lambda < 0$ and thus the corresponding solutions in GR coupled to NLED can not be regular black holes. On the other hand, having set $\beta = 1$, we see that $\gamma \leq -1$ (namely $\lambda \geq 3$) satisfies all three criteria and thus the corresponding solutions in GR coupled to NLED are regular black holes that originate as the endstate of collapse.

12.5.6 Trapped Surfaces

As matter collapses trapped surfaces may form. The equation that implicitly defines the apparent horizon in the interior cloud is (12.31) or

$$1 - \frac{r^2 m_{\rm eff}}{a} = 0\,, \tag{12.160}$$

which in the marginally bound cases reduces to

$$1 - r^2 \dot{a}^2 = 0\,, \tag{12.161}$$

and implicitly gives the apparent horizon curve as

$$r_{\rm ah}(t) = \frac{1}{|\dot{a}|} \,. \tag{12.162}$$

As mentioned previously the apparent horizon is present only for $r_{\rm ah}(t) \leq r_b$ and when $r_{\rm ah}(t) = r_b$ it crosses the boundary and connects with a corresponding horizon in the exterior.

If $\dot{a} \to -\infty$ then $r_{\rm ah} \to 0$ and $r_{\rm ah}(t)$ will cross the boundary only once. Therefore we can not have the formation of an inner horizon during collapse. This is the case of the OSD model. On the other hand if $\dot{a} \to 0$ then $r_{\rm ah} \to +\infty$ and it may cross the boundary twice thus allowing for the possibility of producing the outer and inner horizons. This is the case of the Hayward regular black hole. Interestingly, the regular black hole is not the only possible option for the exterior if $\dot{a} \to 0$. In fact the horizon in the exterior may be closed giving rise to a closed trapped surface that exists for a finite time. These cases may be described with a Vaidya or generalized Vaidya exterior.

To summarize we have four possible scenarios:

1. Only one horizon forms and the singularity forms at the end of collapse. This is the case of OSD collapse. This case is shown in the left panel of Fig. 12.7.
2. A closed trapped surface forms before the bounce and the singularity is averted. If the expansion after the bounce is the time reversal of the collapse case a second closed trapped surface will form, this time describing a white hole instead of a black hole. This is the case of the LQG inspired model [7]. This case is shown in the right panel of Fig. 12.7.
3. An inner and an outer horizon form as matter settles asymptotically and the singularity is averted. The outer spacetime is described by a regular black hole. This is the case discussed in [53]. This case is shown in the left panel of Fig. 12.8.
4. A closed trapped surface forms as matter settles asymptotically and the singularity is averted. The outer spacetime is described by an horizonless compact remnant. This case is shown in the right panel of Fig. 12.8.

12.5.7 Example: Collapse and Bounce in LQG

The LQG inspired model discussed in [7] bounces turning collapse into expansion in a finite time. This model is obtained in the above formalism by taking the (−) case, $k = 0$ and $\gamma = 1$, $\beta = 1$ (i.e. $\lambda = -3$, $\kappa = 3$) in Eq. (12.143). The effective energy density in this model is given by Eq. (12.140) and relates to the effective Misner-Sharp mass via

$$\varepsilon_{\rm eff} = \frac{3m_{\rm eff}}{a^3} \,, \tag{12.163}$$

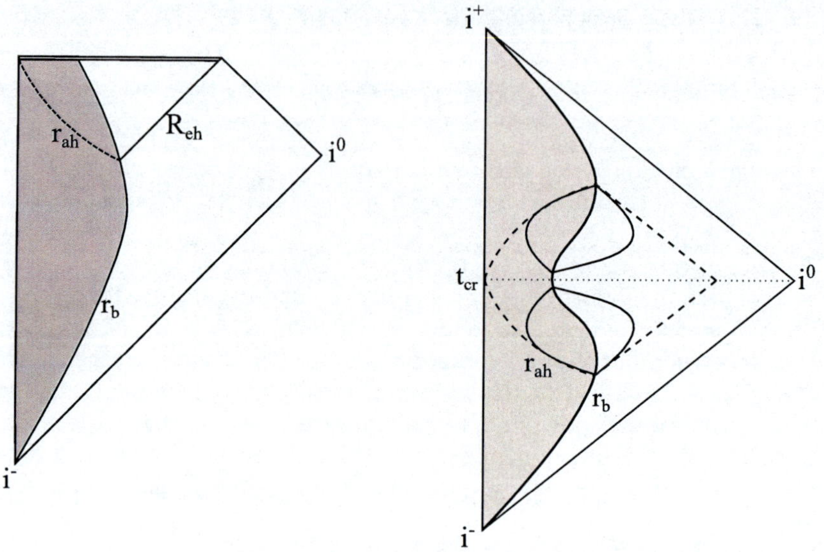

Fig. 12.7 Left panel: Penrose diagram for the OSD collapse. The darker region represents the interior of the cloud with boundary r_b while the double horizontal line represents the singularity. The solid line $R_{\rm eh}$ in the exterior is the event horizon while the dashed line in the interior is the apparent horizon. Right panel: Penrose diagram for the LQG inspired collapse model with dynamical Vaidya-like exterior. The darker region represents the interior of the cloud with boundary r_b. Collapse turns into expansion at the time $t_{\rm cr}$ and the spacetime is regular everywhere. The solid closed lines represent the trapped regions before and after the bounce while the dashed lines represent the event horizon and apparent horizon of the OSD case. The expanding solution for $t > t_{\rm cr}$ is given by the time reversal of the collapsing one

and is shown in the left panel of Fig. 12.9. The resulting effective pressure is given by

$$p_{\rm eff} = -\frac{\dot{m}_{\rm eff}}{a^2\dot{a}} = -\frac{\varepsilon^2}{\varepsilon_{\rm cr}} . \tag{12.164}$$

Notice that the effective pressure is negative and vanishes for $\varepsilon_{\rm cr} \to +\infty$ which corresponds to the OSD case. The effective Misner-Sharp mass $m_{\rm eff}$ still obeys Eq. (12.27), which for homogeneous dust gives Eq. (12.141), which can be written as

$$m_{\rm eff} = m_0\left(1 - \frac{a_{\rm cr}^3}{a^3}\right) = a(\dot{a}^2 - k) . \tag{12.165}$$

Notice that to the critical density parameter $\varepsilon_{\rm cr}$ corresponds a critical scale for collapse $a_{\rm cr}$ from $\varepsilon_{\rm cr} = 3m_0/a_{\rm cr}^3$. To retrieve the OSD model we must consider the limit $\varepsilon_{\rm cr} \to +\infty$, which corresponds to $a_{\rm cr} \to 0$. The solution for the marginally bound case is then easily obtained as

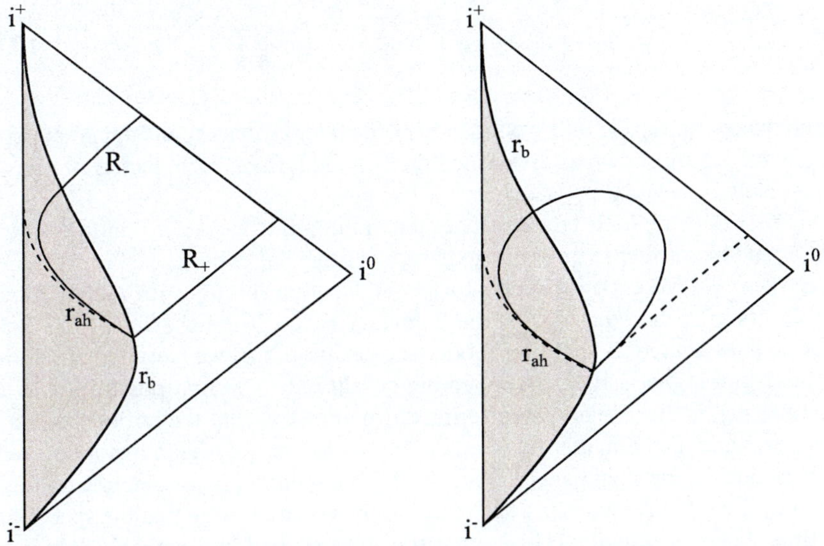

Fig. 12.8 Left panel: Penrose diagram for the collapse model leading to the Hayward black hole. The darker region represents the interior of the cloud with boundary r_b while solid lines represent the horizons, namely the apparent horizon $r_{\rm ah}$ in the interior and the inner R_- and outer R_+ horizons in the exterior. The dashed line in the interior represents the apparent horizon of the OSD case. Right panel: Penrose diagram for collapse model leading to a horizonless remnant. The darker region represents the interior of the cloud with boundary r_b. The solid closed line represents the trapped region while the dashed lines represent the event horizon and apparent horizon of the OSD case

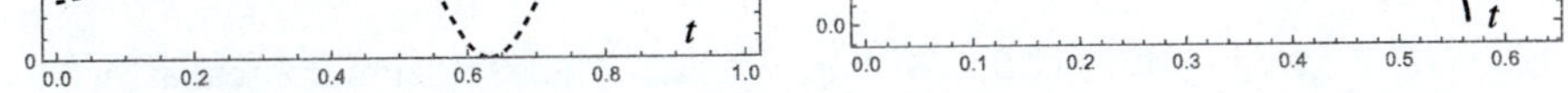

Fig. 12.9 Semi-classical marginally bound collapse, i.e. $k = 0$, with $(-)$ in Eq. (12.158). Left panel: The effective density $\varepsilon_{\rm eff}(t)$ is plotted for $\gamma = 0$ (large dashed), $\gamma = 1$ (short dashed) and $\gamma = -1$ (solid). The effective density is obtained for $q^3 = 0.1$. Right panel: The apparent horizon $r_{\rm ah}(t)$ is plotted for $\gamma = 0$ (large dashed), $\gamma = 1$ (short dashed) and $\gamma = -1$ (solid). The boundary at $r_b = 0.6$ is represented by the horizontal dotted line. The case $\gamma = 0$ corresponds to OSD collapse where $\varepsilon_{\rm eff} = \varepsilon$, while $\gamma = 1$ corresponds to the LQG inspired collapse. The apparent horizon is obtained for $q^3 = 0.01$

$$a(t) = \left(a_{\rm cr}^3 + \left(\sqrt{1 - a_{\rm cr}^3} - \frac{3}{2}\sqrt{m_0} t \right)^2 \right)^{1/3} , \tag{12.166}$$

which reduces to Eq. (12.113) for $a_{\rm cr}^3 = 0$. The effective density and apparent horizon for this model are shown as the dashed lines in the left and right panels of Fig. 12.9, respectively.

In this model the scale factor reaches the minimum value $a_{\rm cr}$ in a finite co-moving time $t_{\rm cr}$ and then collapse turns into expansion as the matter bounces. From the apparent horizon Eq. (12.160) we see that $r_{\rm ah} = r_b$ has two solutions for $t < t_{\rm cr}$ showing that the trapping horizon crosses the boundary twice. The behavior of the interior solution does not tell us anything about the exterior. In fact we have two possibilities for the exterior that satisfy the matching conditions. The first possibility, as suggested in [7], is that a black hole forms during collapse but the horizon eventually disappears and as the matter bounces at $t = t_{\rm cr}$ the central region is not covered by the horizon (see the right panel in Fig. 12.9). This scenario is consistent with the fact that at $t_{\rm cr}$ the effective Misner-Sharp mass vanishes suggesting that the spacetime at the time slice $t = t_{\rm cr}$ is flat. This scenario is not matched to a static black hole such as the ones described by GR coupled to NLED because at $t = t_{\rm cr}$ there must be no horizon anywhere in the spacetime, while we have seen that black holes coupled to NLED must have at least two horizons. In fact it can be shown that the LQG inspired collapse model can be matched to an exterior with variable mass as described by the Vaidya metric [65]. The second possibility, also discussed in [65], is that the interior is matched to a static exterior of the form (12.44). In this case the two roots of $r_{\rm ah} = r_b$ match with the inner and outer horizons of the exterior. However, this scenario does not produce a regular black hole in NLED, as it can also be seen from the fact that $\lambda = -3$.

We may still ask what exterior spacetime would result from the corresponding NLED Lagrangian and consider its properties. It is easy to see that the NLED Lagrangian for this model is

$$\mathscr{L}_{\rm NLED}({\rm F}) = -12\sqrt{\alpha}{\rm F}^{3/2} , \tag{12.167}$$

and $f(R)$ is given by Eq. (12.53) with $q_* = -a_{\rm cr}$. The Kretschmann scalar for this metric diverges for $R \to 0$ as

$$\mathscr{K} = \frac{48M^2}{R^{12}}(39q_*^6 - 10q_*^3R^3 + R^6) . \tag{12.168}$$

Nevertheless, looking the equation for the radial infall of a particle in this spacetime we can easily see that $R = 0$ can not be reached and a test particle on a radial ingoing trajectory must bounce. In fact the equation of motion for a test particle of energy per unit mass E falling radially along a trajectory $R(\tau)$ is

$$E^2 - 1 = \dot{R}^2 - \frac{2M_0}{R} + \frac{2M_0 q_*^3}{R^4}, \tag{12.169}$$

which in the case of a particle with zero initial velocity at spatial infinity, i.e. with $E = 1$, analogous to the marginally bound collapse, has a turning point at $R = q_*$.

12.5.8 Example: Collapse to the Hayward Black Hole

For black hole solutions in GR coupled to NLED we retrieve the Hayward regular black hole [37] by taking $\lambda = \kappa = 3$, which corresponds to $\gamma = -1$, $\beta = 1$ in the (+) case in Eq. (12.143). We can then investigate the corresponding semi-classical marginally bound dust collapse, which was considered in [53]. The effective density is

$$\varepsilon_{\text{eff}} = \varepsilon\left(1 - \frac{\varepsilon}{\varepsilon_{\text{cr}} + \varepsilon}\right). \tag{12.170}$$

Notice that for small densities this collapse model behaves in the same manner as the LQG inspired one. In fact expanding ε_{eff} for $\varepsilon << \varepsilon_{\text{cr}}$ we get

$$\varepsilon_{\text{eff}} = \sum_{n=0}^{\infty} \varepsilon \frac{(-\varepsilon)^n}{\varepsilon_{\text{cr}}^n}, \tag{12.171}$$

which stopping at $n = 1$ is the same as Eq. (12.140). However, while for the LQG inspired model all higher order terms vanish, in this case they are non zero. As a consequence for $\varepsilon = \varepsilon_{\text{cr}}$ we have that $\varepsilon_{\text{eff}} = \varepsilon_{\text{cr}}/2$, while in the LQG case the effective density vanishes at ε_{cr}. Also, for this case the effective density is always positive and it reaches the limiting value ε_{cr} asymptotically for $\varepsilon \to +\infty$. As a consequence the scale factor goes to zero asymptotically, as shown in Fig. 12.1. The effective density for this model is shown in the right panel of Fig. 12.10.

The effective pressure for this model is

$$p_{\text{eff}} = -\frac{\varepsilon^2}{\varepsilon_{\text{cr}} + \varepsilon}\left(1 - \frac{\varepsilon}{\varepsilon_{\text{cr}} + \varepsilon}\right), \tag{12.172}$$

and the equation of motion for the scale factor is

$$\dot{a} = -\frac{\sqrt{m_0}a}{\sqrt{a^3 + q^3}}. \tag{12.173}$$

In this case $a \to 0$ as t goes to infinity and the effective density remains finite and goes to the maximum value $\varepsilon_{\text{eff}} \to 3m_0/q^3$ asymptotically. Notice that $\dot{a} \to 0$ so, if trapped surfaces develop, the apparent horizon Eq. (12.160) must have two times

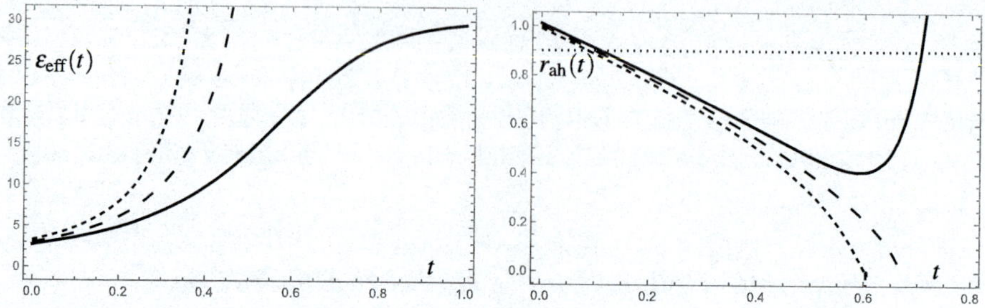

Fig. 12.10 Semi-classical marginally bound collapse, i.e. $k = 0$, with (+) in Eq. (12.158). Left panel: The effective density $\varepsilon_{\rm eff}(t)$ is plotted for $\gamma = 0$ (large dashed), $\gamma = 1$ (short dashed) and $\gamma = -1$ (solid). The effective density is obtained for $q^3 = 0.1$. Right panel: The apparent horizon $r_{\rm ah}(t)$ is plotted for $\gamma = 0$ (large dashed), $\gamma = 1$ (short dashed) and $\gamma = -1$ (solid). The boundary at $r_b = 0.8$ is represented by the horizontal dotted line. The case $\gamma = 0$ corresponds to OSD collapse where $\varepsilon_{\rm eff} = \varepsilon$, while $\gamma = -1$ corresponds to the NLED collapse leading to the Hayward regular black hole. The apparent horizon is obtained for $q^3 = 0.01$

at which $r_{\rm ah} = r_b$, thus forming the outer and inner horizons, as shown in the right panel of Fig. 12.10. Also notice that since $\ddot{a} \to 0$ collapse must halt asymptotically and does not bounce, leaving the Hayward black hole solution as a late time remnant of collapse. The effective density and apparent horizon for this model are shown as the solid lines in the left and right panels of Fig. 12.10, respectively.

Collapse to the Bardeen black hole [5, 10] can be obtained in a similar manner for $\lambda = 3$ and $\kappa = 2$. Other black holes obtained in GR coupled to NLED, such as the ones described in [4], may also be similarly translated into collapse models.

12.6 Conclusions

We know that General Relativity is an incomplete theory. The generic presence of singularities in physically viable solutions of Einstein's equations suggests that the theory needs to be replaced by another, better suited for the description of gravity at large curvature. At present we do not possess such a theory and GR remains the one that best reproduces observations while keeping a simple and beautiful mathematical framework [75]. Also we can not tell for sure at what scales (energy, density, size, etc.) the effects of such a new theory would become relevant [78]. If GR holds unaffected until the Planck scale we may expect it to be replaced by a theory of quantum gravity. Nevertheless there is also the possibility that GR fails before the Planck scale and an alternative classical theory is needed to bridge the gap between GR and quantum gravity. Therefore, in order to investigate the effects that the new theory of gravity would have in the universe we need to look for models of observable phenomena where the curvature becomes large. Then black holes and gravitational collapse are the ideal candidates.

We have considered here semi-classical models of gravitational collapse. Such models are typically used to describe classically the effects arising at Planck scale from a theory of quantum gravity. However the same formalism may be used also to describe corrections to GR coming from alternative theories or the coupling of GR to some other field, such as a theory of electrodynamics. It is well known that the static black hole solution produced by the coupling of GR with Maxwell's electrodynamics is the Reissner-Nordstrom solution. It is also well known that, under some conditions, from the coupling of GR to a theory of non-linear electrodynamics one may obtain regular black hole solutions such as Hayward's [37] or Bardeen's [10]. Here we have shown how to obtain such solutions and more from a semi-classical description of gravitational collapse of homogeneous dust.

The study of the properties of such solutions and their possible observational signatures in astrophysical scenarios may prove to be a valuable guide towards our understanding of the limits of GR and the features that a new theory of gravity must posses.

Acknowledgements DM acknowledges support from Nazarbayev University Faculty Development Competitive Research Grant No. 11022021FD2926.

References

1. A. Ashtekar, T. Pawlowski, P. Singh, Phys. Rev. D **73**, 124038 (2006)
2. A. Ashtekar, T. Pawlowski, P. Singh, Phys. Rev. D **74**, 084003 (2006)
3. E. Ayón-Beato, A. Garcia, Phys. Rev. Lett. **80**, 5056 (1998)
4. E. Ayón-Beato, A. Garcia, Phys. Lett. B **464**, 25 (1999)
5. E. Ayón-Beato, A. Garcia, Phys. Lett. B **493**, 149 (2000)
6. C. Bambi, L. Modesto, Phys. Lett. B **721**, 329 (2013)
7. C. Bambi, D. Malafarina, L. Modesto, Phys. Rev. D **88**, 044009 (2013)
8. C. Barcelò, M. Visser, Int. J. Mod. Phys. D **11**, 1553 (2002)
9. C. Barcelò, S. Liberati, S. Sonego, M. Visser, Phys. Rev. D **77**, 044032 (2008)
10. J.M. Bardeen, in: *Conference Proceedings of GR5*, Tbilisi, USSR, 174 (1968)
11. F. Benìtez, R. Gambini, L. Lehner, S. Liebling, J. Pullin, Phys. Rev. Lett. **124**, 071301 (2020)
12. M. Bojowald, T. Harada, R. Tibrewala, Phys. Rev. D **78**, 064057 (2008)
13. M. Bojowald, Phys. Rev. Lett. **86**, 5227 (2001)
14. M. Bojowald, Phys. Rev. Lett. **95**, 061301 (2005)
15. M. Bojowald, R. Goswami, R. Maartens, P. Singh, Phys. Rev. Lett. **95**, 091302 (2005)
16. H. Bondi, Mon. Not. Astron. Soc. **107**, 343 (1947)
17. I. Booth, Can. J. Phys. **83**, 1073 (2005)
18. S. Brahma, C.-Y. Chen, D.-H. Yeom, Phys. Rev. Lett. **126**, 181301 (2021)
19. K.A. Bronnikov, Phys. Rev. Lett. **85**, 4641 (2000)
20. K.A. Bronnikov, Phys. Rev. D **63**, 044005 (2001)
21. K.A. Bronnikov, Phys. Rev. D **96**, 128501 (2017)
22. S. Chandrasekhar, *An Introduction to the Study of Stellar Structure* (The University of Chicago press, Chicago, USA, 1939)
23. S. Datt, Z. Phys. **108**, 314 (1938)
24. I. Dymnikova Class. Quantum Grav. **19**, 725 (2002)
25. Z.-Y. Fan, X. Wang, Phys. Rev. D **94**, 124027 (2016)
26. F. Fayos, X. Jaen, E. Llanta, J.M.M. Senovilla, Phys. Rev. D **45**, 2732 (1992)

27. F. Fayos, J.M.M. Senovilla, R. Torres, Phys. Rev. D **54**, 4862 (1996)
28. D. Finkelstein, Phys. Rev. **110**, 965 (1958)
29. V.P. Frolov, Phys. Rev. D **94**, 104056 (2016)
30. V.P. Frolov, G.A. Vilkovisky, Phys. Lett. B **106**, 307 (1981)
31. R. Gambini, J. Pullin, Phys. Rev. Lett. **110**, 211301 (2013)
32. D. Garfinkle, J.M.M. Senovilla, Class. Quantum Grav. **32**, 124008 (2015)
33. R. Goswami, P.S. Joshi, P. Singh, Phys. Rev. Lett. **96**, 031302 (2006)
34. S.W. Hawking, R. Penrose, Proc. Roy. Soc. Lond. A **314**, 529 (1970)
35. S.W. Hawking, G.F.R. Ellis, *The Large Scale Structure of Space-Time* (Cambridge University Press, Cambridge, UK, 1973) <error l="305" c="Invalid command: paragraph not started." />
36. S.A. Hayward, Phys. Rev. D **49**, 6467 (1994)
37. S.A. Hayward, Phys. Rev. Lett. **96**, 031103 (2006)
38. C. Hellaby, K. Lake, Astrophys. J. **290**, 381 (1985)
39. C. Hellaby, K. Lake, Astrophys. J. **300**, 461 (1986)
40. A. Helou, I. Musco, J.C. Miller, Class. Quantum Grav. **34**, 135012 (2017)
41. S. Hossenfelder, L. Modesto, I. Prémont-Schwarz, Phys. Rev. D **81**, 044036 (2010)
42. W. Israel, Nuovo Cimento B **44**, 1 (1966); Nuovo Cimento B **48**, 463 (1966)
43. P.S. Joshi, *Gravitational Collapse and Spacetime Singularities* (Cambridge University Press, Cambridge, UK, 2008)
44. P.S. Joshi, I.H. Dwivedi, Class. Quantum Grav. **9**, L69 (1992)
45. P.S. Joshi, I.H. Dwivedi, Phys. Rev. D **47**, 5357 (1993)
46. P.S. Joshi, I.H. Dwivedi, Class. Quantum Grav. **16**, 41 (1999)
47. P.S. Joshi, D. Malafarina, Int. J. Mod. Phys. D **20**, 2641 (2011)
48. J. Kijowski, E. Czuchry, Class. Quantum Grav. **27**, 235007 (2010)
49. G. Lemaître, Ann. Soc. Sci. Bruxelles I, A **53**, 51 (1933); Gen. Rel. Grav. **29**, 641 (1997)
50. Y. Liu, D. Malafarina, L. Modesto, C. Bambi, Phys. Rev. D **90**, 044040 (2014)
51. D. Malafarina, in *Astrophysics of Black Holes. Astrophysics and Space Science Library* ed. by C. Bambi vol. 440, p. 169 (2016)
52. D. Malafarina, Universe **3**, 48 (2017)
53. D. Malafarina, B. Toshmatov, Phys. Rev. D **105**, L121502 (2022)
54. P. Martín-Moruno, M. Visser, J. High Energ. Phys. **2013**, 50 (2013)
55. M.M. May, R.H. White, Phys. Rev. **141**, 1232 (1966)
56. C. Misner, D. Sharp, Phys. Rev. **136**, B571 (1964)
57. J.R. Oppenheimer, G.M. Volkov, Phys. Rev. **56**, 374 (1939)
58. J.R. Oppenheimer, H. Snyder, Phys. Rev. **56**, 455 (1939)
59. R. Pellicer, R.J. Torrence, J. Math. Phys. **10**, 1718 (1969)
60. R. Penrose, Phys. Rev. Lett. **14**, 57 (1965)
61. R. Penrose, Riv. Nuovo Cim. **1**, 252 (1969)
62. A. Peres, Phys. Rev. **122**, 273 (1961)
63. E. Poisson, *A Relativist's Toolkit: The Mathematics of Black-Hole Mechanics* (Cambridge University Press, Cambridge, UK, 2004)
64. T.A. Roman, P.G. Bergmann, Phys. Rev. D **28**, 1265 (1983)
65. T. Schmitz, Phys. Rev. D **103**, 064074 (2021)
66. J.M.M. Senovilla, Gen. Rel. Grav. **30**, 701 (1998)
67. R.C. Tolman, Proc. Natl. Acad. Sci. USA **20**, 410 (1934)
68. R. Tooper, Astrophys. J. **140**, 434 (1964)
69. B. Toshmatov, B. Ahmedov, A. Abdujabbarov, Z. Stuchlìk, Phys. Rev. D **89**, 104017 (2014)
70. B. Toshmatov, Z. Stuchlìk, B. Ahmedov, Phys. Rev. D **98**, 028501 (2018)
71. B. Toshmatov, Z. Stuchlìk, B. Ahmedov, D. Malafarina, Phys. Rev. D **99**, 064043 (2019)
72. P.C. Vaidya, Proc. Indian Acad. Sc. **A33**, 264 (1951)
73. M. Visser, Phys. Rev. D **90**, 127502 (2014)
74. A. Wang, Y. Wu, Gen. Relativ. Grav. **31**, 107 (1999)
75. C.M. Will, Living Rev. Relativ. **17**, 4 (2014)
76. P. Yodzis, H.-J. Seifert, H.M.Z. Hagen, Commun. Math. Phys. **34**, 135 (1973)
77. Y.B. Zel'dovich, L.F. Grishchuk, Mon. Not. R. Astro. Soc. **2**(07), 23 (1984)
78. P. Żenczykowski, Found. Sci. **24**, 287 (2019)

Chapter 13
Gravitational Collapse with Torsion and Universe in a Black Hole

Nikodem Popławski

13.1 Torsion and Regular Black Holes

The general theory of relativity (GR) is a geometric theory of gravitation, in which curved spacetime is described by the metric tensor. The affine connection, which describes differentiation in a curved spacetime, is constrained to be symmetric and given by the Christoffel symbols constructed from the metric tensor and its derivatives [9, 14, 35]. In the presence of a symmetric connection, the orbital angular momentum of a free particle is conserved [38, 51]. However, the Dirac equation of relativistic quantum mechanics predicts the conservation law for the total (orbital plus spin) angular momentum of a free Dirac particle, allowing a spin-orbit interaction [8]. This conservation law arises naturally in theories of gravity with an asymmetric affine connection [28]. The antisymmetric part of the connection is referred to as the torsion tensor [13, 70, 71].

The simplest theory of gravity that extends GR by relaxing the symmetry constraint of the affine connection and allowing torsion is the Einstein–Cartan (EC) theory [23–25, 27, 31, 38, 51, 53, 68, 69, 72–74]. In this theory, expanded by Sciama and Kibble, the Lagrangian density for the gravitational field is the same as in GR: it is proportional to the Ricci scalar constructed from the connection. The field equations, obtained from varying the action for gravity and matter with respect to the torsion tensor, determine the torsion tensor to be proportional to the spin tensor of fermionic matter [27, 31, 38, 51, 74]. Consequently, EC can be rewritten as GR with the symmetric Levi-Civita connection, in which the energy–momentum tensor of matter acquires additional terms that are quadratic in the spin tensor. In vacuum, torsion vanishes and EC reduces to GR. Following the multipole expansion of the

N. Popławski (✉)
Department of Mathematics and Physics, University of New Haven, 300 Boston Post Road, West Haven, CT 06516, USA
e-mail: NPoplawski@newhaven.edu

C. Bambi (ed.), *Regular Black Holes*, Springer Series in Astrophysics and Cosmology,
https://doi.org/10.1007/978-981-99-1596-5_13

conservation law for the spin tensor in EC, fermionic matter has a form of a spin fluid: an ideal fluid with a spin tensor that is linear in the matter four-velocity [40].

Hehl [22], Trautman [79–81], and Kopczyński [32, 33] discovered that torsion can generate gravitational repulsion and prevent the formation of a cosmological singularity in a homogeneous and isotropic universe described by the Friedmann–Lemaître–Robertson–Walker (FLRW) metric [9, 14, 15, 35, 36, 38, 67, 84] when spins of fermions are aligned. A singularity can be also avoided for randomly oriented spins because macroscopic averaging of the quadratic spin terms in the energy–momentum tensor gives nonzero values [26]. The effective energy density $\tilde{\varepsilon}$ and pressure $\tilde{p}$ of a spin fluid with randomly oriented spins are given by

$$\tilde{\varepsilon} = \varepsilon - \alpha n_{\mathrm{f}}^2, \quad \tilde{p} = p - \alpha n_{\mathrm{f}}^2, \tag{13.1}$$

where ε is the thermodynamic energy density, p is its pressure, n_{f} is the number density of fermions, and $\alpha = \kappa(\hbar c)^2/32$ [26, 43, 60, 61, 82] with $\kappa = 8\pi G/c^4$. At densities much lower than nuclear density, the effects of torsion can be neglected and EC is indistinguishable from GR, passing all observational tests. At densities much higher than nuclear density, that exist in black holes or in the very early universe, the negative corrections from the spin-torsion coupling in (13.1) violate the strong energy condition and manifest themselves as repulsive gravity that may prevent the formation of singularities [18, 34, 44, 54, 56, 60, 61].

In the presence of torsion, the collapsing matter in a black hole would avoid a singularity and instead reach a nonsingular bounce, after which it would expand as a new, closed universe [60, 61, 82] whose total energy is zero [58, 62]. A nonsingular bounce can also occur if the spin tensor is completely antisymmetric [6, 39, 57]. If a black hole creates a baby universe on the other side of its event horizon, then such a universe would be connected to the parent universe through an Einstein–Rosen bridge (wormhole) [12, 52]. The formation and subsequent dynamics of such a universe cannot be observed outside the black hole because of the infinite redshift at the horizon. Consequently, if our universe is closed [19, 83], then it might have originated as a baby universe after a bounce in the interior of a parent black hole existing in another universe [11, 16, 17, 21, 42, 50, 52, 54, 60, 61, 65, 66, 75–77].

13.2 Einstein–Cartan Gravity

The EC theory of gravity naturally extends GR to include matter with spin, providing a more complete account of local gauge invariance with respect to the Poincaré group [27, 31, 51, 74]. In this theory, the affine connection $\Gamma_{i\,j}^{\,k}$ has an antisymmetric part: the torsion tensor

$$S^k_{\ ij} = \frac{1}{2}(\Gamma_{i\,j}^{\,k} - \Gamma_{j\,i}^{\,k}). \tag{13.2}$$

The curvature tensor is given by $R^i{}_{mjk} = \partial_j \Gamma^{\,i}_{m\,k} - \partial_k \Gamma^{\,i}_{m\,j} + \Gamma^{\,i}_{l\,j}\Gamma^{\,l}_{m\,k} - \Gamma^{\,i}_{l\,k}\Gamma^{\,l}_{m\,j}$ and its contraction gives the Ricci tensor $R_{ik} = R^j{}_{ijk}$. The metricity condition $\nabla_j g_{ik} = 0$, where ∇_i denotes the covariant derivative for the connection $\Gamma^{\,k}_{i\,j}$, gives

$$\Gamma^{\,k}_{i\,j} = \{^{\,k}_{i\,j}\} + C^k{}_{ij}, \tag{13.3}$$

where $\{^{\,k}_{i\,j}\} = (1/2)g^{km}(\partial_j g_{mi} + \partial_i g_{mj} - \partial_m g_{ij})$ are the Christoffel symbols for the metric tensor g_{ik} and $C^k{}_{ij} = S^k{}_{ij} + S_{ij}{}^k + S_{ji}{}^k$ is the contortion tensor. The curvature tensor can be decomposed as $R^i{}_{klm} = P^i{}_{klm} + D_l C^i{}_{km} - D_m C^i{}_{kl} + C^j{}_{km} C^i{}_{jl} - C^j{}_{kl} C^i{}_{jm}$, where $P^i{}_{klm}$ is the Riemann tensor (the curvature tensor constructed from the Levi-Civita connection $\{^{\,k}_{i\,j}\}$) and D_i denotes the covariant derivative for the connection $\{^{\,k}_{i\,j}\}$ [51].

The EC gravity is based on the Lagrangian density of the gravitational field that is proportional to the Ricci curvature scalar $R = R_{ik}g^{ik}$, similarly to GR which uses the curvature scalar constructed from $\{^{\,k}_{i\,j}\}$. The field equations are obtained from varying the total action for the gravitational field and matter, $(1/c)\int[-(1/2\kappa)R\sqrt{-g} + \mathcal{L}_\mathrm{m}]d^4x$, where $\mathcal{L}_\mathrm{m}$ is the Lagrangian density for matter, and $g = \det(g_{ik})$, with respect to the metric and torsion tensors. Varying the action with respect to the torsion tensor gives the Cartan field equations, which relate algebraically the torsion tensor to the canonical spin tensor of matter $s^{ijk} = 2(\delta\mathcal{L}_\mathrm{m}/\delta C_{ijk})/\sqrt{-g}$ [27, 31, 38, 51, 74]:

$$S_{jik} - S_i g_{jk} + S_k g_{ji} = -\frac{1}{2}\kappa s_{ikj}, \tag{13.4}$$

where $S_i = S^k{}_{ik}$. Varying the action with respect to the metric tensor gives the Einstein field equations, which relate the curvature tensor to the canonical energy–momentum tensor of matter σ_{ik}:

$$R_{ik} - \frac{1}{2}R g_{ik} = \kappa\sigma_{ki}. \tag{13.5}$$

The canonical energy–momentum tensor is related to the metric energy–momentum tensor $T_{ik} = 2(\delta\mathcal{L}_\mathrm{m}/\delta g^{ik})/\sqrt{-g}$ by $T_{ik} = \sigma_{ik} - (1/2)(\nabla_j - 2S_j)(s_{ik}{}^j - s_k{}^j{}_i + s^j{}_{ik})$ [27, 38, 51].

The field Eqs. (13.4) and (13.5) can be combined to give

$$G^{ik} = \kappa T^{ik} + \frac{1}{2}\kappa^2\Big(s^{ij}{}_j s^{kl}{}_l - s^{ij}{}_l s^{kl}{}_j - s^{ijl}s^k{}_{jl} + \frac{1}{2}s^{jli}s_{jl}{}^k$$
$$+\frac{1}{4}g^{ik}(2s_j{}^l{}_m s^{jm}{}_l - 2s_j{}^l{}_l s^{jm}{}_m + s^{jlm}s_{jlm})\Big), \tag{13.6}$$

where $G_{ik} = P_{ik} - (1/2)P g_{ik}$ is the Einstein tensor of general relativity constructed from the contractions of the Riemann tensor, $P_{ik} = P^j{}_{ijk}$ and $P = P_{ik}g^{ik}$ [27, 51].

The second term on the right-hand side of (13.6) is a correction to curvature from spin. The spin tensor also appears in T_{ik} because $\mathcal{L}_m$ depends on torsion.

Quarks and leptons, that compose all stars, are fermions described in relativistic quantum mechanics by the Dirac equation. Since Dirac fields couple minimally to the torsion tensor, torsion at microscopic scales is generated in the presence of fermions [27, 31, 51, 74]. At macroscopic scales, such particles can be averaged and described as a spin fluid [18, 26, 34, 54]. If particles have randomly oriented spins, then the macroscopic spacetime averages of the spin and of the spin gradients vanish. However, the spin terms in the Einstein equations (13.6) are quadratic in the spin tensor and do not vanish after averaging [26]. These contributions from spin to the field equations are significant only at densities of matter much larger than the density of nuclear matter (on the order of the Cartan density [53]) because of the factor κ^2. At nuclear densities and below, the predictions of EC do not differ from the predictions of the metric general relativity, and reduce to them in vacuum, where torsion vanishes.

The Bianchi identity $\nabla_{[l}R^i_{\ |n|jk]} = 2R^i_{\ nm[j}S^m_{\ kl]}$ and the cyclic identity $R^m_{\ [jkl]} = -2\nabla_{[l}S^m_{\ jk]} + 4S^m_{\ n[j}S^n_{\ kl]}$, together with the Einstein and Cartan field equations give the conservation laws for the canonical energy–momentum and spin tensors: $D_j\sigma^{ij} = C^{\ \ i}_{jk\ }\sigma^{jk} + (1/2)s_{klj}R^{klji}$ and $\nabla_k s_{ij}^{\ \ k} - 2S_k s_{ij}^{\ \ k} = \sigma_{ij} - \sigma_{ji}$ [27, 51]. Using the multipole expansion for these laws leads to the formulas for the macroscopic canonical energy–momentum and spin tensors in the point-particle approximation [40, 51]. These tensors define a spin fluid [26]. The canonical energy–momentum tensor of a spin fluid is given by

$$\sigma_{ij} = c\Pi_i u_j - p(g_{ij} - u_i u_j), \tag{13.7}$$

and its canonical spin tensor is given by

$$s_{ij}^{\ \ k} = s_{ij}u^k, \quad s_{ij}u^j = 0, \quad s^2 = \frac{1}{2}s_{ij}s^{ij} > 0, \tag{13.8}$$

where Π_i is the four-momentum density of the fluid, u^i is its four-velocity, and s_{ij} is its spin density.

The field Eqs. (13.6) give [26]

$$G^{ij} = \kappa\Big(\varepsilon - \frac{1}{4}\kappa s^2\Big)u^i u^j - \kappa\Big(p - \frac{1}{4}\kappa s^2\Big)(g^{ij} - u^i u^j)$$
$$-\frac{1}{2}\kappa(\delta^l_k + u_k u^l)D_l(s^{ki}u^j + s^{kj}u^i), \tag{13.9}$$

where $\varepsilon = c\Pi_i u^i$. For randomly oriented spins of particles, the last term on the right-hand side of (13.9) vanishes after averaging. Thus the Einstein–Cartan equations for such a spin fluid are equivalent to the general-relativistic Einstein equations for an ideal fluid with the effective energy density $\tilde{\varepsilon} = \varepsilon - \kappa s^2/4$ and pressure $\tilde{p} = p - \kappa s^2/4$ [26, 27, 51]. The square of the spin density for a fluid consisting of fermions with random spin orientation is given by [43]

$$s^2 = \frac{1}{8}(\hbar c n_{\mathrm{f}})^2. \tag{13.10}$$

Consequently, the effective energy density and pressure of a spin fluid give (13.1).

13.3 Gravitational Collapse of a Homogeneous Sphere

In this section, we consider gravitational collapse of a sphere of a homogeneous spin fluid that is initially at rest, to demonstrate the formation of a new universe in a black hole [65, 66]. The presented work extends the analysis of collapse of a dustlike sphere by Landau and Lifshitz [35], based on the work of Tolman [78] and Oppenheimer and Snyder [45]. This formalism relates the initial scale factor of the universe in a black hole to the initial radius and mass of the black hole. In the absence of pressure gradients, such a collapse can be described in a system of coordinates that is both synchronous and comoving [35].

For a spherically symmetric gravitational field in spacetime filled with an ideal fluid, the geometry is given by the Tolman metric [35, 78]:

$$ds^2 = e^{\nu(\tau,R)}c^2 d\tau^2 - e^{\lambda(\tau,R)}dR^2 - e^{\mu(\tau,R)}(d\theta^2 + \sin^2\theta\, d\phi^2), \tag{13.11}$$

where ν, λ and μ are functions of a time coordinate τ and a radial coordinate R. We can still apply coordinate transformations $\tau \to \tau'(\tau)$ and $R \to R'(R)$ without changing the form of the metric (13.11). The components of the Einstein tensor corresponding to (13.11) that do not vanish identically are [35, 78]:

$$\begin{aligned}
G^0_0 &= -e^{-\lambda}\Big(\mu'' + \frac{3\mu'^2}{4} - \frac{\mu'\lambda'}{2}\Big) + \frac{e^{-\nu}}{2}\Big(\dot{\lambda}\dot{\mu} + \frac{\dot{\mu}^2}{2}\Big)\\
&+e^{-\mu},\\
G^1_1 &= -\frac{e^{-\lambda}}{2}\Big(\frac{\mu'^2}{2} + \mu'\nu'\Big) + e^{-\nu}\Big(\ddot{\mu} - \frac{\dot{\mu}\dot{\nu}}{2} + \frac{3\dot{\mu}^2}{4}\Big)\\
&+e^{-\mu},\\
G^2_2 &= G^3_3 = -\frac{e^{-\nu}}{4}(\dot{\lambda}\dot{\nu} + \dot{\mu}\dot{\nu} - \dot{\lambda}\dot{\mu} - 2\ddot{\lambda} - \dot{\lambda}^2 - 2\ddot{\mu} - \dot{\mu}^2)\\
&-\frac{e^{-\lambda}}{4}(2\nu'' + \nu'^2 + 2\mu'' + \mu'^2 - \mu'\lambda' - \nu'\lambda' + \mu'\nu'),\\
G^1_0 &= \frac{e^{-\lambda}}{2}(2\dot{\mu}' + \dot{\mu}\mu' - \dot{\lambda}\mu' - \dot{\mu}\nu'),
\end{aligned} \tag{13.12}$$

where a dot denotes differentiation with respect to $c\tau$ and a prime denotes differentiation with respect to R.

In a comoving frame of reference, the spatial components of the four-velocity u^i vanish. Accordingly, the nonzero components of the energy–momentum tensor for a spin fluid, $T_{ik} = (\tilde{\varepsilon} + \tilde{p})u_i u_k - \tilde{p} g_{ik}$, are: $T_0^0 = \tilde{\varepsilon}$, $T_1^1 = T_2^2 = T_3^3 = -\tilde{p}$. The Einstein field equations $G_k^i = \kappa T_k^i$ in this frame of reference are:

$$G_0^0 = \kappa\tilde{\varepsilon}, \quad G_1^1 = G_2^2 = G_3^3 = -\kappa\tilde{p}, \quad G_0^1 = 0. \tag{13.13}$$

The covariant conservation of the energy–momentum tensor gives

$$\dot{\lambda} + 2\dot{\mu} = -\frac{2\dot{\tilde{\varepsilon}}}{\tilde{\varepsilon} + \tilde{p}}, \quad \nu' = -\frac{2\tilde{p}'}{\tilde{\varepsilon} + \tilde{p}}, \tag{13.14}$$

where the constants of integration depend on the allowed transformations $\tau \to \tau'(\tau)$ and $R \to R'(R)$.

If the pressure is homogeneous (no pressure gradients), then $p' = 0$ and $p = p(\tau)$. In this case, the second equation in (13.14) gives $\nu' = 0$. Therefore, $\nu = \nu(\tau)$ and a transformation $\tau \to \tau'(\tau)$ can bring ν to zero and $g_{00} = e^\nu$ to 1. The system of coordinates becomes synchronous [35]. Defining $r(\tau, R) = e^{\mu/2}$ turns (13.11) into

$$ds^2 = c^2 d\tau^2 - e^{\lambda(\tau,R)} dR^2 - r^2(\tau, R)(d\theta^2 + \sin^2\theta\, d\phi^2). \tag{13.15}$$

The Einstein equations (13.12) reduce to

$$\begin{aligned}
\kappa\tilde{\varepsilon} &= -\frac{e^{-\lambda}}{r^2}(2rr'' + r'^2 - rr'\lambda') + \frac{1}{r^2}(r\dot{r}\dot{\lambda} + \dot{r}^2 + 1),\\
-\kappa\tilde{p} &= \frac{1}{r^2}(-e^{-\lambda}r'^2 + 2r\ddot{r} + \dot{r}^2 + 1),\\
-2\kappa\tilde{p} &= -\frac{e^{-\lambda}}{r}(2r'' - r'\lambda') + \frac{\dot{r}\dot{\lambda}}{r} + \ddot{\lambda} + \frac{1}{2}\dot{\lambda}^2 + \frac{2\ddot{r}}{r},\\
2\dot{r}' &- \dot{\lambda}r' = 0.
\end{aligned} \tag{13.16}$$

Integrating the last equation in (13.16) gives

$$e^\lambda = \frac{r'^2}{1 + f(R)}, \tag{13.17}$$

where f is a function of R satisfying a condition $1 + f > 0$ [35]. Substituting (13.17) into the second equation in (13.16) gives $2r\ddot{r} + \dot{r}^2 - f = -\kappa\tilde{p}r^2$, which is integrated to

$$\dot{r}^2 = f(R) + \frac{F(R)}{r} - \frac{\kappa}{r}\int \tilde{p}r^2 dr, \tag{13.18}$$

where F is a positive function of R. Substituting (13.17) into the third equation in (13.16) does not give a new relation. Substituting (13.17) into the first equation in (13.16) and using (13.18) gives

$$\kappa(\tilde{\varepsilon} + \tilde{p}) = \frac{F'(R)}{r^2 r'}. \tag{13.19}$$

Combining (13.18) and (13.19) gives

$$\dot{r}^2 = f(R) + \frac{\kappa}{r}\int_0^R \tilde{\varepsilon} r^2 r' dR. \tag{13.20}$$

Every particle in a collapsing fluid sphere is represented by a radial coordinate R that ranges from 0 (at the center of the sphere) to R_0 (at the surface of the sphere). If the mass of the sphere is M, then the Schwarzschild radius $r_g = 2GM/c^2$ of the black hole that forms from the sphere is equal to [35]

$$r_g = \kappa \int_0^{R_0} \tilde{\varepsilon} r^2 r' dR. \tag{13.21}$$

Equations (13.20) and (13.21) give

$$\dot{r}^2(\tau, R_0) = f(R_0) + \frac{r_g}{r(\tau, R_0)}. \tag{13.22}$$

If $r_0 = r(0, R_0)$ is the initial radius of the sphere and the sphere is initially at rest, then $\dot{r}(0, R_0) = 0$. Consequently, (13.22) determines the value of R_0 [65, 66]:

$$f(R_0) = -\frac{r_g}{r_0}. \tag{13.23}$$

13.4 Spinless Dustlike Sphere

Before considering gravitational collapse of a sphere composed of a spin fluid, it is instructive to consider spinless dust, for which the pressure vanishes and thus $\tilde{p} = 0$. Substituting (13.19) into (13.21) gives

$$r_g = F(R_0) - F(0) = F(R_0), \tag{13.24}$$

which determines the value of R_0. If $f < 0$, then (13.18) has a solution

$$r = -\frac{F}{2f}(1 + \cos\eta), \quad \tau - \tau_0(R) = \frac{F}{2(-f)^{3/2}}(\eta + \sin\eta), \tag{13.25}$$

where η is a parameter and $\tau_0(R)$ is a function of R [35, 78]. Choosing

$$f(R) = -\sin^2 R, \quad F(R) = a_0 \sin^3 R, \quad \tau_0(R) = \text{const.} \tag{13.26}$$

gives

$$r = \frac{a_0}{2} \sin R(1 + \cos\eta) \quad \tau - \tau_0 = \frac{a_0}{2}(\eta + \sin\eta), \tag{13.27}$$

where a_0 is a constant [35]. Initially, at $\tau = \tau_0$ and $\eta = 0$, the sphere is at rest: $\dot{r} = 0$. Clearly, a singularity $r = 0$ is reached for all particles in a finite time. The values of a_0 and R_0 can be determined from (13.23), (13.24), and (13.26):

$$\sin R_0 = \left(\frac{r_g}{r_0}\right)^{1/2}, \quad a_0 = \left(\frac{r_0^3}{r_g}\right)^{1/2}. \tag{13.28}$$

An event horizon for the entire sphere forms when $r(\tau, R_0) = r_g$, that is, at $\cos(\eta/2) = \sin R_0$.

Substituting (13.26) and (13.27) into (13.17) gives $e^{\lambda(\tau,R)} = a_0^2(1 + \cos\eta)^2/4$. If we define

$$a(\tau) = \frac{a_0}{2}(1 + \cos\eta), \tag{13.29}$$

then the square of an infinitesimal interval in the interior of a collapsing dust (13.15) turns into [35]

$$ds^2 = c^2 d\tau^2 - a^2(\tau) dR^2 - a^2(\tau) \sin^2 R(d\theta^2 + \sin^2\theta\, d\phi^2). \tag{13.30}$$

The initial value of a is equal to a_0. This metric has a form of the closed FLRW metric and describes a part of a closed universe with $0 \le R \le R_0$.

13.5 Spin-fluid Sphere

We now proceed to the main part of this chapter and consider gravitational collapse of a sphere composed of a spin fluid. We use the Tolman metric [45, 78] and the EC field equations with a relativistic spin fluid as a source, which can be written as the GR field equations for a fluid source with the energy density and pressure (13.1). We use the temperature to represent the energy density, pressure, and fermion number density in a relativistic fluid [60, 61].

Substituting $r = e^{\mu/2}$ and (13.17) into the first equation in (13.14) gives

$$\frac{d}{d\tau}(\tilde{\varepsilon} r^2 r') + \tilde{p} \frac{d}{d\tau}(r^2 r') = 0, \tag{13.31}$$

which has a form of the first law of thermodynamics for the energy density and pressure (13.1) [60, 61]. If we assume that the spin fluid is composed by an ultra-relativistic matter in kinetic equilibrium, then $\varepsilon = h_\star T^4$, $p = \varepsilon/3$, and $n_\mathrm{f} = h_{n\mathrm{f}} T^3$, where T is the temperature of the fluid, $h_\star = (\pi^2/30)(g_\mathrm{b} + (7/8)g_\mathrm{f})k_\mathrm{B}^4/(\hbar c)^3$, and $h_{n\mathrm{f}} = (\zeta(3)/\pi^2)(3/4)g_\mathrm{f} k_\mathrm{B}^3/(\hbar c)^3$ [60, 61, 82]. For standard-model particles, $g_\mathrm{b} = 29$ and $g_\mathrm{f} = 90$. Since $p' = 0$, the temperature does not depend on R: $T = T(\tau)$. Substituting these relations into (13.31) gives

$$r^2 r' T^3 = g(R), \tag{13.32}$$

where g is a function of R. Putting this equation into (13.20) gives

$$\dot{r}^2 = f(R) + \frac{\kappa}{r}(h_\star T^4 - \alpha h_{n\mathrm{f}}^2 T^6)\int_0^R r^2 r' dR. \tag{13.33}$$

Equations (13.32) and (13.33) give the function $r(\tau, R)$, which with (13.17) gives $\lambda(\tau, R)$ [65, 66]. The integration of (13.33) also contains the initial value $\tau_0(R)$. The metric (13.15) depends thus on three arbitrary functions: $f(R)$, $g(R)$, and $\tau_0(R)$.

We seek a solution of (13.32) and (13.33) as

$$f(R) = -\sin^2 R, \quad r(\tau, R) = a(\tau)\sin R, \tag{13.34}$$

where $a(\tau)$ is a nonnegative function of τ. This choice is analogous to a dust sphere: the first equation in (13.26), the first equation in (13.27), and (13.29). Accordingly, (13.32) gives

$$a^3 T^3 \sin^2 R \cos R = g(R), \tag{13.35}$$

in which separation of the variables τ and R leads to

$$g(R) = \mathrm{const}\cdot\sin^2 R\cos R, \quad a^3 T^3 = \mathrm{const}. \tag{13.36}$$

Consequently, we find

$$aT = a_0 T_0, \quad \frac{\dot{T}}{T} + \frac{H}{c} = 0, \tag{13.37}$$

where $a_0 = a(0)$, $T_0 = T(0)$, and $H = c\dot{a}/a$ is the Hubble parameter. Substituting (13.34) into (13.33) gives

$$\dot{a}^2 + 1 = \frac{\kappa}{3}(h_\star T^4 - \alpha h_{n\mathrm{f}}^2 T^6)a^2. \tag{13.38}$$

Using (13.37) in (13.38) yields

$$\dot{a}^2 = -1 + \frac{\kappa}{3}\left(\frac{h_\star T_0^4 a_0^4}{a^2} - \frac{\alpha h_{n\mathrm{f}}^2 T_0^6 a_0^6}{a^4}\right). \tag{13.39}$$

Substituting (13.34) into (13.17) gives $e^{\lambda(\tau,R)} = a^2$. Consequently, the square of an infinitesimal interval in the interior of a collapsing spin fluid (13.15) is also given by (13.30) [65, 66].

The values of a_0 and R_0 can be determined from (13.23) and (13.34), giving (13.28). Substituting them and $\dot{a}(0) = 0$ into (13.38), in which the second term on the right-hand side is negligible, gives $Mc^2 = (4\pi/3)r_0^3 h_\star T_0^4$. This relation indicates the equivalence of mass and energy of a fluid sphere with radius r_0 and determines T_0. An event horizon for the entire sphere forms when $r(\tau, R_0) = r_g$, which is equivalent to $a = (r_g r_0)^{1/2}$. Equation (13.39) has two turning points, $\dot{a} = 0$, if [82]

$$\frac{r_0^3}{r_g} > \frac{3\pi G\hbar^4 h_{nf}^4}{8h_\star^3} \sim l_{\text{Planck}}^2, \tag{13.40}$$

which is satisfied for astrophysical systems that form black holes.

13.6 Nonsingular Bounce and Formation of a New Universe

Equation (13.39) can be solved analytically in terms of an elliptic integral of the second kind [82], giving the function $a(\tau)$ and then $r(\tau, R) = a(\tau)\sin R$. The value of a never reaches zero because as a decreases, the right-hand side of (13.39) becomes negative, contradicting the left-hand side. The change of the sign occurs when $a < (r_g r_0)^{1/2}$, that is, after the event horizon forms. Consequently, all particles with $R > 0$ fall within the event horizon but never reach $r = 0$ (the only particle at the center is the particle that is initially at the center, with $R = 0$). A singularity is therefore avoided and replaced with a regular bounce [65, 66]. Nonzero values of a in (13.30) give finite values of T and therefore finite values of ε, p, and n_{f}.

If the initial mass of a spin-fluid sphere is insufficient to form an event horizon, then the fluid bounces and disperses back to the region of space outside the sphere [20]. When an event horizon forms, the fluid cannot disperse back to the region of space outside the horizon because of the unidirectionality of the motion of matter through a horizon [35]. Moreover, it cannot tend to a static state because the spacetime within an event horizon is not stationary. Consequently, the spin fluid on the other side of the event horizon must expand as a new, growing universe with a closed geometry (constant positive curvature) [54]. This universe can be regarded as the three-dimensional surface of a four-dimensional sphere with radius $a(\tau)$, which is the scale factor of this universe. The new, closed universe is oscillatory: the value of a oscillates between the two turning points. The value of R_0 does not change. A turning point at which $\ddot{a} > 0$ is a bounce, and a turning point at which $\ddot{a} < 0$ is a crunch. The universe has therefore an infinite number of bounces and crunches, and each cycle is alike.

The Raychaudhuri equation for a congruence of geodesics without rotation and four-acceleration is $d\theta/ds = -\theta^2/3 - 2\sigma^2 - P_{ik}u^i u^k$, where θ is the expansion scalar and σ^2 is the shear scalar [51]. For a spin fluid, the last term in this equation is equal to $-\kappa(\tilde{\varepsilon}+3\tilde{p})/2$. Consequently, the necessary and sufficient condition for avoiding a singularity in a black hole is $-\kappa(\tilde{\varepsilon}+3\tilde{p})/2 > 2\sigma^2$. For a relativistic spin fluid, $p=\varepsilon/3$, this condition is equivalent to

$$2\kappa\alpha n_{\mathrm{f}}^2 > 2\sigma^2 + \kappa\varepsilon. \tag{13.41}$$

Without torsion, the left-hand side of (13.41) would be absent and this inequality could not be satisfied, resulting in a singularity. Torsion therefore provides a necessary condition for preventing a singularity. In the absence of shear, this condition is also sufficient.

The presence of shear opposes the effects of torsion. The shear scalar σ^2 grows with decreasing a like $\sim a^{-6}$, which is the same power law as that for n_{f}^2 [32, 33]. Therefore, if the initial shear term dominates over the initial torsion term in (13.41), then it will dominate at later times during contraction and a singularity will form. To avoid a singularity if the shear is present, n_{f}^2 must grow faster than $\sim a^{-6}$. Consequently, fermions must be produced in a black hole during contraction.

13.7 Particle Production

The production rate of particles in a contracting or expanding universe [3, 46–49, 85, 86] can be phenomenologically given by

$$\frac{1}{c\sqrt{-g}}\frac{d(\sqrt{-g}n_{\mathrm{f}})}{dt} = \frac{\beta H^4}{c^4}, \tag{13.42}$$

where $g=-a^6\sin^4 R\sin^2\theta$ is the determinant of the metric tensor in (13.30) and β is a nondimensional production rate [60, 61]. With particle production, the second equation in (13.37) turns into

$$\frac{\dot{T}}{T} = \frac{H}{c}\Big(\frac{\beta H^3}{3c^3 h_{n\mathrm{f}} T^3} - 1\Big). \tag{13.43}$$

Particle production changes the power law $n_{\mathrm{f}}(a)$:

$$n_{\mathrm{f}} \sim a^{-(3+\delta)}, \tag{13.44}$$

where δ varies with τ [65, 66]. Putting this relation into (13.42) gives

$$\delta \sim -a^{\delta}\dot{a}^3. \tag{13.45}$$

During contraction, $\dot{a} < 0$ and thus $\delta > 0$. The term $n_{\mathrm{f}}^2 \sim a^{-6-2\delta}$ grows faster than $\sigma^2 \sim a^{-6}$ and a singularity is avoided [65, 66]. Particle production and torsion act together to reverse the effects of shear, generating a nonsingular bounce. The dynamics of the nonsingular, relativistic universe in a black hole is described by Eqs. (13.38) and (13.43), with the initial conditions $a(0) = (r_0^3/r_g)^{1/2}$ and $\dot{a}(0) = 0$, that give the functions $a(\tau)$ and $T(\tau)$. The shear would enter the right-hand side of (13.38) as an additional positive term that is proportional to a^{-4}. When the universe becomes nonrelativistic, the term $h_\star T^4$ in (13.38) changes into a positive term that is proportional to a^{-1}. The cosmological constant enters (13.38) as a positive term that is proportional to a^2.

Particle production increases the maximum size of the scale factor that is reached at a crunch. Consequently, the new cycle is larger and lasts longer then the previous cycle. According to (13.28), R_0 is given by

$$\sin^3 R_0 = \frac{r_g}{a(0)}, \tag{13.46}$$

where $a(0)$ is the initial scale factor that is equal to the maximum scale factor in the first cycle. Since the maximum scale factor in the next cycle is larger, the value of $\sin R_0$ decreases. As cycles proceed, R_0 approaches π.

Without torsion and particle production, a singularity would be reached and the metric would be described by the interior Schwarzschild solution, which is equivalent to the Kantowski–Sachs metric describing an anisotropic universe with topology $R \times S^2$ [5, 29, 64]. Thanks to torsion, the universe in a black hole becomes closed with topology S^3 (3-sphere).

13.8 Inflation and Oscillations

During contraction, H is negative and the temperature T increases. During expansion, if β is too big, then the right-hand side of (13.43) could become positive. In this case, the temperature would grow with increasing a, which would lead to eternal inflation [60, 61]. Consequently, there is an upper limit to the production rate: the maximum of the function $(\beta H^3)/(3c^3 h_{nf} T^3)$ must be lesser than 1.

If $(\beta H^3)/(3c^3 h_{nf} T^3)$ in (13.43) increases after a bounce to a value that is slightly lesser than 1, then T would become approximately constant. Accordingly, H would be also nearly constant and the scale factor a would grow exponentially, generating inflation. Since the energy density would be also nearly constant, the universe would produce enormous amounts of matter and entropy. Such an expansion would last until the right-hand side of (13.43) drops below 1. Consequently, inflation would last a finite period of time. After this period, the effects of torsion weaken and the universe smoothly enters the radiation-dominated expansion, followed by the matter-dominated expansion.

If the universe during expansion does not reach a critical size at which the cosmological constant is significant, then it recollapses to another bounce and starts a new oscillation cycle [4, 41]. The new cycle is larger and longer then the previous cycle [1, 2, 60, 61]. After a finite series of cycles, the universe reaches the critical size which prevents the next contraction and enters the cosmological-constant-dominated expansion, during which it expands indefinitely. The value of R_0 asymptotically tends to π, which is the maximum value of R in a closed isotropic universe given by (13.30). The last bounce, referred to as the big bounce, is the big bang.

A more realistic scenario of gravitational collapse should involve a fluid sphere that is inhomogeneous and rotating. If the pressure in the sphere is not homogeneous, then the system of coordinates cannot be comoving and synchronous [35, 37]. Consequently, ν and the temperature would depend on R and the equations of the collapse and the subsequent dynamics of the universe would be more complicated. If the sphere were rotating, then further complications would appear [10] and the angular momentum of the forming Kerr black hole would be another parameter in addition to the mass [30]. Nevertheless, the general character of the effects of torsion and particle production in avoiding a singularity and generating a bounce in a black hole would still be valid.

Torsion and particle production act together to reverse gravitational attraction generated by shear and prevent a singularity, to turn the interior of a black hole into a new universe, and to generate inflation in that universe [65, 66]. Torsion may also explain the matter-antimatter asymmetry in the universe [55]. In addition, it could explain the cosmological constant, which is necessary for a closed universe to expand to infinity [59]. Furthermore, torsion may impose a spatial extension of fermions [53] and eliminate the ultraviolet divergence of radiative corrections represented by loop Feynman diagrams in quantum field theory [63].

If every black hole becomes an Einstein–Rosen bridge to a new universe on the other side of its event horizon [12, 52], then our universe might have been born as a baby universe in a parent black hole existing in another universe. This hypothesis, following from the presented analysis of gravitational collapse of a spin fluid [65, 66], naturally solves the black hole information paradox: the information about the initial state of a collapsing matter is not lost but it goes through the event horizon to the new universe [54]. Furthermore, inflation generated by torsion and particle production, is consistent with the Planck observations of the cosmic microwave background radiation [7].

Acknowledgements I am grateful to Francisco Guedes and my Parents, Bożenna Popławska and Janusz Popławski, for inspiring this work.

References

1. J.D. Barrow, M.P. Dąbrowski, Mon. Not. Roy. Astron. Soc. **275**, 850 (1995)
2. J.D. Barrow, C. Ganguly, Int. J. Mod. Phys. D **26**, 1743016 (2017)
3. V.A. Beilin, G.M. Vereshkov, Y.S. Grishkan, N.M. Ivanov, V.A. Nesterenko, A.N. Poltavtsev, J. Exp. Theor. Phys. **51**, 1045 (1980)
4. H. Bondi, *Cosmology* (Cambridge University Press, 1960)
5. R.W. Brehme, Am. J. Phys. **45**, 423 (1977)
6. J.L. Cubero, N.J. Popławski, Class. Quantum Grav. **37**, 025011 (2020)
7. S. Desai, N.J. Popławski, Phys. Lett. B **755**, 183 (2016)
8. P.A.M. Dirac, *The Principles of Quantum Mechanics* (Oxford University Press, 1930)
9. P.A.M. Dirac, *General Theory of Relativity* (Wiley, 1975)
10. A.G. Doroshkevich, Y.B. Zel'dovich, I.D. Novikov, J. Exp. Theor. Phys. **22**, 122 (1966)
11. D.A. Easson, R.H. Brandenberger, J. High Energ. Phys. **06**, 024 (2001)
12. A. Einstein, N. Rosen, Phys. Rev. **48**, 73 (1935)
13. L.P. Eisenhart, *Non-Riemannian Geometry* (American Mathematical Society, 1927)
14. V.A. Fock, *The Theory of Space, Time and Gravitation* (Macmillan, 1964)
15. A. Friedmann, Z. Phys. A **10**, 377 (1922)
16. V.P. Frolov, M.A. Markov, V.F. Mukhanov, Phys. Lett. B **216**, 272 (1989)
17. V.P. Frolov, M.A. Markov, V.F. Mukhanov, Phys. Rev. D **41**, 383 (1990)
18. M. Gasperini, Phys. Rev. Lett. **56**, 2873 (1986)
19. W. Handley, Phys. Rev. D **103**, 041301 (2021)
20. M. Hashemi, S. Jalalzadeh, A.H. Ziaie, Eur. Phys. J. C **75**, 53 (2015)
21. S. Hawking, *Black Holes and Baby Universes and other Essays* (Bantam Dell, 1993)
22. F.W. Hehl, Abh. Braunschw. Wiss. Ges. **18**, 98 (1966)
23. F.W. Hehl, Gen. Relativ. Gravit. **4**, 333 (1973)
24. F.W. Hehl, Gen. Relativ. Gravit. **5**, 491 (1974)
25. F.W. Hehl, B.K. Datta, J. Math. Phys. **12**, 1334 (1971)
26. F.W. Hehl, P. von der Heyde, G.D. Kerlick, Phys. Rev. D **10**, 1066 (1974)
27. F.W. Hehl, P. von der Heyde, G.D. Kerlick, J.M. Nester, Rev. Mod. Phys. **48**, 393 (1976)
28. F.W. Hehl, J.D. McCrea, Found. Phys. **16**, 267 (1986)
29. R. Kantowski, R.K. Sachs, J. Math. Phys. **7**, 443 (1966)
30. R.P. Kerr, Phys. Rev. Lett. **11**, 237 (1963)
31. T.W.B. Kibble, J. Math. Phys. **2**, 212 (1961)
32. W. Kopczyński, Phys. Lett. A **39**, 219 (1972)
33. W. Kopczyński, Phys. Lett. A **43**, 63 (1973)
34. B. Kuchowicz, Gen. Relativ. Gravit. **9**, 511 (1978)
35. L.D. Landau, E.M. Lifshitz, *The Classical Theory of Fields* (Pergamon, 1975)
36. G. Lemaître, Ann. Soc. Sci. Bruxelles A **53**, 51 (1933)
37. E.M. Lifshitz, I.M. Khalatnikov, J. Exp. Theor. Phys. **12**, 108 (1961)
38. E.A. Lord, *Tensors, Relativity and Cosmology* (McGraw-Hill, 1976)
39. J. Magueijo, T.G. Zlosnik, T.W.B. Kibble, Phys. Rev. D **87**, 063504 (2013)
40. K. Nomura, T. Shirafuji, K. Hayashi, Prog. Theor. Phys. **86**, 1239 (1991)
41. J.D. North, *The Measure of the Universe* (Clarendon Press, 1965)
42. I.D. Novikov, J. Exp. Theor. Phys. Lett. **3**, 142 (1966)
43. I.S. Nurgaliev, W.N. Ponomariev, Phys. Lett. B **130**, 378 (1983)
44. Y.N. Obukhov, V.A. Korotky, Class. Quantum Grav. **4**, 1633 (1987)
45. J.R. Oppenheimer, H. Snyder, Phys. Rev. **56**, 455 (1939)
46. L. Parker, Phys. Rev. Lett. **21**, 562 (1968)
47. L. Parker, Phys. Rev. **183**, 1057 (1969)
48. L. Parker, Phys. Rev. D **3**, 346 (1971)
49. L. Parker, Phys. Rev. D **3**, 2546 (1971)
50. R.K. Pathria, Nature **240**, 298 (1972)
51. N. Popławski, *Classical Physics: Spacetime and Fields* (2009). arXiv:0911.0334

52. N.J. Popławski, Phys. Lett. B **687**, 110 (2010)
53. N.J. Popławski, Phys. Lett. B **690**, 73 (2010) [erratum: Phys. Lett. B **727**, 575 (2013)]
54. N.J. Popławski, Phys. Lett. B **694**, 181 (2010) [erratum: Phys. Lett. B **701**, 672 (2011)]
55. N.J. Popławski, Phys. Rev. D **83**, 084033 (2011)
56. N.J. Popławski, Gen. Relativ. Gravit. **44**, 1007 (2012)
57. N. Popławski, Phys. Rev. D **85**, 107502 (2012)
58. N.J. Popławski, Class. Quantum Grav. **31**, 065005 (2014)
59. N. Popławski, Gen. Relativ. Gravit. **46**, 1625 (2014)
60. N. Popławski, Astrophys. J. **832**, 96 (2016)
61. N. Popławski, Int. J. Mod. Phys. D **27**, 1847020 (2018)
62. N. Popławski, Mod. Phys. Lett. A **33**, 1850236 (2018)
63. N. Popławski, Found. Phys. **50**, 900 (2020)
64. N.J. Popławski, Gen. Relativ. Gravit. **53**, 18 (2021)
65. N. Popławski, Zh. Eksp, Teor. Fiz. **159**, 448 (2021)
66. N. Popławski, J. Exp. Theor. Phys. **132**, 374 (2021)
67. H.P. Robertson, Astrophys. J. **82**, 284 (1935)
68. V. de Sabbata, M. Gasperini, *Introduction to Gravitation* (World Scientific, 1985)
69. V. de Sabbata, C. Sivaram, *Spin and Torsion in Gravitation* (World Scientific, 1994)
70. J.A. Schouten, *Ricci-Calculus* (Springer, 1954)
71. E. Schrödinger, *Space-time Structure* (Cambridge University Press, 1954)
72. D.W. Sciama, Proc. Camb. Phil. Soc. **54**, 72 (1958)
73. D.W. Sciama, in *Recent Developments in General Relativity* (Pergamon, 1962), p. 415
74. D.W. Sciama, Rev. Mod. Phys. **36**, 463 (1964) [erratum: Rev. Mod. Phys. **36**, 1103 (1964)]
75. L. Smolin, Class. Quantum Grav. **9**, 173 (1992)
76. J. Smoller, B. Temple, Proc. Natl. Acad. Sci. USA **100**, 11216 (2003)
77. W.M. Stuckey, Am. J. Phys. **62**, 788 (1994)
78. R.A. Tolman, Proc. Natl. Acad. Sci. USA **20**, 169 (1934)
79. A. Trautman, Bull. Acad. Polon. Sci., Serie Sci. Math. Astr. Phys. **20**, 185 (1972)
80. A. Trautman, Symp. Math. **12**, 139 (1973)
81. A. Trautman, Nature Phys. Sci. **242**, 7 (1973)
82. G. Unger, N. Popławski, Astrophys. J. **870**, 78 (2019)
83. E. Di Valentino, A. Melchiorri, J. Silk, Nature Astron. **4**, 196 (2020)
84. A.G. Walker, Proc. London Math. Soc. **42**, 90 (1937)
85. Y.B. Zel'dovich, J. Exp. Theor. Phys. Lett. **12**, 307 (1970)
86. Y.B. Zel'dovich, A.A. Starobinskii, J. Exp. Theor. Phys. Lett. **26**, 252 (1977)